ANALYSIS AND PERFORMANCE OF ENGINEERING MATERIALS

Key Research and Development

ANALYSIS AND PERFORMANCE OF ENGINEERING MATERIALS

Key Research and Development

Edited by

Gennady E. Zaikov, DSc

AAP | APPLE ACADEMIC PRESS

Apple Academic Press Inc.
3333 Mistwell Crescent
Oakville, ON L6L 0A2
Canada

Apple Academic Press Inc.
9 Spinnaker Way
Waretown, NJ 08758
USA

Exclusive worldwide distribution by CRC Press, a member of Taylor & Francis Group

International Standard Book Number-13: 978-1-77188-085-5 (Hardcover)
International Standard Book Number-13: 978-1-4987-0773-2 (eBook)

Library of Congress Control Number: 2015946443

Library and Archives Canada Cataloguing in Publication

Analysis and performance of engineering materials: key research and development / edited by Gennady E. Zaikov, DSc.

Includes bibliographical references and index.
Issued in print and electronic formats.
ISBN 978-1-77188-085-5 (hardcover).--ISBN 978-1-4987-0773-2 (ebook)

1. Polymers--Testing. 2. Materials--Testing. 3. Polymer engineering. 4. Chemistry, Physical and theoretical. I. Zaikov, G. E. (Gennadiĭ Efremovich), 1935-, author, editor

TA455.P58A53 2015   620.1'920287   C2015-905182-7   C2015-905221-1

Apple Academic Press also publishes its books in a variety of electronic formats. Some content that appears in print may not be available in electronic format. For information about Apple Academic Press products, visit our website at **www.appleacademicpress.com** and the CRC Press website at **www.crcpress.com**

CONTENTS

LIST OF CONTRIBUTORS

D. S. Andreev
Volgograd State Architect-Build University Sebryakov Department, Russia

Yurij A. Antonov
N.M. Emanuel Institute of Biochemical Physics, Russian Academy of Sciences, Kosigin Str. 4. 119334 Moscow, Russia

R. Anyszka
Lodz University of Technology, Faculty of Chemistry, Institute of Polymer and Dye Technology, Stefanowskiego 12/16, 90-924 Lodz, Poland

V. A. Babkin
Volgograd State Architect-Build University Sebryakov Department, Russia

Yu. V. Berestneva
Donetsk National University, 24 Universitetskaya Street, 83 055 Donetsk, Ukraine; E-mail: N.Turovskij@donnu.edu.ua

D. M. Bieliński
Institute for Engineering of Polymer Materials and Dyes, Division of Elastomers and Rubber Technology, Harcerska 30, 05-820 Piastow, Poland; Tel.: +4842 6313214; Fax: +4842 6362543; E-mail: dbielin@p.lodz.pl

T. A. Borukaev
Kabardino-Balkarian State University after H.M. Berbekov, 360004, Nalchik, Chernyshevskaya St., 173, Russia; E-mail: boruk-chemical@mail.ru

V. Yu. Dmitriev
Volgograd State Architect-Build University Sebryakov Department, Russia

F. R. Gaisin
Birsk Branch Bashkir State University, Birsk, Russia

M. A. Gastasheva
Kabardino-Balkarian State University after H.M. Berbekov, 360004, Nalchik, Chernyshevskaya St., 173, Russia; E-mail: boruk-chemical@mail.ru

M. P. Gonçalves
CEQUP/Departamento de Engenharia Química, Faculdade de Engenharia, Universidade do Porto, Rua dos Bragas, 4099 Porto Codex, Portugal

A. K. Haghi
University of Guilan, Rasht, Iran

A. V. Ignatov
Volgograd State Architect-Build University Sebryakov Department, Russia

Peter Jurkovič
VIPO, Partizánske, Slovakia; E-mail: upolnovi@savba.sk

V. F. Kablov
Volzhsky Polytechnical Institute (Branch) Volgograd State Technical University, Russia, E-mail: kablov@volpi.ru, www.volpi.ru

Francois Kajzar
Faculty of Applied Chemistry and Materials Science, University Politehnica Bucharest, Bucharest, Romania; Tel/fax: +40203154193; E-mail: frkajzar@yahoo.com

M. Kanafchian
University of Guilan, Rasht, Iran

A. L. Kovarski
N.M. Emanuel Institute of Biochemical Physics, Kosygin st.4., Moscow, Russia

Milan Marônek
Slovak Academy of Sciences, Polymer Institute of the Slovak Academy of Sciences, 845 41 Bratislava, Slovakia

B. S. Mashukova
Kabardino-Balkarian State University after H.M. Berbekov, 360004, Nalchik, Chernyshevskaya St., 173Russia; E-mail: boruk-chemical@mail.ru

Ján Matyašovský
VIPO, Partizánske, Slovakia; E-mail: upolnovi@savba.sk

Ivan Michalec
Slovak Academy of Sciences, Polymer Institute of the Slovak Academy of Sciences, 845 41 Bratislava, Slovakia

V. M. Misin
N.M. Emanuel Institute of Biochemical Physics, 4, Kosygin street, Moscow, 119334, Russian Federation

B. Hadavi Moghadam
University of Guilan, Rasht, Iran

M. S. Mohyeldin
Polymeric Materials Department, Advanced Technologies and New Materials Research Institute, City of Scientific Research and Technological Applications, New Borg El-Arab City 21934, Alexandria, Egypt

Rafit R. Nabiev
Kazan National Research Technological University, 68 Karl Marx Street, 420015 Kazan, Republic of Tatarstan, Russian Federation; Fax: +7 (843) 231-41-56

Ilshat I. Nasyrov
Kazan National Research Technological University, 68 Karl Marx Street, 420015 Kazan, Republic of Tatarstan, Russian Federation; Fax: +7 (843) 231-41-56

Igor Novák
Department of Welding and Foundry, Faculty of Materials Science and Technology in Trnava, 917 24 Trnava, Slovakia

Nail K. Nuriev
Kazan National Research Technological University, 68 Karl Marx Street, 420015 Kazan, Republic of Tatarstan, Russian Federation; Fax: +7 (843) 231-41-56

A. M. Omer
Laboratory of Bioorganic Chemistry of Drugs, Institute of Experimental Pharmacology and Toxicology, Slovak Academy of Sciences, Bratislava, Slovakia; Materials Delivery Group, Polymeric Materials Department, Advanced Technologies and New Materials Research Institute (ATNMRI), City of Scientific Research and Technological Applications (SRTA-City), New Borg El-Arab City 21934, Alexandria, Egypt; E-mail: Ahmedomer_81@yahoo.com

E. V. Raksha
Donetsk National University, 24 Universitetskaya Street, 83 055 Donetsk, Ukraine; E-mail: N.Turovskij@donnu.edu.ua

Ileana Rau
Faculty of Applied Chemistry and Materials Science, University Politehnica Bucharest, Bucharest, Romania; Tel/fax: +40203154193

Daria A. Shiyan
Kazan National Research Technological University, 68 Karl Marx Street, 420015 Kazan, Republic of Tatarstan, Russian Federation; Fax: +7 (843) 231-41-56

Yu. M. Sivergin
N.N. Semenov Institute of Chemical Physics, Kosygin st.4., Moscow, Russia

Ladislav Šoltés
Institute of Experimental Pharmacology of the Slovak Academy of Sciences, 845 41 Bratislava, Slovakia

O. V. Stoyanov
Kazan State Technological University, Russia

T. M. Tamer
Polymeric Materials Department, Advanced Technologies and New Materials Research Institute, City of Scientific Research and Technological Applications, New Borg El-Arab City 21934, Alexandria, Egypt

E. S. Titova
Volgograd State Technical University, Russia

M. A. Tlenkopachev
Kabardino-Balkarian State University after H.M. Berbekov, 360004, Nalchik, Chernyshevskaya St., 173Russia; E-mail: boruk-chemical@mail.ru

V. V. Trifonov
Volgograd State Architect-Build University Sebryakov Department, Russia

Nikolai V. Ulitin
Kazan National Research Technological University, 68 Karl Marx Street, 420015 Kazan, Republic of Tatarstan, Russian Federation; Fax: +7 (843) 231-41-56; E-mail: n.v.ulitin@mail.ru

S. M. Usmanov
Birsk Branch Bashkir State University, Birsk, Russia

R. R. Usmanova
Ufa State technical university of aviation, 12 Karl Marks str., Ufa 450000, Bashkortostan, Russia; E-mail: Usmanovarr@mail.ru

Marian Valentin
Department of Welding and Foundry, Faculty of Materials Science and Technology in Trnava, 917 24 Trnava, Slovakia

N. I. Vatin
Saint-Petersburg State Polytechnical University, 29 Polytechnicheskaya street, Saint-Petersburg 195251, Russia; E-mail: vatin@mail.ru

V. A. Volkov
N.M. Emanuel Institute of Biochemical Physics, 4, Kosygin street, Moscow, 119334, Russian Federation; E-mail: vl.volkov@mail.ru

Gennady E. Zaikov
N.M. Emanuel Institute of Biochemical Physics, Russian Academy of Sciences, 4, Kosygina st., Moscow, Russian Federation, 119334; Fax: +7(499)137-41-01; E-mail: chembio@sky.chph.ras.ru, www.ibcp.chph.ras.ru

N. A. Turovskij
Donetsk National University, 24 Universitetskaya Street, 83 055 Donetsk, Ukraine; E-mail: N.Turovskij@donnu.edu.ua

LIST OF ABBREVIATIONS

AFM	atomic-force microscope
BD	Brownian dynamics
CHC	Cahn–Hilliard–Cook
CNTs	carbon nanotube's
DFT	dynamic density functional theory
DGEBA	diglycidyl ether of bisphenol-A
DMTA	dynamic mechanical thermal analyzer
DPD	dissipative particle dynamics
ERM	effective reinforcing modulus
FEM	finite element method
HA	hexylamine
HMDA	hexamethylenediamine
LB	lattice Boltzmann
MC	Monte Carlo
MD	molecular dynamics
MM	molecular mechanics
MWCNTs	multiwalled CNTs
PMMA	poly(methyl methacrylate)
RVE	representative volume element
SUSHI	Simulation Utilities for Soft and Hard Interfaces
SWCNTs	single-walled CNTs
TDGL	time-dependent Ginsburg–Landau

PREFACE

This book facilitates the study of problematic chemicals in such applications as chemical fate modeling, chemical process design, and experimental design. This volume provides comprehensive coverage of modern physical chemistry and chemical engineering.

In the first chapter, aiming to control by stress birefringence the radio-transparent fiber-glass plastic products based on highly cross-linked polymer matrices, theoretical regularities for mathematical description of this property were developed. Computer physical modeling of topological structure of experimental objects was carried out on epoxy-amine polymers with different cross-link density taken as an example. And constants of this model were specified. The adequacy of the model was demonstrated by comparison of the model-calculated against experimental approach to thermal polarization curves.

Hyaluronan (HA) is a high-molecular weight, naturally occurring linear polysaccharide and found in all tissues and body fluids of higher animals. The excellent properties of HA such as biodegradability, biocompatibility, safety, excellent mucoadhesive capacity and high water retaining ability make it well-qualified for using in various bio-medical applications. In addition, HA is nontoxic, noninflammatory, and nonimmunogenic. Because of all these advantages, HA has received much attention as a matrix for drug delivery system. Chapter 2 will summarize our present knowledge about HA, as well as its properties and its development in some pharmaceutical applications.

The goal of Chapter 3 is to review the adhesive bonding of steel sheets treated by nitrooxidation and to compare the acquired results to the non-treated steel.

In Chapter 4, for the first time in the world practice the results of simulation by the Monte Carlo method of the kinetics of three-dimensional free-radical polymerization of tetrafunctional monomers (TFM) were obtained in the framework of the formation of a unitary three-dimensional structural element (UTDSE) and their structure formation on the simple cubic lattice, depending on the length l of molecules tetrafunctional monomers (l = 1 to 40 ribs of the lattice). Peculiarities of kinetics of changes in parameters such as the degree of polymerization of the Pn UTDSE, the number of radicals, the number of cross-links and cycles, and other characteristics were revealed. It was established that UTDSE are characterized by low levels of P_n for l = 1 and an explanation of this phenomenon was given. The study of the granulometric distribution (GMD) of UTDSE showed that curves of GMDs are bimodal and the probability density of these maximums was calculated.

The research in Chapter 5 is devoted to creation of elastomeric compositions based on systems with functionally active components for extreme conditions. The use of polycondensation capable monomers (PCCM) and other compounds with reactive groups was proposed for generating the stabilizing physical and chemical transformations. Thermodynamic analysis of open polycondensation systems and substantiation of various PCCM application as functionally active components of elastomeric materials have been conducted in the work; research results of polycondensation in an elastomeric matrix have been represented and a possibility of improving heat and corrosion resistance of elastomeric materials with introduction PCCM has been shown; and different ways of applying PCCM have been proposed and experimentally proved.

Aromatic polyimides and polyamides-based 4,4′-diaminothreephenylmethane has been synthesized. Their thermal, rheological properties and solubility in various organic solvents have been studied. In Chapter 6, it is shown that the solubility of the obtained polymers is connected with a free internal rotation triphenylmethan of bridge group and an effect of a surround phenyl substituent in diaminodiphenylmethane.

Possibilities of utilization of biopolymers, and particularly of the deoxyribonucleic acid (DNA) are reviewed and discussed in Chapter 7. The ways of their functionalization with photoresponsive molecules to get desired properties are described and illustrated on several examples as well as the processing of materials into thin films. Their room – and photothermal stability, studied by spectroscopic techniques is reported, together with optical damage thresholds. Physical properties and, more particularly linear, nonlinear and photoluminescent properties of obtained materials are also reviewed and discussed.

In Chapter 8, NMR ^{1}H and ^{13}C spectra of *tert*-butyl hydroperoxide in acetonitrile-d_3, chloroform-d and dimethyl sulfoxide-d_6 have been investigated by the NMR method. The calculation of magnetic shielding tensors and chemical shifts for ^{1}H and ^{13}C nuclei of the *tert*-butyl hydroperoxide molecule in the approximation of an isolated particle and considering the influence of the solvent in the framework of the continuum polarization model was carried out. Comparative analysis of experimental and computer NMR spectroscopy results revealed that the GIAO method with MP2/6–31G (d,p) level of theory and the PCM approach can be used to estimate the parameters of NMR ^{1}H and ^{13}C spectra of *tert*-butyl hydroperoxide.

Air pollution source by manufacture of ceramic materials are emissions of a smoke from refire kilns. Designs on modernization of system of an aspiration of smoke fumes of refire kilns in manufacture ceramic and refractories are devised. In this regard, experimental researches of efficiency of clearing of gas emissions are executed in Chapter 9. Modelling of process of a current of a gas-liquid stream is implemented in the program of computing hydrodynamics Ansys CFX. The ecological result of implementation of system consists highly clearings of a waste-heat and betterment of ecological circumstances in a zone of the factories.

In Chapter 10, an update on the modeling and mechanical properties of CNT/polymer nano-composites is presented. A very comprehensive references and further reading is also provided at the end of this chapter as well. Chapter 11 provides a detailed review on relevant approach of 3D reconstruction from two views of single 2D image and it potential applications in pore analysis of electrospun nanofibrous membrane. The review has concisely demonstrated that 3D reconstruction consists of three steps which is equivalent to the estimation of a specific geometry group. These steps include: estimation of the epipolar geometry existing between the stereo image pair, estimation of the affine geometry, and also camera calibration. The advantage of this system is that the 2D images do not need to be calibrated in order to obtain a reconstruction. Results for both the camera calibration and reconstruction are presented to verify that it is possible to obtain a 3D model directly from features in the images. Finally, the applications of 3D reconstruction in pore structure characterization of electrospun nanofibrous membrane are discussed.

Today, energy is an important requirement for both industrial and daily life, as well as political, economical, and military issues between countries. While the energy demand is constantly increasing every day, existing energy resources are limited and slowly coming to an end. Due to all of these conditions, researchers are directed to develop new energy sources which are abundant, inexpensive, and environmentally friendly. The solar cells, which directly convert sunlight into electrical energy, can meet these needs of mankind. Chapter 12 reviews the efforts in incorporating of solar cells into textile materials.

Quantum-chemical calculation of molecules dekacene, eicocene was done by method MNDO in Chapter 13. And optimized by all parameters geometric and electronic structures of these compounds. Each of these molecular models has a universal factor of acidity equal to 33 (pKa=33). They all pertain to class of very weak H-acids (pKa>14).

In Chapter 14 ceramization (ceramification) of polymer composites is presented as a promising method for gaining flame retardancy of the materials. Because of its passive fire protecting character, ceramization effect can be applied in polymer composites, which are dedicated for work in public places like shopping centers, sport halls, galleries and museums, office buildings, theaters or cinemas and public means of transport. In case of fire, ceramizable polymer composites turn into barrier ceramic materials, ensuring integrity of objects like electrical cables, window frames, doors, ceilings, etc., exposed on flames and heat, preventing from collapsing of materials and making electricity working, enabling effective evacuation. Moreover, ceramization process decreases emission of toxic or harmful gaseous products of polymer matrix degradation as well as its smoke intensity. The chapter describes mechanisms of ceramization for various polymer composites, especially focusing on silicone rubber-based ones, basic characteristics of the materials and ways of their testing.

In Chapter 15, we have attempted to establish relationships between the phase viscosity ratios of liquid gelatin (2%)-LBG (0.8%) systems and their rheological properties. To do that, LBG samples with different degrees of thermodegradation were taken.

In Chapter 16, a kinetic method of analysis of compounds from edible and medicinal plant extracts activity against stable radical 2,2-diphenyl-1-picrylhydrazyl (DPPH) is developed. The initial rate of DPPH's decay in standard conditions is suggested and theoretically explained as a kinetic parameter to compare the extract antiradical activity. A 10–150-fold decrease of the DPPH's reaction rate with plant extract antioxidants is achieved by the addition of acids into the reaction medium. Such results were explained by changes of input of different mechanisms into the whole process of scavenge of DPPH radical. A decrease of the reaction rate for the optimum of added acid's concentration with the acid strength increasing is also observed. The influence of the acids extracted from plant material on the results is excluded by this method because of the stronger acid addition. It is found that the linear interval of the dependence of DPPH's conversion degree after the first 30 min from the start of the reaction vs. the initial antioxidant concentration lies from 0 to 60%.

Chapter 17 discusses the development of a mathematical model for describing stress birefringence of highly cross-linked polymer matrices. Chapter 18 of this book gives an update on modern fibers, fabrics, and clothing.

This book offers a valuable overview and myriad details on current chemical processes, products, and practices. The book serves a spectrum of individuals, from those who are directly involved in the chemical industry to others in related industries and activities. It provides not only the underlying science and technology for important industry sectors, but also broad coverage of critical supporting topics.

ABOUT THE EDITOR

Gennady E. Zaikov, DSc

Gennady E. Zaikov, DSc, is the Head of the Polymer Division at the N. M. Emanuel Institute of Biochemical Physics, Russian Academy of Sciences, Moscow, Russia, and professor at Moscow State Academy of Fine Chemical Technology, Russia, as well as Professor at Kazan National Research Technological University, Kazan, Russia. He is also a prolific author, researcher, and lecturer. He has received several awards for his work, including the Russian Federation Scholarship for Outstanding Scientists. He has been a member of many professional organizations and on the editorial boards of many international science journals. Dr. Zaikov has recently been honored with tributes in several journals and books on the occasion of his 80th birthday for his long and distinguished career and for his mentorship to many scientists over the years.

CHAPTER 1

MODELING OF STRESS BIREFRINGENCE FOR HIGHLY CROSS-LINKED POLYMERS

NIKOLAI V. ULITIN,[1] NAIL K. NURIEV,[1] RAFIT R. NABIEV,[1] ILSHAT I. NASYROV,[1] DARIA A. SHIYAN,[1] and GENNADY E. ZAIKOV[2]

[1]*Kazan National Research Technological University, 68 Karl Marx Street, 420015 Kazan, Republic of Tatarstan, Russian Federation; Fax: +7 (843) 231-41-56; E-mail: n.v.ulitin@mail.ru*

[2]*N.M. Emanuel Institute of Biochemical Physics, Russian Academy of Sciences, 4, Kosygina st., Moscow, Russian Federation, 119334; Fax: +7(499)137-41-01; E-mail: chembio@sky.chph.ras.ru*

CONTENTS

ABSTRACT

Aiming to control by stress birefringence the radiotransparent fiber-glass plastic products based on highly cross-linked polymer matrices, theoretical regularities for mathematical description of this property were developed. Computer physical modeling of topological structure of experimental objects was carried out on epoxy-amine polymers with different cross-link density taken as an example. And constants of this model were specified. The adequacy of the model was demonstrated by comparison of the model-calculated against experimental approach to thermal polarization curves.

1.1 INTRODUCTION

To manufacture protective domes for radar installations, radiotransparent fiber-glass plastics are used, that is, polymer composite materials, consisting of highly cross-linked polymer matrix reinforced by glass fiber. When in use, radiotransparent products are subject to static (in particular, by its own weight) or dynamic loads, and highly cross-linked polymer matrix demonstrates an effect of stress birefringence. The actual challenge that arises during application of radiotransparent fiber-glass plastics is reduction of stress birefringence. Therefore, the aim of this work is to develop a mathematical model for describing stress birefringence of highly cross-linked polymer matrices in all physical states of the ones (glassy state, rubbery state, and transition state between them).

1.2 MATHEMATICAL MODEL

We can define relative deformation of polymer solid body of free shape in the form of tensor as follows [1]:

$$u_{ik} = \frac{1}{2}\mathbf{J}\tau_{ik} - \frac{1}{3}B_{\infty}p\delta_{ik}, \tag{1}$$

where B_{∞} is balanced bulk creep compliance, (MPa^{-1}); $\mathbf{J}$ is relaxation operator of shear compliance (MPa^{-1}), τ_{ik} is tensor of shear stress (MPa);

p is pressure, that is compressing any volume element without changing its shape (MPa); δ_{ik} is Kronecker symbol.

Contributions B_∞ in u_{ik} for highly cross-linked polymer matrices are very small and they are basically ignored [1]. If the deformation of cross-linked polymer matrix is not accompanied by destruction of its chemical structure, then:

$$\mathbf{J} = J_\infty \mathbf{J}_N, \tag{2}$$

where J_∞ is balanced shear compliance at the given temperature (MPa^{-1}); $\mathbf{J}_N$ is a normalized to 1 relaxation Volterra operator.

In this paper, all the discussions were made for highly cross-linked polymer matrices, which topological structure is spatially uniform. Highly cross-linked polymer matrices on supramolecular level of the structure is characterized by microhetero-phasicity: apart from gel-fraction formed by globules and their aggregates, here are present microdispersed formations (sol fraction) which are formed of linear and/or branched macromolecules of low molecular weight. Since different topologies of cross-linked macromolecules forming microgel super-molecular formations are found in the ones equally often over the entire volume, so they are considered statistically equivalent, and highly cross-linked polymer matrices are considered spatially uniform, owing to their topological structure. An important experimental proof of that is only inversely proportional dependence of balanced shear compliance J_∞ (MPa^{-1}) verses temperature (T, K):

$$J_\infty = A_\infty / T, \tag{3}$$

where A_∞ is an independent of T constant of rubbery state, K/MPa.

Since relaxation spectrum of shear compliance consists of β- and α-branches[1], therefore for highly cross-linked polymer matrices with spatially uniform topological structure, operator $\mathbf{J}_N$ is the following:

$$\mathbf{J}_N = w_{J,\beta} + (1 - w_{J,\beta})\mathbf{J}_{N,\alpha}, \tag{4}$$

[1] α-branch reflects cooperative mobility of network nodes, which are not participating in local movements; β-branch is connected with local conformation mobility.

where $w_{J,\infty}$ is weighting coefficient, independent of T and reflecting contribution of β-transitions in J_∞; $\mathbf{J}_{N,\alpha}$ is fractional exponential operator connected to distribution of α-relaxation time $L_{J,\alpha}(\theta)$:

$$\mathbf{J}_{N,\alpha} = \int_{-\infty}^{\infty} L_{J,\alpha}(\theta)[1-\exp(-t/\theta)]\,\mathrm{d}\ln\theta. \tag{5}$$

In Eq. (5): θ is relaxation time, t is current time.

Normalized to 1 α-mode has been described using distribution developed by Rabotnov [3]:

$$L_{J,\alpha}(\theta) = \frac{\sin[\pi(1-\Xi_{J,\alpha})]}{[2\pi\{\mathrm{ch}[(1-\Xi_{J,\alpha})\ln(\theta/\Theta_{J,\alpha})]+\cos[\pi(1-\Xi_{J,\alpha})]\}]}, \tag{6}$$

where $\Theta_{J,\alpha}$ is average α-relaxation time; $\Xi_{J,\alpha}$ is independent of T distribution width ($0 \leq \Xi_{J,\alpha} \leq 1$).

Operator $\mathbf{J}_{N,\alpha}$ in glassy state takes on the value equal to 0, in rubbery state equal to 1, and in transition state between these physical states it is equal to 0 up to 1. That is why Eqs. (1)–(6) will describe **J** of highly cross-linked polymer matrices with spatially uniform topological structure in all their physical states.

Stress birefringence was reduced to **J**, considering that ordered orientation of solid body molecules[2] is occurring by deformational shear. The point of departure for our discussions was the equation linking dielectric permittivity of the deformed polymer dielectric with independent components of relative deformation tensor [4]:

$$\varepsilon_{ik} = \varepsilon_0\delta_{ik} + a_1\gamma_{ik} + \frac{1}{3}\left(a_1 + 3a_2\right)u_{ll}\delta_{ik}, \tag{7}$$

here ε_0 is dielectric permittivity of the nondeformed solid body; γ_{ik} is deformation shear tensor; a_1, a_2 are polarization coefficients; u_{ll} is volumetric compression deformation tensor.

From the Eq. (7), can be obtained the equation of Brewster-Wertheim law:

$$\Delta n = \xi_\infty \Delta\gamma = C_\infty \Delta\tau, \tag{8}$$

[2]This orientation is the purpose of polarization anisotropy.

where Δn is stress birefringence; $\Delta\gamma$ and $\Delta\tau$ are differences of main shear deformations and stresses in the given point; C_∞ is balanced electromagnetic susceptibility (MPa^{-1}) connected with J_∞ equation:

$$C_\infty = 0.5\xi_\infty J_\infty, \tag{9}$$

where ξ_∞ is balanced elastic coefficient of electromagnetic susceptibility.

Let us introduce the relaxation operators for electromagnetic susceptibility $\mathbf{C}$ (MPa^{-1}) and for elastic coefficient of electromagnetic susceptibility ξ, assuming that they are connected with $\mathbf{J}$ in the form of equation below:

$$\mathbf{C} = 0.5\boldsymbol{\xi}\mathbf{J}. \tag{10}$$

The importance of this Eq. (10) lies in the fact that we get a result of coincidence of relaxation spectra ξ and shear module. Operator ξ is as follows:

$$\boldsymbol{\xi} = \xi_\infty(w_{\xi,\beta} + (1 - w_{\xi,\beta})\mathbf{G}_{N,\alpha}), \tag{11}$$

where $w_{\xi,\infty}$ is weighting coefficient independent of T and reflecting contribution of β-transitions in ξ_∞; $\mathbf{G}_{N,\alpha}$ is fractional exponential operator, reverse to $\mathbf{J}_{N,\alpha}$.

Basing on the rule of multiplication of the fractional exponential operators [3] from Eq. (10), we get:

$$\mathbf{C} = C_\infty(w_{C,\beta} + (1 - w_{C,\beta})\mathbf{J}_{N,\alpha}), \tag{12}$$

where $w_{C,\beta}$ is weighting coefficient independent of T and reflecting contribution of β-transitions in C_∞.

The result of Eq. (12) shows coincidence of relaxation spectra $\mathbf{J}$ and C, hereby $\mathbf{C}$, as well as $\mathbf{J}$, will cover all physical states of highly cross-linked polymer matrices. To apply the obtained regularities in practice we must know the temperature dependence of α-relaxation time:

$$\lg(\Theta_{J,\alpha}(T)/\Theta_{J,\alpha}(T_g)) = 40\big((f_g/(f_g + \alpha(1 - f_g)(T - T_g)) - 1)\big),$$
$$\alpha = \begin{cases} \alpha_g, & T < T_g \\ \alpha_\infty, & T > T_g \end{cases}, \tag{13}$$

where T_g is glass transition temperature, K; α_g, α_∞ are coefficients of thermal expansion in glassy and rubbery states, K^{-1}; f_g is fractional free volume at T_g.

Equation (13) is obtained analytically from equation below

$$\lg(\Theta_{J,\alpha}(T)/\Theta_{J,\alpha}(T_g)) = (f_g^{/} + [0.025\alpha_\infty(1-f_g)/f_g](T-T_g))^{-1} - (f_g^{/})^{-1},$$

suggested by Ferry [5] to describe relaxation of linear, branched and lightly cross-linked polymers at higher than T_g temperatures, taking into account the known temperature function f_g of highly cross-linked polymer matrices [6]. It is to be noted that value $f_g^{/} = 0.025$, which Ferry identified with fractional free volume at T_g, is not the same in reality, and for highly cross-linked polymer matrices, it is included in coefficient 40.

So, theoretical regularities of stress birefringence of highly cross-linked polymer matrices, forming under the influence of temperature fields and mechanical stresses, are fully described by an aggregate of Eqs. (2)–(6), (8), (10)–(13). According to these equations, stress birefringence can be assessed basing on such parameters as: T_g, α_g, α_∞, A_∞, ξ_∞, $w_{C,\infty}$, $w_{J,\infty}$, $\Theta_{J,\alpha}$, $\Xi_{J,\alpha}$. In this connection, theoretical and experimental assessment of these values was carried out to demonstrate adequacy of the introduced theoretical representations and their operability for evaluation of stress birefringence for highly cross-linked epoxy amine polymers at different conditions.

1.3 EXPERIMENTAL PART

1.3.1 EXPERIMENTAL OBJECTS

Experimental objects became highly cross-linked polymer matrices with a various of crosslink density on the basis of diglycidyl ether of bisphenol-A (DGEBA – Fig. 1.1 (a)), cured by mixtures of hexylamine (HA – Fig. 1.1(b)) and hexamethylenediamine (HMDA – Fig. 1.1(c)) at a variation of a molar ratio of the ones $x = n$ (HA)/n (HMDA) from 0 to 2 (step 0.5) taking into account a stoichiometry of epoxy groups and hydrogen of the amine group.

Preparation of epoxyamine compositions: DGEBA and HMDA weighted on scales METTLER TOLEDO AB304-S/FACT (up to 0.0001 g)

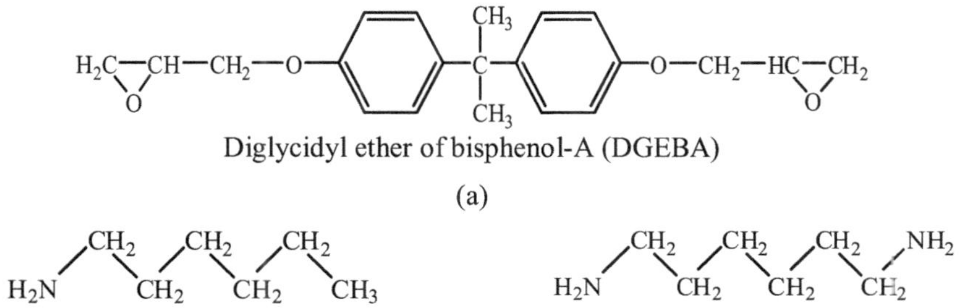

Diglycidyl ether of bisphenol-A (DGEBA)

(a)

Hexylamine (HA)

(b)

Hexamethylenediamine (HMDA)

(c)

FIGURE 1.1 Initial substances.

and heated to 315 K, then molt of HMDA and required amount of HA were mixed with DGEBA; then the mixture was stirred to form a homogeneous mass and was poured into an ampoule and vacuumized for 1 h in the "freeze-thaw" conditions; ampoule filled with argon and sealed. Curing conditions: at 293 K for 72 h, at 323 K for 72 h, at 353 K for 72 h, at 393 K for 72 h (chosen on the basis of the representations set forth in [7]).

1.3.2 *EXPERIMENTS BY PHOTOELASTIC METHOD*

One of the methods of experimental determination of stress birefringence is a photoelastic method. The tests of the experimental objects were conducted on the test facilities [8], designed to measure the relative stress birefringence in the center of the disk and the horizontal diameter relative deformation of the disk. Disc is made of the test material and 18 mm in diameter and 3 mm in thickness. At the test, the disc was compressed by the concentrated forces on the vertical diameter. Measurement error of the relative deformation and the stress birefringence does not exceed 3% and 1%, respectively. To determine the T_g, α_g, α_∞, test facilities were used as a dilatometer. Dilatometric curves $u(T)$ are temperature dependences of the relative deformation of unloaded samples, when cooling of the ones with a constant average speed of 0.4 K/min, were averaged in the results over four measurements. The T_g, α_g, α_∞ were determined by the method of ordinary least squares according to equation [9]:

$$u(T)=\frac{1}{3}\left[\alpha_{\infty}\left(T_g-T_0\right)+\alpha\left(T-T_g\right)\right],\quad \alpha=\begin{cases}\alpha_g, & T<T_g\\ \alpha_{\infty}, & T>T_g\end{cases},$$

where T_0 is initial temperature of the sample (rubbery state), K.

J_{∞} and C_{∞} are dependent of T, and therefore were determined on the basis of measurements of the relative deformation and stress birefringence with four loads for a series of temperatures above T_g+30 K followed by averaging the results of four tests for each load. Substituting the obtained values of J_{∞} and C_{∞} in Eqs. (3) and (9) allows determining of the experimental A_{∞} and ξ_{∞}.

C and **J** in glassy state has been calculated in terms of measuring results of relative deformations and stress birefringence using four loads at 298 K with subsequent averaging of results of four tests for each load. Experimental values $w_{J,\beta}$ and $w_{C,\beta}$ were defined by substitution of **J**, C values (in glassy state), and A_{∞}, ξ_{∞} in Eqs. (2), (4), (12), on condition that in the glassy state $\mathbf{J}_{N,\alpha}=0$. Creep and photocreep curves are the development of relatives deformation and stress birefringence into time at constant T under the influence of constant stress. They were taken for several T of transit state between glassy and rubbery states, $T_g\pm 15$ K, and were averaged over the results of four measurements. Empirical values of creep and photocreep functions were calculated by Eq. (8).

1.4 MODELING OF TOPOLOGICAL STRUCTURE OF EXPERIMENTAL OBJECTS

Modeling of topological structure of experimental objects was carried out in two ways: using computer modeling and graphs theory.

1.4.1 COMPUTER MODELING

The topological structure of highly cross-linked polymer matrices in the length interval of 0.25 to 2 nm is a random fractal [10]. Models of topological structure (Fig. 1.2) of experimental objects were arranged in the

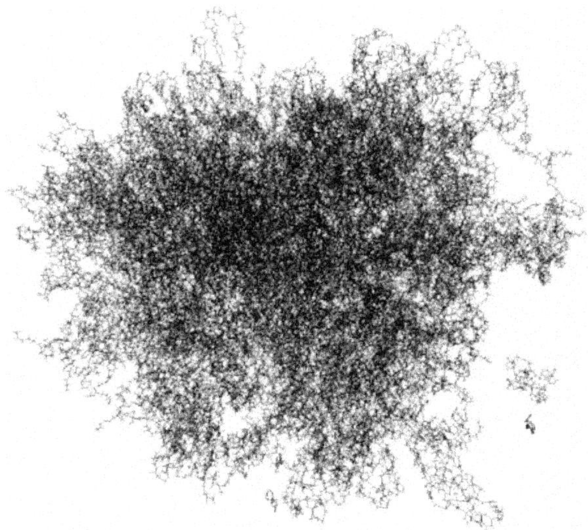

FIGURE 1.2 The model of topological structure of the experimental object composed of the composition $x = 2.0$.

Bullet Physics Library (www.bulletphysics.org) with use of known values of Van-der-Waals volume of the atoms and link lengths between them [6] plus fractal dimension (d_f) (Table 1.1) defined by the Bullet Physics Library from the following expression[10]:

$$N_{st} \propto R_w^{d_f},$$

where N_{st} is a number of random segments within the sphere radius R_w.

The nature of modeling was in polyaddition imitation. For each experimental object, the model was arranged in such a way, so that the topological structure was spatially uniform and the total number of elastically effective nodes comprised minimum 10,000.

1.4.2 *MODELING ON THE BASIS OF GRAPHS THEORY*

Figure 1.3 shows a structure of the repeating fragment common to all polymer series. If we designate the numbers of HA fragments with two methylene groups attached to nitrogen atom as N_{2f}, the elastically effective nodes as N_{3f}, the tetramethylene fragments as N_σ, and DGEBA fragments

TABLE 1.1 The Theoretical and Experimental Values of Glass Transition Temperature, the Constant of Rubbery State, the Balanced Elastic Coefficient of Electromagnetic Susceptibility and Weighting Coefficients

x	$\langle l \rangle$	n_{3f}	d_f	T_g, K			A_∞, K/MПa			ξ_∞			$w_{J,\beta}$			$w_{C,\beta}$		
				theor.	**exp.**	**ε,%**	**theor.**	**exp.**	**ε,%**	**theor.**	**exp.**	**ε,%**	**theor.**	**exp.**	**ε,%**	**theor.**	**exp.**	**ε,%**
0.0	1.00	0.4000	2.63	380	382	2	33.5	35.0	4	0.0240	0.0263	9	0.0231	0.0180	28	0.0407	0.0280	45
0.5	1.25	0.3333	2.65	373	372	1	47.3	53.0	11	0.0230	0.0235	2	0.0164	0.0150	9	0.0298	0.0260	15
1.0	1.50	0.2857	2.69	360	361	1	62.1	69.3	10	0.0220	0.0224	2	0.0125	0.0140	11	0.0232	0.0230	1
1.5	1.75	0.2500	2.72	350	350	0	77.7	81.6	5	0.0220	0.0207	6	0.0100	0.0120	17	0.0188	0.0200	6
2.0	2.0	0.2222	2.74	344	344	0	93.9	93.5	4	0.0220	0.0192	15	0.0083	0.0080	4	0.0160	0.0180	11

* The relative difference of the theoretical value versus the experimental value was calculated by formula $\varepsilon = |(\text{exp.value} - \text{theor.value})/\text{exp. value}| \cdot 100,\%$.

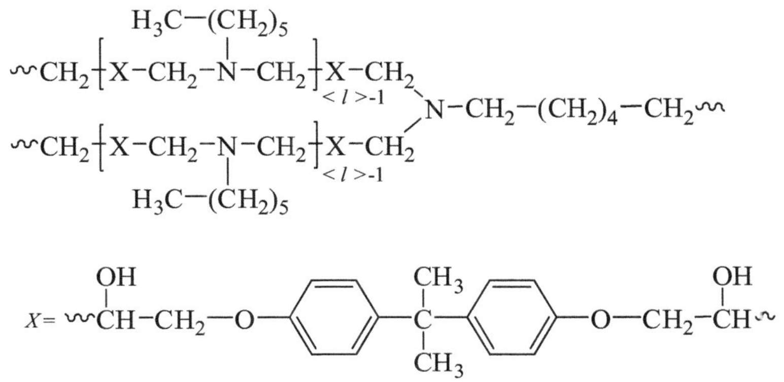

FIGURE 1.3 The repeating fragment of the topological structure covering all experimental objects.

as N_{π}, and N_{tot} being the total number of links, then, according to stoichiometry, we work out as follows:

$$N_{2f} = xn(HMDA), N_{3f} = 2n(HMDA), N_{\sigma} = n(HMDA),$$

$$N_{\pi} = (2+x)n(HMDA),\ N_{tot} = (5+2x)n(HMDA)\cdot$$

In this case, random parameter representing the number of elastically effective nodes n_{3f} (Table 1.1), and the number average degree of polymerization of the intermodal chain $<l>$ (Table 1.1) are as follows:

$$n_{3f} = N_{3f} / N_{tot} = 2/(5+2x), \quad <l> = N_{\pi} / N_{3f} = 1+0.5x\cdot$$

1.5 THEORETICAL ASSESSMENT OF CONSTANTS FOR STRESS BIREFRINGENCE MODEL

1.5.1 GLASS TRANSITION TEMPERATURE

Glass Transition Temperature [10] (Table 1.1):

$$T_g = Cl_{st}^{d_f - d}, K$$

where C = 270 K is a constant, d is Euclidean dimension, l_{st} is an average size of a random segment, nm.

The difference between theoretical and experimental values T_g is maximum 2%.

1.5.2 COEFFICIENTS OF THERMAL EXPANSION

The calculated by the below equations [10] values of α_g and α_∞ for all experimental objects turned out to be approximately equal, and in average were 3.2×10^{-4} K^{-1} and 6.1×10^{-4} K^{-1}, respectively. These values agree with experimental values: $4.3\cdot10^{-4}$ K^{-1} and $7.0\cdot10^{-4}$ K^{-1}, the relative difference between the theory and the experiment is 26% and 13%, respectively.

$$\alpha_g = \frac{-3(T_g\cdot10^{-4}-2d_f+5)-\sqrt{9(T_g\cdot10^{-4}-2d_f+5)^2-12T_g^2\cdot10^{-8}}}{2T_g^2\cdot10^{-4}}, K^{-1},$$

$$\alpha_\infty = \frac{0.106+\alpha_g T_g}{T_g} = \frac{0.106}{T_g}+\alpha_g, K^{-1}.$$

1.5.3 CONSTANT OF RUBBERY STATE AND BALANCED ELASTIC COEFFICIENT OF ELECTROMAGNETIC SUSCEPTIBILITY

Constant of rubbery state is as follows [6]:

$$A_\infty = (0.5f\,C_{tot}n_{3f}RF)^{-1}, \tag{14}$$

where f is functionality of the network nodes; F – front-coefficient; R – gas constant (8.314 J/(mol·K)); C_{tot} – concentration of elemental links of the topological structure at T_g, mol/cm³.

By definition, we have:

$$C_{tot} = N_{tot}/V_{tot}(T_g),$$

where $V_{tot}(T_g)$ is volume of elemental links at T_g (cm³):

$$V_{tot}(T_g) = N_\sigma V_\sigma(T_g)+N_{3f}V_{3f}(T_g)+N_{2f}V_{2f}(T_g)+N_\pi V_\pi(T_g),$$

where, $V_\sigma(T_g)$, $V_{3f}(T_g)$, $V_{2f}(T_g)$, $V_\pi(T_g)$ are molar volumes of elemental links at T_g, cm^3/mol. Then:

$$C_{tot} = (5+2x)d(T_g)/(M_\sigma + 2M_{3f} + xM_{2f} + (2+x)M_\pi), \qquad (15)$$

where M_σ, M_{3f}, M_{2f}, M_π are molar mass of elemental links, g/mol; $d(T_g)$ is polymer density at T_g, g/cm^3.

For highly cross-linked polymer matrices temperature dependence of density is as follows [6]:

$$d(T) = k_g M_{r.f.} / \left(10^{-24}\left[1+\alpha\left(T-T_g\right)\right]N_A\left(\sum_i \Delta V_i\right)_{r.f.}\right), \alpha = \begin{cases} \alpha_g, T < T_g \\ \alpha_\infty, T > T_g \end{cases},$$

where, $d(T)$ is polymer density at T, g/cm^3; kg is molecular packing coefficient at T_g; $M_{r.f.}$ is molar mass of the repeating fragment of network, g/mol; 10^{-24} is conversion coefficient, from $Å^3$ to cm^3; N_A is Avogadro constant; $\left(\sum_i \Delta V_i\right)_{r.f.}$ is Van-der-Waals volume of the repeating fragment of network.

The molar mass of the repeating fragment of network is $M_{r.f.} = 441 < l > - 43$ g/mol.

Taking into account, that for highly cross-linked polymer matrices at T_g, the kg $\approx$ 0.681 [6], we work out:

$$d(T) = 1.13(441 < l > - 43)/\left(\left[1+\alpha\left(T-T_g\right)\right](458.9 < l > - 57.4)\right),$$
$$\alpha = \begin{cases} \alpha_g, T < T_g \\ \alpha_\infty, T > T_g \end{cases}.$$

Thus, substitution of T_g, α_g and α_∞ allows finding the $d(T)$ values at any T values. The molar masses of the elemental links are as follows: $M_\sigma = 56$ g/mol, $M_{3f} = 56$ g/mol, $M_{2f} = 127$ g/mol, $M_\pi = 314$ g/mol.

Substitution of resultant expression for C_{tot}

$$C_{tot} = (1.13(5+2x)(441 < l > - 43))/((796+441x)(458.9 < l > - 57.4))$$

in Eq. (14) yields:

$$A_\infty = \frac{(796+441x)(458.9 < l > -57.4)}{\left(1.695FR(5+2x)(441 < l > -43)n_{3f}\right)}.$$

The F values for highly cross-linked polymer matrices with spatially uniform topological structure are within the range of 0.65 ÷ 0.85 and should increase in a linear fashion as the network density grows up [2]. On this basis we receive:

$$F = 1.125 n_{3f} + 0.4$$

$$A_{\infty} = \frac{(796 + 441x)(458.9 < l > -57.4)}{\left(1.695\left(1.125 n_{3f} + 0.4\right) R (5 + 2x)(441 < l > -43) n_{3f}\right)}. \tag{16}$$

Experimental values of A_{∞} (Table 1.1) can be determined by way of approximation of empirical J_{∞} values (2). Small difference between theory and experiment for A_{∞} (4 ÷ 11%) prove the adequacy of the Eq. (16). Experimental values of F can be determined from A_{∞} values, for that, on the basis of experimental polymer density values, preliminary calculation of concentration of elemental links should be found by Eq. (15).

To assess ξ_{∞}, the following equation was established:

$$\xi_{\infty} = -3K(\partial \delta \varepsilon / \partial T) / (\alpha_{\infty} \sqrt{\varepsilon_0}), \tag{17}$$

where ε_0 is dielectric permittivity of the unstressed polymer at T_g, K is constant, depending on polymer topological structure; $\partial \delta \varepsilon / \partial T$ is one of

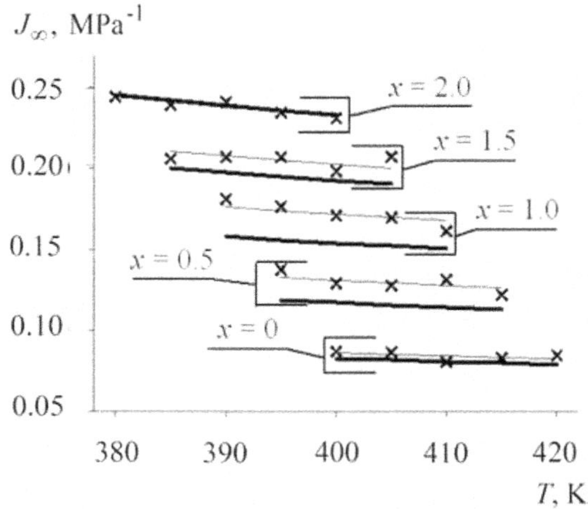

FIGURE 1.4 Temperature dependence of balanced shear compliance (cross-experiment, line-approximation).

components of the derivative ε_0 in T, which relates to electromagnetic anisotropy in stressed polymer, K^{-1}; ε_0 and $\partial\delta\varepsilon/\partial T$ were calculated incrementally according to standard methods [6]. Constant K was determined by the method described in the work [4], its value for experimental objects is 15.7.

Finally (Table 1.1):

$$\xi_\infty = 0.07\left[\frac{212.7362 < l > +5.309}{458.9 < l > -57.4}\right]\sqrt{\frac{186.860314 < l > -24.388346}{455.465782 < l > -54.939368}}.$$

The difference between theoretical and experimental values, ξ_∞, was maximum 15%.

1.5.4 WEIGHTING COEFFICIENTS

Weighting coefficients (Table 1.1):

$$w_{J,\beta} = \frac{J_{\beta,\infty}}{A_\infty}T, \quad w_{C,\beta} = \frac{C_{\beta,\infty}}{0.5\xi_\infty A_\infty}T,$$

where $C_{\beta,\infty} = \frac{J_{\beta,\infty} a_1}{4n_0}$; $J_{\beta,\infty} = \frac{2}{3}\cdot\frac{1+\mu_{\beta,\infty}}{1-2\mu_{\beta,\infty}}B_\infty$ $\mu_{\beta,\infty} = \frac{d_f}{d-1} - 1.$

Here a_1 is coefficient (refer to Eq. (7)); n_0 is a refractive index of the nondeformed polymer dielectric, which is in glassy state. The a_1 and n_0 are calculated incrementally according to standard methods at 298 K [6]. Table 1 shows a comparison of experimental and theoretical determination of the weighting coefficients.

1.5.4 PARAMETERS OF THE RELAXATION SPECTRUM

Parameters of the relaxation spectrum theoretically cannot be assessed neither within the framework of the increments method, nor within the framework of the fractal approach. Empirical values of $\Theta_{J,\alpha}$ (T) and

$\Xi_{J,\alpha}$ were determined from the creep and photocreep curves (method of least squares with regularization of solutions by singular decomposition [11]). It was revealed that, for the highly cross-linked polymer matrices with spatially uniform topological structure $\Xi_{J,\alpha}$ is determined by network topology and it is independent of temperature (no splitting of α-transition). The experimentally found value $\Xi_{J,\alpha}$ increases linearly from 0.4 to 0.65 with an increase $\langle l \rangle$ (for a priori prediction, the value of 0.5 can be used). Example of $\Theta_{J,\alpha}$ (T) dependence of temperature is shown in Fig. 1.5: the Eq. (13) describes both branches of the experimental curve lg$\Theta_{J,\alpha}$ (T) with high accuracy. The obtained value of the share of the fluctuation free volume was averaged over all experimental objects, the average value was f_g = 0.095. This result is consistent with the currently accepted value for the highly cross-linked polymer matrices – 0.09.

1.6 ADEQUACY OF THE MODEL

The adequacy of the model of stress birefringence of highly cross-linked polymer matrices was experimentally demonstrated by comparing of the predicted and the actual course of the thermal polarization curves (Fig. 1.6).

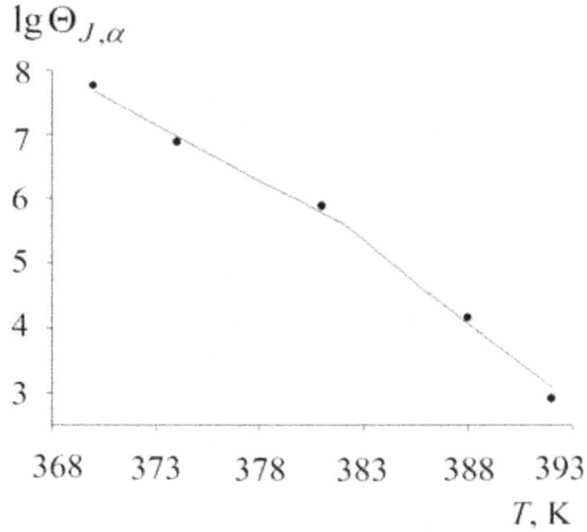

FIGURE 1.5 The dependence of lg$\Theta_{J,\alpha}$ on T for the object of the composition $x = 0$ (dots-experiment, line-approximation).

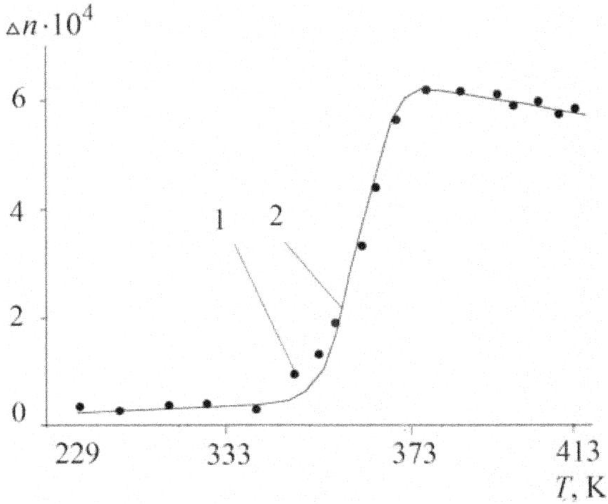

FIGURE 1.6 Thermal polarization curve for the experimental object of the composition $x = 1.5$ (1 – experiment, 2 – calculation based on the proposed model).

1.7 CONCLUSION

Thus, the stress birefringence model and the fractal-incremental approach allow assessing the stress birefringence to be done, even before the experiment starts. So, a priori, we can assess the maximum ultimate value of stress birefringence for which highly cross-linked polymer matrix is capable in the given operating conditions of radio-transparent fiberglass. Knowing the value of stress birefringence of highly cross-linked polymer matrices, one can assess the change of fiberglass radio-transparency coefficient application-wise. Hence, the possibility of highly cross-linked polymer matrix application for fiberglass is being justified.

KEYWORDS

- **fractal analysis of macromolecules**
- **heredity theory**
- **highly cross-linked epoxy-amine polymers**
- **modeling**

REFERENCES

1. Davide S. A., De Focatiis, C. Paul Buckley. Prediction of Frozen-In Birefringence in Oriented Glassy Polymers Using a Molecularly Aware Constitutive Model Allowing for Finite Molecular Extensibility. *Macromolecules*, 2011, 44(8), 3085–3095.
2. Irzhak V. I. Topological structure and relaxation properties of polymers, *Russ. Chem. Rev.* 2005, 74, 937.
3. Rabotnov Y. N. Mechanics of a Deformable Solid Body. 1979. Nauka, M., 744 (in rus).
4. Blythe T., Bloor D. Electrical properties of polymers. Cambridge: Cambridge University Press, 2005, 492.
5. Ferry, J. D. Viscoelastic properties of polymers, 3rd ed. New York-Chichester-Brisbane-Toronto-Singapore: John Wiley & Sons, Inc., 1980, 641.
6. Askadskii A. A. Computational Materials Science of Polymers. Cambridge: Cambridge International Science Publish, 2003, 650.
7. Irzhak V. I., Mezhikovskii S. M. Structural aspects of polymer network formation upon curing of oligomer systems, *Russ. Chem. Rev*. 2009, 78(2), 165–194.
8. Zuev, B. M., Arkhireev, O. S. The initial stage in the fracture of stressed dense-cross-linked polymer systems, 1990, *Polymer Science U.S.S.R.* 32(5), 941–947.
9. Handbook of Thermal Analysis and Calorimetry, Volume 3: Applications to Polymers and Plastics Ed. by S.Z.D. Cheng, Amsterdam: Elsevier Science B. V., 2002, 1–45.
10. Novikov V. U., Kozlov G. V. Structure and properties of polymers in terms of the radical approach, *Rus. Chem. Rev*. 2000, 69(6), 523–549.
11. Tihonov A. N., Arsenin V. Ja. Methods of Decision of Incorrect Problems, 3rd ed. M. Nauka. 1986, 287 (in rus).

CHAPTER 2

HIGH-MOLECULAR WEIGHT BIOPOLYMER

A. M. OMER,[1,2] T. M. TAMER,[1] and M. S. MOHYELDIN[1]

[1]*Polymeric Materials Department, Advanced Technologies and New Materials Research Institute, City of Scientific Research and Technological Applications, New Borg El-Arab City 21934, Alexandria, Egypt; Materials Delivery Group, Polymeric Materials Department, Advanced Technologies and New Materials Research Institute (ATNMRI), City of Scientific Research and Technological Applications (SRTA-City), New Borg El-Arab City 21934, Alexandria, Egypt; E-mail: Ahmedomer_81@yahoo.com*

[2]*Laboratory of Bioorganic Chemistry of Drugs, Institute of Experimental Pharmacology and Toxicology, Slovak Academy of Sciences, Bratislava, Slovakia*

CONTENTS

ABSTRACT

Hyaluronan (HA) is a high-molecular weight, naturally occurring linear polysaccharide and found in all tissues and body fluids of higher animals. The excellent properties of HA such as biodegradability, biocompatibility, safety, excellent mucoadhesive capacity and high water retaining ability make it well-qualified for using in various bio-medical applications. In addition; HA is nontoxic, noninflammatory and nonimmunogenic. Because of all these advantages, HA has received much attention as a matrix for drug delivery system. This review will summarize our present knowledge about HA, properties and its development in some pharmaceutical applications.

2.1 INTRODUCTION

2.1.1 HISTORICAL PERSPECTIVE OF HYALURONAN

Hyaluronan is one of the most interesting and useful natural biopolymer macromolecules and considered as a member of a similar polysaccharides group, and also known as mucopolysaccharides, connective tissue polysaccharides, or glycosaminogylcans [1–3]. The popular name of hyaluronic acid (HA) is derived from "hyalos," which is the Greek word for glass + uronic acid, and it was discovered and investigated in 1934 by Karl Meyer and his colleague John Palmer [4]. Firstly, they isolated a previously unknown chemical substance from the vitreous body of cows' eyes as an acid form but it behaved like a salt in physiological conditions (sodium hyaluronate) [5–7], they solved the chemical structure of HA and found that its composed from two sugar molecules (D-glucuronic acid (known as uronic acid) and D-Nacetyl glucosamine) and they named the molecule "hyaluronic acid" because of the hyaloid appearance of the substance when swollen in water and the probable presence of hexuronic acid as one of the components. Hyaluronan (HA) is the currently used name; hence it represents a combination of "hyaluronic acid" and "hyaluronate," in order to indicate the different charged states of this polysaccharide [8].

In 1942, HA was applied for the first time as a substitute for egg white in bakery products [6], and shortly afterward, in 1950s HA was isolated from umbilical cord and then from rooster combs [7], and finally it was isolated from other sources. HA is present in synovial fluid (SF) with final physiological concentration about 2–3 mg/mL, and the largest amounts of HA are found in the extracellular matrix (ECM) of soft connective tissues [9. 10] and so its widely distributed in vertebrate connective tissues, particularly in umbilical cord, vitreous humor, dermis, cartilage, and intervertebral disc [11, 12]. Also, it was reported that HA is present in the capsules of some bacteria (e.g., strains of Streptococci) but it's absent completely in fungi, plants, and insects [13].

2.1.2 PHYSICOCHEMICAL PROPERTIES OF HYALURONAN

2.1.2.1 Chemical Structure

HA is an un-branched nonsulfated glycosaminoglycan (GAG) composed of repeating disaccharides and present in the acid form [14–16], and composed of repeating units from D-glucuronic acid and N-acetyl-D- glucosamine linked by a glucuronidic β(1–3) bond [17] as shown in Fig. 2.1. Also HA forms specific stable tertiary structures in aqueous solution.

Both sugars are spatially related to glucose which in the β-configuration allows all of its bulky groups (the hydroxyls, the carboxylate moiety, and the anomeric carbon on the adjacent sugar) to be in sterically favorable equatorial positions while all of the small hydrogen atoms occupy the less sterically favorable axial positions. Thus, the structure of the disaccharide is energetically very stable [18]. Several thousand sugar molecules can be

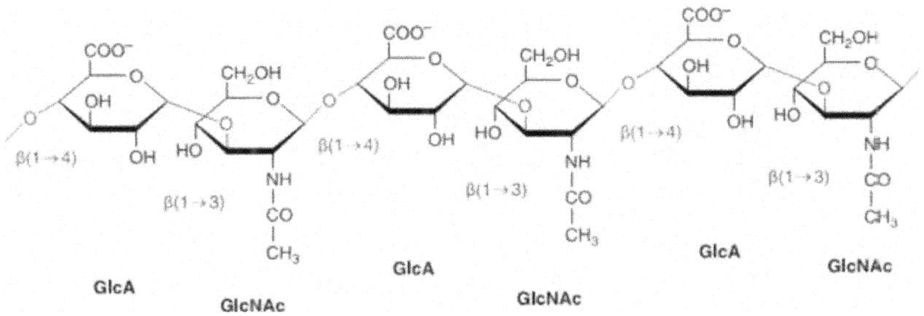

FIGURE 2.1 Hyaluronan is composed of repeating polymeric disaccharides D-glucuronic acid (GlcA) and *N*-acetyl-D-glucosamine (GlcNAc) linked by a glucuronidic (1–3) bond. Three disaccharide GlcA-GlcNAc are shown [17].

included in the backbone of HA. The structure of HA called a coiled structure, and this can attributed to that the equatorial side chains form a more polar face (hydrophilic), while the axial hydrogen atoms form a nonpolar face (relatively hydrophobic), and this, led to a twisted ribbon structure for HA (i.e., a coiled structure) [6].

2.1.2.2 Solubility and Viscosity

According to hygroscopic and homeostatic properties of HA; the molecules of HA can be readily soluble in water and this property prompt the proteoglycans for hydration producing a gel like a lubricant [19. HA also exhibit a strong water retention property and this advantage can be explained by the fact that HA is a natural hydrophilic polymer, (i.e., water soluble polymer), where its contain carboxylic group and also high number of hydroxyl groups which impart hydrophilicity to the molecule, and so increase affinity of water molecules to penetrate in to the HA network and swells the macromolecular chains consequently. The water retention ability of HA can also attributed to the strong anionic nature of HA, where the structure of the HA chains acts to trap water between the coiled chains and giving it a high ability to uptake and retain water molecules. It was stated that stated that HA molecules can retain water up to 1,000 times from own weight [3]. The water holding capacity of HA increases with increasing relative humidity [20] therefore, the hydration parameters are independent of the molecular weight of the HA [21].

On the other hand, the viscosity is one of the most important properties of HA gel, in which several factors affecting the viscosity of this molecule such as the length of the chain, molecular weight, cross-linking, pH and chemical modification [22]. The rotational viscometry is considered one of the successful and simplest instruments which used for identification of the dynamic viscosity and the 'macroscopic' Properties of HA solutions [23, 24]. It was indicated that the viscosity is strongly dependent on the applied shear-stress. At concentrations less than 1 mg/mL HA start to entangle. Morris and his co-workers identified the entanglement point by measuring the viscosity, they confirmed that the viscosity increases rapidly and exponentially with concentration ($\sim c^{3.3}$) beyond the entanglement point, also, the viscosity of a solution with concentration10 g/L at low shear probably equal to10^6 times the viscosity of the solvent [25]. While, at high shear the viscosity may drop as much as ~10^3 times [26]. However; in the synovial fluid (SF), unassociated high molar mass HA confers its unique viscoelastic properties which required for maintaining proper functioning of the synovial joints [27].

2.1.2.3 Viscoelasticity

Viscoelasticity is another characteristic of HA resulting from the entanglement and self-association of HA random coils in solution [5]. Viscoelasticity of HA can be related to the molecular interactions which are also dependent on the concentration and molecular weight of HA. The higher the molecular weight and concentration of HA, the higher the viscoelasticity the solutions possess. In addition, with increasing molecular weight, concentration or shear rate, HA in aqueous solution is undergo a transition from Newtonian to non-Newtonian characteristics [28]. The dynamic viscoelasticity of HA gels was increased relative to HA–HA networks when the network proteoglycan–HA aggregates shift the Newtonian region to lower shear rates [29]. In addition to the previous properties of HA, the shape and viscoelasticity of HA molecule in aqueous solution like a polyanion is undergo the pH sensitivity (i.e., pH dependent) and effected by the ionic strength [30, 31]. Indead, HA has a pKa value of about 3.0 and therefore, the extent of ionization of the HA chains was affected by the change in pH. The intermolecular interactions between the HA molecules may be affected by the shift in ionization, which its rheological properties changes consequently [32].

2.1.3 DEGRADATION OF HA

In principle there are many ways of HA degradation depends on biological (enzymatic) or physical and chemical (non enzymatic) methods.

2.1.3.1 Biological Methods

In the biological methods, the degradation of HA can take place using enzymes. It was reported that there are three types of enzymes which are present in various forms, in the intercellular space and in serum (hyaluronidase, β-D-glucuronidase, and β–N-acetyl-hexosaminidase) are involved in enzymatic degradation [33]. Hyaluronidase (HYAL) is considered as a most powerful degradation enzyme for hyaluronan [34]. Volpi et al. [35] reported that Hyaluronidase cleaves high molecular weight HA into smaller fragments of varying size via hydrolyzing the hexosaminidic β (1–4) linkages between N-acetyl-D-glucosamine and D-glucuronic acid residues in HA [36], while the degradation of fragments via removal of non-reducing terminal sugars can be done by the other two enzymes. However, it was found that the HYAL enzymes are present with very low concentrations and the measuring of its activity, characterization and purification of it are difficult, in addition, measuring their activity, which is high but unstable, so that this family has received little attention until recently [37].

2.1.3.2 Physical Methods

By physical methods, the degradation and depolymerization of HA can be performed by different techniques, it was reported that HA can be degraded using ultrasonication in a nonrandom fashion and the obtained results shows that high molecular weight HA chains degrade slower than low molecular weight HA chains [38]. However, it was noted that the degradation of HA into monomers is not fully completed when using different HA samples under applying different ultrasound energies, and the increasing of absorbance at 232 nm after sonication is not observed [38]. Heat is another type of the physical methods used for HA degradation, in which with increasing temperature the degradation increased consequently and the viscosity strongly decreased [39]. In case of thermal degradation method, it was reported that the treatment of different HA samples at temperatures from 60 to 90°C for 1 h results in only moderate degradation and a small increase of polydispersity [40]. Bottner and his co-workers have proved that that thermal degradation of HA occurs in agreement with the random-scission mechanism during the study of two high-molar-mass HA samples that were extensively degraded at 128°C in an autoclave [41]. HA can also degrade by other physical methods like γ-irradiation [38].

2.1.3.3 Chemical Methods

HA like other polysaccharides can be degraded by acid and alkaline hydrolysis or by a deleterious action of free radicals [38, 42, 43].

Stern et al. [38] reported the degradation of HA by acid and alkaline conditions occurs in a random fashion often resulting in disaccharide fragment production. Where, the glucuronic acid moiety of HA degraded via acidic hydrolysis, while the alkaline hydrolysis occurs on *N*-acetylglucosamine units and giving rise to furan containing species [44]. Also, the oxidation processes can degrade HA via reactive oxygen species (ROS) such as hydroxyl radicals and superoxide anions which generated from cells as a consequence of aerobic respiration [5]. It was found that acceleration of degradation of high-molecular-weight HA occurring under oxidative stress produces an impairment and loss of its viscoelastic properties [45, 46]. Figure 2.2 describes the fragmentation mechanism of HA under free radical stress [47].

The ROS are involved in the degradation of essential tissue or related components such as synovial fluid (SF) of the joint which contains high-molar-mass HA. It is well known that most of rheumatic diseases are resulting from reduction of HA molar mass in the synovial fluid of patients. Numerous studies have been reported to study the effect of various ROS on HA molar mass. Soltes and his team focused their research on the hydroxyl radicals resulting from the reaction mechanism of (H_2O_2 + transitional metal cation H_2O2 in the presence of ascorbic acid as a reducing agent) under aerobic conditions and studied its effect on the degradation of HA molar mass [38, 48–50], in which the system of ascorbate and metal cation as copper(II) ions enduces hydrogen peroxide (H_2O_2) to turn into OH^- radicals by a Fenton-like reaction [51] and this system is called Weiss Berger's oxidative system. They observed a decrease of the dynamic viscosity value of the HA solution, and this indicate the degradation of the HA by the system containing Cu (II) cations [52]. Therefore, agents that could delay the free-radical-catalyzed degradation of HA may be useful in maintaining the integrity of dermal HA in addition to its moisturizing properties [53]. It should be noted that the concentrations of ascorbate and Cu(II) were comparable to those that may occur during an early stage of the acute phase of joint inflammation [12, 54–63].

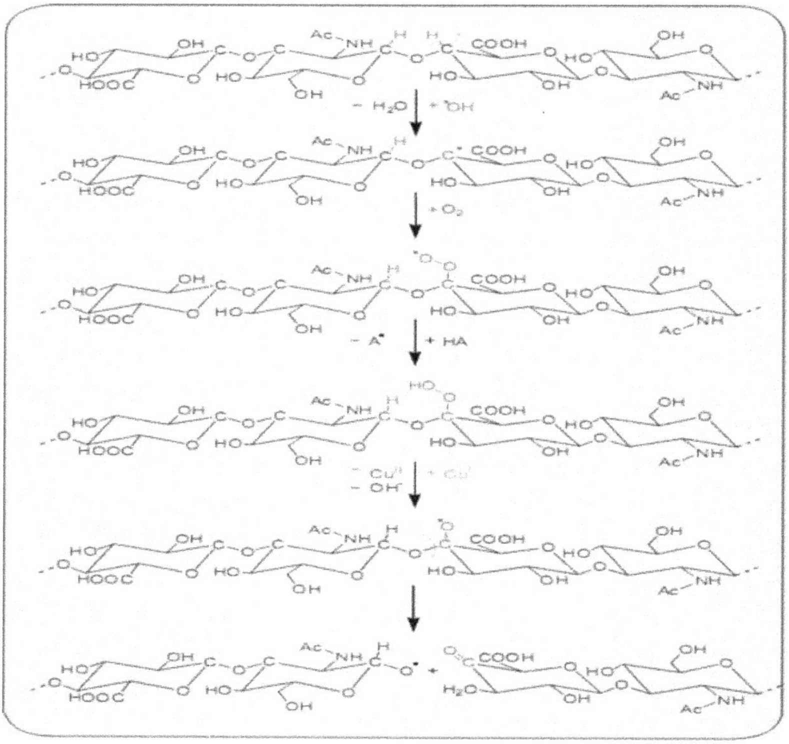

FIGURE 2.2 Schematic degradation of HA under free radical stress [47].

2.2 APPLICATIONS OF HA

2.2.1 PHARMACEUTICAL APPLICATIONS

Indeed, HA and its modified forms have been extensively investigated and widely used for various pharmaceutical applications [64]. In the current review, we presented in brief some pharmaceutical applications of HA biopolymer.

2.2.1.1 HA in Drug Delivery Systems

It is well known that macromolecular drug forms are composed basically of three components: (i) the carrier; (ii) the drug; and (iii) a link between them [34, 65].

It was reported that polysaccharide-based microgels are considered as one class of promising protein carriers due to their large surface area, high water absorption, drug loading ability, injectability, nontoxicity, inherent biodegradability, low cost and biocompactibility. Among various polysaccharides, HA, has been recently most concerned [66].

The physicochemical and biological properties of hyaluronan qualify this macromolecule as a prospective carrier of drugs. This natural anionic polysaccharide has an excellent mucoadhesive capacity and many important applications in formulation of bioadhesive drug delivery systems. It was found that this biopolymer may enhance the absorption of drugs and proteins via mucosal tissues [67]. In addition, it is immunologically inert, safely degraded in lysosomes of many cells [34] and could be an ideal biomaterial for drug and gene delivery [68, 69]. Therefore, HA biopolymer has become the topic of interest for developing sustained drug delivery devices of peptide and protein drugs in subcutaneous formulations. The recent studies also suggested that HA molecules may be used as gel preparations for nasal and ocular drug delivery [70]. Also, HA has been used for targeting specific intracellular delivery of genes or anticancer drugs.

The applications of HA in the above mentioned drug delivery systems and its advantages in formulations for various administration routes delivery were summarized in Table 2.1 [22].

2.2.1.2 Nasal Delivery

Over the last few decades nasal route has been explored as an alternative for drug delivery systems (Nonparenteral). This is due to the large surface area and relatively high blood flow of the nasal cavity and so the rapid absorption is possible [79, 81]. It was reported that viscous solutions of polymer have been shown to increase the residence time of the drug at the nasal mucosa and thereby promote bioavailability [82]. The mucoadhesive properties of HA could promote the drugs and proteins absorption through mucosal tissues [78]. The mucoadhesive property of HA can be increased by conjugating it with other bioadhesive polymers such as Chitosan and polyethylene glycol. Lim and his team prepared biodegradable microparticles using chitosan (CA) and HA by the solvent evaporation method, they used gentamicin used as a model drug for intranasal

TABLE 2.1 Summary of Some Drug Delivery Applications of HA and Its Advantages [22]

Administration route	Advantages	References
Intravenous	• Enhances drug solubility and stability • Promotes tumor targeting via active (CD44 and other cell surface receptors) and passive (EPR) mechanisms • Can decrease clearance, increase AUC and increase circulating half-life	[71, 72]
Dermal	• Surface hydration and film formation enhance the permeability of the skin to topical drugs • Promotes drug retention and localization in the epidermis • Exerts an anti-inflammatory action	[29, 73]
Subcutaneous	• Sustained/controlled release from site of injection • Maintenance of plasma concentrations and more favorable • Pharmacokinetics • Decreases injection frequency	[74, 75]
Intra-articular	• Retention of drug within the joint • Beneficial biological activities include the anti-inflammatory, analgesic and chondroprotective properties of HA	[76, 77]
Ocular	• The shear-thinning properties of HA hydrogels mean minimal effects on visual acuity and minimal resistance to blinking • Mucoadhesion and prolonged retention time increase drug bioavailability to ocular tissues	[50]
Nasal	• Mucoadhesion, prolonged retention time, and increased permeability of mucosal epithelium increase bioavailability	[78, 79]
Oral	• Protects the drug from degradation in the GIT • Promotes oral bioavailability	[6]
Gene	• Dissolution rate modification and protection	[69, 80]

studies in rats and sheep [83]. The results showed that the release of gentamicin is prolonged when formulated in HA, CH and HA/CH and that the resultant microparticles are mucoadhesive in nature [78, 84]. In addition, much attention has received for delivery of drugs to the brain via the olfactory region through nasal route, that is, nose-to-brain transport [85]. Horvat group developed a formulation containing sodium hyaluronate in combination with a nonionic surfactant to enhance the delivery of hydrophilic compounds to the brain via the olfactory route, the results proved that HA, a nontoxic biomolecule used as a excellent mucoadhesive polymer in a nasal formulation, increased the brain penetration of a hydrophilic compound, the size of a peptide, via the nasal route [67].

2.2.1.3 Ocular Delivery

The current goals in the design of new drug delivery systems in ophthalmology are to achieve directly: (a) precorneal contact time lengthening; (b) an increase in drug permeability; and (c) a reduction in the rate of drug elimination [86]. The excellent water-holding capacity of HA makes it capable of retaining moisture in eyes [87]. Also, the viscosity and pseudoplastic behavior of HA providing mucoadhesive property can increase the ocular residence time [22]. Nancy and her group work reported that HA solutions have tremendous ocular compatibility both internally (when used during ophthalmic surgery) and externally, at concentrations of up to 10 mg/ml (1%). Also, topical HA solutions (0.1–0.2%) have been shown to be effective therapy for dry eye syndrome [88]. It was noted also that HA may interact with the corneal surface and tear film to stabilize the tear film and provide effective wetting, lubrication and relief from pain caused by exposed and often damaged corneal epithelium. The ability to interact with and to stabilize the natural tear film is a property unique to hyaluronan [89]. Pilocarpine – HA vehicle is considered the most commonly studied HA delivery system. Camber and his group proved that 1% pilocarpine solution dissolved in HA increased the 2-fold absorption of drug [90], improving the bioavailability, and miotic response while extending the duration of action. In another study gentamicin bioavailability was also reported to be increased when formulated with a 0.25% HA solution [22]. It was found that HA in the form of Healon can be used in artificial tears for the treatment of dry eye syndrome, and its efficacy for the treatment was evaluated [2]. On the other hand, A few studies have reported on the use of HA with contact lenses in different applications [3]. Pustorino and his group were conducted a study to determine whether HA could be used to inhibit bacterial adhesion on the surface of contact lenses. He showed that HA did not act as an inhibitor or a promoter of bacterial adhesion on the contact lens surfaces [91].

Also, in another application, Van Beek and others evaluated the use of HA containing hydrogel contact lenses to determine the effect on protein adsorption. Protein deposition on the contact lens surface can result in reduced vision, reduced lens wettability, inflammatory complications, and reduced comfort. They incorporated releasable and chemically crosslinked HA of different molecular weights as a wetting agent in soft contact lenses. The results showed that the addition of HA had no effect on the modulus or tensile strength of the lens regardless of molecular weight and no effect on the optical transparency of the lens. While, the protein adsorption on the lens did not affected by the releasable HA at either molecular weight [92].

2.2.1.4 Protein Sustained Delivery

Indeed, during the past few decades HA has been shown to be useful for sustained release (SR) formulations of protein and peptide drugs via parenteral delivery [97]. Because of the hydrophilic nature of HA, hydrogels can provide an aqueous environment preventing proteins from denaturation [94]. The swelling properties of hydrogel were shown to be affecting the protein diffusion; hence the diffusion of protein was influenced by the crosslink structure itself. In addition, the sustained delivery of proteins without denaturation is realized by tailoring the crosslink network of HA microgels. Luo and his group studied the sustained delivery of bovine serum albumin (BSA) protein from HA microgels by tailoring the crosslink network. He prepared a series of HA microgels with different crosslink network using an inverse microemulsion method, and studied the effect of different crosslink network in HA microgels on the loading capacity and sustained delivery profile of BSA as a model protein.

The date showed that the BSA loading had no obvious influence on the surface morphology of HA microgels but seemed to induce their aggregation. Increase of crosslink density slowed down the degradation of HA microgels by hyaluronidase and reduced the BSA loading capacity as well, but prolonged the sustained delivery of BSA [66]. However, physically cross-linked hydrogel behave very soft and easily disintegrated, thus, an initial burst and rapid protein release resulted hybrid hyaluronan hydrogel encapsulating nanogel was developed to overcome the above mentioned problems. The nanogels were physically entrapped and well dispersed in a three-dimensional network of chemically cross-linked HA (HA gel) [95].

2.2.1.5 Anticancer Drug Delivery

To date, the potentialities of HA in drug delivery have been investigated as carrier of antitumoral and anti-inflammatory drugs. HA is considered one of the major components of the extracellular matrix (ECM), also it is the main ligand for CD44 and RHAMM, which are overexpressed in a variety of tumor cell surfaces including human breast epithelial cells, colon cancer, lung cancer and acute leukemia cells [22]. In fact, it's essential in treatment and prevention of cancer cell metastasis that the localization of drug not only to the cancerous cells, but also to the surrounding lymph. HA is known as a bioadhesive compound capable of binding with high affinity to both cell-surface and intracellular receptors, to the extracellular matrix (ECM) components and to itself. HA can bind to receptors in cancer cells, and this is involved in tumor growth and spreading [96]. CD44 regulates cancer cells proliferation and metastatic processes. In addition, disruption of HA–CD44 binding was shown to reduce tumor progression. Also, administration of exogenous HA resulted in arrest of tumor spreading [97].

Therefore, anticancer drug solubilization, stabilization, localization and controlled release could be enhanced via coupling with HA [98]. Yang and his team work reported that the degradation of HA by intratumoral administration of hyaluronidases (HYAL) resulted in improved tumor penetration of conventional chemotherapeutic drugs [99]. Also, they stated that high HA level has been detected at the invasive front of growing breast tumors, 3.3-fold higher than in central locations within the tumor [100]. In addition, HA over production is associated with poor prognosis of breast cancer. In women <50 years, breast tumor HA level could predict cancer relapse [101]. It was reported that HA conjugates containing anticancer drugs include sodium butyrate, cisplatin, doxorubicin and mitomycin C. Therefore, depending on the degree of substitution of HA with drugs, these exhibited enhanced targeting ability to the tumor and higher therapeutic efficacy compared to free-anticancer drugs [22]. It is well known that Cisplatin (cis-diaminedichloroplatinum or CDDP) is an extensively employed chemotherapeutic agent for the treatment of a wide spectrum of solid tumors. Xiea and others presented a successful drainage of hyaluronan–cisplatin (HA–Pt) conjugates into the axillary lymph nodes with reduced systemic toxicities after local injection in a breast cancer xenograft model in rodents. They also observed that the pulmonary delivery of the HA–Pt conjugate to the lungs may be useful in the treatment of lung cancer by reducing systemic toxicities and increasing CDDP deposition and retention within lung tumors, surrounding lung tissues, and the mediastinal lymph.

2.2.1.6 Gene Delivery

HA could be an ideal biomaterial for gene delivery. Since it has been introduced as a nanocarrier for gene delivery since 2003 [69, 99]. Yang and his team work demonstrated the utility of HA microspheres for DNA gene delivery, they showed that show that DNA can be easy incorporated before derivatization. Once the HA-DNA microspheres degrade, the released DNA is structurally intact and able to transfect cells in culture and in vivo using the rat hind limb model. In addition, they found that the release of the encapsulated plasmid DNA can be sustained for months and is capable of transfection in vitro or in vivo and concluded that the native HA can be used to delivers DNA at a controlled rate and adaptable for site-specific targeting [99]. Another study for the same group, the DNA-HA matrix crosslinking with adipic dihydrazide (ADH) was able to sustain gene release while protecting the DNA from enzymatic degradation. It has been reported that HA combined with polyethyleneimine (PEI), poly (L-lysine) (PLL) and poly (L-arginine) (PLR) [102, 103]20 Therefore, the biocompatibility of the anionic HA is achieved by shielding the positive surface of gene/polycation complexes and by inhibiting nonspecific binding to serum proteins, thereby reducing the cytotoxicity of cationic polymers. Figure 2.3 show the binding of anionic HA (negatively charged) to the surface of DNA/PEI complexes, where PEI is polycation (positively charged) through electrostatic interactions between them to form ternary complex [22].

2.2.2 *HA FOR EVALUATION THE ACTIVITY OF ANTIOXIDANTS*

It is well known that the fast HA turnover in SF of the joints of healthy individuals can be attributed to the oxidative/degradative action of the reactive oxygen species (ROS), which generated among others by the catalytic effect of transition metal ions on the autoxidation of ascorbate. It has been reported that among the ROS, hydroxyl radical (OH·) represents the most active substance in terms of degradation of HA [104].

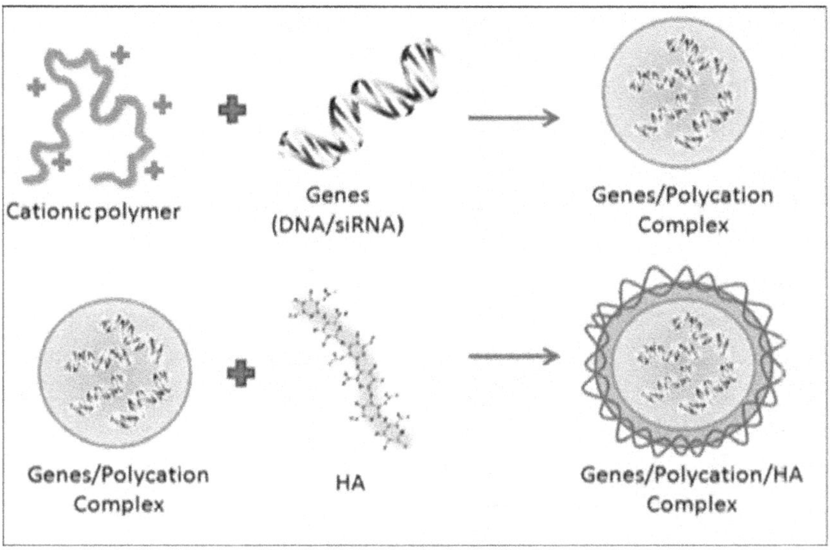

FIGURE 2.3 Schematic representation for the formation of electrostatic complex between negatively charged HA and positively charged polycation to produce gene/polycation (PEI)/HA ternary complex [22].

Therefore, chondrocytes are able to protect against free radical damage by means of endogenous antioxidants such as catalase and the glutathione peroxidase/reductase system, which are both involved in the removal of H_2O_2 [105]. The degradation behavior of HA via ROS has been studied on applying several in vitro models. The team of laboratory of bioorganic chemistry of drugs at the institute of experimental pharmacology and toxicology, Slovak academy of science, focused their attention on studying the effect of various antioxidants on the degradation behavior of HA which used as a model for evaluation the extent of the activity of these antioxidants [106–110]. As it well known that among various thiol compounds, D-penicillamine and L-glutathione are considered the best antioxidants compounds. Valachová and her work group studied monitored the pro- and antioxidative effects of an antirheumatoid drug (D-penicillamine (D-PN)), on the degradation kinetics of high-molar-mass HA using the method of rotational viscometry. They observed that the addition of d-PN dose dependently prolonged the period of complete inhibition of the degradation of HA macromolecules, and the initial antioxidative action of D-PN is followed by induction of prooxidative conditions due to the generation of reactive free radicals. It has been reported that L-glutathione (GSH) composed of L-glutamate (glu), L-cysteine (cys) and glycine (gly) moieties and is commonly named the "mother" of all antioxidants [111].

In addition, GSH is endogenic antioxidant, belongs among the most efficient substances protecting the cells against reactive oxygen species (ROS) escaping from mitochondria and maintains the intracellular reduction oxidation (redox) balance and regulates signaling pathways during oxidative stress/conditions. Valachová et al. studied the antioxidative effect of L-glutathione (GSH) using HA high-molar mass in an oxidative system composed of Cu(II) plus ascorbic acid by using rotational viscometry. Results showed GSH added to the oxidative system in sufficient amount resulted in total inhibition of HA degradation. By the same way, Šoltés et al. [48] used HA solutions as in vitro model for studying the scavenging effect of ibuprofen isomers using H_2O_2 and Cu^{2+} to prove that ibuprofen can be used as an anti-inflammatory drug.

In another study, HA was used for evaluation the activity of stobadine as an antioxidant drug. The protective effect of stobadine·2HCl on ascorbate plus Cu (II)-induced HA degradation was published by Rapta et al. [62]. Surovciková et al. [12] studied the antioxidative effects of two HHPI derivatives, namely SM1dM9dM10·2HCl and SME1i-ProC$_2$·HCl, and compared those effects with that of stobadine·2HCl. From data it was observed that the most effective scavengers of •OH and peroxy-type radicals were recorded to be stobadine·2HCl and SME1i-ProC2·HCl, respectively [112–116]. On the other hand, the most effective scavenger, determined by applying the ABTS assay, was stobadine·2HCl.

ACKNOWLEDGEMENTS

The corresponding author would like to thank the Institute of Experimental Pharmacology and Toxicology for having invited him and oriented him in the field of medical research. He would also like to thank Slovak Academic Information Agency (SAIA) for funding him during his work in the Institute.

KEYWORDS

- **antioxidant**
- **drug delivery system**
- **hyaluronan**
- **hydrogel**

REFERENCES

1. Lapcík, L., De-Smedt, S., Demeester, J., Chabrecek, P., Lapcík L Jr. (1998). Hyaluronan: Preparation, structure, properties, and applications. Chem Rev, 98, 2663–2684.
2. Stuart, C., Linn, G. (1985). Dilute sodium hyaluronate (Healon) in the treatment of ocular surface disorders. Ann Ophthalmol, 17, 190–192.
3. Marjorie, J. (2011). A review of hyaluronan and its ophthalmic applications, Optometry, 82, 38–43.

4. Meyer, K., Palmer, J. W. (1934).The polysaccharide of the vitreous humor. Journal of Biology and Chemistry, 107, 629–634.
5. Garg, G., Hales, A. (2004). Chemistry and biology of hyaluronan. Elsevier Science.
6. Necas, J., Bartosikova, L., Brauner, P., Kolar, J. (2008). Hyaluronic acid (hyaluronan): a review, Veterinarni Medicina, 53(8), 397–411.
7. Falcone, S., Palmeri, D. (2006). Berg, R., editors. Biomedical applications of hyaluronic acid, ACS Publications.
8. Balazs, E. A., Laurent, T. C., Jeanloz, R. W. (1986). Nomencla ture of hyaluronic acid. Biochemical Journal 235, 903.
9. Laurent, T. C., Fraser, J. R.E. (1992). Hyaluronan. FASEB J, 6, 2397–2404.
10. Kogan, G., Soltés, L., Stern, R., Gemeiner, P. (2007a). Hyaluronic acid: A natural biopolymer with a broad range of biomedical and industrial applications. Biotechnol Lett, 29, 17–25.
11. Kongtawelert, P., Ghosh, P. (1989): An enzyme-linked immunosorbent-inhibition assay for quantitation of hyaluronan (hyaluronic acid) in biological fluids. Anal. Biochem, 178, 367–372.
12. Surovciková, L., Valachová, K., Baňasová, M., Snirc, V., Priesolová, E., Nagy, M., Juránek, I., Šoltés, L. (2012). Free-radical degradation of high-molar-mass hyaluronan induced by ascorbate plus cupric ions: Testing of stobadine and its two derivatives in function as antioxidants. General Physiol Biophys, 31, 57–64.
13. Kogan, G., Soltés, L., Stern, R., Mendichi, R. (2007). Hyaluronic acid: A biopolymer with versatile physico-chemical and biological properties. Chapter 31 – in: Handbook of Polymer Research: Monomers, Oligomers, Polymers and Composites. Pethrick, R. A., Ballada A., Zaikov, G. E. (eds.), Nova Science Publishers, New York, 393–439.
14. Valachová, K., Vargová, A., Rapta, P., Hrabárová, E., Drafi, F., Bauerová, K., Juránek, I., Šoltés, L. (2011a). Aurothiomalate as preventive and chain-breaking antioxidant in radical degradation of high-molar-mass hyaluronan. Chemistry & Biodiversity 8, 1274–1283.
15. Kogan, G. (2010). Hyaluronan – A High Molar mass messenger reporting on the status of synovial joints: part 1. Physiological status In: New Steps in Chemical and Biochemical Physics. ISBN: 978-1-61668-923-0. 121–133.
16. Kurisawa, M., Chung, J., Yang, Y., Gao, S., Uyama, H. (2005). Injectable biodegradable hydrogels composed of hyaluronic acid–tyramine conjugates for drug delivery and tissue engineering. Chemical communications.,(34), 4312–4314.
17. Jiang, D., Liang, J., Noble, P. (2011). Hyaluronan as an Immune Regulator in Human Diseases. Physiol Rev, 91, 221–264.
18. Tamer, T. M. (2013). Hyaluronan and synovial joint function, distribution and healing. Interdiscip Toxicol Vol. 6(3), 101–115.
19. Price, R., Berry, M., Navsaria, H. (2007). Hyaluronic acid: the scientific and clinical evidence. Journal of Plastic, Reconstructive & Aesthetic Surgery, 60, 1110–1119.
20. Milas, M., Rinaudo, M. (2005). Characterization and properties of hyaluronic acid (hyaluronan). In: Dumitriu, S., ed. Polysaccharides Structural Diversity and Functional Versatility. New York, NY: Marcel Dekker 535–49.

21. Davies, A., Gormally, J., Wyn-Jones, E. (1982). A study of hydration of sodium hyaluronate from compressibility and high precision densitometric measurements. Int J Biol Macromol, 4, 436.
22. Jin, Y., Ubonvan, T., Kim, D. (2010). Hyaluronic Acid in Drug Delivery Systems. Journal of Pharmaceutical Investigation, 40, 33–43.
23. Šoltés, L., Valachová K, Mendichi, R., Kogan, G., Arnhold, J., Gemeiner, P. (2007). Solution properties of high-molar-mass hyaluronans: the biopolymer degradation by ascorbate. Carbohydr Res, 342, 1071–1077.
24. Bothner, H., Wik, O. (1987). Rheology of hyaluronate. Acta Otolaryngol Suppl, 442, 25–30.
25. Morris, E. R., Rees, D. A., Welsh, E. J. (1980). Conformation and dynamic interactions in hyaluronate solutions. J Mol Biol, 138, 383–400.
26. Gibbs, D. A., Merrill, E. W., Smith, K. A., Balazs EA. (1968). Rheology of hyaluronic acid. Biopolymers, 6, 777–91.
27. Valachova, K., Banasova, M., Machova, L., Juranek, I., Bezek, S., Soltes, L. (2013a). Antioxidant activity of various hexahydropyridoindoles. Journal of Information Intelligence and Knowledge, 5, 15–32.
28. Gribbon, P., Heng, B. C., Hardingham TE. (2000). The analysis of intermolecular interactions in concentrated hyaluronan solutions suggests no evidence for chain–chain association. Biochem, J., 350, 329–335.
29. Brown, M., Jones, S. (2005a). Hyaluronic acid: a unique topical vehicle for the localized delivery of drugs to the skin. Journal of the European Academy of Dermatology and Venereology, 19, 308–318.
30. Kobayashi, Y., Okamoto, A., Nishinari, K. (1994) Viscoelasticity of hyaluronic-acid with different molecular-weights. Biorheology, 31, 235–244.
31. Laurent, T. C., Ryan, M., Pictruszkiewicz, A. (1960). Fractionation of hyaluronic acid. The polydispersity of hyaluronic acid from the vitreous body. Biochim Biophys Acta, 42, 476–85.
32. Brown, M. B., Jones, S. A. (2005b). Hyaluronic acid: a unique topical vehicle for the localized delivery of drugs to the skin. JEADV,19, 308–318.
33. Fakhari, A. (2011). Biomedical Application of Hyaluronic Acid Nanoparticles. PhD. Faculty of the University of Kansas.
34. Drobnik, J. (1991). Hyaluronan in drug delivery. Adv Drug Dev Rev, 7, 295–308.
35. Volpi, N., Schiller, J., Stern, R., Soltés, L. (2009). Role, metabolism, chemical modifications and applications of hyaluronan. Current medicinal chemistry, 16(14), 1718–45.
36. Papakonstantinou, E., Roth, M., Karakiulakis, G. (2012). Hyaluronic acid: A key molecule in skin aging. Dermato-Endocrinology, 4, 3, 1–6.
37. Kreil, G. (1995). Hyaluronidases--a group of neglected enzymes. Protein Sci 4, 1666–9.
38. Stern, R., Kogan, G., Jedrzejas, M. J., Šoltés, L. (2007). The many ways to cleave hyaluronan. Biotechnology advances, 25 (6), 537–57.
39. Šoltés, L., Mislovičová, D., Sebille, B. (1996) Insight into the distribution of molecular weights and higher-order structure of hyaluronans and some β-(1→3)-glucans by size exclusion chromatography. Biomed Chromatogr, 10, 53–9.
40. Reháková, M., Bakoš, D., Soldán M, Vizárová, K. (1994). Depolymerization reactions of hyaluronic acid in solution. Int J Biol Macromol, 16, 121–4.

41. Bottner, H., Waaler, T., Wik, O. (1988). Limiting viscosity number and weight average molecular weight of hyaluronate samples produced by heat degradation. Int J Biol Macromol, 10, 287–91.
42. Šoltés, L., Mendichi, R., Kogan, G., Schiller, J., Stankovska, M., Arnhold, J. (2006). Degradative action of reactive oxygen species on hyaluronan, Biomacromolecules, 7(3), 659–68.
43. Topolska, D., Valachova, K., Hrabárová, E., Rapta, P., Banasova, M., Juránek, I., Soltés, L. (2014). Determination of protective properties of Bardejovske Kupele spa curative waters by rotational viscometry and ABTS assay. Balneo Research Journal, 5 (1), 3–15.
44. BeMiller, J. N., Whistler, R. L. (1962). Alkaline degradation of amino sugars. J Org Chem,7, 1161–4.
45. Hrabárová, E., Gemeiner, P., Šoltés, L. (2007). Peroxynitrite: In vivo and in vitro synthesis and oxidant degradative action on biological systems regarding biomolecular injury and inflammatory processes. Chem Pap, 61, 417–437.
46. Šoltés, L., Stankovská, M., Kogan, G., Gemeiner, P., Stern, R. (2005). Contribution of oxidative-reductive reactions to high-molecular-weight hyaluronan catabolism. Chem Biodivers, 2, 1242–5.
47. Hrabárová, E., Valachova, K., Juránek, I., Soltés, L. (2012). Free-radical degradation of high-molar-mass hyaluronan induced by ascorbate plus cupric ions: evaluation of antioxidative effect of cysteine-derived compounds. Chemistry & Biodiversity, 9, 309–317.
48. Šoltés, L., Lath, D., Mendichi, R., Bystrický, P. (2001). Radical degradation of high molecular weight hyaluronan: Inhibition of the reaction by ibuprofen enantiomers. Meth Find Exp Clin Pharmacol, 23, 65–71.
49. Šoltés, L., Brezová, V., Stankovská, M., Kogan, G., Gemeiner P. (2006a) Degradation of high molecular-weight hyaluronan by hydrogen peroxide in the presence of cupric ions. Carbohydr Res, 341, 639–44.
50. Šoltés, L., Stankovská, M., Brezová, V., Schiller, J., Arnhold, J., Kogan, G., Gemeiner, P. (2006b). Hyaluronan degradation by copper (II) chloride and ascorbate: rotational viscometric, EPR spin-trapping, and MALDI-TOF mass spectrometric investigation. Carbohydr Res, 341, 2826–34.
51. Hrabárová, E., Valachová, K., Rychly, J., Rapta, P., Sasinková, V., Maliková, M., Šoltés, L. (2009). High-molar-mass hyaluronan degradation by Weissberger's system: Pro- and anti-oxidative effects of some thiol compounds. Polymer Degradation and Stability, 94, 1867–1875.
52. Rapta, P., Valachová K, Gemeiner, P., Šoltés, L. (2009). High-molar-mass hyaluronan behavior during testing its radical scavenging capacity in organic and aqueous media: Effects of the presence of Manganese (II) ions. Chem Biodivers, 6, 162–169.
53. Stern, R., Maibach, H. I. (2008). Hyaluronan in skin: aspects of aging and its pharmacologic modulation. Clin Dermatol, 26, 106–122.
54. Baňasová, M., Valachová, K., Hrabárová, E., Priesolová, E., Nagy, M., Juránek, I., Šoltés, L. (2011). Early stage of the acute phase of joint inflammation. In vitro testing of bucillamine and its oxidized metabolite SA981 in the function of antioxidants. 16th Interdisciplinary Czech-Slovak Toxicological Conference in Prague. Interdiscip Toxicol 4(2), 22.

55. Baňasová, M., Valachová, K., Rychly, J., Priesolová, E., Nagy, M., Juránek, I., Šoltés, L. (2011a). Scavenging and chain breaking activity of bucillamine on free-radical mediated degradation of high molar mass hyaluronan. ChemZi, 7, 205–206.
56. Stankovská, M., Šoltés, L., Vikartovská, A., Mendichi r, Lath, D., Molnarová, M., Gemeiner, P. (2004). Study of hyaluronan degradation by means of rotational Viscometry: Contribution of the material of viscometer. Chem Pap, 58, 348–352.
57. Valachová, K., Hrabárová, E., Gemeiner, P., Šoltés, L. (2008). Study of pro- and anti-oxidative properties of d-penicillamine in a system comprising highmolar-mass hyaluronan, ascorbate, and cupric ions. Neuroendocrinol Lett, 29 (5), 697–701.
58. Valachová, K., Rapta, P., Kogan, G., Hrabárová, E., Gemeiner, P., Šoltés, L. (2009). Degradation of high-molar-mass hyaluronan by ascorbate plus cupric ions: effects of D-penicillamine addition. Chem Biodivers 6, 389–395.
59. Valachová, K., Mendichi, R., Šoltés, L. (2010b). Effect of L-glutathione on high-molar-mass hyaluronan degradation by oxidative system Cu(II) plus ascorbate. In: Monomers, Oligomers, Polymers, Composites, and Nanocomposites, Ed: R. A. Pethrick, P. Petkov, A. Zlatarov, G. E. Zaikov, S. K. Rakovsky, Nova Science Publishers, N.Y, Chapter 6, 101–111.
60. Valachová, K., Hrabárová, E., Priesolová, E., Nagy, M., Baňasová, M., Juránek, I., Šoltés, L. (2011). Free-radical degradation of high-molecular-weight hyaluronan induced by ascorbate plus cupric ions. Testing of bucillamine and its SA981-metabolite as antioxidants. J Pharma & Biomedical Analysis, 56, 664–670.
61. Valachová, K., Hrabárová, E., Juránek, I., Šoltés, L. (2011b). Radical degradation of high-molar-mass hyaluronan induced by Weissberger oxidative system. Testing of thiol compounds in the function of antioxidants. 16th Interdisciplinary Slovak-Czech Toxicological Conference in Prague. Interdiscip Toxicol, 4(2), 65.
62. Rapta, P., Valachová, K., Zalibera, M., Šnirc, V., Šoltés, L. (2010). Hyaluronan degradation by reactive oxygen species: scavenging eggect of the hexapyridoindole stobadine and two of its derivatives. In Monomers, Oligomers, Polymers, Composites, and Nanocomposites, Ed: R. A.
63. Dráfi, F., Valachová, K., Hrabárová, E., Juránek, I., Bauerová, K., Šoltés, L. (2010). Study of methotrexate and β-alanyl-L-histidine in comparison with L-glutathione on high-molar-mass hyaluronan degradation induced by ascorbate plus Cu (II) ions via rotational viscometry. 60th Pharmacological Days in Hradec Králové. Acta Medica, 53(3), 170.
64. Balazs, E., Denlinger, J. (1993). Viscosupplementation: a new concept in the treatment of osteoarthritis. Journal of rheumatology Supplement, 20, 3–9.
65. Kalal, J., Drobnik, J. Rypacek, F. (1982). Affinity chromatography and affinity therapy. In: T.C.J. Gribnau, J. Visser and R.J.F. Nivard (Eds.), Affinity Chromatography and Related Techniques, Elsevier, Amsterdam.
66. Luo, C., Zhao, J., Tua, M., Zenga, R., Rong, J. (2014). Hyaluronan microgel as a potential carrier for protein sustained delivery by tailoring the crosslink network. Materials Science and Engineering, C., 36, 301–308.
67. Horvát, S., Fehér, A., Wolburg, H., Sipos, P., Veszelka, S., Tóth, A., Kis, L., Kurunczi, A., Balogh, G., Kürti, L., Eros, I., Szabó-Révész, P., Deli, M. (2009). Sodium hyaluronate as a mucoadhesive component in nasal formulation enhances delivery of molecules to brain tissue. European Journal of Pharmaceutics and Biopharmaceutics, 72, 252–259.

68. Langer, R. (2003). Biomaterials in drug delivery and tissue engineering: one laboratory's experience. Acc Chem Res, 33, 94–101.
69. Kim, A., Checkla, D. M., Chen, W. (2003). Characterization of DNA hyaluronan matrix for sustained gene transfer. J Control Release, 90, 81–95.
70. Le Bourlais, C., Acar, L., Zia, H., Sado, P., Needham, T., Leverge, R. (1998). Ophthalmic drug delivery systems—recent advances. Progress in retinal and eye research, 17, 33–58.
71. Bourguignon, L., Zhu, H., Shao, L., Chen, Y. (2000). CD44 interaction with tiam1 promotes Rac1 signaling and hyaluronic acid-mediated breast tumor cell migration. Journal of Biological Chemistry, 275 (3), 1829–38.
72. Dollo, G., Malinovsky J, Peron, A., Chevanne F, Pinaud, M., Verge, R., Corre P. (2004). Prolongation of epidural bupivacaine effects with hyaluronic acid in rabbits. International Journal of Pharmaceutics, 272, 109–119.
73. Tammi, R., Ripellino, J., Margolis, R., Tammi, M. (1988). Localizationof epidermal hyaluronic acid using the hyaluronate binding region of cartilage proteoglycan as a specific probe. Journal of Investigative Dermatology, 90, 412–414.
74. Prisell, P., Camber, O., Hiselius, J., Norstedt, G. (1992). Evaluation of hyaluronan as a vehicle for peptide growth factors. International Journal of Pharmaceutics, 85, 51–56.
75. Esposito, E., Menegatti, E., Cortesi, R. (2005). Hyaluronan-based microspheres as tools for drug delivery: a comparative study. International Journal of Pharmaceutics, 288, 35–49.
76. Rydell, N., Balazs, E. (1971). Effect of intra-articular injection of hyaluronic acid on the clinical symptoms of osteoarthritis and on granulation tissue formation. Clinical Orthopaedics and Related Research, 80, 25–32.
77. Marshall, K. (2000). Intra-articular hyaluronan therapy. Current opinion in rheumatology, 12, 468–474.
78. Lim, S., Martin, G., Berry, D., Brown, M. (2000). Preparation and evaluation of the in vitro drug release properties and mucoadhesion of novel microspheres of hyaluronic acid and chitosan. Journal of Controlled Release, 66, 281–292.
79. Ugwoke, M., Agu, R., Verbeke, N., Kinget, R. (2005). Nasal mucoadhesive drug delivery: Background, applications, trends and future perspectives. Advanced drug delivery reviews, 57, 1640–1665.
80. Yun, Y. H., Goetz, D. J., Yellen, P., Chen, W. (2004). Hyaluronan microspheres for sustained gene delivery and site-specific targeting. Biomaterials, 25, 147–157.
81. Turker, S., Onur, E., Ozer, Y. (2004). Nasal route and drug delivery systems. Pharmacy World & Science, 26, 137–142.
82. Morimoto, K., Morisaka, K., Kamada, A. (1985). Enhancement of nasal absorption of insulin and calcitonin using polyacrylic acid gel. J. Pharm. Pharmacol., 37, 134–136.
83. Illum, L., Farraj, N. F., Critchley, H., Davis, S. S. (1988). Nasal administration of gentamicin using a novel microsphere delivery system. Int. J. Pharm., 46, 261–265.
84. Lim, T., Forbes, B., Berry, J., Martin, G., Brown, M. (2002). In vivo evaluation of novel hyaluronan/Chitosan microparticulate delivery systems for the nasal delivery of gentamicin in rabbits. International Journal of Pharmaceutics, 231, 73–82.

85. Illum, L. (2000).Transport of drugs from the nasal cavity to the central nervous system, Eur. J. Pharm. Sci,11, 1–18.
86. Calles, J. A., Tártara, L. I., Lopez-García, A., Diebold, Y., Palma, S. D., Vallés, E. M. (2013). Novel bioadhesive hyaluronan–itaconic acid crosslinked films for ocular therapy, International Journal of Pharmaceutics, 455, 48– 56.
87. Robert, L., Robert, A. M., Renard, G. (2010). Biological effects of hyaluro-nan in connective tissues, eye, skin, venous wall. Role Aging Pathol. Biol. 58 (3), 187–198,.
88. Nancy, E., Larsen, Endre, A. Balazs. (1991). Drug delivery systems using hyaluronan and its derivatives, Advanced Drug Delivery Reviews, 7, 279–293.
89. Gurny, R., Ibrahim, H., Aebi, A., Buri, P., Wilson, C. G., Washington, N., Edman, P., Camber, O. (1987). Design and evaluation of controlled release systems for the eye, J. Controlled Release, 6, 367 373.
90. Camber, O., Edman, P., Gurny, R. (1987). Influence of sodium hyaluronate on the meiotic effect of pilocarpine in rabbits. Current eye research, 6, 779–784.
91. Pustorino, R., Nicosia, R., Sessa, R. (1996). Effect of bovine serum, Hyaluronic acid and netilmicine on the in vitro adhesion of bacteria isolated from human-worn disposable soft contact lenses. Ann Ig,8, 469–75.
92. Van Beek, M., Jones, L., Sheardown, H. (2008) Hyaluronic acid containing hydrogels for the reduction of protein adsorption. Biomaterials, 29, 780–9.
93. Elbert, D., Pratt, A., Lutolf, M., Halstenberg, S., Hubbell, J. (2001). Protein delivery from materials formed by self-selective conjugate addition reactions. Journal of Controlled Release, 76, 11–25.
94. Hirakura, T., Yasugi, K., Nemoto, T., Sato, M., Shimoboji, T., Aso, Y., Morimoto, N., Akiyoshi, K. (2009). Hybrid hyaluronan hydrogel encapsulating nanogel as a protein nanocarrier: New system for sustained delivery of protein with a chaperone-like function. Journal of Controlled Release.
95. Jeong, B., Bae, Y., Kim, S. (2000). Drug release from biodegradable injectable thermosensitive hydrogel of PEG-PLGA-PEG triblock copolymers. Journal of Controlled Release, 63, 155–163.
96. Luo, C., Zhao, J., Tu, M., Zeng, R., Rong, J. (2013). Hyaluronan microgel as a potential carrier for protein sustained delivery by tailoring the crosslink network, Materials Science and Engineering, C., 36, 301–308.
97. Naor, D., Wallach-Dayan, S. B., Zahalka, M. A., Sionov, R. V. (2008). Involvement of CD44, a molecule with a thousand faces, in cancer dissemination, Semin. Cancer Biol, 18, 260–267.
98. Maeda, H., Seymour, L., Miyamoto, Y. (1992). Conjugates of anticancer agents and polymers: advantages of macromolecular therapeutics in vivo. Bioconjugate chemistry, 3, 351–362.
99. Yang, H. Yuna, Douglas, J. Goetzb, Paige Yellena, Weiliam Chen, (2004) Hyaluronan microspheres for sustained gene delivery and site-specific targeting, Biomaterials, 25, 147–157.
100. Bertrand, P., Girard, N., Delpech, B., Duval, C., d'Anjou, J., Dauce, J. (1992). Hyaluronan (hyaluronic acid) and hyaluronectin in the extracellular matrix of human breast carcinomas: comparison between invasive and noninvasive areas. Int J Cancer,52(1), 1–6.

101. Casalini, P., Carcangiu, M. L., Tammi, R., Auvinen, P., Kosma, V. M., Valagussa P. (2008). Two distinct local relapse subtypes in invasive breast cancer: effect on their prognostic impact. Clin Cancer Res, 14(1), 25–31.
102. Ito, T., Iidatanaka, N., Niidome, T., Kawano, T., Kubo, K., Yoshikawa, K., Sato, T., Yang, Z., Koyama, Y. (2006). Hyaluronic acid and its derivative as a multi-functional gene expression enhancer: Protection from non-specific interactions, adhesion to targeted cells, and transcriptional activation. Journal of Controlled Release,112, 382–388.
103. Takei, Y., Maruyama, A., Ferdous, A., Nishimura, Y., Kawano, S., Ikejima, K., Okumura, S., Asayama, S., Nogawa, M., Hashimoto M. (2004). Targeted gene delivery to sinusoidal endothelial cells: DNA nanoassociate bearing hyaluronanglycocalyx. The FASAB Journal, 18, 699–701.
104. Cortivo, R., Brun, P., Cardarelli, L., O'Regan, M., Radice, M., Abatangelo G. (1996). Antioxidant Effects of Hyaluronan and Its a-Methyl-Prednisolone Derivative in Chondrocyte and Cartilage Cultures. Seminars in Arthritis and Rheumatism, 26 (1), 492–501.
105. Baker, M., Feigan, J., Lowther, D. (1988). Chondrocyte antioxidant defences: the roles of catalase and glutathione peroxidase in protection against H202 dependent inhibition of PG biosynthesis. J Rheumatol, 15, 670–677.
106. Valachová, K., Hrabárová, E., Dráfi, F., Juránek, I., Bauerová, K., Priesolová, E., Nagy, M., Šoltés, L. (2010). Ascorbate and Cu(II) induced oxidative degradation of high-molar-mass hyaluronan. Pro- and antioxidative effects of some thiols. Neuroendocrinol Lett, 31(2), 101–104.
107. Valachová, K., Kogan, G., Gemeiner, P., Šoltés, L. (2008a). Hyaluronan degradation by ascorbate: Protective effects of manganese (II). Cellulose Chem. Technol, 42(9–10), 473–483.
108. Valachová, K., Kogan, G., Gemeiner, P., Šoltés, L. (2009a). Hyaluronan degradation by ascorbate: protective effects of manganese (II) chloride. In: Progress in Chemistry and Biochemistry. Kinetics, Thermodynamics, Synthesis, Properties and Application, Nova Science Publishers, N.Y, Chapter 20, 201–215.
109. Valachová, K., Rapta, P., Slováková, M., Priesolová, E., Nagy, M., Mislovičová, D., Dráfi, F., Bauerová, K., Šoltés, L. (2013). Radical degradation of high-molar-mass hyaluronan induced by ascorbate plus cupric ions. Testing of arbutin in the function of antioxidant. In: Advances in Kinetics and Mechanism of Chemical Reactions, G. E. Zaikov, A. J. M. Valente, A. L. Iordanskii (eds), Apple Academic Press, Waretown, NJ, USA, 1–19.
110. Valachová, K., Šoltés, L. (2010a). Effects of biogenic transition metal ions Zn(II) and Mn(II) on hyaluronan degradation by action of ascorbate plus Cu(II) ions. In: New Steps in Chemical and Biochemical Physics. Pure and Applied Science, Nova Science Publishers, Ed: E. M. Pearce, G. Kirshenbaum, G.E. Zaikov, Nova Science Publishers, N.Y, Chapter 10, 153–160.
111. Hrabárová, E., Valachová, K., Rapta, P., Šoltés, L. (2010). An alternative standard for trolox-equivalent antioxidant-capacity estimation based on thiol antioxidants. Comparative 2,2'-azinobis[3-ethylbenzothiazoline-6-sulfonic acid] decolorization and rotational viscometry study regarding hyaluronan degradation. Chemistry & Biodiversity, 7(9), 2191–2200.

112. Cuixia Yang, Yiwen Liu, Yiqing He, Yan Du, Wenjuan Wang, Xiaoxing Shi, Feng Gao. (2013). The use of HA oligosaccharide-loaded nanoparticles to breach the endogenous hyaluronan glycocalyx for breast cancer therapy, Biomaterials, 34, 6829–6838.
113. Dalit Landesman-Milo, Meir Goldsmith, Shani Leviatan Ben-Arye, Bruria Witenberg, Emily Brown, Sigalit Leibovitch, Shalhevet Azriel, Sarit Tabak, Vered Morad, Dan Peer. (2013). Hyaluronan grafted lipid-based nanoparticles as RNAi carriers for cancer cells, Cancer Letters, 334, 221–227.
114. Kogan, G. (2010). Hyaluronan – A High Molar mass messenger reporting on the status of synovial joints: part 1. Physiological status In: New Steps in Chemical and Biochemical Physics. ISBN: 978-1-61668-923-0. 121–133.
115. Pethrick, P., Petkov, A., Zlatarov, G. E., Zaikov S K, Rakovsk. Nova Science Publishers, NY, Chapter 7, 113–126.
116. Yumei Xie, Kristin L Aillon, Shuang Cai, Jason, M. Christian, Neal, M. Davies, Cory, J. Berkland, M. Laird Forrest. (2010). Pulmonary delivery of cisplatin–hyaluronan conjugates via endotracheal instillation for the treatment of lung cancer, International Journal of Pharmaceutics, 392, 156–163.

CHAPTER 3

STEEL SURFACE MODIFICATION

IGOR NOVÁK,[1] IVAN MICHALEC,[2] MARIAN VALENTIN,[1] MILAN MARÔNEK,[2] LADISLAV ŠOLTÉS,[3] JÁN MATYAŠOVSKÝ,[4] and PETER JURKOVIČ[4]

[1]*Department of Welding and Foundry, Faculty of Materials Science and Technology in Trnava, 917 24 Trnava, Slovakia*

[2]*Slovak Academy of Sciences, Polymer Institute of the Slovak Academy of Sciences, 845 41 Bratislava, Slovakia*

[3]*Institute of Experimental Pharmacology of the Slovak Academy of Sciences, 845 41 Bratislava, Slovakia*

[4]*VIPO, Partizánske, Slovakia*

E-mail: upolnovi@savba.sk

CONTENTS

3.1 INTRODUCTION

The surface treatment of steel surface is often used, especially in an automotive industry, that creates the motive power for research, design and production. New methods of surface treatment are also developed having major influence on improvement the surface properties of steel sheets while keeping the price at reasonable level [1–3].

The nitrooxidation is one of the nonconventional surface treatment methods which combine the advantages of nitridation and oxidation processes. The improvement of the mechanical properties (Tensile Strength, Yield Strength) together with the corrosion resistance (up to level 10) can be achieved [5–8]. The fatigue characteristics of the nitrooxidized material can be also raised [6].

Steel sheets with surface treatment are more often used, especially in an automotive industry which creates the motive power for research, design and production. New methods of surface treatment are also developed having major influence on improvement the surface properties of steel sheets while keeping the price at reasonable level.

Previous outcomes [1, 3, 4, 9, 10] dealt with the welding of steel sheets treated by the process of nitrooxidation by various arc and beam welding methods. Due to high oxygen and nitrogen content in the surface layer, problems with high level of porosity had occurred in every each method. The best results were achieved by the solid-state laser beam welding, by which the defect-free joints were created. Due to high initial cost of the laser equipment, the further research was directed to the joining method that has not been tested. Therefore the adhesive bonding was chosen, because the joints are not thermally affected, they have uniform stress distribution and good corrosion resistance.

The goal of the paper is to review the adhesive bonding of steel sheets treated by nitrooxidation and to compare the acquired results to the non-treated steel.

3.2 EXPERIMENTAL

For the experiments, low carbon deep drawing steel DC 01 EN 10130/91 of 1 mm in thickness was used. The chemical composition of steel DC 01 is documented in Table 3.1.

TABLE 3.1 Chemical Composition of Steel DC 01 EN 10130/91

EN designation	C[%]	Mn[%]	P[%]	S[%]	Si[%]	Al[%]
DC 01 10130/91	0.10	0.45	0.03	0.03	0.01	–

3.2.1 CHEMICAL MODIFICATION

The base material was consequently treated by the process of nitrooxidation in fluidized bed. The nitridation fluid environment consisted of the Al_2O_3 with granularity of 120 μm. The fluid environment was wafted by the gaseous ammonia. After the process of nitridation, the oxidation process started immediately. The oxidation itself was performed in the vapors of distilled water. Processes parameters are referred in Table 3.2.

3.2.2 ADHESIVES

In the experiments, the four types of two-component epoxy adhesives made by Loctite Company (Hysol 9466, Hysol 9455, Hysol 9492 and Hysol 9497) were used. The properties of the adhesives are documented in Table 3.3.

TABLE 3.2 Process of Nitrooxidation Parameters

	Nitridation	Oxidation
Time [min]	45	5
Temperature [°C]	580	380

TABLE 3.3 The Characterization of the Adhesives

	Hysol 9466	Hysol 9455	Hysol 9492	Hysol 9497
Resin type	Epoxy	Epoxy	Epoxy	Epoxy
Hardener type	Amin	Methanethiol	Modified Amin	
Mixing ratio	2:1	1:1	2:1	2:1
Elongation [%]	3	80	0.8	2.9
Shore hardness	60	50	80	83

3.2.3 METHODS

The experiments were done at the Faculty of Materials Science and Technology, Department of Welding and Foundry in Trnava. The adhesive bonding was applied on the grinded as well as nongrinded surfaces of the material to determine the grinding effect on total adhesion of the material so as on ultimate shear strength of the joints. The grinded material was prepared by grinding with silicone carbide paper up to 240 grit.

Before the adhesive bonding, the bonding surfaces (both grinded as well as nongrinded) were decreased with aerosol cleaner. The overlap area was 30 mm. To ensure the maximum strength of the joints, the continuous layer of the adhesive was coated on the overlap area of both bonded materials. The thickness of the adhesive layer was 0.1 mm and it was measured by a caliper. The joints were cured under fixed stress for 48 h at the room temperature. The dimensions of the joints are referred in Fig. 3.1.

The mechanical properties of the joints were examined by the static shear tests. As a device, the LaborTech LabTest SP1 was used. The conditions of the tests were set in accordance with STN EN 10002–1. The static shear tests were repeated on three separate samples and an average value was calculated.

The fracture areas were observed in order to obtain the fracture character of the joints. The JEOL JSM-7600F scanning electron microscope was used as a measuring device.

The differential scanning calorimetry (DSC) was performed on Netzsch STA 409 C/CD equipment. As the shielding gas, Helium with purity of

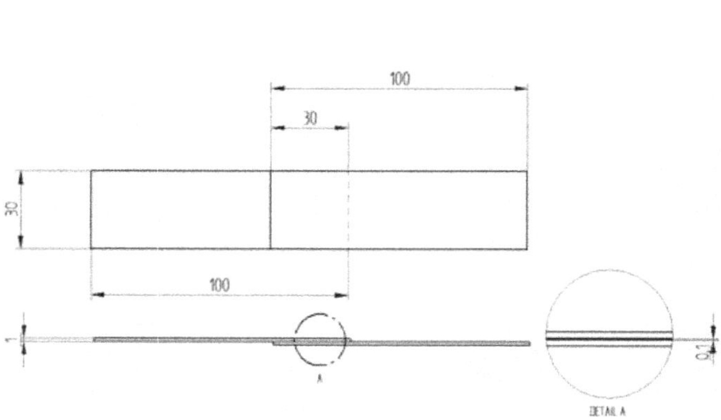

FIGURE 3.1 The dimension of the bonded joints.

99.999% was used. The heating process starts at the room temperature and continued up to 400°C with heating rate 10°C/min. The DSC analysis on Hysol 9455 was done on Diamond DSC Perkin Elmer, capable of doing analyzes from – 70°C.

3.3 RESULTS AND DISCUSSION

The material analysis represents the first step of evaluation. There are many factors having an influence on the joint quality. The properties of the nitrooxidized material depends on the treatment process parameters. For the adhesive bonding, the surface layer properties are important because of that, the high adhesion is needed to ensure the high strength of the joint.

The overall view on the microstructure of the nitrooxidized material surface layer is referred in Fig 3.2a. On the top of the surface, the oxide layer (see Fig. 3.2b) was created. This layer had a thickness of approx. 700 μm. Beneath the oxide layer, the continuous layer of ε-phase, consisting of nitrides $Fe_{2-3}N$ and with the thickness of 8–10 μm was observed.

The surface energy measurements were performed due to obtain the properties of the material, which are important for adhesive bonding. For observing the grinding effect on the total surface energy, the measurements were done on the base as well as on the grinded material.

To determine the surface energy, the portable computer-based instrument SeeSystem was used. Four different liquids (distilled water,

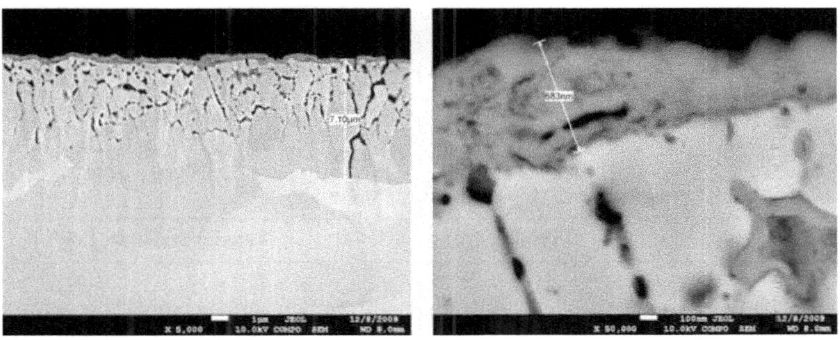

FIGURE 3.2 The microstructure of the surface layer. (a) overall view; (b) detail view on the oxide layer.

formamide, diiodomethan and ethylene glycol) were instilled on the material surface and contact angle was measured. The Owens-Wendt regression model was used for the surface energy calculation. The total amount of six droplets were analyzed of each liquid. The results (see Table 3.4) proved that the nitrooxidation treatment had a strong affect on material surface energy, where the decrease by 28% in comparison to nonnitrooxidized material had occurred. The surface energies of grinded and nongrinded material without nitrooxidation were very similar, while the increase of surface energy of grinded nitrooxidized material by 35% in comparison to nongrinded material was observed. In the case of barrier plasma modified steel is the surface energy higher compared unmodified material and namely its polar component is significantly higher than polar component of surface energy for unmodified sample as well as for sample modified by nitrooxidation.

The mechanical properties of the material were obtained by the static tensile test. Total amount of three measurements were done and the average values are documented in Table 3.5. By the results, it can be stated,

TABLE 3.4 The Surface Energy of Materials

Material type	Total surface energy [mJ/m^2]	Dispersion component [mJ/m^2]	Acid-base component [mJ/m^2]
DC 01	38.20	35.62	2.58
DC 01 grinded	38.80	33.66	5.14
Nitrooxidized	27.60	25.79	1.80
Nitrooxidized grinded	37.31	32.54	4.77
Barrier plasma treated	39.99	33.69	6.30

TABLE 3.5 Mechanical Properties of the Base Materials

	Yield Strength [MPa]	Tensile Strength [MPa]
DC 01	200	270
Nitrooxidized	310	380
Barrier plasma treated	200	270

that after the process of nitrooxidation, the increase of Yield Strength by 55% and Tensile Strength by 40 % were observed. The barrier plasma did not influence the mechanical properties after surface modification of steel, that remained the same as for unmodified sample.

The tensile test of the adhesives were carried out on the specimens, which were created by curing of the adhesives in special designed polyethylene forms for 48 h. The results are shown in Table 3.6. In three of the adhesives (Hysol 9466, Hysol 9492 and Hysol 9497), very similar values were observed while in case of Hysol 9455, only tensile strength of 1 MPa was observed.

The differential scanning calorimetry was performed due to obtain the glass transition temperature as well as the melting points of the adhesives. The results are in Table 3.7. To measure the glass transition temperature of Hysol 9455, the measurements had to be started from the cryogenic temperatures. The results of such a low glass transition temperature explained the low tensile strength of the Hysol 9455, where at the room temperature, the mechanical behavior changed from rigid to rubbery state. The results of DSC analyzes are given in Fig. 3.3–3.6.

The adhesive joint evaluation consisted of observing the mechanical properties and fracture surface respectively.

In order to obtain mechanical properties of the joints, the static shear tests were carried out. Results (Table 3.8) showed that the highest shear strength was observed in grinded nitrooxidized joints. The Hysol 9466 provided joints with the highest shear strength.

TABLE 3.6 The Mechanical Properties of the Adhesives

	Hysol 9466	Hysol 9455	Hysol 9492	Hysol 9497
Tensile Strength [MPa]	60	1	58	65

TABLE 3.7 The Glass Transition Temperatures of Adhesives

	Hysol 9466	Hysol 9455	Hysol 9492	Hysol 9497
Glass transition temperature [°C]	52.6	5.0	61.2	62.4
Melting point [°C]	315.7	337.0	351.8	327.3

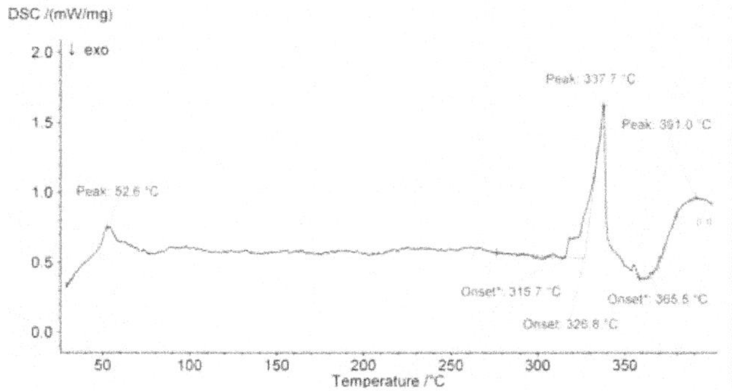

FIGURE 3.3 The DSC analysis of the Hysol 9466.

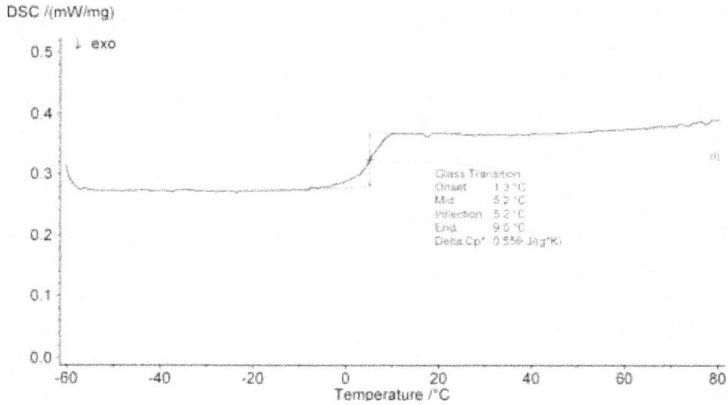

FIGURE 3.4 The DSC analysis of the Hysol 9455.

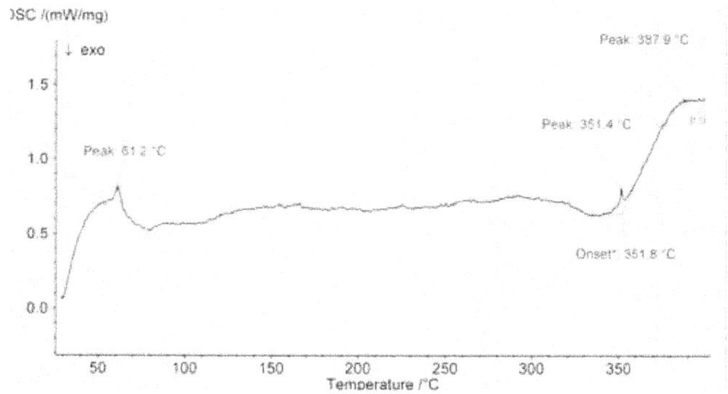

FIGURE 3.5 The DSC analysis of the Hysol 9492.

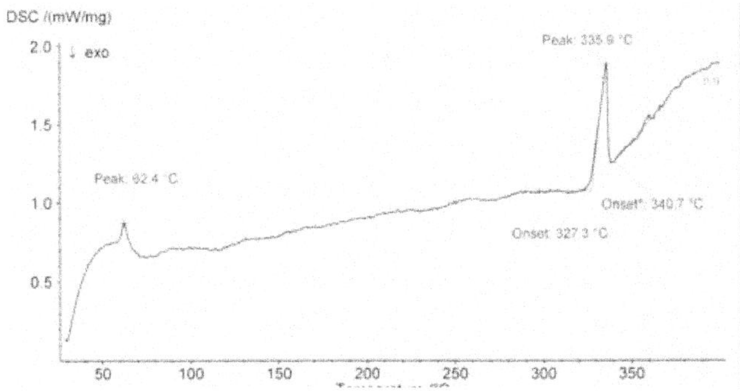

FIGURE 3.6 The DSC analysis of the Hysol 9497.

TABLE 3.8 The results of shear test of the joints

Material	Shear strength [MPa]			
	Hysol 9466	Hysol 9455	Hysol 9492	Hysol 9497
DC 01	9.0	2.8	7.1	6.0
DC 01 grinded	8.9	2.9	7.2	6.1
Nitrooxidized	12.9	3.1	7.8	5.4
Nitrooxidized grinded	12.9	5.9	12.7	7.0
Barrier plasma treated	13.8	6.1	14.2	7.8

The mechanical properties of the joints made on nonnitrooxidized material did not depended on the surface grinding. The mechanical properties of the joints made on nitrooxidized material, in comparison to DC 01, were higher by 43% in case of Hysol 9466, 11% in case of Hysol 9455 and 10% in case of Hysol 9492. In the case of adhesive Hysol 9497, the decrease of the shear strength had occurred. The joints produced on grinded nitrooxidized material, in comparison to DC 01, had a higher shear strength by 43% in case of Hysol 9466, 110% in case of Hysol 9455, 79% in case of Hysol 9492 and 15% in case of Hysol 9497. The shear strength of adhesive joint was for barrier plasma modified steel for all kinds of adhesive Hysol higher compared unmodified and nitrooxidized steel.

Results of fracture morphology of the joints made of nonnitrooxidized material are shown in Fig. 3.7a. Only the adhesive type of fracture morphology (see Fig. 3.7b) was observed in every type of the adhesive. Cohesive and combined fracture type were not observed. It can be stated that the adhesion forces were not strong enough so that the joints were fractured between the material and adhesive.

The results of the fractographic analysis of the nongrinded nitro-oxidized steel are documented in Fig. 3.8. No oxide layer peeling was observed. The cleavage fracture pattern was observed only as well as adhesive fracture morphology. The close-up view on the cleavage fracture is shown in Fig 3.8b.

Differential Scanning Calorimetry revealed that three of the adhesives had a very similar glass transition temperature, so the meshing of the adhesives will start in the same way.

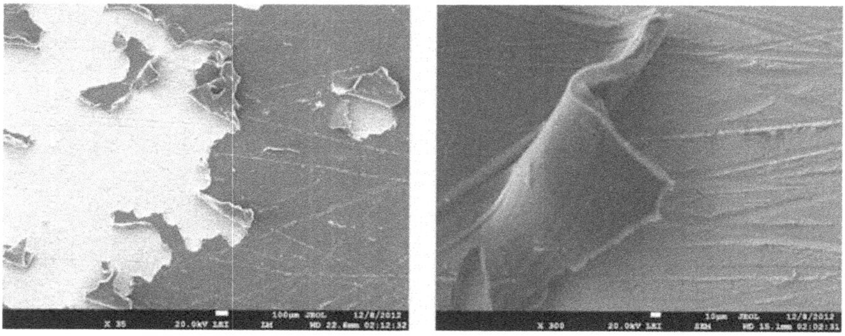

FIGURE 3.7 The fractographic analysis of the fractured adhesive joint of nonnitrooxidized material. (a) overall view (b) close-up view.

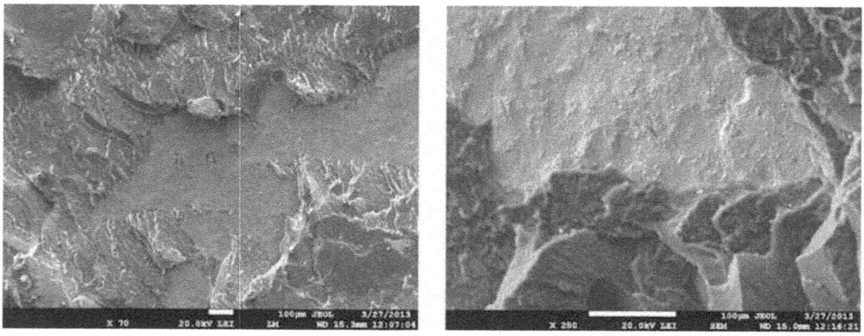

FIGURE 3.8 The fractographic analysis of the fractured adhesive joint of nitrooxidized material. (a) overall view (b) close-up view.

The results of mechanical properties evaluation of the joints proved that the material after the nitrooxidation process had a better adhesion to the epoxy adhesives than plain material DC01. Due to this fact the higher shear strength was achieved. It can be explained by the surface oxide layer porosity, which helped the adhesive to leak in.

On the other hand, increase of mechanical properties of joints prepaired from grinded nitrooxidized material can be explained by removing the surface oxide layer and thus resulting into rapid increase of surface energy.

The only adhesive type of fracture was observed and the fractographic analysis showed, that only cleavage type of fractures has been created. It can be stated, that the surface energy of the materials was not appropriate for the cohesive fracture pattern.

3.4 CONCLUSION

Joining of steel sheets treated by the process of nitrooxidation represents an interesting technical as well as technological problem. The fusion welding methods with high energy concentration, for example, laser beam welding are one of the possible options, however even with high effort of minimizing the surface layer deterioration, it's not possible to completely avoid it.

Adhesive bonding of nitrooxidized steels presents thus the second alternative, when the surface layer is not damaged and the adhesive joint keeps its properties after it has been cured. Adhesive bonding of metallic substrates, often requires removing the surface oxide layer from the areas to be bonded. In case of materials treated by nitrooxidation, this is possible, however the damage of the formed surface layer will occur. The goal of this paper was to review the effect of surface layer, created by the process of nitrooxidation and or by barrier discharge plasma treatment, on final mechanical properties of the joints, evaluation of fracture morphology and results comparison for both, treated and untreated material.

The acquired results have been revealed, that the presence of nitrooxidation surface layer caused decrease of free surface energy by 28%. The surface energy and namely its polar component were for barrier plasma modified steel higher than for unmodified and nitrooxidized steel. On the other hand, this surface layer brings on the joints shear strength increase by 10–43% in dependence of the adhesive used. In case of Hysol 9497, the decrease of the joint shear strength by 10% was observed. In the case of barrier plasma modified steel were the strength of adhesive joint higher compared unmodified and nitrooxidized material. We can presume that the increase of the shear strength was mainly due to porous structure of the surface layer, which enabled the adhesive to leak in.

The adhesive type of fracture morphology was observed during fractographic analysis. Regarding the characteristics of used adhesives, the cleavage fracture morphology of the joints was occurred.

There can be concluded on the base of received results, that the epoxy adhesive bonding represents the suitable alternative of creating the high quality joints of steel sheets treated by nitrooxidation as well as treated by barrier discharge plasma.

ACKNOWLEDGEMENTS

This paper was prepared within the support of Slovak Research and Development Agency, grant No. 0057–07 and Scientific Grant Agency VEGA, grant No. 1/0203/11 and 2/0199/14.

This publication was prepared as an output of the project 2013–14547/39694:1–11, Research and Development of Hi-Tech Integrated Technological and Machinery Systems for Tyre Production – PROTYRE" cofunded by the Ministry of Education, Science, Research and Sport of the Slovak Republic pursuant to Stimuli for Research and Development Act No. 185/2009 Coll.

KEYWORDS

- **automotive industry**
- **improvement**
- **mechanical properties**
- **modification treatment**
- **steel**
- **surface**

REFERENCES

1. Michalec, I. CMT Technology Exploitation for Welding of Steel Sheets Treated by Nitrooxidation. Diploma thesis, Trnava 2010.
2. Konjatić, Pejo; Kozak, Dražan; Gubeljak, Nenad. The Influence of the Weld Width on Fracture Behavior of the Heterogeneous Welded Joint. Key Engineering Materials. 488–489 (2012); 367–370.
3. Bárta, J. Welding of special treated thin steel sheets Dissertation thesis, Trnava, 2010.
4. Marônek, M. et al. Laser beam welding of steel sheets treated by nitrooxidation, 61st Annual Assembly and International Conference of the International Institute of Welding, Graz, Austria, 6–11 July 2008.
5. Lazar, R., Marônek, M., Dománková, M. Low carbon steel sheets treated by nitrooxidation process, Engineering extra, 2007, No. 4, p. 86.
6. Palček, P et al. Change of fatigue characteristics of deep-drawing sheets by nitrooxidation. Master Journal List, Scopus. In Chemické listy. – ISSN 0009–2770. – Vol. 105, Iss. 16, Spec. Iss (2011), 539–541.
7. Bárta, J. et al. Joining of thin steel sheets treated by nitrooxidation, Proceeding of lectures of 15th seminary of ESAB + MTF-STU in the scope of seminars about welding and weldability. Trnava, Alumni Press, 2011, 57–67.
8. Marônek, M. et al. Welding of steel sheets treated by nitrooxidation, JOM-16 16-th International Conference On the Joining of Materials and 7-th International Conference on Education in Welding ICEW-7, May 10–13th, Tisvildeleje, Denmark, ISBN 87–89582–19–5.
9. Viňáš, J. Quality evaluation of laser welded sheets for cars body. In Mat/tech automobilového priemyslu Zborník prác vt-seminára s medzinárodnou účasťou Košice, 25.11.2005. Košice TU, 2005. 119–124. ISBN 80–8073–400–3.
10. Michalec, I. et al. Resistance welding of steel sheets treated by nitrooxidation. In TEAM 2011 Proceedings of the 3rd International Scientific and Expert Conference with simultaneously organized 17th International Scientific Conference CO-MAT-TECH 2011, 19th–21st October 2011, Trnava Slovakia. Slavonski Brod University of Applied Sciences of Slavonski Brod, 2011. ISBN 978-953-55970-4-9. 47–50.

CHAPTER 4

MONTE CARLO SIMULATION OF THE THREE-DIMENSIONAL FREE-RADICAL POLYMERIZATION OF TETRAFUNCTIONAL MONOMERS

YU. M. SIVERGIN,[1] S. M. USMANOV,[2] F. R. GAISIN,[2] and A. L. KOVARSKI[3]

[1]*N.N. Semenov Institute of Chemical Physics, Kosygin st.4., Moscow, Russia*

[2]*Birsk Branch Bashkir State University, Birsk, Russia*

[3]*N.M. Emanuel Institute of Biochemical Physics, Kosygin st.4., Moscow, Russia*

CONTENTS

ABSTRACT

For the first time in the world practice the results of simulation by Monte Carlo method of the kinetics of three-dimensional free-radical polymerization of tetrafunctional monomers (TFM) were obtained in the framework of the formation of a unitary three-dimensional structural element (UTDSE) and their structure formation on the simple cubic lattice, depending on the length l of molecules tetrafunctional monomers (l = 1 to 40 ribs of the lattice). Peculiarities of kinetics of changes in parameters such as the degree of polymerization of the P_n UTDSE, the number of radicals, the number of cross-links and cycles, and other characteristics were revealed. It was established that UTDSE are characterized by low levels of P_n for l = 1 and an explanation of this phenomenon was given. The study of the granulometric distribution (GMD) of UTDSE showed that curves of GMDs are bimodal and the probability density of these maximums was calculated.

4.1 INTRODUCTION

Kinetics of three-dimensional free-radical polymerization of tetrafunctional monomer (TFM) have been studied superficially using speculative assumptions that caused by complexity of this problem. It is proved that for three-dimensional polymers (TDP) microheterogeneous globular morphology is typical [1–3]. Considering the primary globular (or nanogel) structure of TDP as a unitary three-dimensional structural element (UTDSE) we developed a model of the UTDSE- MC for studying the kinetics of three-dimensional free-radical polymerization of TFM within the framework of formation of UTDSE [4, 5].

In the works of Refs. [4, 5], we studied this problem mainly using the simple cubic lattices of small dimension (from 3×3×3 to 10×10×10). The present work is devoted to identify impact of TFM molecules length *l* in a wide range of *l* (*l* = 1 to 40 ribs of the lattice). Experimental investigation of the kinetics of a block of free-radical polymerization of oligoesteracrylates (OEA) with oligomeric block of varying length revealed the influence of the block length on the gross kinetics of polymerization of OEA [6, 7]. It was obtained the kinetic curves of the degree of conversion of OEA and the reaction rate, the position of which were changed by varying the length of the oligomeric block in the direction of acceleration of the process of polymerization. These regularities were explained by the increase of viscosity of the reaction media (viscosity increases with the length of OEA molecules) as well as by the ratio of the rate constants of propagation and termination of chains [6, 7].

There are no literature data for the study of kinetics of three-dimensional free-radical polymerization of polyfunctional monomers within the framework of forming a unitary three-dimensional structural element (UTDSE) except our works [4, 5, 8]. The reason is the lack of experimental equipment with the necessary sensitivity which would enable to solve the problem.

In this chapter, we analyze the impact of the length of tetrafunctional monomers molecules (TFM) on the kinetics of formation of UTDSE, which is the primary structural element (nanogel) macrobody in a three-dimensional polymer (TDP) [1–3]. According to the offered scheme of structure formation of macrobodies of TDP, this process is realized in the following order: monomer (oligomer → nanogel (UTDSE) → microgel (many of the UTDSE) → macrobody TDP (multitude of microgels).

In the works of Refs. [5, 8] a detailed analysis of the literature data related to the problem of the kinetics of three-dimensional free-radical polymerization of polyfunctional monomers was performed, including a clear distinction between processes of structurization from the positions of percolation in conditions of physical and chemical structure formation. The analysis frees us from the need to perform the literature review in this work.

4.2 SETTING UP A PROBLEM AND THE EXPERIMENTAL PART

Modeling of three-dimensional free-radical polymerization TFM carried out by the statistical Monte-Carlo (MC) method using the simple cubic lattice with the dimensions 40x40x40 with inert walls. Randomly on the lattice molecule (M) of TFM were positioned and then initiated one of the double bonds (randomly selected) of TFM. In the framework of formation of UTDSE initiation reaction of the double bond M is implemented only once – at the time of the initial act of chain initiating.

At the subsequent stages double bonds of PG in the side groups of chains are initiated (the share of these groups we denotes by *D*). Consider the reactivity of double bonds in M and PG is the same [9, 10]. When modeling we took into account the reactions of initiation:

$$v_i = k_i[R_o][M] \quad (1)$$

$$v'_i = k_i[R_o][PG] \quad (1a).$$

After initiation the reactions of chain propagation

$$v_p = k_p[R][M] \quad (2)$$

$$v'_p = k'_p[R'][M] \quad (3)$$

and termination become possible

$$v_t = k_t[R]^2 \quad (4)$$

$$v'_t = k_t[R][R'] \quad (5)$$

$$v''_t = k_t[R']^2, \quad (6)$$

where R the is radical in the form of pendant groups of the following type 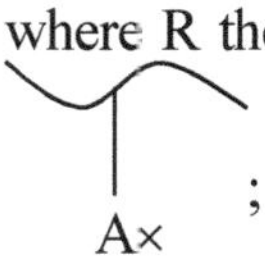 ; *R'*- active radical of A  type and he is characterized by the less reactivity compared to *R* due to the greater steric difficulties in *R'*; it is assumed for it that $k'_p = 0.1\ k_p$. In these equations v_i, v'_i, v_p, v'_p, v_t, v'_t, v''_t and k_i, k_p, k'_p, k_t – the rates and rate constants of the reactions of initiation, propagation, and termination of the chain; R_0 – primary initiating radical, that in the reactions of chain termination is not involved, as in this case, the model is not work.

In the process of implementing of a three-dimensional polymerization of TFM reactions of chain cross-linking (7–11) and the cycles formation (12–16) occurs by mechanisms for *R-PG, R'-PG, R-R, R-R'* and *R'-R'*:

$$v_{cr} = k_{cr}[R][PG] \quad (7)$$

$$v'_{cr} = k'_{cr} [R'][PG] \tag{8}$$

$$v_{tcr} = k_{tcr}[R]^2 \tag{9}$$

$$v'_{tcr} = k_{tcr}[R][R'] \tag{10}$$

$$v''_{tcr} = k_{tcr}[R']^2 \tag{11}$$

$$v_{cy} = k_{cy}[R][PG] \tag{12}$$

$$v'_{cy} = k'_{cy}[R'][PG] \tag{13}$$

$$v_{tcy} = k_{tcy}[R]^2 \tag{14}$$

$$v'_{tcy} = k_{tcy}[R][R'] \tag{15}$$

$$v''_{tcy} = k_{tcy}[R']^2, \tag{16}$$

Here: v_{cr}, v'_{cr}, v_{tcr}, v'_{tcr}, v''_{tcr}, v_{cy}, v'_{cy}, v_{tcy}, v'_{tcy}, v''_{tcy} are the rates of chain cross-linking and the cycles formation, k_{cr}, k'_{cr}, k_{tcr}, k_{cy}, k'_{cy}, k_{tcy} are the rate constants of these reactions. As the reactivity of double bonds in M and PG equal and it was accepted the equality of reaction constant rates $k_{cr} = k_{cy} = k_p$, $k'_{cr} = k'_{cy} = k'_p$, reactions (7, 8, 12, 13) are among reactions (2, 3); in view of the equality of reactivity of radicals *R* and *R'* in reactions (4–6) and in (9–11, 14–16) and equality of the rate constants of reactions $k_{tcr} = k_{tcy} = k_t$, reactions (9–11, 14–16) are included in the reactions (4–6). The possibility of occurrence of this or that elementary reaction of the above mentioned will depend on the probability of encounters (contact) of appropriate reactive centers.

For the above reasons, in the calculation of the probability reactions p_i we took into account only six elementary stages $j = \overline{1..6}$ At each stage of modeling, one needs to identify which of these six reactions will be run. The stage number j ($j = \overline{1..6}$) is found from the condition:

$$\sum_{i=1}^{j-1} p_i < \xi_1 < \sum_{i=1}^{i=j} p_i, \tag{17}$$

where $P_i = \frac{V_i}{\sum_{i=1}^{n} V_i}$ is the probability of occurrence of the i-th elementary stage (or p_i is the statistical weight of the i-th reactions); designating by the V_i – the rate of i-the reaction one can obtains $V = \sum V_i = v_i + v_p + v'_p + v_t + v'_t + v''_t$; total probability $P = \sum p_i = 1$.

Waiting time of specific elementary stage of polymerization is found from the relation:

$$\tau_k = \frac{-\ln \xi_2}{\sum_{i=1}^{6} V_i}, \tag{18}$$

here ξ_1 и ξ_2 – are random numbers uniformly distributed in the interval (0, 1). In the product tk_p the parameter k_p is the constant rate of pseudo-first order of chain propagation. A total polymerization time (T_{pol}) is determined by the sum of time waiting for all stages

$$T_{пол} = \sum \tau_k. \tag{19}$$

In the initial stages of three-dimensional polymerization of TFM is a purely stochastic process (selection of possible reaction, etc.), but from a certain degree of conservation (G) an element of determinism interfere and its contribution increases with the growth of G with the transition to the final stages of the process primarily in deterministic mode.

In addition to of reactions quadratic termination of the radicals R and R' there is linear radicals termination in reaction system with their immobilization (automatically considered as residual radicals R and R'):

$$R + Y \rightarrow RY \qquad v_{tl} = k_{tl}[R][Y] \tag{20}$$

$$R' + Y \rightarrow R'Y \qquad v'_{tl} = k'_{tl}[R'][Y], \tag{21}$$

where v_{tl}, v'_{tl}, k_{tl}, k'_{tl} – the rates and rate constants of linear termination of radicals.

From the reaction schemes it is clear that the growth of the molecular mass of UTDSE frame is performed only by the reactions of chain propagation (2, 3). Reactions of cross-linking and cycle formation do not lead to growth of molecular mass of UTDSE, and cause only topological changes in the frame of UTDSE – macroradical cross-linking between each other either the formation of cyclic structures of different sizes. The process was conducted before the possibility of a reaction proceeding (exhaustion of available double bonds, and so on). Re-visiting the nodes was forbidden (self-intersections are absent).

When modeling of TFM polymerization by MC method at every stage of the simulation we randomly determined the choice of elementary reactions (random choice of conditions in which the reaction system will convert), its direction, location of TFM molecule, the choice of functional group, the waiting time of transition [2, 8].

The following values for the rate constants of reactions were adopted: $k_i = k_p = 400$ l/mol•c, $k_t = 10^6$ l/mol•c, $k'_p = 40$ l/mol•c. The length l of TFM molecules varied in a range from 1 to 40 ribs (2 to 41 nodes). If you take the density of TFM equal to 1.05 g/ml and the molar mass to 200 g/mol, the simulated volume for this lattice and TFM molecule with $l = 2$ ribs is 13179 nm^3 (the lattice 40^3 consist of 21333 this monomer molecules).

Accordingly, for monomers with another length l this reaction volume will be different for the same amount of monomer, as well as the reaction volume was adopted constant, then different number of monomer molecules (from 32,000 at $l = 1$ to 1560 monomer molecules at $l = 40$). Averaging of obtaining parameters to the current moment of time we spent 5000 experimental implementations; it is clear that every experiment did not depend on others.

4.3 RESULTS AND DISCUSSION

The kinetic curves of free-radical polymerization of TFM in coordinates: the degree of polymerization P_n – time tk_p (Fig. 4.1) have the form of sigma-shaped curves (except the case of TFM with $l = 1$, for which the process is not managed to grow by average values P_n up to high values of P_n), to be the exact form of the experimental curve G(t) obtained in the case of block polymerization OEA [6, 7]. Quantitatively limit values P_n decrease in the row: $l = 2 \rightarrow 40$, which is quite natural and expected, because at the same row reduces the number of molecules that located in a

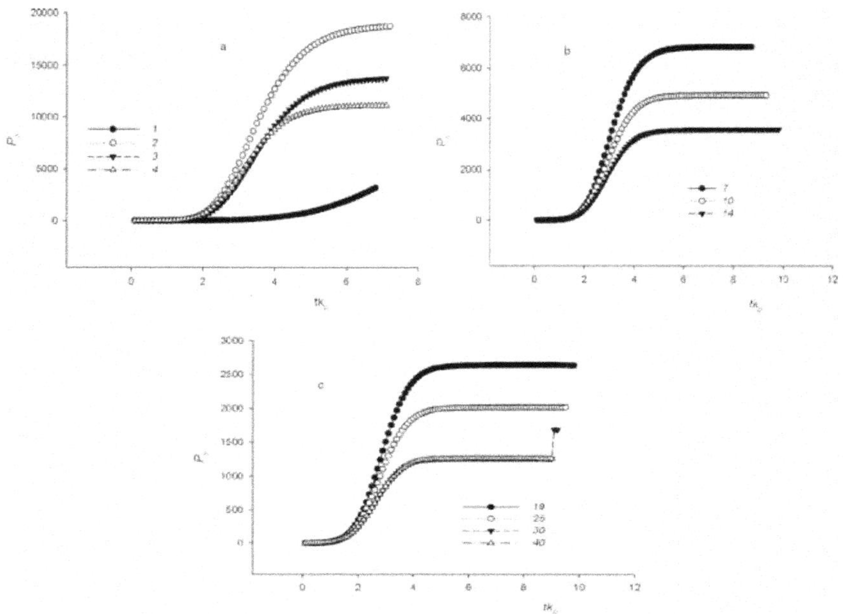

FIGURE 4.1 Kinetic dependence of the degree of polymerization UTDSE at different length l (in the ribs of the lattice, numbers in front bracket) for TFM molecules: a – 1 (1), 2 (2), 3 (3), 4 (4); b – 7 (1), 10 (2), 14 (3); c – 19 (1), 25 (2), 30 (3), 40 (4).

given constant volume of the reaction (Table 4.1, Fig. 4.1). At appropriate G values curves $P_n(t)$ to some extent shifted towards shorter times with the elongation of TFM molecules, which is in agreement with experiment for the case of block polymerization of OEA [6, 7]. This result is predictable, if you put the emerging UTDSE is a nano-reactor in which the polymerization of TFM realized. In the regime of block polymerization of TFM there are numerous of such of nano-reactors, located in the reaction volume.

The essential difference between the curve $P_n(t)$ for monomer with $l = 1$ rib is due to the fact that such a short-chained monomer is characterized by a high probability to cycle formation arising in the process of polymerization. Due to the presence of a high share of cyclic structures when the number of parallel experiments is 5000 the averaging of parameters causes the observed shape of the curve $P_n(t)$ for this monomer (Fig. 4.1, curve 1). Of course, among 5000 structures UTDSE at $l = 1$ one can find UTDSE with high values of $P_n \approx 23{,}665–23{,}840$, but the share of such UTDSE,

TABLE 4.1 Limit Values of Kinetic Parameters at Different Lengths l of TFM Molecules

l, Number of grids	P_n	N_R	$N_{R'}$	$N_R+N_{R'}$	$N_R/N_{R'}$	D	N_{PG}	k_{pack}
1	3226	124	129	253	0.96	0.124	457	0.25
2	18,696	3580	491	4071	7.3	0.012	227	0.786
3	13,716	3568	464	4032	7.69	0.018	251	0.765
4	11,119	3884	320	4204	12.1	0.009	107	0.822
7	6818	3556	186	3742	19.1	0.004	24	0.837
10	4919	3116	109	3225	28.6	0.0008	4	0.836
14	3565	2574	66	2640	39	0.0006	2	0.831
19	2637	2073	40	2113	51.8	0.003	6.9	0.825
25	2019	1686	25	1711	67.4	0.001	2	0.81
30	1684	1451	18	1469	80.6	0.001	1.9	0.808
40	1262	1130	9.4	1139.4	120.2	0.001	1.3	0.773

as will be shown below, significantly less than the share of UTDSE with low values of P_n.

As we can see from the Table 4.1, the difference between the P_n for monomers with $l = 2$ and $l = 40$ reaches 15.5, but by the values of achieved molar mass m UTDSE this difference significantly less: the ratio $mL_{=2}/m_{l=40} = 0.75$, and for the whole range of lengths l the value of the ratio $mL_{=2}/m_l$ varies from 0.91 to 0.74, falling down with increasing of l.

Equations (22)–(22b) allow to estimate the importance of the limit value of polymerization degree of UTDSE P_n at the desired length and conditions under consideration of the three-dimensional polymerization of TFM:

$$P_n = 30925\cdot\exp(-0.026l) \quad (l = 2\div4) \tag{22}$$

$$P_n = 12735\cdot\exp(-0.092l) \quad (l = 7\div14) \tag{22a}$$

$$P_n = 7215\cdot\exp(-0.051l) \quad (l = 14\div25) \tag{22б}$$

$$P_n = 4335\cdot\exp(-0.031l) \quad (l \geq 25) \tag{22в}$$

The change in the number of radicals N_R and $N_{R'}$ in the course of polymerization process of TFM illustrated in Fig. 2: the number of radicals N_R decreases in a row TFM with $l = 4 > 2 > 3 > 7 > 10 > 14 > 19 > 25 > 30 > 40 > 1$, and for change of the number of radicals $N_{R'}$ we got the dependence of the l type: $l = 2 > 3 > 4 > 7 > 1 > 10 > 14 > 19 > 25 > 30 > 40$. Limit values of numbers N_R and $N_{R'}$ fit in the same dependence, and the dependence of the sum of the limit values ($N_R + N_{R'}$) is similar to the dependence of N_R from l (Table 4.1). Except the variation of $l = 1$, all dependencies N_R (t) have the form of sigma shaped curves. The curves $N_{R'}(t)$ has similar character, but expressed less clear for $l = 2, 3, 4, 7$ (Fig. 4.2). Out of these regularities is the case for TFM with $l = 1$. The number of N_R in this version is significantly less compared with other l and the number $N_{R'}$, were almost

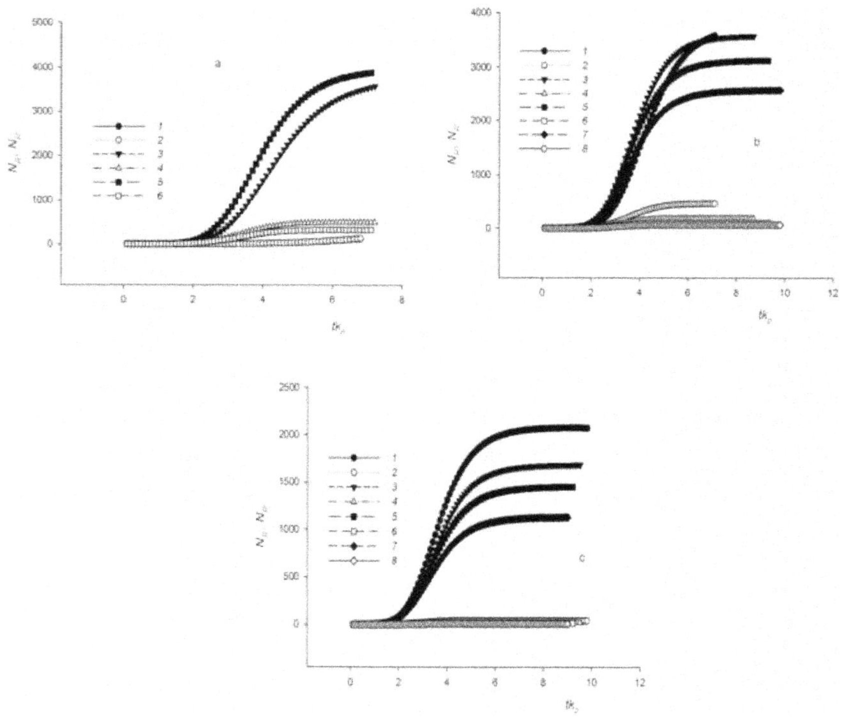

FIGURE 4.2 The change in the number of radicals N_R (1, 3, 5, 7) и $N_{R'}$ (2, 4, 6, 8) with time for UTDSE, obtained at: a. $l = 1$ (1, 2), 2 (3, 4), 4 (5, 6); b. $l = 3$ (1, 2), 7 (3, 4), 10 (5, 6), 14 (7, 8); c. $l = 19$ (1,2), 25 (3, 4), 30 (5, 6), 40 (7, 8).

equal to N_{R}, less when $l = 2$ to 7, but more than at $l > 10$ (Fig. 4.2, Table 1). The dependences of limited values of N_R and $N_{R'}$ on l describe by exponential functions (23) and (24) allowing to assess the magnitude of N_R and $N_{R'}$, for other values of l. Functions converge at a limit → 0:

$$N_R = 4715 \cdot \exp(-0.043l) \quad (l = 4 \div 22) \tag{23}$$

$$N_R = 3240 \cdot \exp(-0.026l) \quad (l = 22 \div 40) \tag{23a}$$

$$N_{R'} = 750 \cdot \exp(-0.195l) \quad (l = 2 \div 11) \tag{24}$$

$$N_{R'} = 210 \cdot \exp(-0.084l) \quad (l = 12 \div 40), \tag{24a}$$

From dependencies N_R of and $N_{R'}$ it follows that the rate of change of radicals in the process of three-dimensional free-radical polymerization of TFM are not equal to zero, that is, to use the approach of steady-state condition of the process relative to the radicals in this case, is unauthorized. This conclusion is very useful and should be taken into account in the analysis of experimental kinetic study of TFM polymerization [6, 7]. From the relations of the limit values of $N_R/N_{R'}$ (Table 4.1) it is clear that with increasing l this attitude increases, that is, with increasing the length of TFM molecules the number of generated radicals of R' type falls, including due to decrease of the contribution of the shares of the process of cycle formation with increasing of l.

In all cases for l the number of cross-linked N_{cr} exceeds the number of cycles N_{cy} ($N_{cr} > N_{cy}$), and in a series of increasing l, as earlier, from the patterns of growth in the N_{cr} and N_{cy} with the rise of l the case $l = 1$ drops out (Fig. 4.3, Table 4.2). As it was expected, with increasing of l the contribution of cycle formation process should decrease, that we observe in the example of the relationship N_{cr}/N_{cy}.

It is known that with chain elongation the probability of encounters of two ends of this chain in any microvolume decreases. For molecules of TFM with $l = 1$, the probability of two ends of this molecule encounters is equal to zero, but in the course of polymerization process there

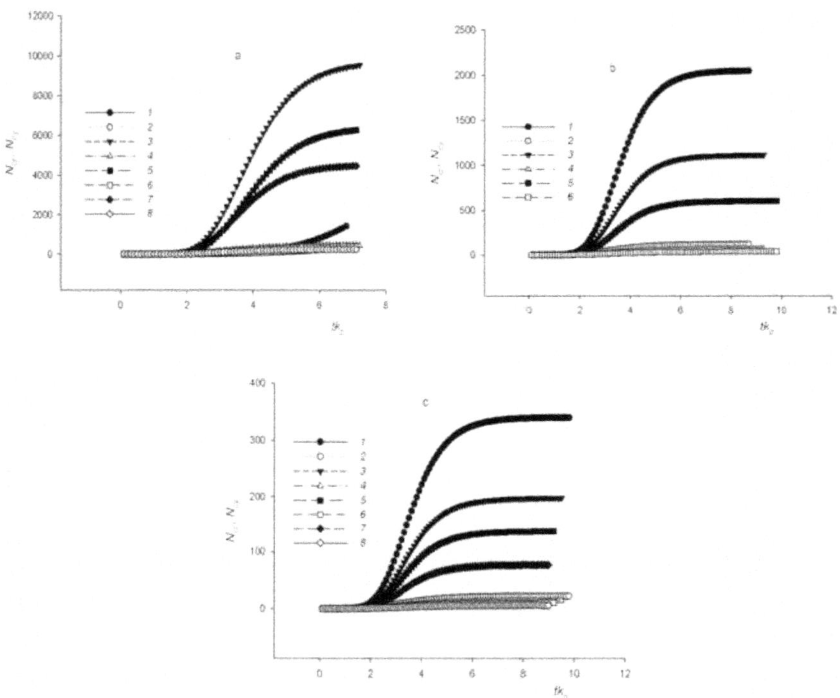

FIGURE 4.3 The dependence of the number of cross-links N_{cr} (1, 3, 5, 7) and the number of cycles N_{cy} (2, 4, 6, 8) on time for UTDSE obtained at: a. $l = 1$ (1, 2), 2 (3, 4), 3 (5, 6), 4 (7, 8); b. $l = 7$ (1, 2), 10 (3, 4), 14 (5, 6); c. $l = 19$ (1, 2), 25 (3, 4), 30 (5, 6), 40 (7, 8).

are situations when the two ends of the growing macrochain come into contact with each other, causing a cycle that is facilitated by the high probability of the reaction of chain termination at radicals contact. These situations arise often when $l = 1$, as evidenced by achieving $N_{cy} \sim 950$–1050 in a hypothetical variant of the achievement of $k_{pack} \sim 0.76$–0.83 for TFM with $l = 1$.

Such values of N_{cy} noticeably exceed N_{cy} even for TFM with $l = 2$. The number of N_{cy} found in a single experiments during which the P_n of UTDSE was maximal confirm the following dependencies of $N_{cy}(l)$ and $N_{cr}(l)$ (where the figures at N_{cy} mean: I – the average UTDSE (see Table 4.2), II – single maximal UTDSE, III – per 100 chain units (average UTDSE), IV – per 100 chain units (single maximal UTDSE). In the

TABLE 4.2 Limit Values of N_{cr} and N_{cy}, the Amount of Cycles ($N_{cr}+N_{cy}$), Relations (N_{cr}/N_{cy}) and the Number of Reactions R-PG, R′-PG, R-R, R-R′, R′-R′ at Different Lengths l of TFM Molecules

Number of grids, l	N_{cr}	N_{cy}	$N_{cr}+N_{cy}$	N_{cr}/N_{cy}	N_{R-PG}	$N_{R'-PG}$	N_{R-R}	$N_{R-R'}$	$N_{R'-R'}$
1	1421	313	1734	4.54	1133	74	188	219	120
2	9542	451	9993	21.16	5208	602	2046	1338	799
3	6278	382	6660	16.43	3670	376	1561	604	449
4	4482	250	4732	17.93	2472	282	1356	314	308
7	2048	118	2166	17.4	1185	100	711	69	101
10	1115	64	1179	17.42	629	43	441	24	42
14	605	37	642	16.35	344	17	256	9.0	16
19	339	22	361	15.41	198	7	146	3.6	6.4
25	198	14	212	14.14	116	3	89	1.6	2.4
30	137	10	147	13.7	78	1.9	64.4	1.1	1.6
40	77	6	83	12.83	43	1	38	0.4	0.6

Table 2 also shown the number of cross-links N_{cr} per 100 units (V – average UTDSE, VI – single maximal UTDSE)):

l	1	2	3	4	7	10	14	19	25	30	40
N_{cy} (II)	1487	565	405	286	133	63	34	19	15	13	8
N_{cy} (III)	9.7	2.5	2.78	2.25	1.72	1.31	1.04	0.84	0.69	0.61	0.47
N_{cy} (IV)	6.8	3.04	2.8	2.16	1.86	1.43	0.99	0.75	0.58	0.8	0.63
N_{cr} (V)	44	51	45.8	40.3	30	22.7	17	12.9	9.8	8.1	6.1
N_{cr} (VI)	58	50.4	45.6	40.7	29.6	22.6	17.1	12.6	9.6	8.1	5.4

As one can see, N_{cy} was the greatest (case 1) at $l = 1$, indicating a very significant private contribution to the process of cycle formation for this variant of TFM. This case probably extends to other TFM which molecules longer that $l = 1$, but they are also hard (simbat to $l = 1$), excluding the possibility of encounters of macromolecule ends, and yet the share of process of cycle formation can be very significant for them in the course of construction of UTDSE for the reasons mentioned for molecule with $l = 1$.

The comparison of the number of N_{cy} per 100 units of UTDSE (cases III and IV) also confirms the highest probability of cycle formation for TFM with $l = 1$ (calculation of N_{cy} made on the basis of average (5000 experiments) values of N_{cy} both with the value of N_{cy} for a single UTDSE). Assessing the number of cycles N_{cy} for these different options may be carried out using the following equations:

I. UTDSE with the averaged P_n

$$N_{cy} = 737 \cdot \exp(-0.251l) \quad (l = 2 \div 7) \tag{25}$$

$$N_{cy} = 123 \cdot \exp(-0.088l) \quad (l = 10 \div 19) \tag{25a}$$

$$N_{cy} = 55 \cdot \exp(-0.056l) \quad (l \geq 25) \tag{25б}$$

II. Single UTDSE with the maximal value P_n

$$N_{cy} = 910 \cdot \exp(-0.271l) \quad (l = 2 \div 10) \tag{26}$$

$$N_{cy} = 230 \cdot \exp(-0.133l) \quad (l = 10 \div 19) \tag{26a}$$

$$N_{cy} = 42 \cdot \exp(-0.041l) \quad (l \geq 19) \tag{26b}$$

III. Per 100 units of UTDSE (averaged UTDSE)

$$N_{cy} = 3.15 \cdot \exp(-0.082l) \quad (l = 2 \div 14) \tag{27}$$

$$N_{cy} = 1.52 \cdot \exp(-0.03l) \quad (l = 14 \div 40) \tag{27a}$$

IV. Per 100 units (single UTDSE with maximal value of P_n)

$$N_{cy} = 3.5 \cdot \exp(-0.09l) \quad (l = 2 \div 19) \tag{28}$$

$$N_{cy} = 0.92 \cdot \exp(-0.008l) \quad (l \geq 19) \tag{28a}$$

In the equations of the form $N_{cy} = N_{0cy} \bullet \exp(-\Delta \cdot l)$ preexponential factor N_{0cy} indicates the number of cycles for chains with self-intersection, and the coefficient Δ is the variations of cycles in real chain.

Dependence N_{cy} (l) is shown in Fig. 4.4 for options I and III, IV. It gives vivid example about this dependence. Experimental verification of the number of cycles in a three-dimensional polymer is rather scanty and executed, for example, for diallyl derivatives of dicarboxylic acids [11–13]. It was revealed changes in the number of cycles for polymers of diallyl sebacinate (DAS) (N_{cy} = 3.8–26.9 for fractions of pregel stage with P_n = 28.4–191, when P_n = 100 N_{cy} = 10–18) and for diallyl isophtalate (DAIPh) (N_{cy} = 2.7–68.4 for fractions pregel stage with P_n = 7–407, when P_n = 100 N_{cy} = 16–34). During polymerization of diallyl derivatives great contribution to the formation of structural units makes the reaction of chain transfer to monomer, including influencing the process of cycle formation. The length of the DAS molecule greater than the length of the DAIPh molecule, that determines the increase in the number of cycles in DAIPh macrochain. These results correlate with the regularities which we established in this work and that testifies in favor of a model of the UTDSE-MC, developed by us. Earlier [5, 8] we have shown that the process of cycle formation in the three-dimensional polymerization of TFM affect the rate of initiation of the chain, the size of reaction volume (V), the geometry of lattices (g.l. or reaction volume), the activity of reactor walls (arw.), the chemical nature of TFM (c.n.). In this work we have established the influence of the length of molecules TFM on this process. Thus, one can write that $N_{cy} = f\,(v_i, V, g.l., a.w., c.n., l)$.

The variation of the number of cycles N_{cy} per 100 units of UTDSE has a drop-down type with l rising and that is to be expected (the reaction

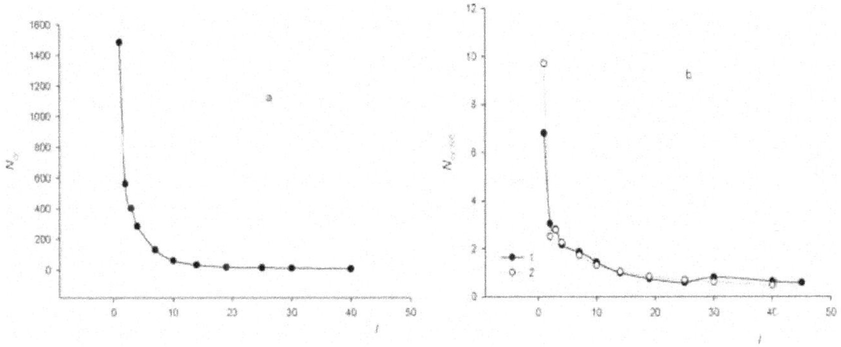

FIGURE 4.4 Change the number of cycles N_{cy} with elongation of TFM molecules for options II (a) and III, IV (b).

volume is constant, and l is growing). An estimate of the number of N_{cy} 100 units of UTDSE allow one to sign the expressions:

V. per 100 units of UTDSE (averaged UTDSE)

$$N_{cr} = 61.6 \cdot \exp(-0.1l) \quad (l = 2 \div 14) \tag{29}$$

$$N_{cr} = 27.5 \cdot \exp(-0.04l) \quad (l = 14\text{–}40) \tag{29a}$$

VI. per 100 units of UTDSE (single UTDSE with maximal P_n)

$$N_{cr} = 62.6 \cdot \exp(-0.104l) \quad (l = 1 \div 14) \tag{30}$$

$$N_{cr} = 29.7 \cdot \exp(-0.043l) \quad (l = 14\text{–}40) \tag{30a}$$

$$N_{cr} = 49.7 \cdot \exp(-0.062l) \quad (l = 1\text{–}40) \tag{30b}$$

In all cases, the correlation coefficient r = 0.98–0.999. In Eqs. (29) and (30), (29a) and (30a) the values of numerical coefficients close to each other, preexponential factor N_{0cr} is the number of cross-linking in the case of chains with self-intersection, and coefficient Δ is a change in the number of cross-linking in a real chain. Eq. (30b) illustrates the admissibility of the use of more general function for l = (1–40). In Refs. [11–13] it was estimated the number of branching points in chains of prepolymers of diallyl derivatives. Because each cross-link has two branch points, for cases V and VI we shall receive from 10 to 116 branching points in a row l = 1–40, which significantly exceeds the number of branch points (6–7) to prepolymers diallyl derivatives, but this fact should not come as a surprise, because in the latter case considered pregel stage, the reactions of chain transfer and other. In a series with $l = (2 \rightarrow 40)$ the number of cross-links regularly decreases, as well as the attitude N_{cr}/N_{cy}, and the number of cross-links significantly greater than the value N_{cy} (Table 4.2, Fig. 4.3). The dependence N_{cy} on P_n is predominantly linear, and dependence of N_{cr} on P_n – is an nonlinear increasing curve [8].

The processes of cross-linking and cycle formation during three-dimensional polymerization of TFM include reactions R-PG, R′-PG, R-R, R-R′ and R′-R.’ The ratio of the contributions of these reactions in the cross-linking of macrochains and cycle formation is clear from the Table 4.2 and Fig. 4.5, that is, the main share of these reactions is cross-linking.

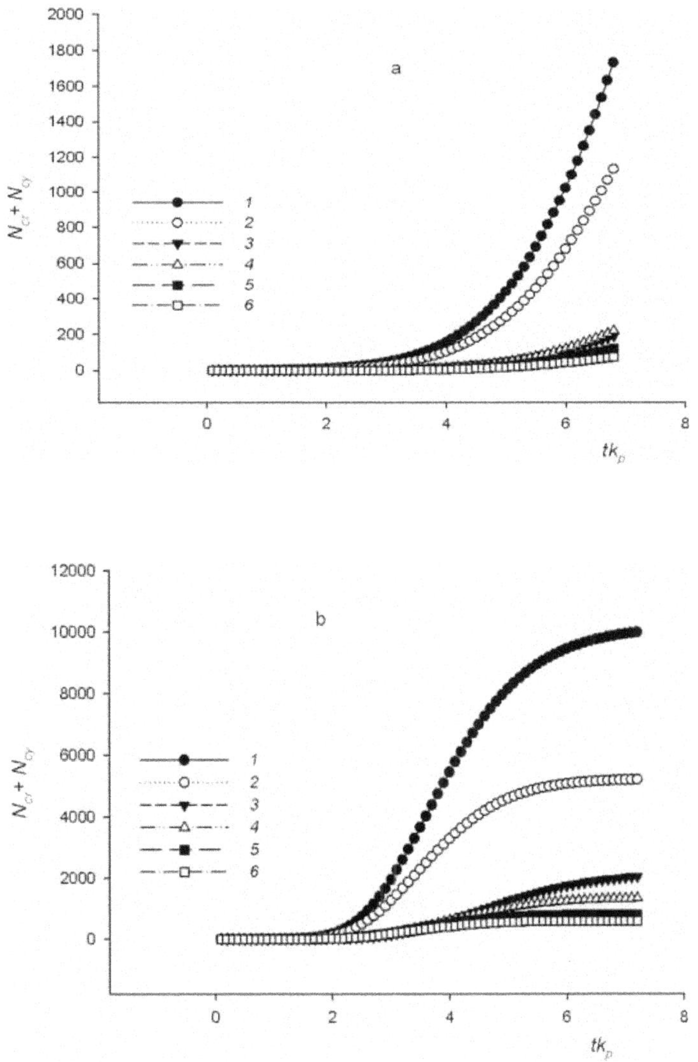

FIGURE 4.5 The impact of the mechanism of reactions R-PG (curve 2), R-R (3), R-R′(4), (R′-R′ (5), R′-PG (6) on the number of cross-links and cycles ($N_{cr} + N_{cy}$, curve 1) with $l = 1$ (a), 2 (b), 3 (c), 4 (d), 7 (e), 10 (f), 14 (g), 19 (h), 25 (i), 30 (j), 40 (k).

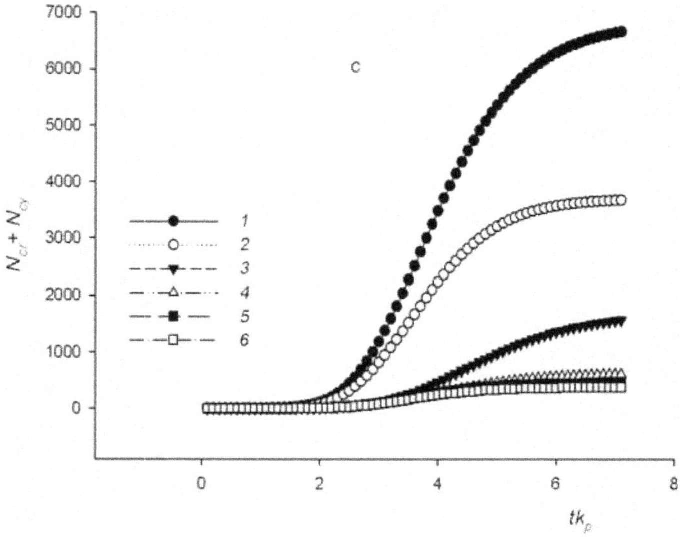

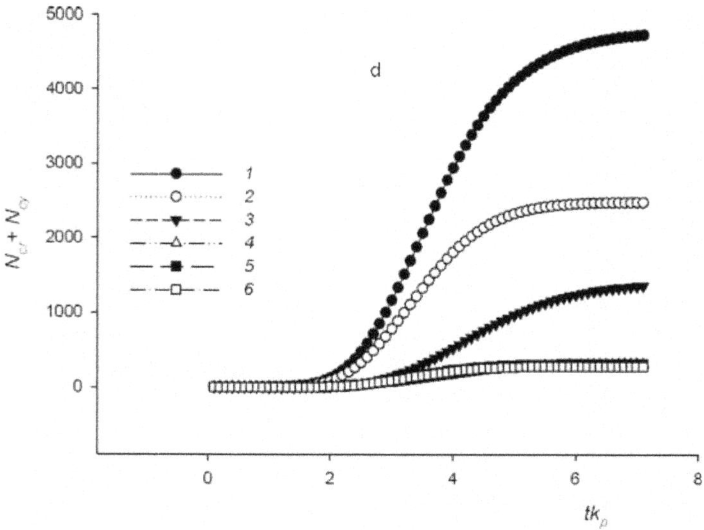

FIGURE 4.5 (Continued)

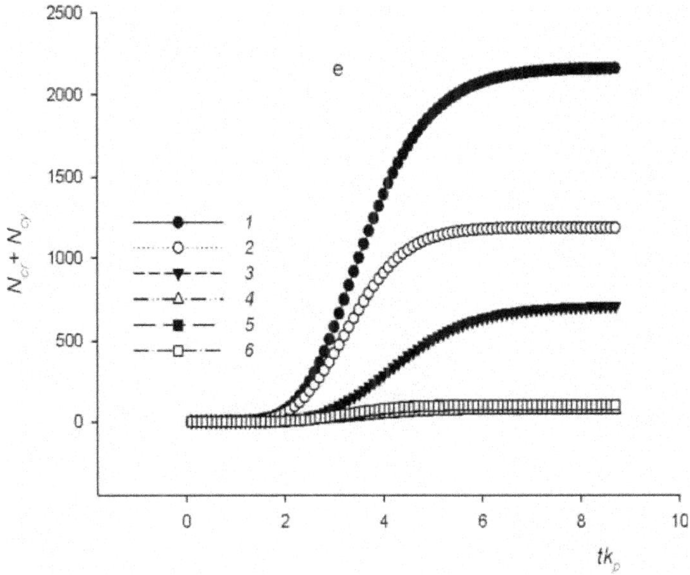

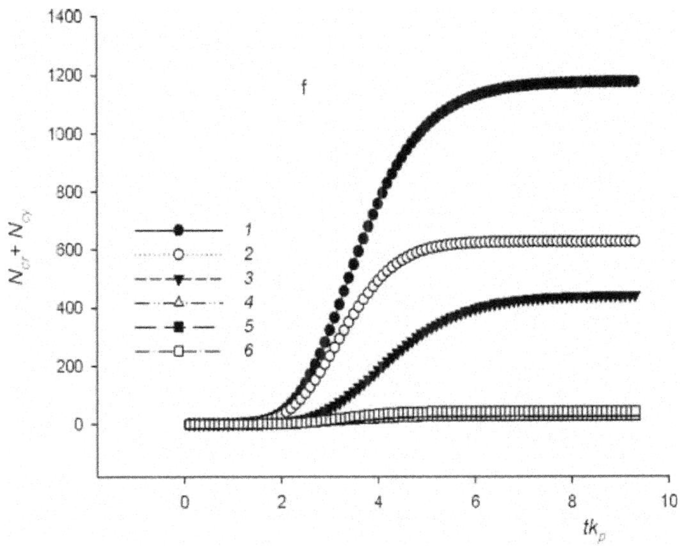

FIGURE 4.5 (Continued)

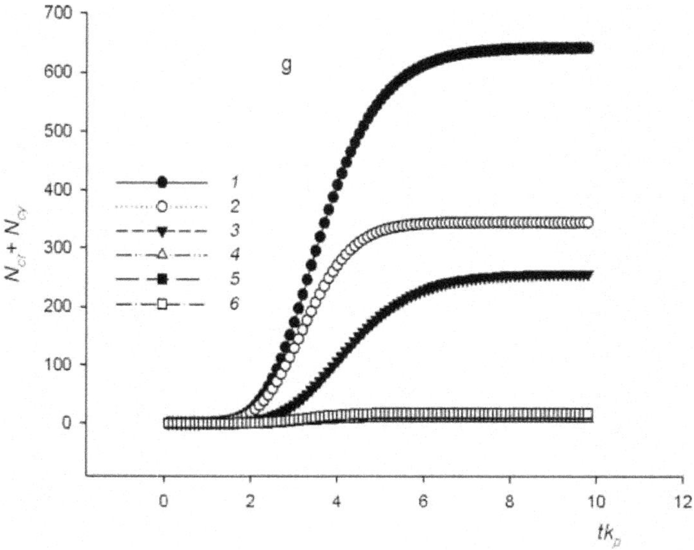

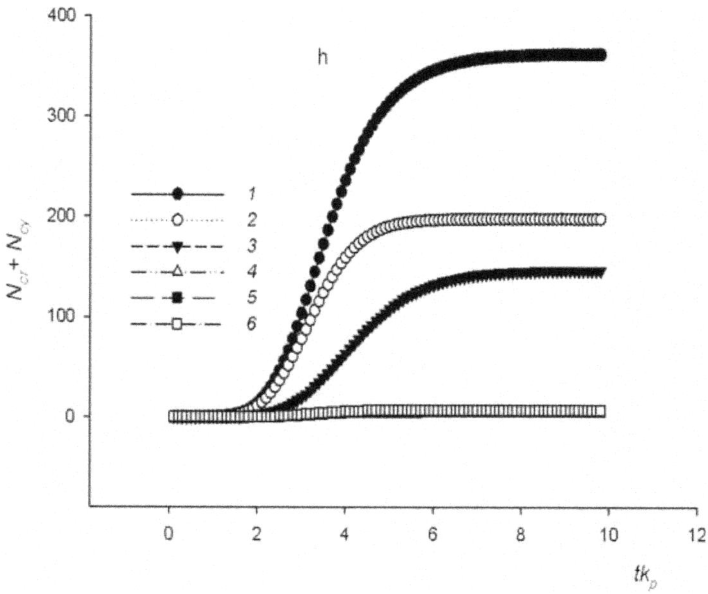

FIGURE 4.5 (Continued)

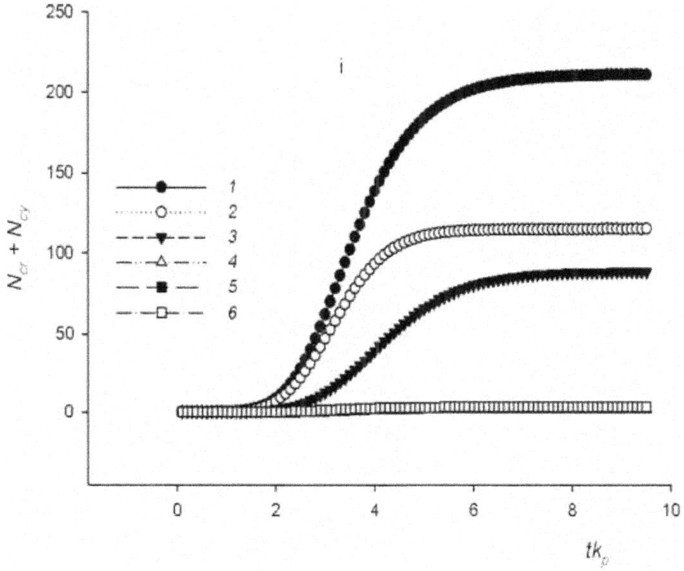

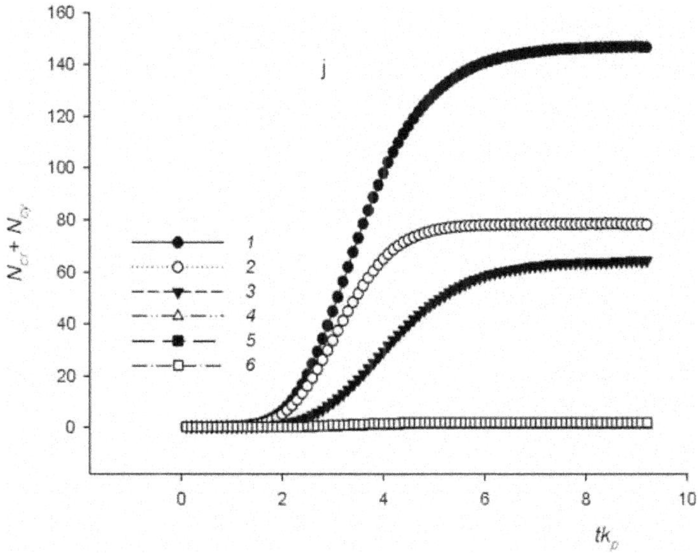

FIGURE 4.5 (Continued)

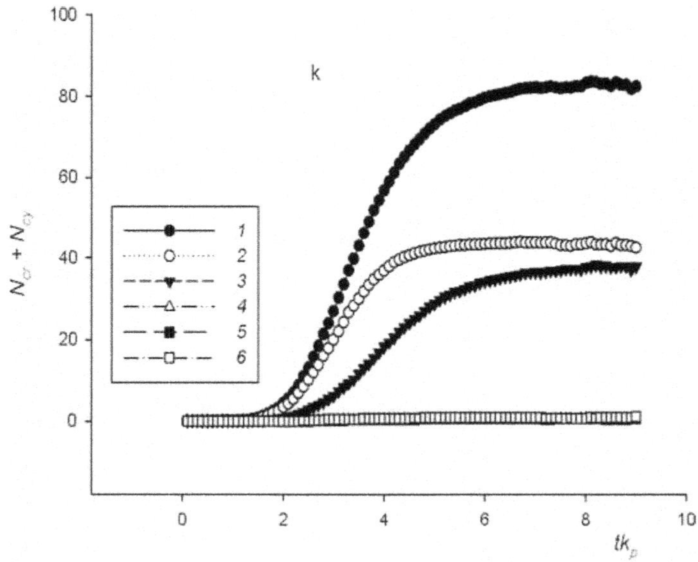

FIGURE 4.5 (Continued)

We do not separate in this work the share of each of the reactions of cross-linking and cycles formation (although in principle these is possible to realize). On a basis of the competition of these reactions it is obvious that the prevailing reaction is R-PG (Fig. 4.5, Table 4.2). Excluding TMF with $l = 1$, we see that considered reactions compose a row $N_{R-PG} > N_{R-R} > N_{R-R'} > N_{R'-R'} > N_{R'-PG}$ at $l = 2$–4; at $l \geq 7$ the row is $N_{R-PG} > N_{R-R} > N_{R'-PG} > N_{R'-R'} > N_{R-R'}$, reflecting the impact on these reactions the length of TFM molecules. In the case with $l = 1$ the row of these reactions is $N_{R-PG} > N_{R-R'} > N_{R-R} > N_{R'-R'} > N_{R'-PG}$ and this dependence reflects the fact that $N_{R'} > N_R$ at $l = 1$ (Table 4.1).

From comparison of the limiting values of the number of reactions follows that the row $l = 2$–40 the ratio N_{R-PG}/N_{R-R} decreases (relative contribution of reaction R-R grows), and the ratios $N_{R-PG} > N_{R-R'}$, $N_{R-PG} > N_{R'-PG}$ and $N_{R-PG} > N_{R'-R'}$ increase (relative contribution of reaction R-PG decreases) (Table 4.2). Curves of the dependence of the number of reactions on t are sigma-shaped curves in a row $l \to$ (2 to 40), and at $l = 1$ averaged curves have the shape of the curves with acceleration (but in the case of single UTDSE with maximal P_n the curves also are sigma-shaped) (Fig. 4.5). The obtained regularities are unique and is absent in the literature, excluding our works [4, 5, 8]. Our kinetic dependences (N_{cr} + N_{cy}) (Fig. 4.5) reflect the character of changes of such topological parameters such as the number of cross-links and the number of cycles in the framework of formation of UTDSE, and the change of N_{cr} and N_{cy} with increasing P_n (or the degree of conversion of TFM) allows to understand the nature of their variation during the process of polymerization.

The fall of residual unsaturation in UTDSE (or share of pendant to a carcass of UTDSE double bonds PG) at $l = 1$ has the appearance of a smooth drop-down curve (Fig. 4.6), and with increasing l curves *D(t)* has a patch sharp drop *D*, turning to the curve of sigma-shaped type the initial part of which is located in the area $D \approx 0.34–0.36$, and the end part located in the field of limit values of $D \approx 0.001$ to 0.018 (Table 4.2,

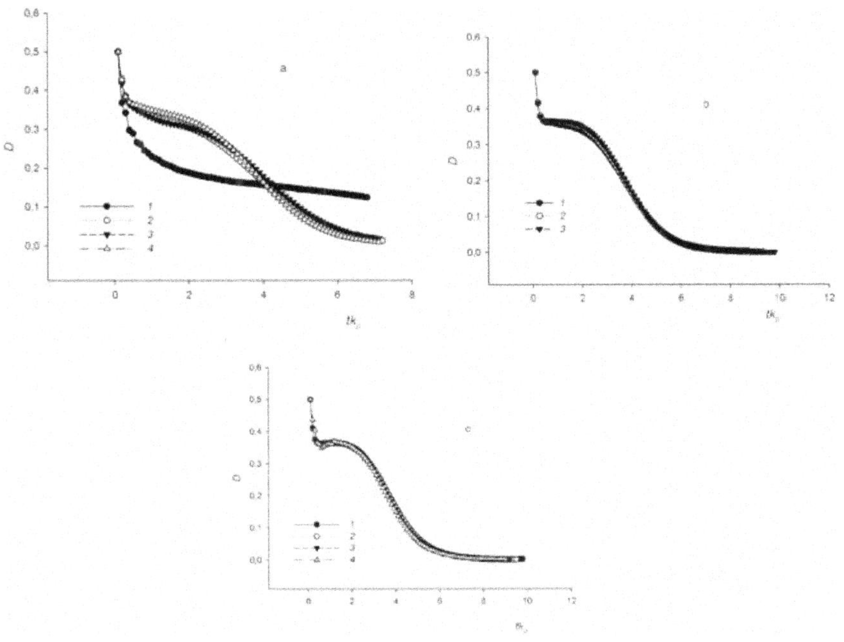

FIGURE 4.6 Kinetic curves of the residual instauration *D* for UTDSE: a. $l = 1$ (1), 2 (2), 3 (3), 4 (4); b. $l = 7$ (1), 10 (2), 14 (3); c. $l = 19$ (1), 25 (2), 30 (3), 40 (4).

Fig. 4.6) with the number of $N_{PG} \approx 1.3\%$ to 251. In a series $l \rightarrow$ (1–15) limit values fall, when $l = 19$, then rise and then fall again in the row $l \rightarrow$ (19–40) (table. 1, Fig. 4.6).

In terms of statistics 5000 experiences reliably established the existence of bimodal granulometric distribution (GMD) of UTDSE (Fig. 4.7),

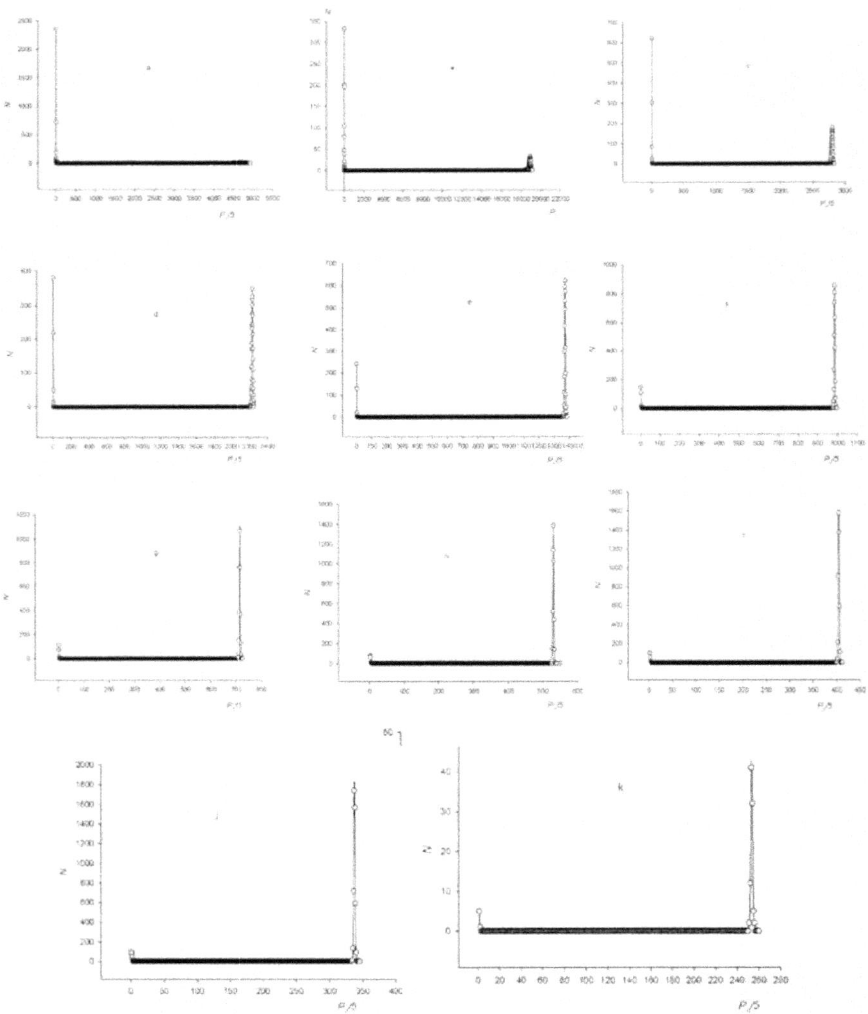

FIGURE 4.7 Curves GMD of UTDSE at: $l = 1$ (a), 2 (b), 3 (c), 4 (d), 7 (e), 10 (f), 14 (g), 19 (h), 25 (i), 30 (j), 40 (k).

which characterized by a marked increase of the density of probability (q) in the field of higher P_n with increasing l:

l, number of ribs	1	2	3	4	7	10	14	19	25	30	40
Peak 1, q	0.715	0.207	0.211	0.134	0.082	0.057	0.042	0.031	0.035	0.029	0.04
Peak 2, q	0.285	0.793	0.789	0.866	0.918	0.943	0.958	0.969	0.965	0.971	0.96
$P_{n\,max}$	~23500	18890	14010	11131	6855	4920	3565	2650	2020	1685	1265

Maximum of peak 1 is located in the area $P_{n\,max} \approx 3–5$. Peak 2 for $l = 1$ is weak (smeared). Obviously, if $l = 1$ $q_1/q_2 = 2.51$, then, for example, when $l = 30$ the attitude $q_1/q_2 = 0.03$, i.e., with increasing l increases the formation of UTDSE with higher values of P_n. The formation of a significant proportion of low-molecular UTDSE at $l = 1$ confirm the foregoing features of UTDSE formation for TFM with $l = 1$. The results of the study GMD of UTDSE are important for solving some practical problems of material science. Note that in Ref. [14] was also observed bimodal GMD for the case of three-dimensional polymerization of 1,2-dimethacryloiloxyethylene.

During the three-dimensional polymerization of TFM with $l = 1$ the coefficient of packing (filling of the reaction volume by structural elements) $k_{pack} = 0.25$, indicating low filling of reaction volume that is correlated with the data of GMD-research (high share of low-molecular UTDSE). With the elongation of TFM molecules at $l = 2–7$ ribs k_{pack} grows from 0.786 to 0.837, and when $l > 10$ k_{pack} again slightly reduced to value ≈ 0.773 at $l = 40$ (Table 4.1).

4.4 CONCLUSION

Thus, the performed research allowed identifying a number of the foregoing specific peculiarities and regularities of UTDSE formation which obeyed with increasing of the length of TFM molecules when carrying out the three-dimensional free-radical polymerization of TFM. Namely, it were revealed the features of change of the kinetic dependences of polymerization degree P_n, the number of radicals N_R and $N_{R'}$, the number of cross-links N_{cr} and cycles N_{cy}, share of pendant double bonds D and regularities of changes in these topological parameters of UTDSE depending on the length of TFM molecules. Formulated: (a) the molecular mass of UTDSE sceleton arise because of the reactions of chain growth and topological framing of the sceleton with radicals and cycles is implemented by the reactions of chains initiating, R-R, R-R′, R′-R′, R-PG, R′-PG; (b) the cross-linking of j-mentioned macrochains is carried out according to the last five reactions; (c) structure formation of macrobodies of three-dimensional polymer proceeds according to the scheme: monomer → nanogel (through branched macromolecule to UTDSE) → microgel (maltitude of nanogels) → macrobody (maltitude of microgels). In accordance with this scheme of macrobodies formation of TDP we have to pay attention to the fact that the stage of formation (synthesis) of UTDSE, assigns to nanochemistry, and the next stage, the stage of formation microgel, assigns to mesoscopic chemistry; (d) it was discovered bimodal grainulometric distribution of UTDSE and it was calculated the ratio of peaks intensity in the fields of low- and high-molecular UTDSE; (e) it was shown a tendency to increase the occupancy lattices with increasing length of TFM molecules (including due to a change of complementarity of UTDSE with the lattice). The obtained results are useful for understanding the nature of the three-dimensional free-radical polymerization of TFM in the manufacture of industrial products, the development of gel-technology, technology of integrated circuits fabrications, etc.

The authors hope that this work will attract the attention of researchers to this problem and they will create new models and programs for solving this problem, that will deepen understanding of the processes in a three dimensional polymerization of TFM. The authors are skeptical about the creation of experimental techniques to study the formation of UTDSE, at least in the near future.

We emphasize that our series of little-known works [4, 5, 8, 15–17] far promoted understanding of the problems of kinetics and structure formation in the case of three-dimensional polymers in comparison with the level of knowledge's in twentieth century and in the beginning of twenty-first century.

REFERENCES

1. Sivergin Yu. M., Perniks R.Ya., Kireeva S. M. Polycarbonatmethakrylates. Zinatne, Riga, 1988.
2. Sivergin Yu. M., Kireeva S. M., Berlin A. A. Moscow, 1972. 39 PP. – VINITI, 09.11.72, No 5033.
3. Berlin A. A., Kireeva S. M., Sivergin Yu. M. Moscow, 1974. 162 PP. – VINITI, 17.06.74, No 1660.
4. Usmanov S. M., Gaisin F. R., Sivergin Yu. M. Russian plastics technology journal. 2005. No 8. P.19 (in Rus.).
5. Gaisin F. R., Sivergin Yu. M., Usmanov S. M. Monte Carlo Simulation of the Three-dimensional Free-radical Polymerization of Ttetraphunktional Monomers (On Lattices of Different Dimensions and Geometry as a Part of the Formation of a Single Three-Dimensional Structural Eelement). Gilem, Ufa., Russia. 2009.
6. Berlin A. A., Kefely T. Ya., Korolev G. V. Polyesteracrylates. "Nauka" ("Science," in Rus.) Publishing House, Moscow, Russia. 1967.
7. Sivergin Yu. M., Usmanov S. M. Synthesis and Properties of Oligoestermethacrylates. "Khimiya" ("Chemistry," in Rus.) Publishing House. Moscow, Russia. 2000.
8. Gaisin F. R., Sivergin Yu. M., Usmanov S. M. Monte Carlo Simulation of the Three-dimensional Free-radical Polymerization of Ttetraphunktional Monomers (On Lattices of Different Dimensions and Geometry as a Part of the Formation of a Single Three-Dimensional Structural Eelement). Al'tair, Moscow, Russia. 2010, Part 3.
9. Sivergin Yu. M., Stankevitch I. V. Chemical Physics. 2006. V. 25. No 1. P.89 (in Rus.).
10. Sivergin Yu. M., Stankevitch I. V. Plastics. 2005. No 5. p. 35 (in Rus.).
11. Pavlova O. V., Kireeva S. M., Sivergin Yu. M. Ploymer Science. 1987. V. 29A. P.1777 (in Rus.)
12. Pavlova O. V., Kireeva S. M., Sivergin Yu. M. Vysokomol. Soed. 1990. V.32A. P.1256 (in Rus.)
13. Pavlova O. V., Kireeva S. M., Sivergin Yu. M. Acta Polymer. 1992. V.43. P.114 (in Rus.).
14. Chiu Y. Y., Lee L. J. J. Polym. Sci.: Part A: Polymer Chemistry. 1995. V. 33. No 2. p. 269.
15. Gaisin F. R., Sivergin Yu. M., Usmanov S. M. Monte Carlo Simulation of the Three-dimensional Free-radical Polymerization of Ttetraphunktional Monomers (On Lattices of Different Dimensions and Geometry as a Part of the Formation of a Single Three-Dimensional Structural Eelement). Al'tair, Moscow, Russia. 2010, Part 2, Section 1.
16. Gaisin F. R., Sivergin Yu. M., Usmanov S. M. Monte Carlo Simulation of the Three-dimensional Free-radical Polymerization of Ttetraphunktional Monomers (On Lattices of Different Dimensions and Geometry as a Part of the Formation of a Single Three-Dimensional Structural Eelement). Al'tair, Moscow, Russia. 2010, Part 2, Section 2.
17. Gaisin F. R., Sivergin Yu. M., Usmanov S. M. Monte Carlo Simulation of the Three-dimensional Free-radical Polymerization of Ttetraphunktional Monomers (On Lattices of Different Dimensions and Geometry as a Part of the Formation of a Single Three-Dimensional Structural Eelement). Al'tair, Moscow, Russia. 2010, Part 4.

CHAPTER 5

APPLICATION OF POLYCONDENSATION CAPABLE MONOMERS FOR PRODUCTION OF ELASTOMERIC MATERIALS WORKING IN EXTREME CONDITIONS

V. F. KABLOV[1] and G. E. ZAIKOV[2]

[1] *Volzhsky Polytechnical Institute (Branch) Volgograd State Technical University, Russia, E-mail: kablov@volpi.ru, www.volpi.ru*

[2] *N.M.Emanuel Institute of Biochemical Physics Russian Academy of Sciences, Russia, E-mail: chembio@sky.chph.ras.ru, www.ibcp.chph.ras.ru*

CONTENTS

ABSTRACT

The research is devoted to creation of elastomeric compositions based on systems with functionally active components for extreme conditions. The use of polycondensation capable monomers (PCCM) and other compounds with reactive groups was proposed for generating the stabilizing physical and chemical transformations. Thermodynamic analysis of open polycondensation systems and substantiation of various PCCM application as functionally active components of elastomeric materials have been conducted in the work; research results of polycondensation in an elastomeric matrix have been represented and a possibility of improving heat and corrosion resistance of elastomeric materials with introduction PCCM has been shown; different ways of applying PCCM have been proposed and experimentally proved.

5.1 INTRODUCTION

A polymer operated at high temperatures, in reactive environments, under intense friction, and other intensive effects can be regarded as a nonequilibrium open system, that is, a system exchanging the medium substance (or energy). Energy and substance exchange is carried out by heat conduction, diffusion of the medium in the material, low molecular additives and products of thermal and chemical destruction of the polymer matrix. Under these conditions, the material as a system is not in equilibrium state, and the further it from equilibrium, the greater the intensity of exposure to the material and, consequently, the more intensive mass and energy exchange with the medium is. The presence of reactive components in a material increases an equilibrium deviation and leads to greater intensification of mass and energy exchange.

Obtaining of elastomeric materials based on systems with functionally active components aimed to work in high temperatures, pressure and reactive environments is essential when rubber technical products are produced for oil drilling equipment, geophysical instruments and articles for chemical industry.

Theoretical analysis of capabilities to apply principles of the nonequilibrium thermodynamics and theory of open systems for production of elastomeric materials with physical and chemical transformations under operational exposures represents the prospects of the direction [1–3].

In the research the use of polycondensation capable monomers (PCCM) for surface modification and obtainment of gradient nonequilibrium systems with optimal organized spatial structure when the concentration of functionally active components is increased toward the surface of the article have been proposed and experimentally substantiated.

The method for producing gradient systems is that the first condensation monomer is introduced in rubber, then the rubber is cured, and, finally, processing in the second monomer at a temperature providing polycondensation is carried out. Herewith, due to the diffusion of the second monomer into the product the generation of a gradient structure is provided. An important advantage of the gradient systems of is the absence of the phase boundary, and, thus, the layering during operation.

5.2 RESULTS AND DISCUSSION

Functionally active components include the components capable of chemical reactions and physical and chemical transformations in bulk or on a surface of a material under external exposures (heat, mechanical, reactive environments). In addition, either protective agents are formed or physical effects are realized that allow for increasing service durability of rubbers.

For generating the stabilizing physical and chemical transformations, it was proposed in the work to use polycondensation capable monomers (PCCM) and other compounds with reactive groups. PCCM in an elastomeric matrix can react with formation of a new polymeric phase and heat absorption.

Thermodynamic analysis of open polycondensation systems and substantiation of various PCCM application as functionally active components of elastomeric materials have been conducted in the work; research results of polycondensation in an elastomeric matrix have been represented and a possibility of improving heat and corrosion resistance of elastomeric materials with introduction PCCM has been shown; different ways of applying PCCM have been proposed and experimentally proved [4–6].

Polycondensation processes allow for obtaining a large variety of chemical structures and, consequently, a possible wide range of properties when PCCM are applied as modifying agents. A distinctive feature of the developing direction is the thermodynamic nonequilibrium of polycondensation systems. In this case, nonequilibrium means that polycondensation systems have functionally active groups capable of further transformations (both the growth of macromolecules and a reversible reaction) nearly at any stage of a transformation. Low molecular product recovery, exo- and endothermic effects enable to classify polycondensation systems as open systems. Since the heat effects of the polycondensation are not too large, and the obtained low molecular product (usually water) takes much heat away, then, when an open system is considered, the process runs as endothermic. Besides, when the low molecular product is removed, an additional negative entropy flow arises resulting in thermodynamic assumptions of self-organization within the system.

According to Le Chatelier's principle, an increase in temperature shifts endothermic reactions towards heat absorption, that is, towards the polymer formation. In this regard, PCCM introduction to rubbers is of interest in terms of creating a kind of self-cooling rubbers for heat protective materials.

The following reactions have been studied in the research: polyetherification, polytransesterification, polyamidation, the formation of polyhydrazides and their polycyclization into polyoxidiazoles, polyurethane formation and production of poly 1-acylhydrazides, three-dimensional polymerization, and catalytic polycondensation.

Diffusive and kinetic features of polycondensation in the high viscosity rubber matrix are very specific because of the difficult removal of the obtained product, thermodynamic incompatibility of a rubber to monomers and the formed polymer. All this required a study of polycondensation kinetics in a matrix of different elastomers.

One of the most interesting phenomena found at polycondensation in elastomeric matrices is acceleration of the reaction compared to the reaction in a melt. Kinetic constants of reactions running in rubbers exceed greatly the rates of reactions running in a melt, and the rates of reactions in rubbers of different nature are very different. It has been shown that such dependence of the reaction rate on the rubber nature is connected with a type of the phase structure in systems "rubber-monomer" and "rubber-reaction products." Electron-microscopic analysis has revealed that the reaction rate is maximum, if system "rubber-monomer" is a system with emulsion of one monomer in a rubber with the second monomer dissolved in the rubber. In this case, polycondensation runs on the interfacial mechanism with the product recovery into an individual phase. The rate of the interfacial polycondensation goes up with an increase in the interfacial surface and increase in activating effect of free surface energy, respectively.

The essential thing is that polycondensation runs also in vulcanizates at time-temperature aging modes.

That allows for realizing the proposed concept concerned to the production of materials with physical and chemical structure which is nonequilibrium in operating conditions. Before the certain values of conversion degree of a polycondensation system, the stabilization of rubber characteristics takes place, which is related to inhibition of residual functional groups and endothermic effects of polycondensation. The results of differential thermal analysis represent a flowing of polycondensation and a slow-down of thermal-oxidative processes in rubbers modified with PCCM.

The features of flowing polycondensation in bulk and surface layers have been investigated while application of mechanical loads and high pressure.

The observed activation of polycondensation under mechanical loading of the elastomeric matrix is an occurrence of coupling chemical and mechanical processes in accordance with the Onsager principle. The possibility of using high pressure (up to 1000 MPa) to selectively target polycondensation in an elastomeric matrix has been shown.

The opportunity to improve heat and corrosion elastomeric materials by using physical and chemical effects running at operational exposures into monomer containing nonequillibrium and open elastomeric systems has been displayed (Table 5.1). PCCM application is especially effective for some rubbers exploited in extreme conditions (high temperatures, aggressive oxidative environments based on mixes of

TABLE 5.1 Coefficient of Properties Variation for Monomer Containing Rubbers at Aging

PCCM	Δfp	ΔEp	PCCM	Δfp	ΔEp
Vulcanized rubbers based on butadiene-nitrile rubber SKN-40					
Aging at 398 K x 72 hrs			**Aging at 423 K x 72 hrs**		
without PCCM	0.30	0.44	without PCCM	0.29	0.07
PA – Glyc.	0.45	0.30	HFDP – AA	0.69	0.59
PA – Glyc. – DEG	0.70	0.40	PFAA – AA	0.88	0.82
PA – DEG	0.60	0.30	**Aging at 473 K x 60 min**		
AA– DEG	0.51	0.35	without PCM	0.52	0.44
DHAA – AA	0.64	0.68	PA– Glyc.	0.84	0.78
unfilled PA– Glyc.	1.20	0.70	PA– Glyc. – DEG	1.50	0.83
unfilled PA– Glyc. – DEG	0.90	0.80	PA– DEG	1.70	0.66
unfilled PA– DEG	1.25	0.80			
Vulcanized rubbers based on ethylenepropylene rubber SKEP					
Aging at 423 K x 72 hrs			**Aging at 443 K x 72 hrs**		
without PCCM	0.70	0.62	without PCCM	0.15	0.13
AA– DEG	0.87	0.88	DCA–MPDA	0.52	0.40
1,3 ADA– DAA	1.40	1.50			

inorganic and organic acids). It has been demonstrated that PCCM and polycondensation products form some unique buffer system inside a rubber.

An increase in corrosion resistance of rubbers is also possible in elastomeric compositions based on PVC. In the designed compositions for work in nitrating environments a number of proposed concepts are implemented, in particular the use of aggressive nitrating mixture to seal and structure a surface layer and creation of internal functional subsystems of physical and chemical transformations based PCCM and other components. All this avoids surface gumming.

Figure 5.1 demonstrates the efficiency of PCCM application.

HFDP–hexafluorodiphenylolpropane; Glyc.–glycerin; PA–phthalic anhydride; AA – adipic acid, DHAA- dihydrazine of adipic acid, 1,3 ADA-1,3 adamantanedicarboxylic acid, PFAA-bis(monoethanolamide)-perfluoroadipic

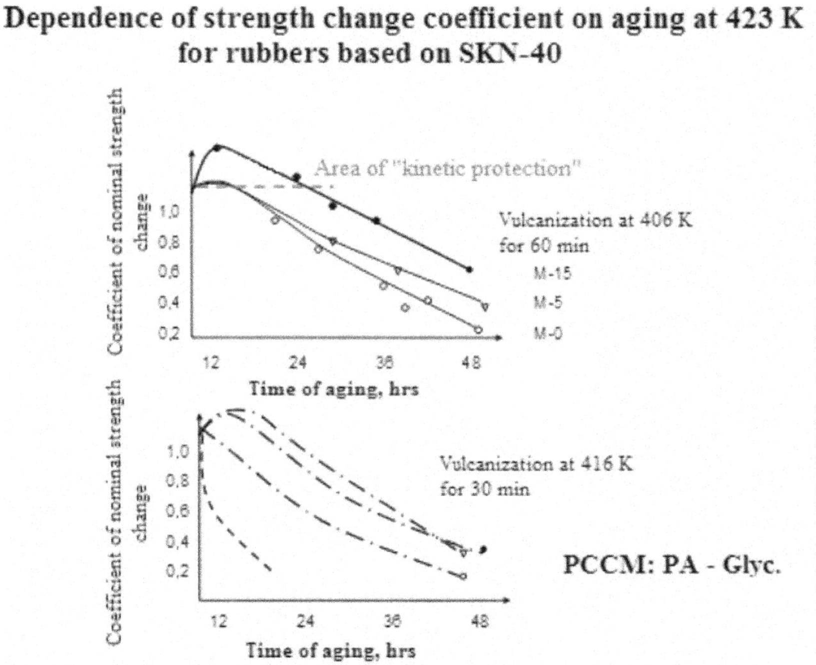

FIGURE 5.1 Dependence of a change in properties of rubbers containing PCCM at aging.

acid, DCA – 1,3-dimethyl-5,7-dicarboxyladamantane, MPDA – m-phenylenediamine, f – the aging coefficient on strength, E– the aging coefficient on elongation.

5.3 CONCLUSION

Thus, the conducted investigation has shown that there is a possibility to create heat and aggressive resistant elastomeric materials based on systems with functionally active ingredients for different operating conditions.

The research has been carried out with the support of the project "Development of modifiers and functional fillers for fire and heat protective polymeric materials" performed by the institute within the state task of Ministry of Education and Science of the Russian Federation.

KEYWORDS

- **components**
- **elastomeric material**
- **monomers**
- **polycondensation**
- **production**
- **thermodynamics**

REFERENCES

1. Kablov, V.F. Regulation of properties of elastomeric materials with functionally active components by conjugation of thermodynamic forces and flows. II Russian scientific-practical conference "Raw materials and materials for rubber industry: Today and Tomorrow": Proceedings. Moscow, STC"NIISHP," 1995.
2. Kablov V.F. Sychev, N.V., Ogrel A.M. Using thermodynamic forces and flows to create composite materials. International conference on rubber, Beijing, 1992.
3. Kablov V.F. Creation of elastomeric materials based on nonequilibrium systems with functionally active components. Raw materials and materials for rubber industry. Today and Tomorrow (SM. RKR-99): Proceedings of 5 Anniversary Russian scientific-practical conference, May 1998. Moscow, 1998, 385–386.
4. Effect dicarboxylic acid and their dihydrazides on curing characteristics of EPDM. V.E. Derbisher, V.F. Kablov, A.M. Koroteeva, A.M. Ogrel. Kauchuk i Rezina, 1983, №1, 24–26.
5. Kablov V.F. Verwendung polymerization sfahiger Verbindungen zur Regulierung der Eigenschaften von Gummi. V.F. Kablov, V.E. Derbisher, A.M. Ogrel. Plaste und Kautschuk. 1985, №5, 163–167.
6. Kablov V.F. Kinetic features of polycondensation of monomers introduced in elastomeric matrix. V.F. Kablov, A.M. Ogrel, A.M. Koroteeva. Bulletin of Higher Professional Institutions. Series: Chemistry and Chemical Technology. 1985. Vol. 28, Issue 5, 96–98.

CHAPTER 6

AROMATIC POLYAMIDES AND POLYIMIDES OF TRIARILMETHANE FRAGMENTS IN MAIN CHAIN*

T. A. BORUKAEV,[1] M. A. GASTASHEVA,[1] M. A. TLENKOPACHEV,[1] B. S. MASHUKOVA,[1] and G.E. ZAIKOV[2]

[1]*Kabardino-Balkarian State University after H.M. Berbekov, 360004, Nalchik, Chernyshevskaya St., 173Russia; E-mail: boruk-chemical@mail.ru*

[2]*N.M. Emanuel Institute of Biochemical Physics, Russia, 117334, Moscow, 4, Kosygin St., E-mail: chembio@sky.chph.ras.ru*

**This is work is carried out with the financial support of the Ministry of Education and Science as part of the state task (project 2199).*

CONTENTS

ABSTRACT

Aromatic polyimides and polyamides-based 4,4′-diaminothreephenylmethane has been synthesized. Their thermal, rheological properties and solubility in various organic solvents has been studied. It is shown that the solubility of the obtained polymers is connected with a free internal rotation triphenylmethan of bridge group and an effect of a surround phenyl substituent in diaminodiphenylmethane.

6.1 INTRODUCTION

Aromatic polyamides and polyimides are widely used in modern industry due to its excellent mechanical properties, high thermal and chemical stability. However, many of these polymers are difficult objects for processing because of their infusibility and insolubility in organic solvents. In this regard, great interest of polymer chemistry to new, fully aromatic structures that would preserve their inherent high level of physical-mechanical properties and at the same time would be fusible and easily soluble in organic solvents is understandable.

One of monomers, which may be of interest from this point of view, is 4,4′-diaminothreephenylmethane. Although the details of this monomer synthesis were published in 1928 [1], there are very few works on its use in the polycondensation reactions [2, 3]. Meanwhile, it is shown that the aromatic polyimides and polyamides on the basis of aromatic diamines of similar structure, such as N,N'-diaminothreephenylmethane [4,5] and N,N'-diaminotetraphenylmethane [5] can be easily soluble in organic solvents, and have a range of interesting properties. However, it should be noted that the synthesis of these monomers is quite complex. Unlike the above monomers, 4,4′-diaminothreephenylmethane can easily be obtained in one stage of available connections with high output.

The present research is devoted to the synthesis of aromatic polyimides and polyamides-based 4,4′-diaminothreephenylmethane and study their main properties as solubility and mechanical characteristics.

6.2 EXPERIMENTAL PART

All reagents were obtained from ALDRICH Company. Aniline was surpassed in vacuum (87°C, 1 mm Hg.) before its use. Benzaldehyde was used without treatment. Iso- and tetraphthalic chloroanhydrides was distilled at 170°C (57 mm Hg) and at 165°C (50 mm Hg), respectively. Pyromellitic of dianhydride was double-sublimated at 245°C (10^{-4} mm Hg.). Dichlorohydrin of diphenylcarbazone acid of firm ALDRICH (the degree of purity 98%) was used without treatment. N-methylpyrrolidone was ferried over barium oxide BaO at 98°C (20 mm Hg) and was kept over $CaCl_2$.

6.2.1 *SYNTHESIS 4,4′-DIAMINOTHREEPHENYLMETHANE*

4,4′-diaminothreephenylmethane was synthesized by the reaction of aniline with benzaldehyde at 140°C in nitrogen atmosphere. The details of its synthesis is described in works [1, 2]. Received 4,4′-diaminothreephenylmethane was purified by recrystallization from benzene followed by sublimation at 110°C (10^{-4} mm Hg.). Elemental analysis results agree well with the structure of the monomer. 83.15; N – 6.47; N – 10.17 was found (%). C – 83.17; H – 6.61; N – 10.21 was calculated (%).

6.2.2 *SYNTHESIS OF POLYAMIDE-BASED ON 4,4′-DIAMINOTHREEPHENYLMETHANE*

4,4′-diaminothreephenylmethane was dissolved in a dry N – methylpyrrolidone and cooled to 0°C. Stoichiometric amount of dichlorohydrin Iso – or tetraphthalic acid were added to the solution in the mixing and were raised to the room temperature (reaction time 4–5 h, the concentration of polymer in the solution 15% weight). Received polyamides were undercooled from N-methylpyrrolidone. The solutions of treated polymers in N-methylpyrrolidone (20% weight) were put on glass plates and were dried at 150°C in vacuum (50 mm Hg) for 10 h.

6.2.3 SYNTHESIS OF POLYIMIDES BASED ON 4,4′-DIAMINOTHREEPHENYLMETHANE

Polyimide films were synthesized in two stages by chemical or thermal imidization of prepolymers-polyamide acids (PAA) [4, 7, 8]. The degree of imidization we controlled with IR-spectroscopy [7]. The output of polyimides is close to quantitative (about 100%).

6.2.4 MEASUREMENTS

The Intrinsic viscosity (η_{in}, dL/g) we were determining in the solution of dimethylformamide (0.5 g per 100 mL of dimethylformamide) at 25°C. IR-spectra were taken on the spectrometer NICOLET 510 FT-IR. X-ray spectra of polymers were received on the diffractometer Siemen's D-500 c CuK. Data of thermogravimetric analysis were obtained using the device Du Pont 2950 (in nitrogen atmosphere, the heating rate is 5°C/min). Glass transition temperature polymers (TD) was determined from the results of thermomechanical analysis on the device Du Pont 2950 in nitrogen atmosphere at heating rate 50°C/min using technology oriented films [9]. Mechanical properties of polymer films (samples 30 × 40 mm^2) was determined on the device INSTRON 111 at speeds stretching 50 mm/min.

6.2.5 THEORETICAL CALCULATIONS

Structural parameters and the rotation barriers of phenyl rings around the Central carbon atom of 4,4′-diaminothreephenylmethane were calculated using supercomputer SKOU. After full minimization according to molecular method MMX-89 [10] the structure were optimized by semi empirical method of molecular orbitals AM1 [11].

6.3 RESULTS AND DISCUSSION

Structural parameters and energy barriers of rotation on the Central carbon atom were calculated for 4,4′-diaminothreephenylmethane to evaluate the possibility of rotation of fragments of a polymer chain around the bridged group of 4,4′-diaminothreephenylmethane. The data obtained show that SP^3-hybridization of the Central carbon atom leads to the pyramidal structure of the molecule 4,4′-diaminothreephenylmethane. The estimated value of the angle C^2 C^1 C^3 in the molecule is 112,07°, which is very close to the calculated 4,4′-diaminothreephenylmethane [12]. It is found that the energy barrier of rotation around the Central atom in 4,4′-diaminothreephenylmethane is of 5.89 kcal/mol, which is rather close to those of 4,4′-diaminodiphenylmethane or 4,4′-diaminodiphenylamine [13].

The results obtained agree well with the experimental data on the flexibility of polyimide circuit-based 4,4′-diaminothreephenylmethane [5]. Equilibrium flexibility (σ) of these polymeric structures amounted to no more than 1.10 indicating almost free internal rotation (from the point of view of thermodynamics) polymer chains. Some of these polyimides have a good solubility in organic solvents.

Table 6.1 presents data of solubility aromatic polyimides and polyamides-based 4,4′-diaminothreephenylmethane in various organic solvents at room temperature. It should be noted that the polymers of such structures on the basis of known diamines type 4,4′-diaminothreephenylmethane or 4,4′-diaminothreephenylmethane are insoluble in organic solvents, and solubility of polyimides (polyamides) on the basis of 4,4′-diaminothreephenylmethane can be explained by the effect of volume phenyl rings in the monomer molecule. This monomer, apparently, can be considered as a representative of the so-called "cardoided diamines" [14], which is known for obtaining soluble polyamides and polyimides.

Polyamides-based on 4,4′-diaminothreephenylmethane have better solubility than polyimides on the basis of the same monomer. The latter gives the possibility to prepare solutions polyamides with concentration up to 30–35% wt. in such amide solvents as the N- methylpyrrolidone or dimethylformamide. Solubility of received polyamides in pyridine and tetrahydrofuran is much lower. Polyamides are insoluble in acetone, and in the presence of even small amounts of solvent polyamide films become extremely fragile. The observed fact, apparently, is the result of weakening inside of the sample molecules acetone intermolecular by hydrogen bonds between the amide groups in polymers, responsible for the mechanical properties of the ground state.

TABLE 6.1 Solubility* Polyamides and Polyimides Based on 4,4′-diaminothreephenylmethane

Polymer	solvent				
	dimethyl-formamide	N-methyl-pyrrolidone	Pyridin	Tetrahy-drogue-ran	Acetone
	33h. 1d.7d	3h. 1d. 7d.	33h. 1d. 7d.	3h. 1d. 7d.	3h. 1d. 7d
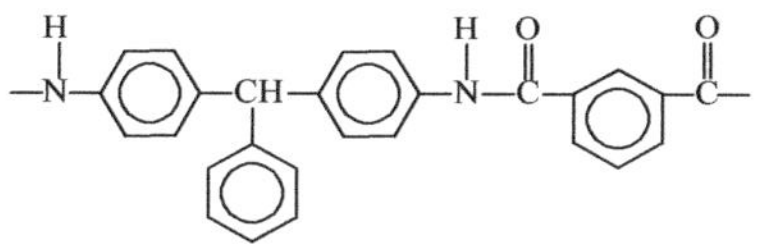	++++	+++	++++	– p.s +	– –
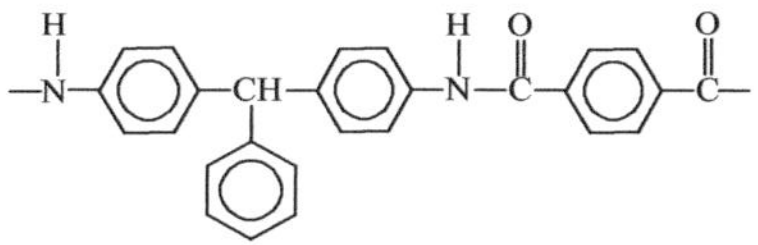	++++	+++	++++	– p.s +	– –
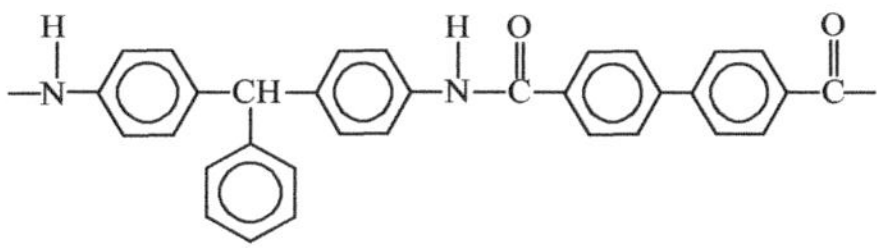	++++	+++	++++	– – p.s	– p.s
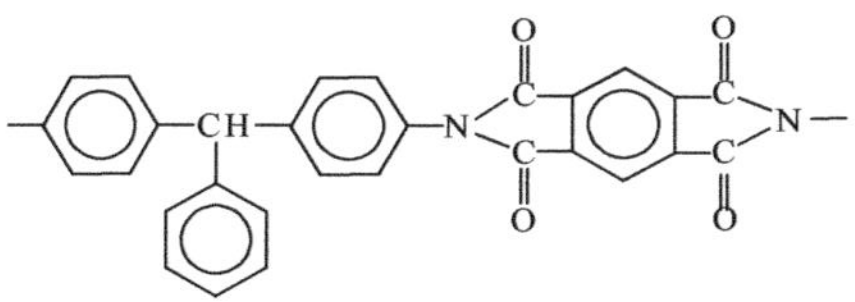	++++	+++	++++	– – –	– p.s

*Solubility were checked after 3 hours, 1 day and 7 days at room temperature; (+) – completely soluble, (–) – insoluble, (p.s) – partially soluble.

In the case of polyimides synthesized by chemical imidization the samples are easily soluble in amide solvents (maximum concentration 15–20 weight. %), in pyridine and swell in the dichloroethane. It should be noted that previously it was reported about insolubility of polytriphenylamine pyromellitics [5] and, it seems that polyimide synthesized by us which based on 4,4′-diaminothreephenylmethane is one of the few which is soluble in organic solvents.

Solubility of synthesized polyimides by thermal imidization is much worse. Samples of polyimide films which were prepared at a temperature of 270°C for 30 min (further increase in the temperature does not increase the degree of imization, which is close to 100%) are only partially soluble in dimethylformamide or N-methylpyrrolidone and inert in other solvents. The number of insoluble fraction increases with increasing temperature or increase time of thermal processing.

A similar effect was observed after annealing higher than the temperature of 270°C of completely soluble polyimides, synthesized by chemical imidization at room temperature. Thus, we can conclude that the polyimides based on 4,4′-diaminothreephenylmethane are heat-stitching at temperatures above 270°C, just as previously we observed for other polyimides [8,12].

Table 6.2 shows the thermal properties of the synthesized polymers on the basis of 4,4′-diaminothreephenylmethane.

The results of thermomechanical analysis show that all polymers show distinct differences in glass- transition temperature. In particular, the values of glass transition temperature increase in the range of dichlorohydrin isophthalic acid – 4,4′-diaminothreephenylmethane (I), dichlorohydrin terephthalic acid-4,4′-diaminothreephenylmethane (II), 4,4′-dichlorodiphenyldichloroethane acid – 4,4′-diaminothreephenylmethane (III), pyromellitic of dianhydride-4,4′- diaminothreephenylmethane (IV). Glass transition temperature of the synthesized polyimides was determined on prepared by chemical imidizaion samples. The values of glass transition temperature T_g are very close to values of glass transition temperature T_g of polyimides, obtained on the basis of known diamines type 4,4′-diaminothreephenylmethane [12, 15] previously published in the literature.

According to the DTA and TGA the thermal destruction of polymers which synthesized on the basis of 4,4′-diaminodiphenylmethane in nitrogen atmosphere is beginning at temperatures above 400°C (Table 6.2). The greatest resistance of all investigated polymers is

TABLE 6.2 Thermal Properties of Polyamides and Polyimides Based on 4,4′-diaminothreephenylmethane

n/n	Polymer	glass transition temperature, T_g,°C	weight loss*, 5%	weight loss*, 10%
I	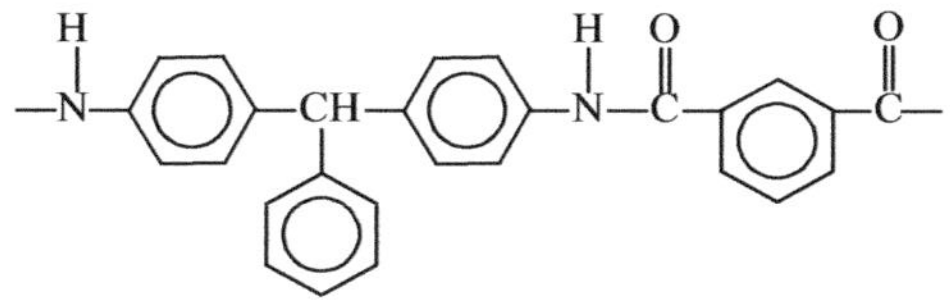	177	238	470
II	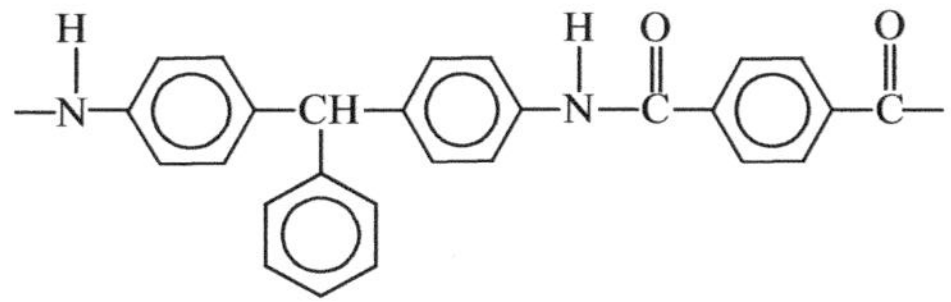	231	449	474
III	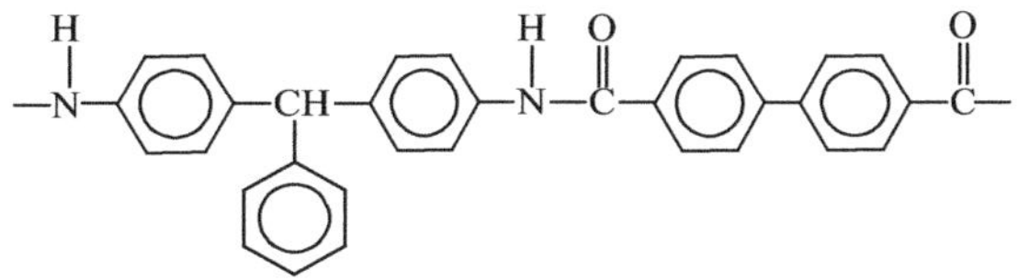	288	516	538
IV	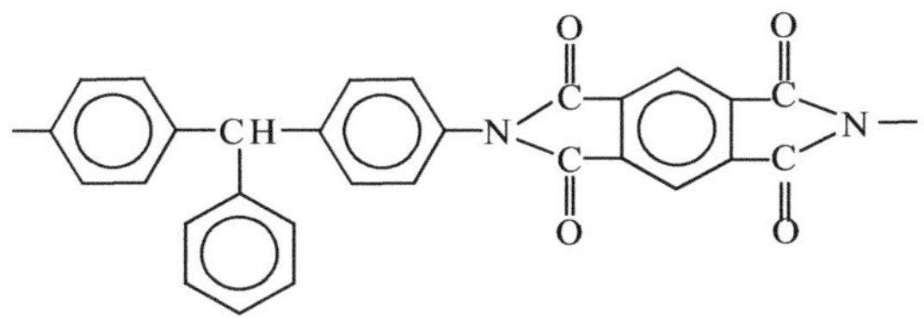	356	538	564

*Weight loss polymers were identified in nitrogen atmosphere at heating rates of 5°C/min.

TABLE 6.3 Mechanical Properties of Polyamides and Polyimides Based on4,4′-diaminothreephenylmethane (Film Thickness 25 mkm)

n/n	Polymer	viscosity, η_{in}, dL/g	Modulus of elasticity, E, HPa	rupture stress, σ_r, MPa	deformation, ε, %
I	isophthalic acid dichlorohydrin – 4,4′-diaminothreephenyl-methane	1,2	1,1	125	67
II	terephthalic acid dichlorohydrin – 4,4′-diaminothreephenyl-methane	0,9	1,3	147	39
III	diphenylcarbazone acid 4,4′-dichlorohydrin – 4,4′- diaminothreephenyl-methane	1,4	1,5	135	75
IV	pyromellitic of dianhydride – 4,4′-diaminothreephenyl-methane	1,1	1,8	158	28

observing for polyimide based on benzene tetracarboxylic dianhydride and 4,4′-diaminothreephenylmethane (IV). The values of temperatures given in Table 6.2 in which there is a 5 and 10% weight loss of polymers in an inert atmosphere, only slightly lower than industrial polyimide film KAPTON HN (For KAPTON these values are 545°C and 573°C, respectively) [16].

Results of mechanical tests of polymer films with thickness of 25 microns are shown in table 3. Studies have shown that the obtained polymers have a fairly high mechanical characteristic. From polyamides to polyimides values of elastic modulus and destructive voltage are growing. This is due to the stringency of the chain macromolecules. In turn, this increase of rigidity of structures chain macromolecules leads to decreasing their opportunities to high values of deformation, which is what we observe (table 3).

The results of x-ray analysis of films based on synthetic polyamide and polyimides suggest that all polymers are amorphous. Significant differences in the position and form of amorphous halo in polyimides were observed between the materials prepared by chemical and thermal imidization. A similar effect of imidization method on packaging aromatic polyimides was noted in previous messages for polypyromellitimide based on 4,4′- diaminothreephenylmethane and 4,4′- diaminodiphenyl [17].

6.4 CONCLUSION

Series of completely aromatic polyamides and polyimides based on 4,4′- diaminothreephenylmethane was synthesized and characterized. All studied polymers have good solubility in organic solvents. Solubility of polymer-based on 4,4′-diaminothreephenylmethane, obviously, is connected with the freedom of internal rotation triphenylmethanol bridge group and a surround deputy in the monomer. Solubility of polyimides based on 4,4′-diaminothreephenylmethane decreases sharply after heat treatment at a temperature 270°C and above (or samples, which was prepared by thermal imidization), which may be explained by the course of the process of cross-linking of polymers. It is found that studied polymers by thermal and mechanical properties are close to polyamides and polyimides on the basis of other monomers with bridge group.

KEYWORDS

- **4,4′-diaminothreephenylmethane**
- **aromatic polyamides**
- **polyamide acids**
- **polymers**

REFERENCES

1. Weil H., Sapper E., Kramer E., Kloter K., Selberg H. Uber Diaminotriphenylmethan und Ahnliches. Ber. 1928. Bd.61. №6. S. 1294–1307.
2. Vishnevaya N.A., Borukaev T.A., Tlenkopachev M.A., Vasilieva O.V., Mikitaev A.K. Synthesis of aromatic polyazomethines based on 4,4′-diaminodiphenylmethane. Polymer Science (in Rus.), Series A. 1993. Vol. 35. № 9. 1418–1420.
3. Likhatchev D.Yu., Tlenkopachev M. A., Vilar R., Salfedo R., Ogawa T., Borukaev T.A, Vishnevaya N.A., Bekanov M.Ch., Mikitaev A. K. International Symposium on Polymers "POLYMEX-93," Cancun. 1993. 176.
4. Kardash I.E., Likhachev D.Yu., Krotovich M.B., Kozlova N.V., Zhuravlev I.L., Bogachev Yu.S., Pravednikov A.N. IR – and NMR-spectroscopic investigation of substituted N-phenylmaleimide as model compounds of aromatic polyimides. Polymer Science (in Rus.), Series A. 1987. Vol. 29. № 7. 1364–1369.
5. Vasilenko N.A., Ahmetova E.I., Sviridov E.B., Berendyaev V.I., Rogozhkina E.D., Alkaeva O.F., Koshelev K.K., Izyumnikova A.L., Kotov B.V. Soluble polyimides based 4,4′-diaminodiphenylmethane. Synthesis, molecular-mass characteristics, properties of solutions. Polymer Science (in Rus.), Series A. 1991. Vol. 33. № 7. 1549–1560.
6. Cvetkov V.N., Stennikova I.N., Lavrenko P.N. Konformation und Gleichgewichts-Flexibilität von Poly(tetraphenylmethan-terephthalamid)-Molekülen in Lōsung. Acta Polymerica – Weinheim. 1980. Vol. 31. № 7. 434–438.
7. Likhachev D.Yu., Arzhakov M.S., Chvalun S.N., Sinevich E.A., Zubov Yu.A., Kardash I.E., Pravednikov A.N. The effect of the chemical structure on properties of aromatic polyimides, obtained by the method of chemical cyclization. Polymer Science (in Rus.), Series B. 1985. Vol. 27. № 10. 723–728.
8. Kardash I.E., Likhachev D.Yu., Nikitin N.V., Ardashnikov A.Ya., Kozlova N.V., Pravednikov A.N. Plasticizing effect of the solvent in the process of solid-phase thermal cyclization of aromatic polyamidation in the polyimides. Polymer Science (in Rus.), Series A. 1985. Vol. 27. № 8. 1747–1751.
9. Clair T. L. st., Clair A. K., st., Smith E. N. Structure- Solubility Re Lationships in Polymers. Eds. Harris F. W., Academic Press. 1997. 199.

10. Sprague J.T., Tai J.C., Yuh Y., Allinger N.L., Stewart J.P. The MMP2 calculational method. J. Comput. Chem. 1987. P. 581–603.
11. Dewar M.J.S., Zoebisch E.G., Healy E.F., Stewart J.P. The development and use of quantum mechanical molecular models a new general purpose quantum mechanical molecular model. J. Amer. Chem. Soc. 1985. Vol. 107. 3902–3909.
12. Bessonov M.I., Kotov M.M., Kudryavtsev V.V., Laius L.A. Polyimides, Thermally Stable Polymers. Consultants Bureau: New York, 1987.
13. Birshtein T.M., Goryunov A.I. Theoretical analysis of flexibility polyimides and polyamidation. Polymer Science (in Rus.), Series A. 1979. Vol. 21. № 9. 1990–1997.
14. Vinogradova S.V., Vigodsky A.S., Vorobev V.D., Churochkina N.A., Chudina L.I., Spirina T.N., Korshak V.V. Investigation of chemical cyclization of polyamidation in solution. Polymer Science (in Rus.), Series A. 1974. Vol. 16. № 3. 506–510.
15. Clair T.I., Wllson D., Stenzenbergen H.D. St. Polyamides. Blackie: New York, 1990. 297.
16. Du Pont High Perfomance Films: Summury of Properties 1993. 231302 A. USA.
17. Likhachev D.Yu., Chvalun S.N., Zubov Yu.A., Nurmukhametov R.N., Kardash I.E. The Influence of defects chemical structure on the morphology of polyimide films. Polymer Science (in Rus.), Series A. 1991. Vol. 33. № 9. 2010–2019.

CHAPTER 7

BIOPOLYMERS FOR APPLICATION IN PHOTONICS

ILEANA RAU and FRANCOIS KAJZAR*

Faculty of Applied Chemistry and Materials Science, University Politehnica Bucharest, Bucharest, Romania; Tel/fax: +40203154193; E-mail: frkajzar@yahoo.com

CONTENTS

ABSTRACT

Possibilities of utilization of biopolymers, and particularly of the deoxyribonucleic acid (DNA) are reviewed and discussed. The ways of their functionalization with photoresponsive molecules to get desired properties are described and illustrated on several examples as well as the processing of materials into thin films. Their room- and photo-thermal stability, studied by spectroscopic techniques is reported, together with optical damage thresholds. Physical properties, and more particularly linear, nonlinear and photoluminescent properties of obtained materials are also reviewed and discussed.

7.1 INTRODUCTION

In last 30 years time the synthetic polymers have found large applications in almost each domain of human activity, and particularly in construction, car industry, medicine, textile and more recently in advanced technologies. These polymers are obtained principally from coal and from oil by chemical transformation and synthesis. However, due to the fact that the coal and oil resources are limited on one hand and contribute to an important pollution of the planet on the other the scientists turns their attention to nature produced biopolymers. Indeed, the decay time for a thin foil of polyethylene (PET), used largely in fabrication of plastic bottles, is of 5–10 years. Also the largely used polystyrene (PS) decomposes in 50 years, low-density polyethylene (LDPE) in 500 – 1000 years. Polypropylene (PP), used in clothing and rope fabrication, practically does not degrade [1]. The fabrication of some polymers, like polyvinyl chloride (PVC), used largely in construction and in fabrication of toys is done with the use of toxic dioxin. Its degradation is associated with the production of unhealthy subproducts. These facts explain well the already mentioned switch of the scientists interest to natural biopolymers, originating from renewable resources and biodegradable.

One of these polymers which attracted recently some interest is chitosan which is a polysaccharide, occurring in the exoskeleton of invertebrates and in their internal structures. It was shown that it has some interesting optical properties [2, 3]. There are two biopolymers produced in a very large amount by nature which are the deoxyribonucleic acid (DNA) and collagen. Both are biodegradable, abundant and can be obtained from, for example, the waste of food producing industry.

DNA, called also "molecule of live," is present in all living species, being responsible for its development and heritage not only of humans and animals but also of the vegetal ones. Since the discovery of its molecular structure by Watson and Crick [4, 5] (cf. Fig. 7.1) in 1953 the deoxyribonucleic acid (DNA) attracted a lot of interest of biologists, chemists, and later, of physicists. Indeed, this supramolecule exhibits a peculiar double helix structure, consisting of stacked base pairs of molecules arranged as rungs of the ladder. The pairs consist always of adenine with thymine and of guanine with cytosine (cf. Fig. 7.2). The two helix backbones are made of sugar and phosphate groups, joined internally by the ester bonds. The base pairs are linked together by the strong hydrogen bonds (cf. Fig. 7.3). Because the outside groups are phosphates, the DNA macromolecule presents a net negative charge, compensated by sodium ions, which are nonlocalized counter ions. They can move freely along the macromolecular chain surface [7].

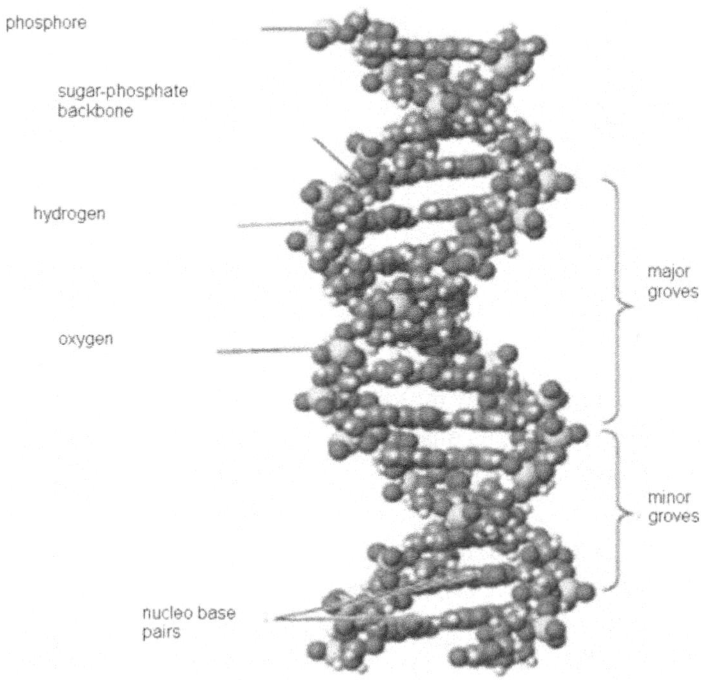

FIGURE 7.1 Chemical structure of a segment of DNA molecule (adapted from Ref. [6]).

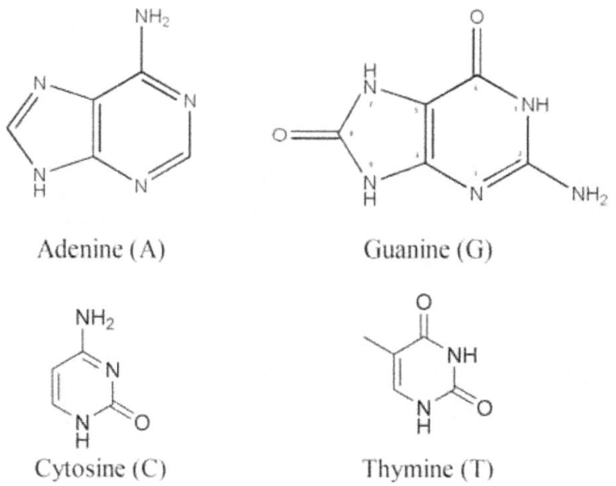

FIGURE 7.2 Chemical structure of four nucleobases.

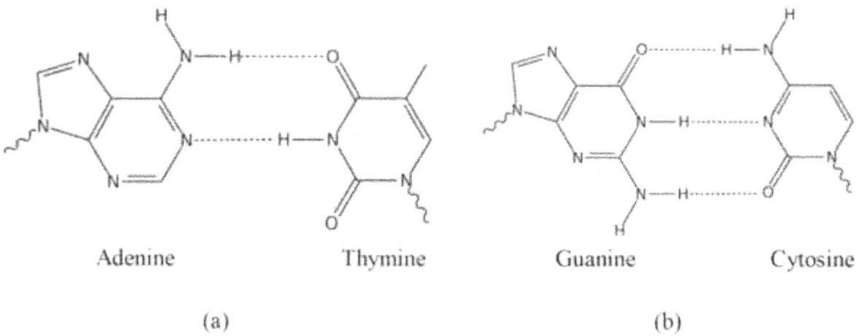

FIGURE 7.3 Nucleobase pairs: Adenine–Thymine (a) and Guanine–Cytosine (b) and the hydrogen bonds between the nucleotides.

As it was found originally by X-ray studies by Watson and Crick the pitch of the helix is of 3.4 nm, its diameter of 1 nm, respectively, and the distance between two neighboring nucleotides of 0.34 nm (cf. Fig. 7.1). In solution these dimensions may be a little different as reported by Mandelkern et al. [8], ranging from 2.2 to 2.6 nm for the helix radius, 3.3 nm for the pitch, and 0.34 nm for the distance between two nucleotides.

The double stranded helix form major and minor grooves, wide, respectively, of 2.2 nm and 1.2 nm [9]. Their presence is important for the functionalization of DNA as it will be discussed later.

The size of DNA depends on the level of development of a given specie. Usually it is expressed in the number of base pairs (bp) and spans from several tens of bp, as for *Escherichia coli* bacteria (76 bp), to 3 000 Mbp for Human DNA [10]. There exist a lot of programs available on internet which transform the base pairs number into molecular mass (Daltons). A base pair has molecular mass of about 660 Da.

One of the important arguments used in favor of biopolymers for replacing the synthetic polymers in photonics and in electronics, and particularly by DNA, is its abundance and renewability. DNA is usually obtained from the waste produced by food processing industry. Thus it can be cheap. In contrary to synthetic polymers, if not protected, biopolymers are biodegradable. Thus their use should permit to decrease the pollution by slowly decomposing synthetic polymers. This is comforted by the present scientific policy related to the humanity problem of creating a sustainable society with durable development, disposing renewable resources and minimizing the environment pollution.

Besides the above mentioned advantages, there are other important properties of DNA, which are in favor of its in photonics and in electronics, as it will be shown and discussed later in this paper. It concerns, in particular the versatility, thin film processability and the possibility of tailoring optical and electrical properties by DNA functionalization. Its specific double strand helical structure, with minor and major groves, provides a large free volume for doping molecules as well as a good protection against photo thermal degradation.

Another potentially very interesting biopolymer is collagen. Its name originates from Greek word *kola* meaning glue. It was indeed used as glue in antic times. This biopolymer is the most abundant in human and animal organisms supramolecule, making *ca* 25% of all their proteins, that is, 5% of their mass. It plays an important role in assuring connectivity of mammals tissues [11] and in their structuring. Its molecular mass is of *ca* 325,000 Da. A collagen fiber with 1 mm diameter may withstand a 10 kg load.

The collagen molecule is composed of three associated alpha polypeptide chains, as shown in Fig. 7.4, linked by hydrogen bonds between hydroxylysine and l'hydroxyproline and by covalent bonds. An alpha chain is constituted of 1055 amino acids. They may combine in different

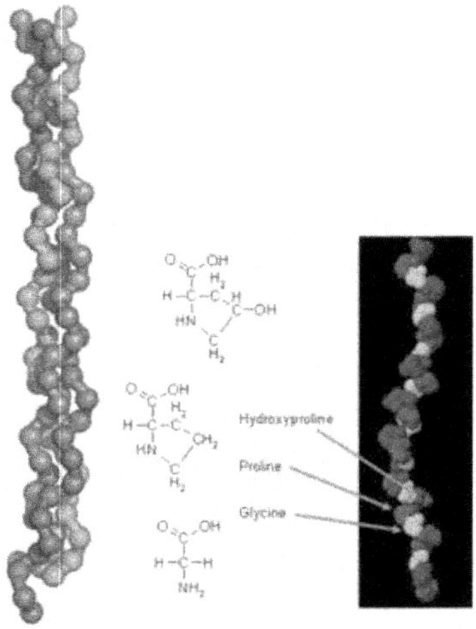

(a) (b)

FIGURE 7.4 Chemical structure of collagen (a) and of tropocollagen (b).

ways and form a large variety of different collagens. Each of them is characterized by an appropriate structure and exerts a particular role in a given organ. For example collagen I is present in cornea, skin and bones while collagen III can be found in the cardiovascular system.

The building unit of collagen is the tropocollagen. It is a noncentrosymmetric molecule, exhibiting second harmonic generation as observed by several research groups. The length of tropocollagen is of *ca* 280 nm and the diameter of 1.5 nm, respectively. The already mentioned constituent elements amino acids comprise glycine, proline, hydroxylysine and 4-hydroxyproline (cf. Fig. 7.4). There are several types of molecular chains, which are composed of repetitive sequences of these amino acids. Glycine is repeated throughout the molecule. The carbohydrates are attached to hydroxylysine. The cohesion of tropocollagen is ensured by strong hydrogen bonds between glycine and hydroxyproline (Fig. 7.4).

The deoxyribonucleic acid exhibits little π electron conjugation which is interesting for application in photonics. It is mainly to the presence of double C=C bands, in nucleobases. Such conjugation is absent in collagen. Therefore, to obtain desired properties necessary for practical applications both biopolymers

In this chapter, we review and discuss the results of our recent studies on functionalization, photo – thermal stability of collagen and DNA and collagen based complexes, together with stability in high intensity laser beams. Some properties of obtained complexes as well as their practical applications also reviewed and discussed.

7.2 MATERIALS

The deoxyribonucleic acid we are using in our studies was purchased at Ogata Photonics Laboratory, Chitose, Hokkaido, Japan. It is obtained from the waste produced in salmon processing [12, 13], particularly from roe and milt. The DNA extraction process is shown schematically in Fig. 7.5. Frozen roe and milt are first grinded. Then the grinded product is homogenized. Then starts the important and difficult process of protein elimination. The homogenized product is treated with enzymes DNA, dissolved

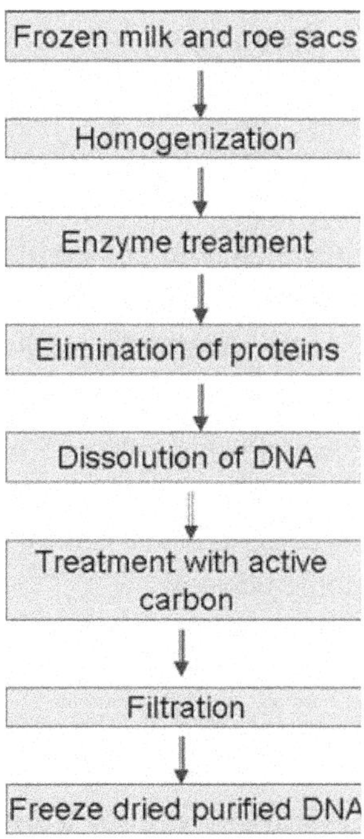

FIGURE 7.5 Successive steps in obtaining pure DNA from the salmon processing waste.

in water and decolorized with active carbon. Finally the product is filtered and freeze-dried. The most delicate and difficult step in purification process is the separation of proteins. The final product contains usually *ca* 98% of DNA and *ca* 2% of proteins [13].

Collagen is also obtained from the waste produced in meat processing. It is usually obtained from skin and bones of animals, principally such as beefs and porcs as well as from fish [14, 15]. The collagen used in our study was obtained at University Politehnica of Bucharest from beef skin using an original procedure described in Refs. [16, 17]. DNA is known to denature at around 90°C, changing its helical structure from double stranded to single stranded [18, 19], limiting in this way the temperature range of applicability. Also thin film processing and water solubility only limits the possible range of its applications.

Collagen can be irreversibly hydrolyzed giving gelatin, which is largely used in food industry.

7.3 FUNCTIONALIZATION

As already mentioned pure DNA and collagen represent a limited interest for applications in, particularly, photonics. They exhibit low π electron conjugation, which is principally present in DNA only owing to the –C=C- conjugated bonds in nucleobases.

Pure DNA has a limited potential for applications in photonics. This biopolymer is soluble in water only, a solvent which doesn't belong to the preferred ones in the device fabrication technologies, although some electronic devices containing water were already described [20, 21]. Also a week π electron conjugation, only in phenyl rings, provides limited hyperpolarizabilities to this compound. Therefore, the only possible practical use of this biopolymer in photonics is as an optically inactive material, except if its chirality can be exploited in some ways.

DNA is an anionic polyelectrolyte [22, 23] with Na^+ ion being a counterion. Therefore, the first possible approach to functionalize is through the electrostatic interaction by substituting Na^+ by a positively charged molecule, which will be bond to the DNA helix by the electrostatic force, changing in this way its properties. Several approaches were done in this direction. A significant DNA material improvement was obtained by functionalizing it chemically with ionic liquids, as shown by Iijiro and Okahata [24], Okahata [25, 26], Serguev et al. [27], Kimura et al. [28], and more particularly by Ogata and co-workers from Chitose Institute of Technology [29–41]. They have shown that the counterion Na+ can be substituted by an amphiphilic cation, leading to a more stable compound, soluble in polar organic solvents and generally insoluble in water. They used several cationic surfactants which react with DNA *via* electrostatic forces and they succeeded in making several stable complexes using surfactants such as: cetyltrimethylammonium (CTMA) (Wang et al. [42] and an aromatic one the benzyldimethylammonium (CBDA) [37] Watanyuki), whose chemical structures are shown in Fig. 7.6. Recently some other surfactants were proposed, such as aromatic ones: benzalkonium chloride (BA), and a linear amphiphilic one didecyldimethyl ammonium chloride (DDCA), whose chemical structures are also shown in Figure 7.6. The complexes formed with the new surfactants are soluble in a larger number of solvents making possible DNA functionalization with a greater class of molecules.

Figure 7.7 shows schematically the reaction of a surfactant (with a counterion, usually Cl^- or ClO_4^-) with DNA (counterion Na^+). As result of this reaction a stable DNA-surfactant complex is formed, which precipitates in

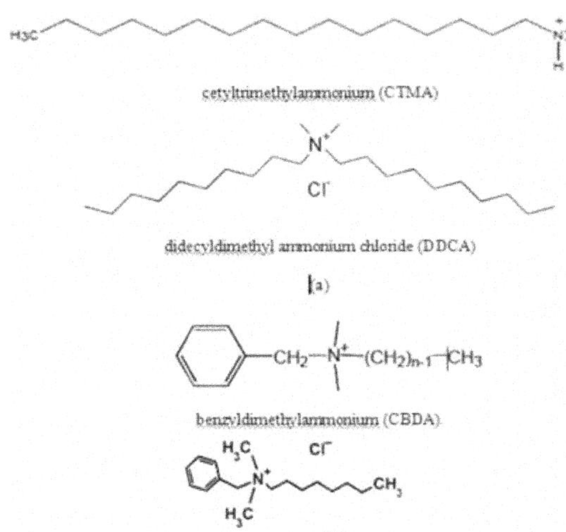

FIGURE 7.6 Chemical structures of several linear amphiphilic (a) and aromatic (b) surfactants.

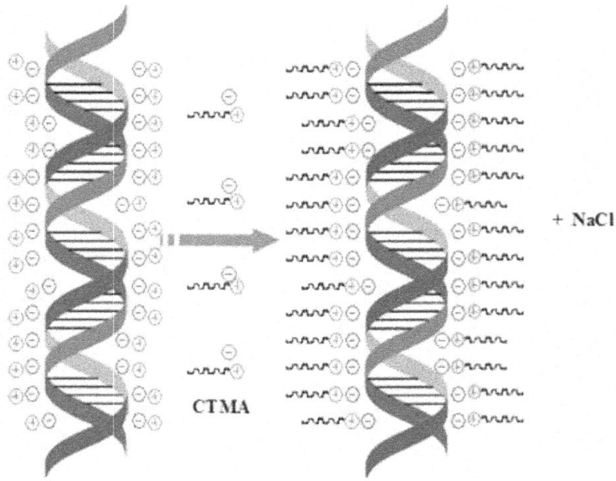

FIGURE 7.7 Electrostatic interaction between DNA and surfactant leading to the formation of a stable complex (courtesy of J. G. Grote, UA Air Force Wright Patterson Research Labs, Dayton, OH, USA).

reaction solvent (water) and can be easily recovered. The resulting from the reaction sodium ion forms a salt with the surfactant counterion which remain in water. The DNA surfactant complexes are well stable. Their thermal degradation takes place at around 230°C [43–45].

As already mentioned DNA undergoes a denaturation process when heated to 90°C. It consists on transformation from double strand to single strand helix. In contrary, the DNA-surfactant complex show a better stability, maintaining its double stranded helical structure to temperatures overpassing 100°C [13], what is sufficient for majority of practical applications.

Functionalization of DNA with the above cited surfactants does not provide them the required in photonics photosensitivity, as the molecules used show also a little (aromatic surfactants) or none (amphiphilic surfactants) π electron delocalization. Figure 7.7 compares optical absorption spectra of thin films of DNA with those made of two complexes: All show absorptions around 260 nm, which are due to conjugated π electrons of nucleobases and of phenyl ring in the case of aromatic surfactants. Therefore, it is necessary to functionalize the biopolymer or its complex with a surfactant with a photosensitive molecule. This can be done in three different ways:

(i) intercalation,
(ii) random doping, as in the case of synthetic polymers,

(iii) doping through molecules inclusion in minor or major groves,
(iv) covalent attachment to DNA chains.

Intercalation consists on introduction of a doping molecule between nucleobase pairs stacks, as shown in Fig. 7.8(a). As the space is limited, only small, ring type, flat molecules can intercalate. A large π electron overlap between the doping molecules and the nucleobase pairs is expected in this case.

Recently Pawlik et al. [46] (see also Refs. [47, 48]) have proposed another intercalation mechanism for DNA-surfactant complexes they proposed to call "semiintercalation." In that case the doping molecules are partly inserted between the surfactant molecules, as shown in Fig 9 for DR1 in DNA–CTMA matrix. Thus the dopants do not bind directly to the DNA backbone but are separated from, keeping in this way a larger conformational mobility. The Monte Carlo simulation calculations performed using this model allowed to explain the photochromic properties of DNA–CTMA-DR1complex. They reproduced accurately the main experimental results of laser dynamic inscription of diffraction gratings in this photochromic material: short response time, low diffraction efficiency, single-exponential kinetics and flat wavelength dependence. It allowed also to explain also the origin of memory effect upon light excitation observed in DNA–CTMA-DR1 complexes.

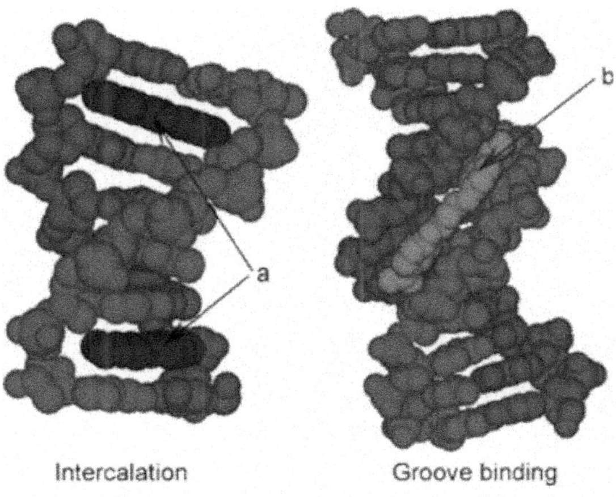

FIGURE 7.8 Intercalation (left) and groove binding (right) of chromophores in DNA [49].

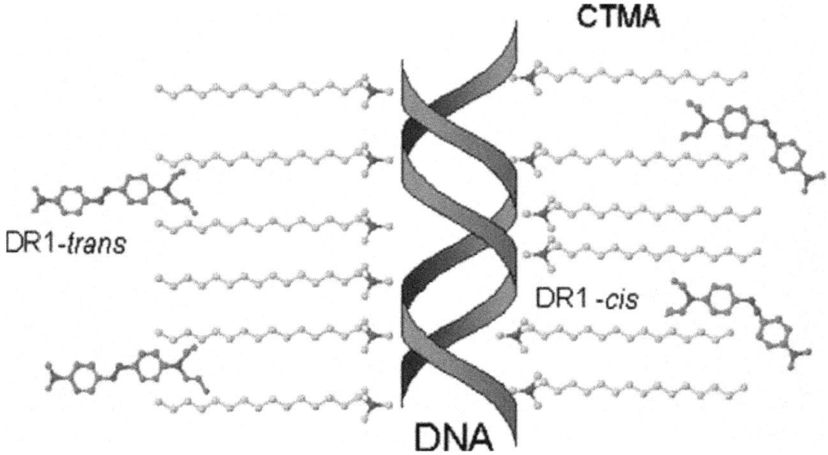

FIGURE 7.9 Illustration of the semiintercalation in *cis* and *trans* forms of DR1 molecule embedded in DNA–CTMA complex [46].

The simplest way of DNA or DNA-surfactant functionalization is by making solid solutions, as it is frequently done with synthetic polymers, and as it was proposed originally by Havinga and Pelt [50] to improve electroluminescent properties of some organic dyes. This is done in solution, while existence of a common solvent is required. It is difficult to be done with DNA and collagen as these biopolymers are soluble in water only. The DNA-surfactant complexes are insoluble in water and soluble, depending on surfactant used, in a large variety of solvents. Indeed, lot of different complexes, particularly DNA-surfactant-photosensitive chromophore complexes were successfully made in this way [see, for example, Grote [13], Rau et al. [51], Derkowska et al. [52]). Photos of some examples of solutions of photosensitive chromophores are reported in Figure 7.10(a).

Some, particularly linear molecules, like so called Hoechst molecule 33258 (cf. Fig. 7.11(a)) fits into the minor or major groves of DNA (cf. Fig. 7.9(b)), without a Van der Walls contact. This type of molecules, if showing photoluminescence, is used to stain DNA, are important and find large practical applications, particularly in criminology.

There are several attempts to attach covalently the photosensitive molecules to DNA by using different synthetic methodologies. In particular a lot of efforts was done with planar molecules, such as porphyrins through

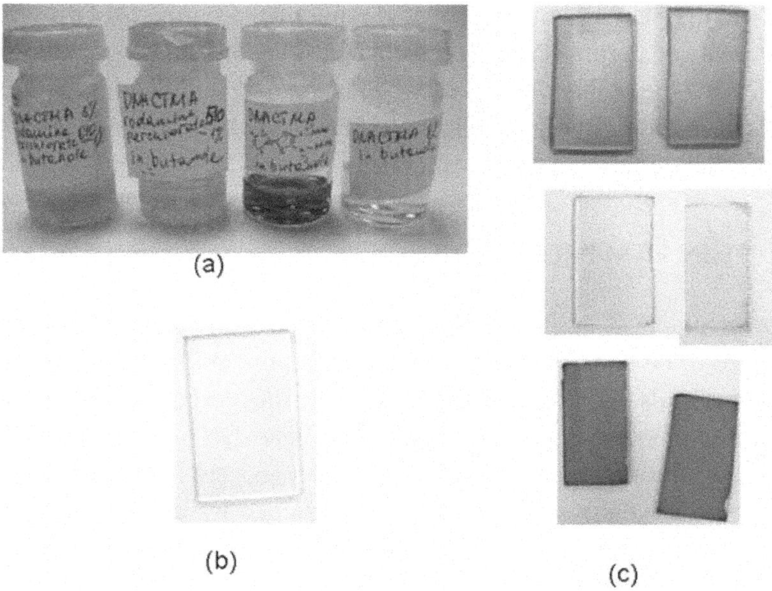

FIGURE 7.10 Photos of solutions in DNA-CTMA matrix (from left to right: first two Rhodamine 610, next – TTF molecule, the last – undoped DNA-CTMA (a), thin films of DNA-CTMA (b) and thin films DNA-CTMA chromophore complexes: from bottom: Rhodamine 590, Nile Blue and Disperse Red 1, respectively (c).

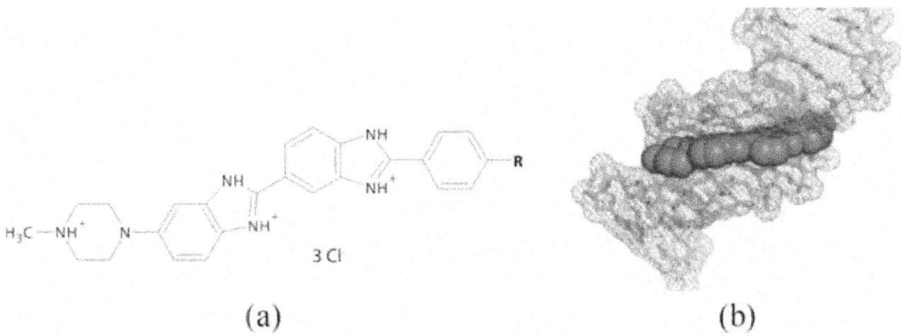

FIGURE 7.11 Chemical structure of Hoechst molecule 33258 ($C_{25}H_{26}N_6R$) chloride (a) and its binding to the minor grove of DNA (after Ref. [53]) (b).

a direct modification of nucleobases [54] or using acyclic linkers [55]. Attempts were also done in replacing nucleobases in the middle of the helix by porphyrin molecules [56] for solar energy conversion [57–58] as well as in photodynamic therapy applications [59].

Recently Stephenson et al. [60, 61] reported the synthesis of β-pyrrolic-functionalised porphyrins and their covalent attachment to 2′-deoxyuridine and DNA. The authors observed a better thermal stabilization of parallel porphyrin-modified triplex-forming oligonucleotide strands, while the antiparallel duplexes were destabilized.

7.4 THIN FILM PROCESSING

A lot of practical applications of photonic and electronic materials are expected to be done in thin films. Therefore, the ability of processing of given materials into this form is important. In the present case all discussed here biopolymers: DNA, DNA-chromophore DNA-surfactant, DNA-surfactant chromophore as well as collagen and collagen chromophore. Good optical quality thin films can be obtained by usual solution casting techniques. For DNA-CTMA we use commonly butanol as solvents. But some other, like methanol, ethanol, isopropanol, cyclohexanone and methyl isobutyl ketone can be also used. For complexes with other surfactants more solvents can be used. Thin films of DNA-CTMA complex, obtained by spin coating, were found to be partly ordered, with measured anisotropy of refractive index [62].

Some examples of thin films deposited by spinning on glass substrates are given in Fig. 7.10 (b-c). An AFM picture of collagen thin films, showing the already mentioned fibrilar structure of this biopolymer is shown in Fig. 7.12.

7.5 STABILITY OF THIN FILMS IN AIR

7.5.1 KINETICS OF MOLECULES DEGRADATION

One of the important problems with organic molecules is their chemical, thermal and photothermal stability. Although some molecules like benzene and one of the allotropic forms of carbon-diamond belong to the most stable molecules/compounds, a lot of others oxidize or decompose when heated to higher temperatures. This is essentially due to the reaction with oxygen which is omnipresent in the nature.

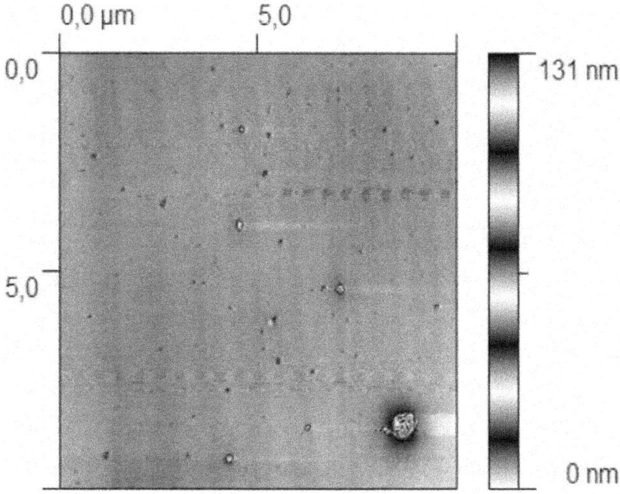

FIGURE 7.12 AFM image of the collagen thin films surface. The mentioned fibrilar structure is well seen.

The material stability issue is one of the most important problems encountered with active molecules and materials. It determines whether or not they may be used in practical devices, as they have to operate in a given temperature range and for a given time duration. More strict requirements exist for materials to be applied in military devices. Therefore, a lot of attention is devoted to the behavior of materials under action of different factors, such as chemical environment, heating, light, and more particularly UV action. Some of these factors can be present simultaneously, influencing strongly the material behavior. The materials have to support large electric optical fields, light illumination, particularly the UV light, as well as action of reactive molecules, such as, for example, oxygen. This problem is of primary importance when using synthetic polymers too.

One of the solutions envisaged to slow the degradation process is encapsulation protecting the material from the action of ambient oxygen. This appeared to be highly efficient in increasing the time of live of organic light emitting diodes (OLEDs), organic field effect transistors (FETs) and organic light emitting field effect transistors (OFETs), which are expected to revolutionize future light generation ways and lighting. Therefore, the studies of the chemical and photothermal stability of new molecules and complexes envisaged to be used in such devices are important from the point of view of their practical utilization.

Moldoveanu et al. [63] and Popescu et al. [64] have studied the stability of several chromophores, such as Rhodamines 590 and 610, DR1, DCM, LDS698, Nile Blue (for chemical structures see Table 1.), dissolved in DNA, DNA-CTMA and in collagen matrices. For the sake of comparison they used for the same chromophores a few synthetic polymers, such as polyethylene glycol (PEG), polycarbonate (PC) and polymethyl methacrylate (PMMA). The chromophores degradation studies were performed in thin films by monitoring variation of their optical absorption spectra with time, at room temperature, at elevated temperature (85°C) and under the action of UV A and B light (312 and 365 nm). The variation of optical absorption spectra under the action of different agents is the usually used technique for such kind of studies (see, for example, Refs [65–68]).

For these purposes the thin film absorption spectra of studied chromophores were monitored in visible light (the used matrices absorb in UV so their absorption does not interfere with the chromophore spectrum variation) at various time intervals. The films were obtained by solution casting (spinning) on transparent substrates (BK7 glass). The observed temporal variation of optical densities of studied films, for a given external degrading factor, were used to calculate the degradation constants, by assuming a pseudofirst order kinetic mechanism.

The advantage of this approach is that according to the Lambert–Beer's law the absorbance (optical density), at low light intensities and for an isotropic medium, like a solution or an isotropic thin film, is directly proportional to the medium thickness l, that is, to the number of molecules in optical beam, provided that the probing light beam is not completely absorbed. At high light intensities, where multiphoton absorptions take place this assumption is no longer valid [69].

This linear relationship between the absorbance and the number of molecules allows thus to determine and to follow the number of absorbing species in a solution, or in a thin film, provided that molecules are arranged in an isotropic way, as it is the case of solid solutions considered here. For ordered systems or the partly ordered thins films the absorption measurements give information on the degree of orientation (see, for

TABLE 7.1 Chemical Structure of Several Chromophores Discussed in this Chapter

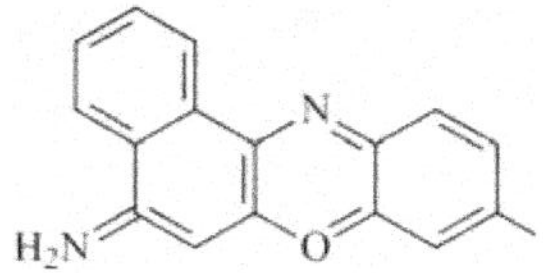 Nile Blue	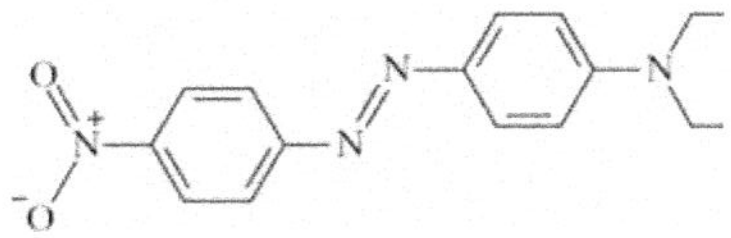 Disperse Red 1	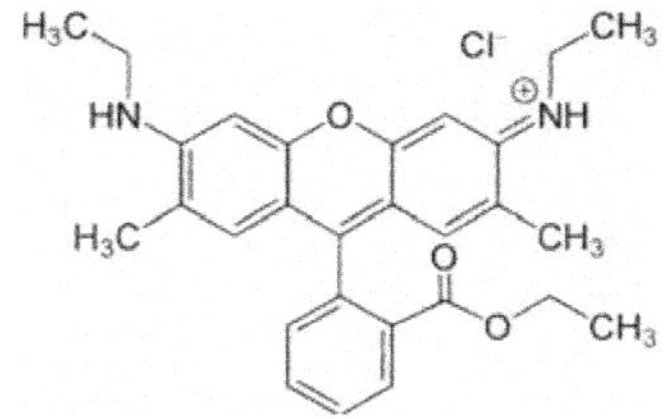 Rhodamine 590 chloride
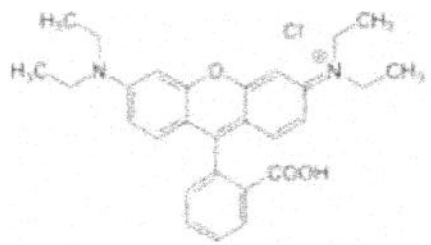 Rhodamine 610 chloride	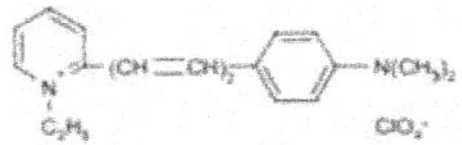 LDS 698 perchlorate	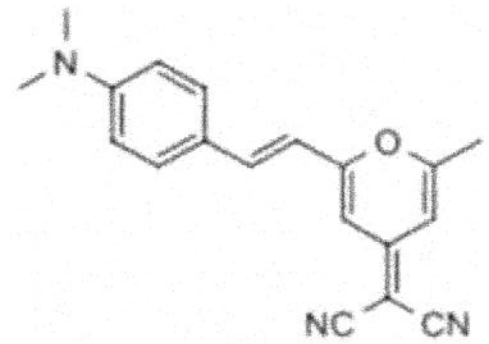 DCM
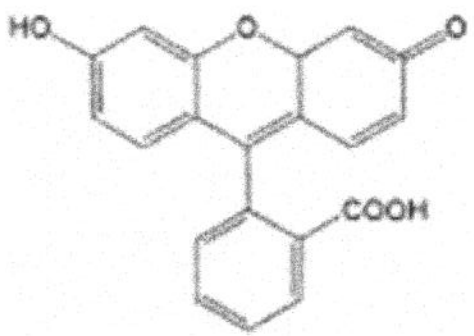 Fluorescein	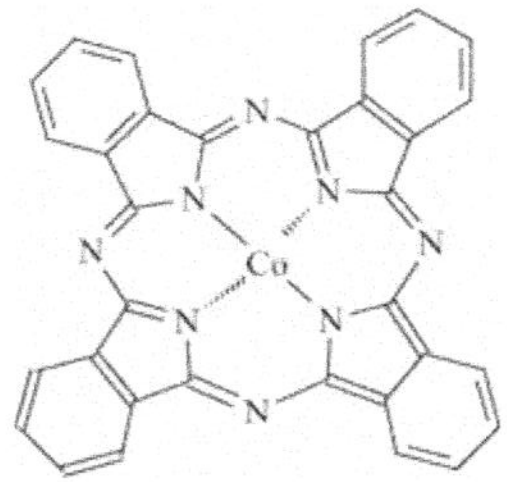 CoPc	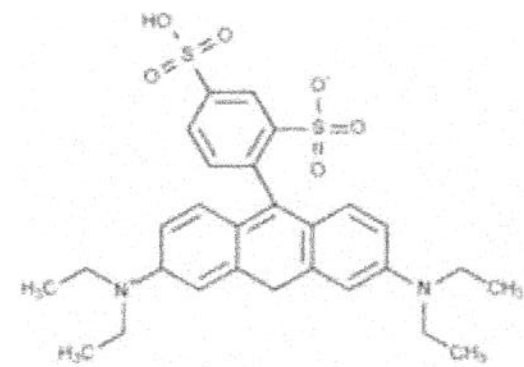 Sulphorodamin B

example, Page et al. [70]). Thus following the absorbance variation of a given material allows monitoring the decay of molecules due to the action of external stimuli, such as: light, temperature, presence of reactive chemical agents, etc.

The kinetics of temporal degradation of a thin film can be described by the first order law:

$$\frac{dc}{dt} = -\,kc \tag{1}$$

where c is the concentration of active species and k is the first order kinetic degradation constant.

It means that the concentration c(t) varies with time t as

$$c(t) = c(t=0)e^{-kt} \tag{2}$$

where c(t = 0) is the initial concentration of absorbing species.

On the other hand, as it follows from the Lambert–Beer's law, the optical absorption of a medium is proportional to the concentration c(t) of absorbing species. The temporal variation of the optical absorption can be represented by the temporal variation of the optical density (absorbance) A(t) at the maximum absorption wavelength. Thus, Eq. (2) can be rewritten as follows

$$A(t) = A(t=0)e^{-kt} \tag{3}$$

where A(t=0) is the initial optical density.

The kinetic degradation constant can be obtained from linear regression of measured temporal variation of optical density (Eq. (3))

$$\ln A(t) = -\,k\,t + \text{const} \tag{4}$$

It may happen that several phenomena contribute to the material degradation. In that case the degradation process is described by several degradation kinetics constants: k_1, k_2, k_3, ….. They can be determined by fitting the temporal variation of the optical density A(t) by two, or more exponential functions

$$A(t) = A_1 e^{-k_1 t} + A_2 e^{-k_2 t} + A_3 e^{-k_3 t} + \ldots \tag{5}$$

with

$$A_1 + A_2 + A_3 + \ldots = A(t = 0) \tag{6}$$

7.5.2 CHEMICAL DEGRADATION AT ROOM TEMPERATURE

The discussed here thermal and photodegradation studies of spin coated thin films, deposited using Laurell–Model WS – 400B – 6NPP/LITE spin coater, were performed with a JASCO UV – VIS – NIR spectrophotometer, model V 670. Figure 7.13 shows, as an example, the variation of the optical absorption spectrum of DNA-Rh 590 thin film as function of time. Over the period of 39 days a slow, monotonic decrease of the optical density of studied film is observed.

Figure 7.14 compares the least squares fit of Eq. (4) to the temporal, experimental variation of the maximum of optical density with time for Rhodamine 590 embedded in DNA (a) and in a synthetic polymer polyethylene glycol (PEG) (b). The observed much larger decay slope in the case of PEG matrix indicates a significantly faster (about 3 orders of magnitude difference) decay of Rhodamine 590 in the synthetic polymer as compared with the DNA –CTMA matrix.

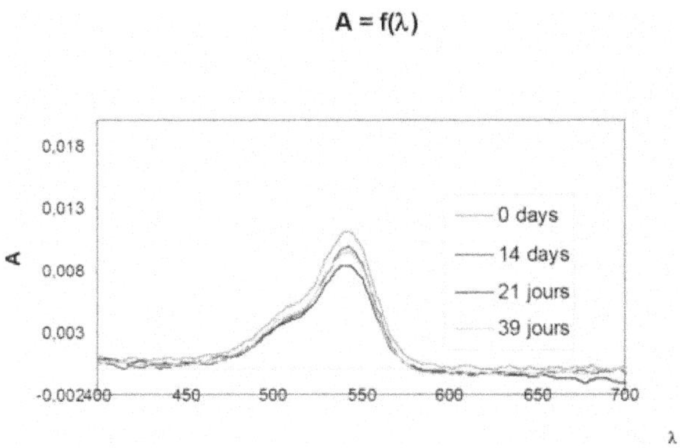

FIGURE 7.13 Variation of the optical absorption spectrum of DNA-Rh 590 thin film as function of time (in days).

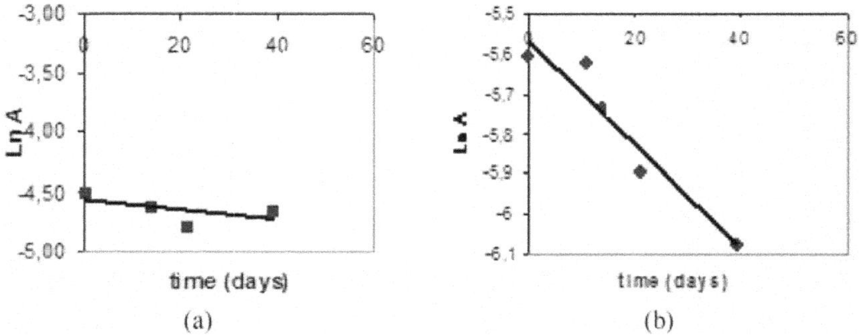

FIGURE 7.14 Least squares fit of Eq. (4) to neperian logarithm of experimental absorbances at room temperature for DNA-Rh590 and PEG-Rh590 thin films (b). Squares and diamonds show measured values while solid lines the fitted ones.

The determined in this way the room temperature kinetic degradation constants for Rhodamine 590 (cf. Table 7.1) embedded in different matrices are given in Table 2. In all cases the chromophore decay can be described by a single exponential function. Although at room temperature the kinetic degradation constants are of the same order of magnitude for different matrices however, the lowest stability is observed in PEG ($k_1 = 9.03\times10^{-6}$ min^{-1}). Adding biopolymers improves significantly the chromophore stability (adding DNA: $k_1 = 6.57\times10^{-6}$ min^{-1}, with collagen one obtains the best stability for this chromophore $k_1 = 1.05\times10^{-6}$ min^{-1}, respectively). The stability of Rhodamine 590 in DNA is very similar to that in other two matrices: collagen and polycarbonate (PC), being close to the value of $k_1 = \sim 3\times10^{-6}$ min^{-1}.

TABLE 7.2 Room and High (85°C) Temperature Kinetic Degradation Constant of Rhodamine 590 at Different Matrices (After Moldoveanu [63])

Host material	Room temperature kinetic degradation constant k_1 (mins^{-1})	Kinetic degradation constant k_1 at 85°C (mins^{-1})
DNA	2.78×10^{-6}	6.68×10^{-6}
DNA+PEG	6.57×10^{-6}	51.7×10^{-6}
Collagen	2.09×10^{-6}	35×10^{-6}
Collagen+PEG	1.05×10^{-6}	55×10^{-6}
PC	3.13×10^{-6}	$11,000\times10^{-6}$
PEG	9.03×10^{-6}	$89,000\times10^{-6}$

Table 7.3 lists the room temperature decay constants for several chromophores embedded in different matrices and for different concentrations. Nile Blue, DCM and LDS 698 show very good stability, with negligible first order decay constants in biopolymers DNA-CTMA and in collagen. A good stability shows also DCM in PC.

7.5.3 CHEMICAL DEGRADATION AT ELEVATED (85°C) TEMPERATURE

Since the room temperature degradation (20°C) are usually very low, the measured values of kinetic degradation constants are affected by a large error because of high uncertainties in the measurements. Therefore, the degradation process was accelerated by heating the films to higher temperature (85°C), however, below the stability limit temperature of DNA (*ca.* 90°C).

Tables 7.2 and 7.3 gives also first order kinetic decay constants measured at elevated temperature (85°C) and compare them with room temperature data Only in the case of LDS 698 chromophore (cf. Table 7.1), embedded in DNA-CTMA matrix, a double exponential decay was observed.. In the case of biopolymer matrices and Rhodamine 590 chromophore the kinetic decay constants are less than one order of magnitude larger than at room temperature (Table 7.2). However, in the synthetic polymer matrices: PC and PEG this increase is by *ca* three orders of magnitude. Also adding of PC or PEG to DNA or to collagen increases the kinetic decay constants.

Table 7.3 lists the kinetic decay constants measured at 85°C for different chromophores and their concentrations, embedded in various matrices. Their large increase, compared to room temperature data, is observed. The chromophore LDS 698, which shows an excellent stability at room temperature, decays rapidly at 85°C in both used matrices: DNA-CTMA and in PC. Moreover, in biopolymer the decay of this chromophore is described by two kinetic decay constants, what means that there are two different processes behind. A very peculiar is also the concentration dependence of kinetic decay parameter observed for Nile Blue embedded in DNA-CTMA matrix. It decreases with chromophore concentration up to 15 w% and increases at 20 w%. This may be due t intercalation or groves inclusion of this chromophore. Indeed Nile Blue is known to stain DNA and is used for this purpose.

TABLE 7.3 Room ($k_{20°C}$) and Elevated ($k_{85°C}$) Temperature Kinetic Degradation Constants, in min^{-1}, for Studied Chromophores Embedded in Different Matrices

Chromphore	Concentrtion w%	Host	k_1(20°C) 10^{-6}(min^{-1})	k_1(85°C) (min^{-1})
Rh590[a]	5	DNA	2.78	6.68
Nile Blue	2	DNA-CTMA	NG	2600
	5			1300
	7			1200
	10			700
	15			400
	20			2600
Nile Blue	7	Collagen	NG	600
	10			400
	15			500
Rh590[a]	10	DNA-CTMA	2.57	40.0
	20		2.78	50.0
	5	Collagen	2.09	35
		Collagen + PEG	1.05	55
		PC	3.13	11,000
		PEG	9.03	89,000
LDS 698	5	DNA-CTMA	NG	k_1=27,400[a]
	10			k_2=3600
				k_1=29,000[a]
				k_2=5200
	5	PC	NG	2600
	10			1900
	15			2400
DCM	5	DNA-CTMA	NG	6400
	10			6300
	15			5300
	5			1400
	10	PC	NG	1700
	15			1500

[a] two decay constants observed.
NG – negligible.

7.5.4 KINETICS OF PHOTODEGRADATION

The described here photodegradation measurements were performed using a commercial Vilber Urmat apparatus equipped with two irradiation sources: UVA at 365 nm and UVB at 312 nm. The illumination intensity was of 6.5 mW/cm^2 for UVA and 2.5 mW/cm^2 for UVB. It means that the ratio of photons illuminating the sample at UVA to that at UVB $n_{UVA}/n_{UVB} \approx 2.6$. All measurements were done at room temperature. Table 7.4 lists first order kinetic degradation constants under UVB for Rhodamine 610 in 3 biopolymers: DNA, DNA-CTMA and collagen. No significant dependence on the matrix and on chromophore concentration is observed within experimental efficiency. However, the chromophore degradation is fast with kinetic degradation constants about three orders of magnitude larger as compared to room temperature values.

Table 7.5 shows the room temperature photodegradation constants for selected chromophores in bio- and synthetic polymer matrices for different chromophore concentrations. The kinetic photodegradation constants are, generally, three orders of magnitude larger than in dark; Interesting is a very good stability of pure Nile Blue in DNA. This molecule is known to stain DNA, forming most likely, a stable complex with this biopolymer. It is less stable in two other matrices PC and DNA-CTMA, with a worse stability in the second. LDS also shows a better stability in PC than in DNA-CTMA. In DNA-CTMA it exhibits two kinetic photodegradation decay constants, similarly as when heating. The kinetic degradation constants increase with increasing NB concentration most likely due to its aggregation. Obviously the photodegradation of chromophores depends

TABLE 7.4 Kinetic Degradation Constants k1 (in min–1) Under UV Irradiation for Rh610 in Different Matrices, Irradiated at 312 nm

Concentration	DNA-Rh610	DNA-CTMA-Rh610	Collagen-Rh610
1%	0.0019	0.0022	0.001
2%	0.0021	0.0029	0.0009
5%	0.0014	0.002	0.0006
7%	0.0002	0.0013	0.001
15%	0.0047	0.0008	0.0013

TABLE 7.5 Photodegradation Kinetic Constants, in min^{-1}, For Selected Chromophores Embedded in Different Matrices and Under Both UVB (312 nm) and UVA (365 nm) Irradiation

Chromophore	Concentration w%	Host material	k_{UVB} $10^{-6}(min^{-1})$	k_{UVA} $10^{-6}(min^{-1})$
Rhodamine 590		DNA	3 800	2000
		Collagen	1600	2200
		PC	8900	2800
	5	PEG	5000	4500
		DNA + PEG	6100	4100
		Collagen + PEG	3330	2100
	10	DNA-CTMA	1000	2300
	20		800	1900
Rhodamine 610	7	DNA	NG	200
	15			470
	7	DNA-CTMA		1300
	15			800
	7	Collagen		1000
	15			1300
	1	PMMA		500
	15			400
DR1	10	DNA-CTMA	880	2200
	20		1000	1800
Nile Blue	20	DNA		NG
	2	DNA-CTMA		2800
	5			1800
	7			1200
	10			1000
	15			900
	20			2300
	7	Collagen		600
	15			500

TABLE 7.5 (Continued)

Chromophore	Concentration w%	Host material	k_{UVB} 10^{-6}(min^{-1})	k_{UVA} 10^{-6} (min^{-1})
LDS698	5	DNA-CTMA		k_1=27400[a]
				k_2=3600
				k_1=29000[a]
	10			k_2=5200[a]
	5			2600
	10	PC		1900
	15			2400
DCM	5			6400
	10	DNA-CTMA		6300
	15			5300
	5			1400
	10	PC		1700
	15			1500

[a] two decay constants observed.
NG – negligible.

on the presence or lack of absorption bands at the illumination wavelength why the results shown in Table 7.5 will be different if shining with another wavelength. Therefore, comparison at one wavelength of kinetic photodegradation constants may be misleading.

Similarly as in the case of thermal degradation the kinetic degradation constants, within experimental accuracy, do not depend significantly on the chromophore concentration for a given matrix.

The kinetic degradation depends on the composition. Heating and UV light accelerates the degradation, as expected. Among the studied chromophores the less stable is the luminophore LDS 698, which exhibits even two photo and thermal degradation processes, described by two distinct kinetic degradation constants. As previously observed these two degradation processes, within experimental, accuracy, do not show dependence on chromophore concentration.

The values listed in Tables 7.2–7.5 for kinetic photodegradation constants correspond to the specific experimental conditions, that is, temperature, ambient atmosphere, etc. The measurements were performed in air and on thin films. In solution, bulk material and in neutral atmosphere they will be different. However, they allow a comparison of the chemical stability of different chromophores in various matrices at the same conditions, although in the case of photodegradation the kinetic degradation constants are expected to depend strongly on the illumination wavelength.

7.5.5 OPTICAL DAMAGE THRESHOLD

Another important parameter determining the usefulness of a given material in devices working in extreme conditions, like high intensity light sources delivered by lasers is the optical damage threshold. It is understood as the highest light intensity the material withstand without a permanent damage. For higher light intensities a permanent damage to the illuminated material takes place. It is caused by very high optical fields leading to the dielectric breakdown in material. Figure 7.15 shows speckles created in a transparent biopolymer thin film by a high intensity laser beam.

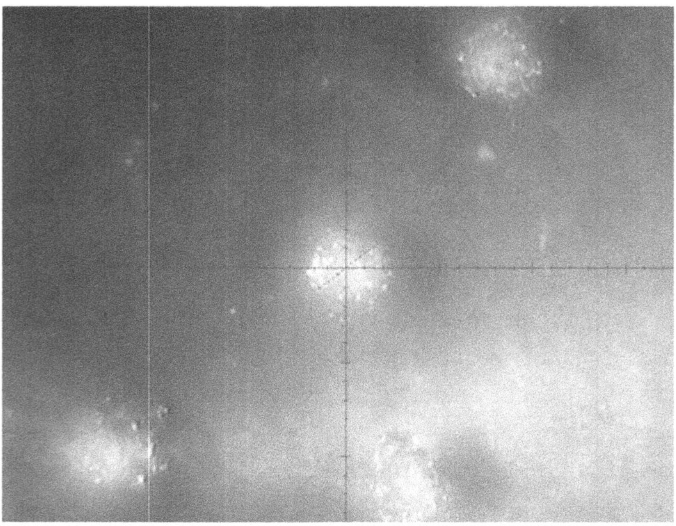

FIGURE 7.15 Optical microscope image of speckles created in a DNA thin film during the optical damage threshold experiment.

High and ultrahigh light intensities are created by pulsed Q switched and mode locked lasers, although the optical damage to a material can be caused also by low intensity cw lasers if the light beam is absorbed. The light absorption causes increase of the temperature and decay of material through melting, photochemical reaction or even evaporation. These processed are not considered here, assuming material with no absorption. The only damage possible, taken under consideration, is that produced through the dielectric breakdown or multiphoton NLO absorptions, thus heating.

The measured optical damage thresholds for studied biopolymers are listed in Table 7.6. They correspond to the specific experimental conditions, that is, wavelength of 1064.2 nm; repetition rate of 10 Hz and pulse duration of 6 ns. At another wavelengths, pulse duration times and repetition rates they will be different. For shorter laser pulses the optical damage threshold will be higher. It will be similar for lower repetition rates. Higher repetition rates will induce a faster material degradation due to the heat accumulation. Also absorption will decrease the damage threshold. However, they allow comparing the behavior of different materials at the same conditions. The data shows, that compared to the thin films of studied synthetic polymers (polycarbonate (PC) and polyethylene glycol (PEG), the films three biopolymers: collagen, DNA and DNA– CTMA exhibits about one order of magnitude higher damage thresholds. Adding DR1 to chromophore decrease only a little the thin

TABLE 7.6 Thin Film Optical Damage Thresholds for Thin Films of Collagen and Selected DNA Based Complexes and for Thin Films of Polycarbonate (PC) and Polyethylene Glycol (PEG)

Material	Optical damage threshold I_{th} GW/cm^2
DNA-CTMA-DR1 – 5 w%	3.3
DNA-CTMA-DR1 – 10 w%	3.6
DNA-CTMA-DR1 – 20 w%	4.8
DNA-CTMA	5.2
DNA	5.3
collagen	4.4
PC	0.30
PEG	0.78

film DNA-CTMA complex damage threshold. This is a very interesting result showing a high potential for using biopolymers in photonics

7.6 OPTICAL THIN FILM PROPERTIES

7.6.1 LINEAR OPTICAL PROPERTIES

DNA and DNA-CTMA molecules exhibit a large transparency range, the lowest energy absorption being that of phenyl rings, located around 260 nm (cf. Fig. 7.16). The absorption UV cut-off is around 325 nm whereas in near IR only the absorption of harmonics of high energy CH and OH vibrations are present.

Adding surfactant does not alter the transparency band as only the aromatic ones add absorption at 260 nm where absorb nucleobases. The refractive index of DNA-CTMA thin films varies between 1.582 in UV (300 nm) and 1.482 in NIR (1000 nm) [70]. Compared to silica it is slightly higher in UV because of the absorption of nucleobases and lower in NIR.

DNA-CTMA thin films exhibit low propagation losses, particularly in the telecommunication wavelength range (1.3–1.55 μm). At 1.3 μm the propagation they are of 0.2 dB/cm, a value comparable to that observed in best synthetic polymers. It is even lower (0.1 dB/cm) at 800 nm, the wavelength used for interconnects and local area networks (LAN's).

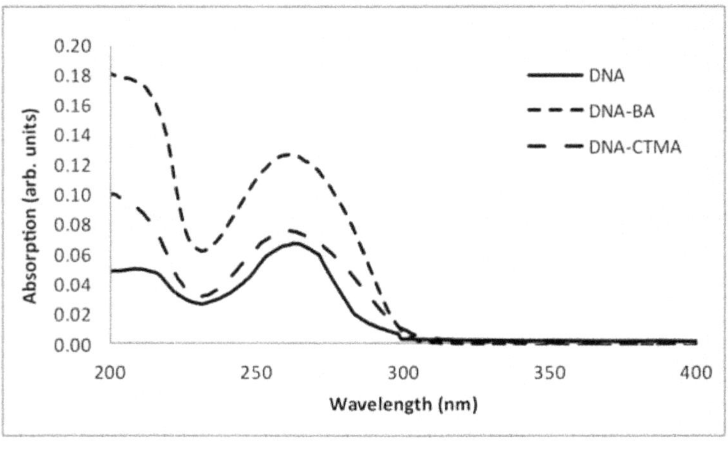

FIGURE 7.16 Optical absorption spectra of DNA and two DNA-surfactant complexes.

The relatively low index of refraction and good transparency make these complexes interesting for application as cladding layers in Mach – Zehnder interferometer type in wave-guiding configuration electro-optic modulator (EOM) [72, 73]. Indeed, in that case a better distribution of electric field in the EOM structure: buffer layer/active layer/buffer layer is obtained with its larger value in active layer. At allows to obtain a better orientation of chromophores in active layer.

Another interesting result concerning the temperature variation of refractive index of DNA-CTMA and DNA-CTMA doped with Nile Blue was reported by Hebda et al. [67 Hebda]. The refractive index of thin film DNA-CTMA alone increase between 0 and 40°C showing negative thermal expansion in this temperature range. Adding 5 w% of Nile Blue compensates this negative expansion. It means that one can get a polymer with no thermal expansion. Such property is very important for practical applications.

7.6.2 NONLINEAR OPTICAL PROPERTIES

7.6.2.1 Second Order Nonlinear Optical Properties

As already mentioned collagen lacks center of symmetry. Thus it is expected it will exhibit, not very important because of lack of conjugated p electrons, second – order nonlinear optical properties (NLO) properties. Indeed Vasilenko et al. [74] reported already in 1965 the first observation of second harmonic generation (SHG) in protein – collagen complex. This observation was confirmed later (1971) by Fine and Hansen [75]. Two other polypeptides, tubulin and myosin [76–78] were shown also to exhibit the second order NLO properties. Much more recently SHG measurements were reported for polysaccharides, such as starch and cellulose [79–85].

The ability of collagen and other biological species to exhibit SHG has attracted a lot of interest (for a review see Knoesen [86]), not only because of its interest for basic knowledge, but also because of the possibility it gives for imaging. Indeed a lot of research which followed was devoted to the use of nonlinear optical techniques for imaging of biological tissues (see, for example, Lee et al. [87]).

There are very little data on second-order NLO properties of DNA alone, although this chiral molecule should exhibit such effects, as observed in this type of materials (see, for example, Ostroverkhov et al. [88], Sioncke et al. [89], Iwamoto et al. [90]) and the already mentioned for collagen. These authors have observed in chiral structure an intensity dependence of SHG signal on the polarization of incident laser beam with respect to the helix.

In several recent papers the second harmonic generation technique was used as a tool for studying the interaction of DNA with environment or for detection of its modifications. In fact, SHG is a very sensitive technique to study the interfaces [91], as the bulk centrosymmetry is broken there and the observation of frequency doubling is no more forbidden by symmetry, as it is the case of centrosymmetric structures. In particular Boman et al. [92] reported on using SHG to study the formation of the DNA double helix at the quartz surface due to the pairing of adenine and thymine nucleobases. Zhuang Zheng-Fei et al. [93] used this technique as a detection tool of the very early malignancy in prostate glandular epithelial cells. Williamson et al. [94] reported observation of a humidity dependent optical SHG signal from spun films of DNA. Note that Yamada Satoru et al. [95].performed theoretical calculations of NLO properties of modified guanine bases with a NLO group.

Using the experimental set up described in Ref. [96] we have attempted poling of spin deposited thin films of DNA-CTMA complex, doped with a highly responsive, noncentrosymmetric (3-(1,1-dicyanothenyl)-1-phenyl-4, 5-dihydro-1H-pyrazole) (DCNP) molecule. This is the commonly used way to create noncentrosymmetry in primarily centrosymmetric thin films by applying high DC electric field, and particularly in so called electro-optic polymers [97]. The polar orientation of chromophores is usually done at higher temperatures where the polymer chain mobility becomes relatively large and facilitates their orientation. Then it is by cooling the polymer thin film down to room temperature. The obtained degree of polar order depends on the used polymer matrix and on the strength of the applied electric field E_{DC} as well as on the ground state dipole moment of chromophore. In the present case it was done by in situ corona poling technique. Figure 7.17 presents the temporal variation of the in situ measured SHG intensities at different temperatures. At 60°C, that is, well below the DNA denaturation temperature (ca 90°C)) we observe after switching on the poling field an increase of SHG signal (triangles, black).

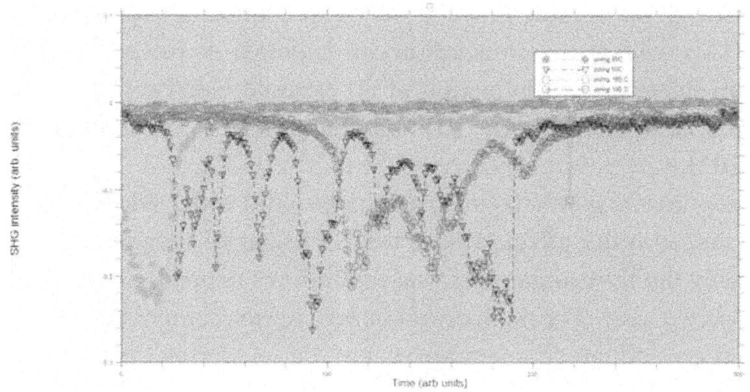

FIGURE 7.17 Temporal variation of SHG intensity during the poling process at different temperatures.

But after reaching a certain value the SHG intensity drops to a low value and starts to increase again, drops, increase. These oscillation don't have a fixed period. Similar behavior was observed in Electric Field Induced Second Harmonic Generation measurements on a polydiacetylene thin film and were explained by formation of electrets in studied material [98] After some number of oscillations no more the polar order formation is observed. This behavior may be due to the existence of deep traps in DNA, a phenomenon we observed already with BioLED [99]. The poling process, accompanied by a flow of charges through the poled films is associated with formation of electrets, that is, pairs of electron holes, because of electron trapping. Such pairs obviously create an electric field opposed to the applied poling field, canceling it. This explains decrease of SHG signal. The charges are obviously created by two and/or more photon absorption [67] process, and/or by SHG photons. The increase after is most likely due to the detraping of charges via a two (and/or more) photon absorption and/or harmonique photons.

7.6.2.2 Third-Order Nonlinear Optical Properties

There is a little research of third-order NLO properties of pure DNA. One of the first problems treated by NLO technique was the mobility of DNA helix under the applied electric field. This study is difficult for two main reasons: DNA is soluble in water only and exhibits a large ionic conductivity. This solvent is also necessary to maintain its integrity. DNA itself is a polyelectrolyte. Application of an external electric field induces an ionic dipole moment [100–103] and changes the conformation of DNA [104, 105].

There are no symmetry restrictions on third order NLO effects, such as for second order effects where lack of center of inversion is required. Obviously the third-order NLO susceptibilities depend on the order, as it is always the case. The number of nonzero tensor components depends on material crystallographic symmetry, but the NLO effect is present in any material medium.

Samoc et al. [106] have measured real and imaginary parts of the nonlinear index of refraction of DNA in solution by wave dispersed femtosecond *z*-scan technique. They have found that it varies between 2×10^{-15} and 10^{-14} cm^2/W in the wavelength range 530–1300 nm. They report also observation of a weak two photon absorption (TPA) below 600 nm, with nonlinear absorption coefficient equal to 0.2 cm/GW at 530 nm. Apparently it corresponds to two photon transition to the first excited level of nucleobases phenyl rings. Such two photon transitions are allowed because of chiral structure of DNA. There is lot of experimental techniques allowing to measure third-order NLO susceptibilities. Hereafter we will describe few of them; particularly those which were used in studying NLO response of DNA based materials.

Rau et al. [108] measured third-order NLO properties of pure DNA-CTMA and DNA-CTMA-DR1 complexes by optical third harmonic generation (THG) technique (see, for example, Kajzar [107]). THG is an NLO process in which three photons with frequency ω generate a photon with triple frequency 3ω *via* interaction with matter. In other words a light beam with wavelength λ will be transformed into a beam with shorter wavelength $\lambda/3$. The main advantage of this technique is the fact that the response time is very fast as the harmonic field oscillations have to follow that of fundamental beam. Thus the response time is in the 10^{-15}s time domain when working in the visible and near IR electromagnetic field range. Therefore, the THG technique is very interesting to measure the fast, electronic origin, third-order NLO susceptibilities of different materials as the other technique, for example, *z*-scan, degenerate and nondegenerate two (TWM), four wave mixing techniques (FWM) give the values which may contain large orientation contributions [108]. Also the thermal contributions can be important when not using very short (fs) laser pulses, particularly when the excitation wavelength is within the material absorption band or a multiphoton absorption [67] is present.

The THG data obtained for thin films of DAN-CTMA and DNA-CTMA complexes doped with DR1 chromophore are reported in Table 7.7. They are calibrated with recently determined THG susceptibility $\chi^{(3)}(-3\omega;\omega,\omega,\omega)$ value [109] of silica used as standard and are compared with the value obtained for polymethyl methacryalate (PMMA) thin film [110]. In the screening procedure the refractive indices for DNA-CTMA complex thin films, reported by Grote et al. [111] were used. We used the same refractive indices for DNA-CTMA-DR1 complexes as for pure DNA-CTMA. Indeed their expected modification is not large and the harmonic intensities do not depend on their difference but on their sum [112]. Therefore, the error in $\chi^{(3)}(-3\omega;\omega,\omega,\omega)$ susceptibility of these complexes is expected to be small.

The DNA based thin films exhibits about one order of magnitude larger cubic susceptibility than silica and PMMA. This is due to the already mentioned contributions from the polarizability of π electrons in nucleobases,

TABLE 7.7 DNA-CTMA and DNA-CTMA – DR1 Thin Film Thicknesses (in μm), Fundamental Wavelength Refractive Indices, Coherence Lengths and Third Order Nonlinear Optical Susceptibilities

Sample	Thickness [μm]	Refractive index n_ω	Coherence length [μm]	$\chi^{(3)}(-3\omega;\omega,\omega,\omega)$ in 10^{-14} [esu]
DNA-CTMA	0.367	1.488[a]	7.4 [a]	11.5±1.2
DNA-CTMA-DR1 (5%)	3.484	1.488[b]	7.4	155±16
DNA-CTMA-DR1 (10%)	4.060	1.488[b]	7.4	69±7
DNA-CTMA-DR1 (15%)	3.484	1.488[b]	7.4	85±9
silica	1010	1.44967	6.71	1.43±0.14[c]
PMMA		1.4795[d]	8.28[d]	3.2[d]

[a] Grote et al. [111].

[b] assumed (see text).

[c] Gubler and Bosshard [109].

[d] Morichere et al. [110].

absent in these materials. Interesting is behavior of $\chi^{(3)}(-3\omega;\omega,\omega,\omega)$ susceptibility with dopant concentration (DR1). At 5 w% of DR1 one observes a large increase, up to (155±16)x10^{-14} esu from (11.5±1.2)x10^{-14} esu for pure DNA-CTMA thin film. At 10 w% of dopant $\chi^{(3)}(-3\omega;\omega,\omega,\omega)$ decreases to (69±7)x10^{-14} esu, to increase again at 15 w% of DR1 to (85±9) x10^{-14} esu, respectively. This strange behavior was tentatively explained by Rau et al. [108] within a three level quantum model as due to the blue shift of DR1 absorption band caused by aggregation of these molecules. As consequences the two photon resonant contribution of charge transfer (CT) band, predominant in NLO susceptibility is varying with the dopant concentration [113].

Samoc et al. [114] determined the real and imaginary parts of the non-linear index of refraction of DNA in solution by wave dispersed femto-second z-scan technique. They found that it varies between 2×10^{-15} and 10^{-14} cm^2/W in the wavelength range 530–1300 nm. They have reported also an observation of a weak two photon absorption (TPA) below 600 nm, with nonlinear absorption coefficient equal to 0.2 cm/GW at 530 nm. Apparently it corresponds to the two photon transition in phenyl rings.

Derkowska et al. [50] have reported the degenerate four wave mixing (DFWM) and nonlinear transmission measurements on DNA-CTMA complex, doped with several complexes, such as DR1, cobalt phthalo-cyanine (CoPc) and fullerene C_{60} (cf. Table 7.1 for chemical structure). A different behavior of DFWM susceptibility than when these mole-cules are dissolved in other solvents was observed, indicating the influ-ence of ionic environment of DNA on electronic structure of embedded chromophores.

Very interesting results on behavior of guest molecules were reported by Mysliwiec et al. [115] by optical phase conjugation performed on DNA-CTMA-DR1 thin films which shows and about four orders faster response, due to the trans-cis-trans isomerization, than if the same chromophore is embedded in PMMA matrix [116]. It is even faster in the case of disperse orange (DO3) molecule [117], which isalso an azodye, but smaller than DR1. In that case the response time is in μs range, depending on beam intensity, about three orders of magnitude shorter than in the case of DR1. in both cases the peculiar double strand structure of DNA and the large free volume are believed to be behind this short response time.

7.6.3 PHOTOLUMINESCENT PROPERTIES OF FUNCTIONALIZED DNA AND DNA-SURFACTANT COMPLEXES

DNA and DNA-CTMA complexes appear also as an interesting matrix for luminophores. Several research groups reported enhancement of photoluminescence of certain luminophores when embedded in DNA or DNA-surfactant (usually CTMA) matrix. In particular Wang et al. [118], reported enhanced photoluminescence of rare earths ions 6,6,7,7,8,8,8-heptafluoro-2,2-dimethyl-3,5,-octanedionate, (Eu^{3+}-FOD) in DNA-CTMA matrix both in thin films and in optical fibers, with fluorescence lifetime of 750 μs. Both the fluorescence lifetime and the emission quantum efficiency were found to be larger in DNA-CTMA matrix than in the reference material PMMA.

Yu et al. [119, 120] observed a 17 times larger photoluminescence from Sulforhodamine (SRh) embedded in DNA matrix as compared when the matrix used was PMMA. The maximum of emission was realized with 1 wt % concentration of SRh in DNA. With a distributed feedback structure with 437 nm period they observed amplified spontaneous emission (ASE) at 650 nm wavelength and lasing with threshold of 30 $\mu J/cm^2$.

Massin et al. [121] compared photoluminescence quantum efficiency of three different luminophores (cf. Fig. 7.18(a)), dissolving them in solid PMMA and in DNA-CTMA matrices. For two of them (for details see Ref. [118]} the photoluminescence quenching takes place at higher concentration when using DNA-CTMA as matrix, as compared with PMMA. Also a blue shift of the fluorescence spectra in the case of DNA-CTMA matrix is observed, showing influence of DNA environment. Figure 7.18 (b) displays the concentration variation of the photoluminescence quantum yield ratio in DNA-CTMA/PMMA for these three chromophores. In the chase of Chr 1 this ratio increases with concentration showing higher quenching limit when the chromophore is embedded in biopolymer matrix. For the other two the ratio is almost constant, being larger than 1.

Koyama et al. [122] reported enhancement of photoluminescence of fluorescein (cf. Table 1) in DNA-CTMA matrix, as compared to PMMA. Increase of photoluminescence efficiency was also observed for natural chromophores: green tea extract (GTE) in DNA [123] and

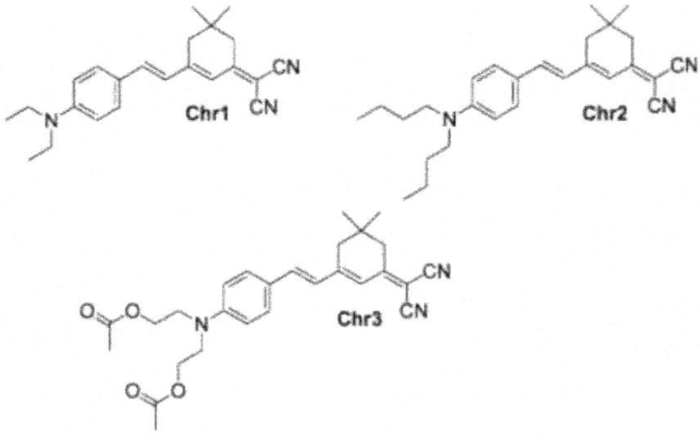

(a)

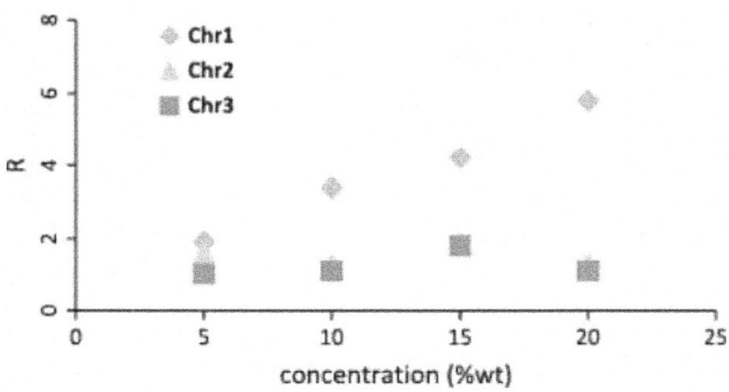

(b)

FIGURE 7.18 Chemical structures of the studied dyes and the concentration variations of their quantum yields ratios when embedded in DNA–CTMA or in PMMA matrices (after Massin et al. [121], for details see text).

sea buckthorn extract (SBE) in DNA-CTMA [124] as compared when using PMMA matrix.

Very recently Kobayashi and co-workers have demonstrated an electric field steered, large emission wavelength range of a DNA based BioLED with AlQ_3 and $Ru(bpy)_3^{2+}$ as active molecules [125, 126].

DNA [127] and DNA-surfactant [43,128–130] thin layers were also used as electron blocking layers in Bio Light Emitting Diodes (BioLEDs) allowing to get a significant increase of luminance efficiencies.

DNA and DNA-CTMA complexes are also interesting matrices for lasing [131–133] Amplified spontaneous emission (ASE) was observed by several research groups [88,134–136] (for a recent review see Ref.

[137]). In particular Mysliwiec et al. [138] have observed photoluminescence and ASE from MR isomer of spiropyrane embedded in DNA-CTMA matrix. These effects are not observed when using synthetic polymers as matrix, such as, for example, polymethyl methacrylate (PMMA). A two photon lasing was also demonstrated in a DNA-CTMA-chromophore complex [139].

Random lasing in luminophore doped DNA based complexes was also recently demonstrated [140–143].

7.7 CONCLUSIONS

In this chapter, we review and discuss the recent work on two biopolymers: collagen and deoxyribonucleic acid in view of their application in photonics. Both are abundant, renewable, biodegradable and nature fabricated macromolecules. They can be obtained from the waste produced in food processing industry. That used in our studies originates from the waste produced in salmon processing industry. Thus they can be cheap and the renewable resources are practically unlimited. It can be used, at least partly, to replace synthetic polymers as matrix for photosensitive molecules, offering an interesting, ecologically friendly, alternative material for applications in photonics and in electronics.

Owing to its peculiar double stranded helical structure DNA offers more than synthetic polymers. Particularly there is a larger free volume offering faster conformational processes, as discussed here, and at the same time a better stability of embedded molecules. Indeed, as shown by described and discussed here the recent photothermal degradation studies performed on a series of DNA based complexes significantly larger first order decay constants for several chromophores embedded in collagen or DNA matrices than when these molecules are dissolved in the commonly used synthetic polymers like the polymethyl methacrylate (PMMA). Also, DNA exhibits a higher optical damage threshold.

DNA can be functionalized with surfactants, leading to a temperature stable material, processable into good optical quality thin films by solution casting techniques. The structure of DNA and the ionic environment it offers changes the physicochemical properties of embedded active molecules in a favorable way as shown and discussed in this paper.

Different doping mechanisms were indicated. The richness of the possible functionalization of DNA gives possibility of obtaining an important class of materials with new properties and new functionalities. They may be cheap, as obtained from the waste produced in the food processing industries and they are environmentally friendly, as already mentioned. As photoactive materials natural chromophores, like anthocyanines, can be, perceptively, used too [144]. We note too that the liquid crystalline phase was also isolated in DNA and investigated [145] as well as in a DNA-surfactant complex [146].

We have described and discussed also linear and nonlinear properties of DNA based systems, One of the important observed property of this biopolymer is the fast response time, corresponding to the conformational transformations of a photoizomerizable molecule It is a few orders of magnitude faster than in PMMA matrix. This effect is due to a large free volume offered by DNA matrix. Also the specific environment of charged DNA modifies electronic structure of embedded molecule and as consequence its physical properties.

There are also a large number of studies and application of DNA based complexes for their electrical properties. *Par excellence* DNA itself is a negatively charged anionic polyelectrolyte, with sodium ions Na^+ as counter ions. DNA. Application of an external electric field induces an ionic dipole moment [147–149] and changes the conformation of DNA molecule [150–152].

The electric conductivity of DNA, not discussed here, was subject of a continuous research interest and of controversies from the early sixties [153–155] with the first theoretical suggestion by Eley and Spivey [156] that the delocalization of π electrons in nucleobases may lead to an efficient electron transport along the DNA stacks [157].

A very interesting output here is obtaining of conducting, solid membranes as it was shown recently [158], with good ionic conductivities, controllable by an appropriate doping. It is getting by plasticizing DNA with glycerol and introducing doping molecules to the system (for details see Refs. [159–161]) of transparent membranes (in visible) with good ionic conductivities.

These membranes can applied in electrochromic cells for displays and in "smart windows." They are also potentially interesting materials for application in solar energy conversion as well.

The research on practical utilization od DNA is still at its beginning. But more and more researchers and laboratories join people already active in this field and a rapid progress is expected with new discoveries' and new practical applications.

ACKNOWLEDGEMENTS

The authors acknowledge the financial support of Romanian Ministry of Education, Research, Youth and Sports, through the UEFISCDI organism, under Contract Number 279/7.10.2011, Code Project PN-II-ID-PCE-2011-3-0505.

KEYWORDS

- **collagen**
- **deoxyribonucleic acid**
- **DNA**
- **DNA-CTMA complex**
- **linear optical properties**
- **nonlinear optical properties**
- **optical damage threshold**
- **photo-thermal stability**
- **photoluminescence**
- **thin films**

REFERENCES

1. http://www.brighthub.com/environment/green-living/articles/107380.aspx.
2. J. P. Vigneron, J.F. Colomer, N. Vigneron, V. Lousse, Natural layer-by-layer photonic structure in the squamae of Hoplia coerulea (Coleoptera,) Phys. Rev. E, 72 (2005), 061904–061906.
3. D. E. Azofeifa, H. J. Arguedas, W. E. Vargas, Optical properties of chitin and chitosan biopolymers with application to structural color analysis, Opt. Mat.,35(2), 175–183 (2012); DOI: http://dx.doi.org/10.1016/j.optmat.2012.07.024.

4. J. D. Watson and F. H. C. Crick, Molecular structure of nucleic acids. A structure for deoxyribose nucleic acid, Nature, 171, 737–738 (1953).
5. F. H. C. Crick and J. D. Watson, The complementary structure of deoxyribonucleic acid. Proc. Royal Soc. (London), 223, 80–96 (1954).
6. L A. Pray, Ph.D. Discovery of DNA structure and function: Watson and Crick, Nature Education, Nature Education, 1(1) (2008).
7. G. S. Manning, The molecular theory of polyelectrolyte solutions with applications to the electrostatic properties of polynucleotides, Q. Rev. Biophys., 11(2), 179–246 (1978).
8. M. Mandelkern, J. Elias, D. Eden, D. Crothers The dimensions of DNA in solution. J Mol Biol., 152 (1), 153–61 (1981)., doi:10.1016/0022–2836(81)90099–1.
9. H. Clausen-Schaumann, M. Rief, C. Tolksdorf, H. Gaub, Mechanical stability of single DNA molecules. Biophys J, 78 (4): 1997–2007(2000), doi:10.1016/S0006–3495(00)76747–6.
10. http://dwb4.unl.edu/Chem/CHEM869N/CHEM869NLinks/chemistry.about.com/science/chemistry/library/weekly/aa061598a.htm.
11. http://fr.wikipedia.org/wiki/Collag%C3%A8ne.
12. L. Wang, J. Yoshida, N. Ogata, S. Sasaki, and T. Kajiyama, Self-assembled supramolecular films derived from marine deoxyribonucleic acid (DNA)-cationic surfactant-complexes: large-scale preparation and optical and thermal properties, Chem. Mater. 13 (4), pp. 1273–1281, 2001.
13. J. Grote, Biopolymer materials show promise for electronics and photonics applications, SPIE newsroom, DOI 10.1117/2.1200805.1082 (2008).
14. M. H. Uriarte-Montoyaa, J. L. Arias-Moscosoa, M. Plascencia-Jatomea, H. Santacruz-Ortega, O. Rouzaud-Sández, J. L. Cardenas-Lopez, E. Marquez-Rios, J. M. Ezquerra-Brauer, Jumbo squid (Dosidicus gigas) mantle collagen: Extraction, characterization, and potential application in the preparation of chitosan–collagen biofilms, Bioresource Technology, 101 4212–4219 (2010).
15. L. Wang, Q. Liang, Z. Wang, J. Xu, Y. Liu, H. Ma, Preparation and characterization of type I and V collagens from the skin of Amur sturgeon (Acipenser schrenckii), Food Chemistry, 148, 410–414 (2014).
16. V. Trandafir, G. Popescu, M. G. Albu, H. Iovu, M. Georgescu, Bioproduse pe baza de colagen, Editura Ars Docendi, Bucuresti, 2007, ISBN: 978–973–558–291–3.
17. M. G. Albu, Collagen gels and matrices for biomedical applications, Lambert Academic Publishing, Saarbrücken. 2011. ISBN 978–3-8443–3057–1.
18. D. L. Vizard, R. White A and A. T. Ansevin, Comparison of theory to experiment for DNA thermal denaturation, Nature, 275, 250–251(1978).
19. J. SantaLucia, Jr., A unified view of polymer, dumbbell, and oligonucleotide DNA nearest-neighbor thermodynamics. Proc. Natl. Acad. Sci. USA, 95(4), 1460–1465 (1998). doi:10.1073/pnas.95.4.1460. PMID 9465037.
20. R. A. Hayes, B. J. Feenstra, Video-speed electronic paper based on electrowetting. Nature, 425, 6956, 383–385 (2003).
21. M. Nogi and H. Yano, Transparent Nanocomposites Based on Cellulose Produced by Bacteria Offer Potential Innovation in the Electronics Device Industry. Adv. Mat., 20, 1849–1852 (2008).

22. M. C. J. Large, D. T. Croke, W. J. Blau, P. McWilliam, F., Kajzar, EFISH in electrolyte and polyelectrolyte systems, Mol. Cryst. Liq. Cryst,. Sc. & Technol., Section B: Nonl. Opt.,12(3), 225–238 (1995).
23. M. C. J. Large, D. T. Croke, W. J. Blau, P. McWilliam, F. Kajzar, Molecular Length Dependent Type Polarizability, in Nonlinear Optical Properties of Organic Molecules IX,G. Mohlmann Ed., Proc. SPIE, vol. 2852, 36 (1996).
24. K. Ijiro, Y. Okahata, A DNA–lipid complex soluble in organic solvents, J. Chem. Soc., Chem. Commun, 18, 1339–1341 (1992).
25. Y. Hoshino, H. Nakayama, Y. J. Okahata, Preparations of a RNA-lipid complex filmand its physical properties, Nucleic Acids Res. Suppl. No. 1, 61–62 (2001).
26. K. Tanaka, Y. J. Okahata, A DNA−Lipid Complex in Organic Media and Formation of an Aligned Cast Film, J. Am. Chem. Soc., 118(44), 10679–10683 (1996).
27. V. G. Sergeyev, O. A. Pyshkina, A. V.Lezov, A. B. Mel'nikov, E. I. Ryumtsev, A. B. Zezin. V. A. Kabanov, DNA Complexed with Oppositely Charged Amphiphile in Low-Polar Organic Solvents, Langmuir, 15, 4434–4440 (1999).
28. H. Kimura, S. Machida, K. Hone, Y. Okahata, Effect of Lipid Molecules on Twisting Motions of DNA Helix Studied by Fluorescence Polarization Anisotropy, Polymer J., 30, 708–712 (1998).
29. Watanuki, J. Yoshida, S. Kobayashi, H. Ikeda, N. Ogata, Optical and photochromic properties of spiropyran-doped marine-biopolymer DNA-surfactant complex films. Proc.SPIE, 5724, 234–241 (2005).
30. Y. Kawabe, L. Wang, T. Koyama, S. Horinouchi, N. Ogata, Light amplification in dye doped DNA-surfactant complex films, Proc. SPIE, 4106, 369–376 (2000).
31. L. Wang, G. Zhang, S. Horinouchi, J. Yoshida, N. Ogata, Optoelectronic materials derived from salmon deoxyribonucleic acid, Nonl. Opt., 24, 63–68 (2000).
32. Y. Kawabe, L. Wang, S. Horinouchi, N. Ogata, Amplified spontaneous emission from fluorescent dye-doped DNA-surfactant films, Adv. Mater. 12, 1281–1283 (2000).
33. T. Koyama, Y. Kawabe, N. Ogata, Electroluminescence as a probe for electrical and optical properties of deoxyribonucleic acid, Proc. SPIE, 4464, 248–255 (2002).
34. L. Wang, M. Fukushima, J. Yoshida, N. Ogata, A novel photochromic film materials derived from supramolecules, DNA-surfactant complex: Spiropyran-DNA-CTMA complex, Nanotechnology toward theorganic photonics, GooTech Ltd., Chitose-shi, Japan, 379–384 (2002).
35. J. Yoshida, L. Wang, S. Kobayashi, G. Zhang, H. Ikeda, N. Ogata, Optical properties of photochromic-compound-doped marine-biopolymer DNA-surfactant complex films for switching applications, Proc. SPIE, 5351, 260 (2004).
36. L. Wang K. Ishihara, H. Izumi, M. Wada, G. Zhang, T. Ishikawa, A. Watanabe, S. Horinouchi, N. Ogata, Strongly luminescent rare-earth ion-doped DNA-CTMA complex film and fiber materials, Proc. SPIE, 4905, 143–152 (2002).
37. Watanuki, J. Yoshida, S. Kobayashi, H. Ikeda, N. Ogata, Optical and photochromic properties of spiropyran-doped marine-biopolymer DNA-surfactant complex films. Proc.SPIE, 5724, 234–241 (2005).
38. Y. Kawabe, L. Wang, T. Koyama, S. Horinouchi, N. Ogata, Light amplification in dye doped DNA-surfactant complex films, Proc. SPIE, 4106, 369–376 (2000).

39. L. Wang, G. Zhang, S. Horinouchi, J. Yoshida, N. Ogata, Optoelectronic materials derived from salmon deoxyribonucleic acid, Nonl. Opt., 24, 63–68 (2000).
40. T. Koyama, Y. Kawabe, N. Ogata, Electroluminescence as a probe for electrical and optical properties of deoxyribonucleic acid, Proc. SPIE, 4464, 248–255 (2002).
41. J. Yoshida, L. Wang, S. Kobayashi, G. Zhang, H. Ikeda, N. Ogata, Optical properties of photochromic-compound-doped marine-biopolymer DNA-surfactant complex films for switching applications, Proc. SPIE, 5351, 260 (2004).
42. L. Wang, J. Yoshida, N. Ogata, S. Sasaki, T. Kamiyama, Self-Assembled Supramolecular Films Derived from Marine Deoxyribonucleic Acid (DNA)-Cationic Surfactant Complexes: Large-Scale Preparation and Optical and Thermal Properties, Chem. Mat., 13(4), 1273–1281 (2001).
43. E. Bajer, Modyfikacja DNA dla zastosowań w optyce nieliniowej (Modification of DNA for application in nonlinear optics), Master thesis, Cracow University of Technology, Poland, 2010.
44. Y.C. Hung, W. T.-Y. Lin Hsu, Y.-W. Chiu, Y.-S. Wang and L Fruk, Functional DNA biopolymers and nanocomposite for optoelectronic applications, Opt. Mat., 34, 1208–1213 (2012).
45. J. Niziol, M. Sniechowski,E. Hebda, M. Jancia, J. Pielichowski, Properties of DNA complexes with new cationic surfactants, Chem. Chem. Technol., 5(4), 397–402 (2011).
46. G. Pawlik, A. C. Mitus, J. Mysliwiec, A. Miniewicz, J. G. Grote, Photochromic dye semiintercalation into DNA-based polymeric matrix: Computer modeling and experiment, Chem. Phys. Lett. 484, 321–323(2010).
47. G. Pawlik, W. Radosz, A. C. Mitus, J. Myśliwiec, A. Miniewicz, F. Kajzar, I. Rau, Grating inscription in DR1:DNA-CTMA thin films: theory and experiment, Proc. SPIE, 8817, 8170D-1–6 (2013).
48. G. Pawlik, W. Radosz, A. C. Mitus, J. Myśliwiec, A. Miniewicz, F. Kajzar, I. Rau, J. G. Grote, Kinetics of grating inscription in DR1:DNA-CTMA thin film: experiment and semiintercalation approach, Proc. SPIE, 8464, 846404(2012).
49. https://www.google.ro/search?q=dna+groove+binding&tbm=isch&tbo=u&source=univ&sa=X&ei=iCDNUsb1OZLioATe-YKACA&ved=0CEUQsAQ&biw=911&bih=361#facrc=_&imgdii=_&imgrc=OhEsBsXoyD0awM%3A%3BiAY0dgn2AAa0gM%3Bhttp%253A%252F%252Fwww.rsc.org%252Fej%252FCS%252F2008%252Fb801433 g%252Fb801433 g-f11.gif%3Bhttp%253A%252F%252Fpubs.rsc.org%252Fen%252Fcontent%252Farticlehtml%252F2008%252Fcs%252Fb801433g%3B331%3B256)
50. E. E. Havinga and P. Van Pelt, Electrochromism of organic dyes in polymer matrices, in Electrochromism of organic dyes in polymer matrices, B. R. Jennings (Ed.), Plenum Press, New York 1979, pp. 89–97.
51. Rau, R. Czaplicki, B. Derkowska, J. G. Grote, F. Kajzar, O. Krupka and B. Sahraoui, Nonlinear Optical Properties of Functionalized DNA-CTMA Complexes, Nonl. Opt. Quant. Opt., 42, 283–324 (2011).
52. B. Derkowska, M. Wojdyla, W. Bala, K. Jaworowicz, M. Karpierz, J.G. Grote, O. Krupka, F. Kajzar and B. Sahraoui, Influence of different peripheral substituents on the nonlinear optical properties of cobalt phthalocyanine core, J. Appl. Phys., 101, (8), 083112(1–8), (2007)

53. http://en.wikipedia.org/wiki/File:Structure_of_Hoechst_dyes.svg.
54. Bouamaied, L.-A. Fendt, D. Hussinger, M. Wiesner, S. Thöni, N. Amiot, E. Stulz, Nucleosides Nucleotides, Nucleic Acids, 26, 1533 –1538 (2007).
55. Onoda, M. Igarashi, S. Naganawa, K. Sasaki, S. Ariyasu, T. Yamamura, Bull. Chem. Soc. Jpn., 82, 1280–1286 (2009).
56. K. Berlin, R. K. Jain, M. D. Simon, C. Richert, A porphyrin embedded in DNA, J. Org. Chem., 63,1527–1535 (1998).
57. W. M. Campbell, K. W. Jolley, P. Wagner, K. Wagner, P. J. Walsh, K. C. Gordon, L. Schmidt-Mende, M. K. Nazeeruddin, Q. Wang, M. Gratzel, D. L. Officer, Highly efficient porphyrin sensitizers for dye-sensitized solar cells, J. Phys. Chem. C, 111, 11760–11762 (2007).
58. W. M. Campbell, A. K. Burrell, D. L. Officer, K. W. Jolley, Porphyrins as light harvesters in the dye-sensitized TiO2 solar cell, Coord. Chem. Rev., 248, 1363–1379 (2004).
59. L. M. Moreira, F. V. dos Santos, J. P. Lyon, M. Maftoum-Costa, C. Pacheco Soares, N. S. da Silva, Photodynamic therapy: porphyrins and phthalocyanines as photosensitizers, Aust. J. Chem., 61, 741 –754 (2008).
60. W. I. Stephenson, N. Bomholt, A. C. Partridge, V. V. Filichev, Significantly Enhanced DNA Thermal Stability Resulting from Porphyrin H-Aggregate Formation in the Minor Groove of the Duplex, Chem. Bio. Chem., 11, 1833 – 1839 (2010); DOI: 10.1002/cbic.201000326.
61. W. I. Stephenson, A. C. Partridge, V. V. Filichev, Synthesis of b-Pyrrolic-Modified Porphyrins and Their Incorporation into DNA, Chem. Eur. J., 17, 6227 – 6238 (2011); DOI: 10.1002/chem.201003200.
62. Samoc, M. Samoc, J. G. Grote, A. Miniewicz and B. Luther-Davies, Optical properties of deoxyribonucleic acid (DNA) polymer host, Proc. SPIE, 6401, 6401–1-6 (2006).
63. M. Moldoveanu, R. Popescu, C. Pîrvu, J. G. Grote, F. Kajzar, I. Rau I., On the stability and degradation of dna based thin films, Mol. Cryst. Liq. Cryst, 522, 182–190 (2010).
64. R. Popescu, C. Pirvu, M. Moldoveanu, J. G. Grote, F. Kajzar, I. Rau I., Biopolymer Thin Films for Optoelectronics Applications, Mol. Cryst. Liq. Cryst., 522, 229–237 (2010).
65. D. Rezzonico, Kwon Seong-Ji, H. Figi, O-Pil Kwon, M. Jazbinsek, P. Günter, Photochemical stability of nonlinear optical chromophores in polymeric and crystalline materials, J. Chem. Phys. 128, 124713 (2008); http://dx.doi.org/10.1063/1.2890964.
66. Dubois, M. Canva, A. Brun, F. Brun, F. Chaput, J. P. Boilot, Photostability of dye molecules trapped in solid matrices, Appl. Opt. 35, 3193(1996).
67. Galvan-Gonzalez, K. D. Belfield, G. I. Stegeman, M. Canva, K.-P. Chan, K. Park, L. Sukhomlinova, R. J. Twieg, Photostability enhancement of an azobenzene photonic polymer, Appl. Phys. Lett. 77, 2083–2086 (2000); http://dx.doi.org/10.1063/1.1313809.
68. M. E. DeRosa, M. Q. He, J. S. Cites, S. M. Garner, Y. R. Tang, Photostability of High μβ Electro-Optic Chromophores at 1550 nm, J. Phys. Chem. B, 108, 8725–8730 (2004).

69. Rau and F. Kajzar, Multiphoton processes in organic materials and their application, Edition des Archives Contemporaines and Old City Publishing, Paris & Philadelphia, 2012.
70. R. H. Page, M. C. Jurich, B. Reck, A. Sen, R. J. Twieg, J. D. Swalen, G. C. Bjorklund, C. G. Willson, Electrochromic and optical waveguide studies of corona-poled electro-optic polymer films, J. Opt. Soc. Am. B, 7(7), 1239–1250 (1990).
71. E. Hebda, M. Jancia, F. Kajzar, J. Niziol, J., Pielichowski, I., Rau I. and A. Tane, Optical Properties of Thin Films of DNA-CTMA and DNA-CTMA Doped with Nile Blue, Mol. Cryst. Liq. Cryst., 556(1), 309–316 (2012), DOI: 10.1080/15421406.2012.642734.
72. J. G. Grote, N. Ogata, D. E. Diggs and F. K. Hopkins, Deoxyribonucleic acid (DNA) cladding layers for nonlinear-optic-polymer-based electro-optic devices, Proc. SPIE, 4991, 621 (2003).
73. J. Grote, D. Diggs, R. Nelson, J. Zetts, F. Hopkins, N. Ogata, J. Hagen, E. Heckman, P. Yaney, M. Stone and L. Dalton, DNA photonics [deoxyribonucleic acid], Mol. Cryst. Liq. Cryst., 3426 (2005).
74. L. S. Vasilenko, V. P. Chebotaev, Y. V. Troitski, Visual observation of infrared laser emission, Soviet Physics JETP, 21(3), 513 (1965).
75. S. Fine, W. P. Hansen, Optical second harmonic generation in biological systems, Appl. Opt., 10(10), 2350–2353 (1971).
76. M. Kim, J. Eichler, L. B. Da Silva, Frequency doubling of ultrashort laser pulses in biological tissues, Appl. Opt., 38(34), 7145–7150 (1999).
77. Y. C. Guo, P. P. Ho, H. Savage, D. Harris, P. Sacks, S. Schantz, F. Liu, N. Zhadin, R. R. Alfano, Second-harmonic tomography of tissues. Opt. Lett., 22(17), 1323–1325 (1997).
78. P. J. Campagnola, A. C. Millard, M. Terasaki, P. E. Hoppe, C. J. Malone, W. A. Mohler, Three-dimensional high-resolution second-harmonic generation imaging of endogenouss structural proteins in biological tissues. Biophys. J., 82(1), 493–508 (2002).
79. G. Cox, N. Moreno, J. Feijo, Second-harmonic imaging of plant polysaccharides,J Biomed. Opt, 10(2), 024013 (2005).
80. S. W. Chu, I. H. Chen, T. M. Liu, P. C. Chen, C. K. Sun, B. L. Lin, Multimodal nonlinear spectral microscopy based on a femtosecond cr:forsterite laser. Opt Lett, 26(23),1909–11 (2001).
81. G. Mizutani, Y. Sonoda, H. Sano, M. Sakamoto, T. Takahashi, S. Ushioda, Detection of starch granules in a living plant by optical second harmonic microscopy, Journal of Luminescence, 87(9), 824–826 (2000).
82. S. W. Chu, I. H. Chen, T. M. Liu, C. K. Sun, S. P. Lee, B. L. Lin, P. C. Kuo, M. X. Cheng, D. J. Lin, H. L. Liu, Nonlinear bio-photonic crystal effects revealed with multimodal nonlinear microscopy. J. Microsc.-Oxford, 208, 190–200 (2002).
83. R. M. Brown, Jr, A. C. Millard, P. J. Campagnola, Macromolecular structure of cellulose studied by second-harmonic generation imaging microscopy. Opt Lett, 28(22),2207–2209 (2003).
84. O. Nadiarnykh, R. B. Lacomb, P. J. Campagnola, W. A. Mohler, Coherent and incoherent second harmonic generation in fibrillar cellulose matrices. Opt Express, 15(6), 3348–3360 (2007).

85. Y. Marubashi, T. Higashi, S. Hirakawa, S. Tani, T. Erata, M. Takai, J. Kawamata, Second harmonic generation measurements for biomacromolecules: Celluloses, Opt. Rev., 11(6), 385–387 (2004).
86. Knoesen, Second order optical nonlinearity in single and triple helical protein supramolecular assemblies, Nonl. Opt. Quant. Opt.: Concepts in Modern Optics, 38(3–4), 213–225 (2009).
87. H. Lee, M. J. Huttunen, K.-J. Hsu, M. Partanen, G.-Y. Zhuo, M. Kauranen, andS.-W. Chu, Chiral imaging of collagen by second-harmonic generation circular dichroism, Biomed Opt Express. 4(6), 909–916 (2013).; doi: 10.1364/BOE.4.000909.
88. V. Ostroverkhov, O. Ostroverkhova, R.G. Petschek, K.D. Singer, L. Sukhomlinova, R. J. Twieg, S.-X. Wang, and L.C. Chien, Optimization of the Molecular Hyperpolarizability for Second Harmonic Generation in Chiral Media, Chem. Phys. 257, 263–274 (2000).
89. S. Sioncke, T. Verbiest and A. Persoons, Second-order nonlinear optical properties of chiral materials, Mat. Sc. Engin., R 42, 115–155 (2003).
90. M. Iwamoto, F. Liu, O.-Y. Zhong-canc, Polarization-dependence of optical second harmonic generation for chiral cylindrical structure and explanation for nonlinear optical imaging of cholesteric liquid crystals, Chem. Phys. Lett., 511, 455–460 (2011); doi.org/10.1016/j.cplett.2011.06.056
91. Y. R. Shen, The principles of nonlinear optics; New York: Wiley, 2003.
92. F. C. Boman,J. M. Gibbs-Davis, L. M. Heckman, B. R. Stepp, S. T. Nguyen, F. M. Geiger, DNA at Aqueous/Solid Interfaces: Chirality-Based Detection via Second Harmonic Generation Activity, J. Am. Chem. Soc., 844–848 (2009).
93. Zhuang Zheng-Fei, Liu Han-Ping, Guo Zhou-Yi, Zhuo Shuang-Mu, Yu Bi-Ying, Deng Xiao-Yuan, Second-harmonic generation as a DNA malignancy indicator of prostate glandular epithelial cells, Chinese Phys. B, 19(5), 4950 (2010).
94. W. Williamson, Y. Wang, S., S. A. Lee, H. J. Simon, A. Rupprecht, Observation of optical second harmonic generation in wet-spun films of Na-DNA, Spectrosc. Lett., 26(5),849–858 (1993).
95. Y. Satoru, N. Masayoshi, Kishi Ryohei, Nakagawa Nozomi, Nitta Tomoshige and Yamaguchi Kizashi, Theoretical study on nonlinear optical (NLO) properties of modified guanine bases having a NLO group, Nippon Kagakkai Koen Yokoshu, 85(1), 302 (2005).
96. M. Manea, I. Rau, A. Tane, F. Kajzar, L. Sznitko and A. Miniewicz, Poling kinetics and second order NLO properties of DCNP doped PMMA based thin film, Opt. Mat., 36 (1), 69–74: http://dx.doi.org/10.1016/j.optmat.2013.05.012
97. F. KAjzar, A. Jen and K. S. Lee Polymeric Materials and Their Orientation Techniques for Second-Order Nonlinear Optics, in Polymers for Photonics Applications II: Nonlinear Optical, Photorefractive and Two-Photon Absorption Polymers, K. S. Lee and G. Wegner Eds, Springer Verlag,. Advances in Polymer Sc, vol. 161, 1–85 (2003).
98. P.A. Chollet, F. Kajzar, J. Messier, Electric Field Induced Optical Second Harmonic Generation and Polarization Effects in Polydiacetylene Langmuir-Blodgett Multilayers, Thin Sol. Films, 132, 1 (1985).
99. R. Grykien, B. Luszczynska, I. Glowacki, J. Ulanski, F. Kajzar, R. Zgarian, I. Rau, A significant improvement of luminence vs current density efficiency of a BioLED,

Opt. Mat., 36(6), 1027–1033(2014); DOI: 10.1016/j.optmat.2014.01.018Opt. Mat. (2014).
100. D. Porschke in Molecular electro-optic properties of Macromolecules and colloids in solution, S. Krause Ed., Plenum Press, New York 1981.
101. K. Yamaoka, K. Fukudome, Electric field orientation of nucleic acids in aqueous solutions. J. Phys. Chem., 92, 4994–5001 (1988).
102. K. Yamaoka, K. Fukudome, Electric field orientation of nucleic acids in aqueous solutions. 2. Dependence of the intrinsic electric dichroism and electric dipole moments of rod-like DNA on molecular weight and ionic strength, J. Phys. Chem., 92, 6896–6903 (1990); DOI: 10.1021/j100380a066.
103. K. Yamaoka, K. Fukudome, K. Matsuda, Electric field orientation of nucleic acids in aqueous solutions. 3. Non-Kerr-law behavior of high molecular weight DNA at weak fields as revealed by electric birefringence and electric dichroism, J. Phys. Chem., 92, 7131 –7136 (1992); DOI: 10.1021/j100196a055.
104. T. O. Konski, N. C. Stellwagen, structural transition produced by electric fields in aqueous sodium deoxyribonucleate, Biophys. J., 5, 607–613 (1965).
105. Neumann, E. Werner, A. Spratke, K. Kruger, in Colloid and Molecular Electro-Optics, B. R. Jennings and S. P. Stoyov, Eds, Institute of Physics Publ., Bristol 1993.
106. M. Samoc, A. Samoc; J. G. Grote, Complex nonlinear refractive index of DNA, Chem. Phys. Lett., 431(1–3), 132–134 (2006).
107. Kajzar, Third Harmonic Generation, in Characterization Techniques and Tabulations for Organic Nonlinear Optical Materials, M. G. Kuzyk, C. W. Dirk Eds, Marcel Dekker, Inc., New York, 1998, pp. 767–839.
108. Rau, F. Kajzar, J. Luc, B. Sahraoui and G. Boudebs, Comparison of Z-scan and THG derived nonlinear index of refraction in selected organic solvents, J. Opt. Soc. Am. B, 25, No. 10, 1738–47 (2008).
109. U. Gubler, C, Bosshard Optical third-harmonic generation of fused silica in gas atmosphere: Absolute value chi(3), Phys Rev. B, 61, 10702 (2000).
110. D. Morichere, M. Dumont, Y. Levy, G. Gadret, F. Kajzar Nonlinear properties of poled polymer films: SHG and electrooptic measurements, Proc. SPIE, 1560, 214 (1991).
111. Grote, D. Diggs, R. Nelson, J. Zetts, F. Hopkins, N. Ogata, J. Hagen, E. Heckman, P. Yaney, M. Stone and L. Dalton, DNA photonics [deoxyribonucleic acid], Mol. Cryst. Liq. Cryst., 3426 (2005).
112. Kajzar, J. Messier, C. Rosilio, Nonlinear Optical Properties of Thin Films of Polysilane, J. Appl. Phys., 60, 3040–3044 (1986).
113. Rau., J. G. Grote,. F. Kajzar., A. Pawlicka., DNA-novel nanomaterial for applications in photonics and in electronics,. Comptes Rendus Physique, 13, 853–864 (2012).
114. M. Samoc, A. Samoc, J. G. Grote, Complex nonlinear refractive index of DNA, Chem. Phys. Lett., 431(1–3), 132–134 (2006).
115. Mysliwiec, A. Miniewicz, I. Rau, O. Krupka, B. Sahraoui, F. Kajzar, J. Grote, Biopolymer-based material for optical phase conjugation, J. Optoel. Adv. Mat., 10(8), 2146 – 2150 (2008).
116. Rodríguez, G. Vitrant, P. A. Chollet, F. Kajzar, Optical control of an integrated interferometer using a photochromic polymer, Appl. Phys. Lett., 79, 461–3 (2001).

117. Myśliwiec, M. Ziemienczuk and A. Miniewicz, Pulsed laser induced birefringence switching in a biopolymer matrix containing azo-dye molecules, Opt. Mat. 33 1382–1386 (2011).
118. L.Wang, K. Ishihara, H. Izumi, M. Wada, G. Zhang, T. Ishikawa, A. Watanabe, S. Horinouchi, N. Ogata, Strongly luminescent rare-earth ion-doped DNA-CTMA complex film and fiber materials, Proc.. SPIE, vol. 4905, 143–152 (2002).
119. Z Yu, J. Hagen, Y. Zhou, D. Klotzkin, J. Grote, and A. Steckl, Photoluminescence and stimulated emission from deoxyribonucleic acid thin films doped with sulforhodamine, Appl. Opt. 46 (9), pp. 1507–1513, 2006.
120. Z. Yu, Y. Zhou, D. Klotzkin, J. Grote, and A. Steckl, Stimulated emission of sulforhodamine 640 doped DNA distributed feedback (DFB) laser devices, Proc. SPIE 6470, 64700 V, 2007.
121. Massin, S. Parola, C. Andraud, F. Kajzar, I. Rau, Enhanced fluorescence of isophorone derivatives in DNA based materials, Opt. Mater. (2013), accessible online: http://dx.doi.org/10.1016/j.optmat.2013.04.021.
122. T. Koyama, Y. Kawabe, and N. Ogata, Electroluminescence as a probe for electrical and optical properties of deoxyribonucleic acid, Proc. SPIE, 4464, 248–255 (2002).
123. A.-M. Manea, I. Rau, F. Kajzar, A. Meghea, Fluorescence, spectroscopic and NLO properties of green tea extract in deoxyribonucleic acid, Opt. Mat., 36 (1) 140–145 (2013); http://dx.doi.org/10.1016/j.optmat.2013.04.016.
124. A.M. Manea, I. Rau, F. Kajzar and A. Meghea, Preparation, Linear and NLO properties of DNA-CTMA-SBE complexes, Proceed. SPIE, vol. 8901 (2013).
125. Nakamura, T. Ishikawa, D. Nishioka, T. Ushikubo and N. Kobayashi, Color-tunable multilayer organic light emitting diode composed of DNA complex and tris 8-hydroxyquinolinato aluminum, Appl. Phys. Lett., 97, 193301 (2010).
126. Kobayashi, Bioled with dna/conducting polymer complex as active layer, Nonl. Opt. Quant. Opt., 43, 233–251 (2011).
127. R. Grykien, B. Luszczynska, I. Glowacki, J. Ulanski, F. Kajzar, R. Zgarian, I. Rau, A significant improvement of luminence vs current density efficiency of a BioLED, Opt. Mat., 36(6), 1027–1033 (2014); DOI: 10.1016/j.optmat.2014.01.018Opt. Mat. (2014).
128. A. Hagen, W. Li, A. J. Steckl, J. G. Grote, Enhanced emission efficiency in organic light-emitting diodes using deoxyribonucleic acid complex as an electron blocking layer, Appl. Phys. Lett., 88, 171109 (2006).
129. J. G. Grote, E. M. Heckman, J. A. Hagen, P. P. Yaney, G. Diggs, G., Subramanyam, R. L. Nelson, J. S. Zetts, D. Y. Zang, B. Singh, N. S. Sariciftci, F. K. Hopkins, DNA: new class of polymer, in Organic Photonic Materials and Devices VIII, Proc. SPIE, 6117, 61170J–6 (2006).
130. J. Steckl, DNA-a new material for photonics?, Nature Photonics,1, 3–5 (2007).
131. Y. Kawabe, L. Wang, T. Koyama, S. Horinouchi and N. Ogata, Light amplification in dye doped DNA-surfactant complex films, Proc. SPIE, 4106, 369–376 (2000).
132. Y. Kawabe, L. Wang, S. Horinouchi, N. Ogata, Amplified spontaneous emission from fluorescent dye-doped DNA-surfactant films, Adv. Mater. 12, 1281–1283 (2000).
133. J. Myśliwiec, L. Sznitko, A. M. Sobolewska, S. Bartkiewicz and A. Miniewicz, Lasing effect in a hybrid dye-doped biopolymer and photochromic polymer system, Appl. Phys. Lett., 96, 141106–1-3 (2010).

134. J. Mysliwiec, L. Sznitko, A. Miniewicz, F. Kajzar and B. Sahraoui B., Study of the amplified spontaneous emission in a dye-doped biopolymer-based material, J. Phys. D: Appl. Phys., 42(8), 085101 (2009).
135. Leonetti, R. Sapienza, M. Ibisate, C. Conti and C. Lopez, Optical gain in DNA-DCM, for lasing in photonic materials, Opt. Lett., 34(24), 3764–3766 (2009), doi:10.1364/OL.34.003764.
136. Sznitko, J. Myśliwiec, P. Karpiński, K. Palewska, K. Parafiniuk, S. Bartkiewicz, I. Rau, F. Kajzar, A. Miniewicz., Biopolymer based system doped with nonlinear optical dye as a medium for amplified spontaneous emission and lasing, Appl. Phys. Lett., 99(3), 031107_1–3 (2011).
137. Y. Kawabe and K.-I. Sakai, DNA Based Solid-State Dye Lasers, Nonl. Opt. Quant. Opt., 43, 273–282 (2011).
138. J. Mysliwiec, L. Sznitko, S. Bartkiewicz, A. Miniewicz, Z. Essaidi, F. Kajzar and B. Sahraoui, Amplified spontaneous emission in the spiropyran-biopolymer based system, Appl. Phys. Lett. 94, 241106 _1–3 (2009).
139. G. S. He, Q. Zheng, P. N. Prasad, J. G. Grote,F. K. Hopkins. Infrared two-photon-excited visible lasing from a DNA-surfactant-chromophore complex. Opt. Lett., 31, 359–361 (2006).
140. L. Sznitko, A. Szukalski, K. Cyprych, P. Karpiński, A. Miniewicz, J. Myśliwiec, Surface roughness induced random lasing in bio-polymeric dye doped film, Chem. Phys. Lett. (579), 31–34 (2013).
141. L. Sznitko, K. Cyprych, A. Szukalski, A. Miniewicz, I. Rau, F. Kajzar, J. Myśliwiec,Lasing and random lasing based on organic molecules, Proc. SPIE, Vol. 8901, p. 89010Y-1–9 2013.
142. J. Mysliwiec; L. Sznitko, K. Cyprych, A. Szukalski, A. Miniewicz; F. Kajzar, I. Rau, Random lasing in bio-polymeric dye-doped systems, Nanobiosystems: Processing, Characterization, and Applications VI, Proc. SPIE, 8817, 88170A (2013); doi:10.1117/12.2025692.
143. L. Sznitko, K. Cyprych, A. Szukalski, A. Miniewicz, J. Myśliwiec, Coherent-incoherent random lasing based on nano-rubbing induced cavities, Laser Physics Letters, 11(4), 1–5 (2014).
144. Iosub, F. Kajzar, M. Makowska-Janusik, A. Meghea, A. Tane A., Rau, I. Electronic structure and optical properties of some anthocyanins extracted from grapes, Opt. Mat., 34(10),1644–1650 (2012).
145. H. Mojzisova, J. Olesiak, M. Zielinski, K. Matczyszyn, D. Chauvat, J. Zyss J., Polarization-Sensitive Two-Photon Microscopy Study of the Organization of Liquid-Crystalline DNA, Biophys. J. 97, 2348–2357 (2009).
146. Ogata and K. Yamaoka, DNA-lipid hybrid films derived from chiral lipids, Polymer J., 40(3), 186–191 (2008).
147. D. Pörschke, in Molecular electro-optic properties of macromolecules and colloids in solution, S. Krause Ed., Plenum Press, New York 1981.
148. K. Yamaoka, K. Fukudome, Electric field orientation of nucleic acids in aqueous solutions. 1. Dependence of steady-state electric birefringence of rodlike DNA on field strength and the comparison with new theoretical orientation functions, J. Phys. Chem., 92, 4994–5001 (1988), ibidem 94, 6896–6903 (1990).

149. K. Yamaoka, K. Fukudome, K. Matsuda, Electric field orientation of nucleic acids in aqueous solutions. 3. Non-Kerr-law behavior of high molecular weight DNA at weak fields as revealed by electric birefringence and electric dichroism, J. Phys. Chem., 96(17), 7131–7136 (1992).
150. T. O'Konski and N. C. Stellwagen, Structural transition produced by electric fields in aqueous sodium deoxyribonucleate, Biophys. J., 5(4), 607–613 (1965).
151. Pörschke, A conformation change of single stranded polyriboadenyiate induced by an electric field, Nucl. Acids Res., 1, 1601–1618 (1974).
152. Neumann, E. Werner E., A. Spratke and K. Krüger, in Colloid and Molecular Electro-Optics, B. R. Jennings and S. P. Stoyov Eds, Institute of Physics Publ., Bristol 1993.
153. J. Duchesne, J. Depireux, A. Bertinchamps, N. Cornet and J. M. Vanderkaa, Nature, 188, 405–406 (1960).
154. J. C. Genereux and J. K. Barton J. K., Mechanism for DNA charge transport. Chem. Rev., 110, 1642–1662 (2010).
155. V. D. Lakhno, The problem of DNA conductivity, Pisma ETSCHAIA, 5(3), 400 –406 (2008).
156. D. Eley and D. I. Spivey, Semiconductivity of organic substances.9. nucleic acid in dry state, Trans. Faraday Soc., 58(470), 411–417 (1962).
157. D. D. Eley., in Organic Semiconducting Polymers, J. E. Katon, Ed., Marcel Dekker, New York, USA, 1968, p. 259.
158. Pawlicka A., Firmino A., Vieira D., Grote J.G., Kajzar F., Gelatin- and DNA-based ionic conducting membranes for electrochromic devices. Proceed. SPIE 7487, 74870J-1 – 74870J-10 (2009).
159. Pawlicka, F. Sentanin, A. Firmino, J. G. Grote, F. Kajzar and I. Rau, Ionically conducting DNA-based membranes for eletrochromic devices, Synt. Met., 161, 2329–2334 (2011)
160. Firmino, J. G. Grote, F. Kajzar, J.-C. M'Peko, A. Pawlicka, DNA-based ionic conducting membranes, J. Appl. Phys., 10, 033704–5 (2011).
161. Pawlicka, J. G. Grote, F. Kajzar, I. Rau, Agar and DNA bio-membranes for electrochromic devices applications. Nonl. Opt., Quant. Opt., 45(1–2), 113–129 (2013).

CHAPTER 8

MOLECULAR MODELING OF THE TERT-BUTYL HYDROPEROXIDE NMR ^{1}H and ^{13}C SPECTRA

N. A. TUROVSKIJ,[1] YU. V. BERESTNEVA,[1] E. V. RAKSHA,[1] N. I. VATIN,[2] and G. E. ZAIKOV[3]

[1]*Donetsk National University, 24 Universitetskaya Street, 83 055 Donetsk, Ukraine;*
E-mail: N.Turovskij@donnu.edu.ua

[2]*Saint-Petersburg State Polytechnical University, 29 Polytechnicheskaya street, Saint-Petersburg 195251, Russia;*
E-mail: vatin@mail.ru

[3]*N.M. Emanuel Institute of Biochemical Physics of RAS, 4 Kosygin Str., Moscow, Russia,*
E-mail: chembio@sky.chph.ras.ru

CONTENTS

ABSTRACT

NMR ^1H and ^{13}C spectra of *tert*-butyl hydroperoxide in acetonitrile-d3, chloroform-d and dimethyl sulfoxide-d6 have been investigated by the NMR method. The calculation of magnetic shielding tensors and chemical shifts for ^1H and ^{13}C nuclei of the *tert*-butyl hydroperoxide molecule in the approximation of an isolated particle and considering the influence of the solvent in the framework of the continuum polarization model was carried out. Comparative analysis of experimental and computer NMR spectroscopy results revealed that the GIAO method with MP2/6–31G (d,p) level of theory and the PCM approach can be used to estimate the parameters of NMR ^1H and ^{13}C spectra of *tert*-butyl hydroperoxide.

8.1 INTRODUCTION

Quantum chemical methods are widely used for investigation the structural features of the organic peroxides and their associates with various classes of compounds, and to study the reactivity of hydroperoxides [1–4]. Computational chemistry is effective tool to get more structural information about peroxide bond activation by enzymes [5], transition metals compounds [6], amines and sulfide [3, 7], quaternary ammonium salts [1, 8]. The existing semiempirical, ab initio and DFT – methods can reproduce the peroxides molecular geometry with sufficient accuracy [9]. Molecular modeling of the peroxide bond homolytic decomposition and hydroperoxides association processes is an additional information source of the structural effects that accompany these reactions. One of the criteria for the quantum-chemical method selection to study the hydroperoxides reactivity can be NMR ^1H and ^{13}C spectra parameters reproduction with sufficient accuracy. It should be noted that the parameters of the NMR spectra are very sensitive to slight changes in spatial and electronic structure of the molecule. The joint use of experimental and computational methods can be an information source of the structural features of the molecule caused by intra and intermolecular interactions [10]. Calculations within the density functional theory (DFT) and perturbation theory (MP2) in the GIAO (Gauge Including Atomic Orbital) approximation for NMR spectra modeling of organic compounds are most often used, since they provided a good relationship between the computational cost and accuracy [10, 11].

The aim of this work is a substantiation of the method and basis for quantum-chemical calculation of the *tert*-butyl hydroperoxide NMR ^1H and ^{13}C spectra.

8.2 EXPERIMENTAL

Tert-butyl hydroperoxide ($(CH_3)_2$C-O-OH) was purified according to Ref. [12], its purity (99%) was controlled by iodometry method. Experimental NMR ^1H and ^{13}C spectra of the hydroperoxide solutions were obtained by using the Bruker Avance II 400 spectrometer (NMR ^1H – 400 MHz, NMR ^{13}C – 100 MHz) at 298 K. Solvents – acetonitrile-d_3 (CD_3CN), chloroform-d ($CDCl_3$), and dimethyl sulfoxide-d_6 (DMSO-d_6) were Sigma-Aldrich reagents and used without additional purification but were stored above molecular sieves before using. Tetramethylsilane (TMS) was used as internal standard. The hydroperoxide concentration in solution was 0.03 mol·dm^{-3}.

Molecular geometry and electronic structure parameters, thermodynamic characteristics of the *tert*-butyl hydroperoxide molecule were calculated using the GAUSSIAN03 [13] software package. The hydroperoxide molecular geometry optimization and frequency harmonic vibrations calculation were carried out on the first step of investigations. The nature of the stationary points obtained was verified by calculating the vibrational frequencies at the same theory level.

To choose the optimal method for the $(CH_3)_2$C-O-OH geometry calculation the hydroperoxide structure parameters were estimated by Hartree-Fock (HF), DFT (B3LYP and BH and HLYP) methods as well as on the MP2 theory level. The following basis sets were used in calculations: 6–31G, 6–31G(d), and 6–31G(d,p). The solvent effect was considered in the PCM approximation [14, 15]. The magnetic shielding tensors (χ, ppm) for ^1H and ^{13}C nuclei of the hydroperoxide molecule were calculated with the MP2/6–31G(d,p) and MP2/6–31G(d,p)/PCM optimized geometries by standard GIAO (Gauge-Independent Atomic Orbital) approach [16]. Optimization of the molecular geometry in the same approximation and calculation of magnetic shielding tensors in the framework of the GIAO approach were also performed for the TMS molecule. Obtained χ values for TMS (Table 8.1) were used for the hydroperoxide ^1H and ^{13}C nuclei chemical shifts calculations.

TABLE 8.1 The Magnetic Shielding Tensors for ^{1}H and ^{13}C Nuclei of the Tetramethylsilane Molecule Calculated Within GIAO/MP2/6–31G(d,p) Approach

Solvent	χ_H, ppm	χ_C, ppm
-	31.958	207.541
CH_3CN	31.952	207.925
$CHCl_3$	31.945	207.815
DMSO	31.940	207.858

Note: (χ vales are averaged over corresponding nuclei)

Inspecting the overall agreement between experimental and theoretical spectra RMS errors (σ) were used to consider the quality of the ^{1}H and ^{13}C nuclei chemical shifts calculations.

8.3 RESULTS AND DISCUSSIONS

8.3.1 EXPERIMENTAL NMR ^{1}H AND ^{13}C SPECTRA OF TERT-BUTYL HYDROPEROXIDE

Experimental NMR ^{1}H and ^{13}C studies of *tert*-butyl hydroperoxide were carried out in the following solvents: acetonitrile-d_3, chloroform-d and dimethyl sulfoxide-d_6 at 298 K. The concentration of the hydroperoxide in all samples was 0.03 mol·dm^{-3}. Tetramethylsilane was used as the internal standard. Parameters of the experimental NMR ^{1}H and ^{13}C spectra of the $(CH_3)_3C$-O-OH are listed in Table 8.2.

Table 8.2 reveals the following features. The shift of the -CH_3 and-COOH groups signals in the NMR ^{1}H spectrum of the hydroperoxide with

TABLE 8.2 Experimental Parameters of the *Tert*-Butyl Hydroperoxide NMR ^{1}H and ^{13}C Spectra

Solvent	ε [17]	δ (^{1}H), ppm.		δ (^{13}C), ppm	
		-CH_3	-COOH	-CH_3	-COOH
$CDCl_3$	4.8	1.27	7.24	25.71	80.87
CD_3CN	37.5	1.18	8.80	26.27	80.53
DMSO-d_6	46.7	1.12	10.73	26.01	76.56

the solvent polarity increasing is observed. The signal of the hydroperoxide group proton appears at 7.24 ppm in chloroform-d, and in the more polar dimethyl sulfoxide-d_6 it was found at 10.73 ppm. When passing from $CDCl_3$ to DMSO-d_6 the signal of the -CH_3 groups protons are shifted to the strong field. Another effect is observed in the case of the NMR ^{13}C spectrum, the signal of a tertiary carbon atom is shifted to a strong field on 4.34 ppm in DMSO-d_6 as compared to $CDCl_3$.

8.3.1.1 Equilibrium Configuration of the Tert-Butyl Hydroperoxide in the HF, MP2, and DFT Methods Approximation

The first step of the $(CH_3)_3C$-O-OH NMR spectra calculation was the estimation of the hydroperoxide molecular geometry parameters and electronic structure to choose method and basis set for the further investigations. The $(CH_3)_3C$-O-OH molecular geometry optimization was carried out by HF, DFT (B3LYP и BH and HLYP) and MP2 methods, the following basis sets were used in calculations: 6–31G, 6–31G(d), and 6–31G(d,p). Peroxide bond O-O is a reaction center in this type of chemical initiators thus the main attention was focused on the geometry of -CO-OH fragment. Table 8.3 illustrates the influence of theory level as well as basis set on the -CO-OH fragment geometry parameters. The calculation results were compared with experimental values [18]. The best agreement between calculated and experimental parameters can be seen in the case of MP2/6–31G(d,p) method. Thus it was used in the further calculations.

Calculations with solvent effect accounting were also carried out for the $(CH_3)_3C$-O-OH molecule by MP2/6–31G(d,p) method within PCM approximation. Obtained geometry parameters are listed in Table 8.4. The solvent does not affect the O-O bond length. Solvent effect is noticeable for O-H and C-O bonds, O-O-H and C-O-O bond angles, and C-O-O-H torsion angle. Significant changes of O-H and C-O bonds, O-O-H and C-O-O bond angles are observed in the case of CH_3CN and DMSO, while CHC_{13} has less influence.

Thus MP2/6–31G(d,p) and PCM/MP2/6–31G(d,p) equilibrium configuration of the *tert*-butyl hydroperoxide will be used for the further calculations.

TABLE 8.3 Molecular Geometry Parameters of -COOH Moiety of the *Tert*-Butyl Hydroperoxide Molecule

Method/basis set	Bond length, Å			Angle, °		
	C–O	O–O	O–H	COO	OOH	COOH
HF/6–31G	1.464	1.459	0.954	108.8	101.2	166.9
HF/6–31G(d)	1.426	1.393	0.948	110.3	101.9	117.6
HF/6–31G(d,p)	1.426	1.394	0.945	110.2	102.3	115.9
B3LYP/6–31G	1.489	1.529	0.984	107.7	98.4	134.4
B3LYP/6–31G(d)	1.450	1.457	0.973	109.2	99.7	110.9
B3LYP/6–31G(d,p)	1.450	1.458	0.971	109.1	99.8	110.7
BHandHLYP/6–31G	1.468	1.487	0.968	107.8	99.8	144.7
BHandHLYP/6–31G(d)	1.431	1.419	0.960	109.4	100.9	113.4
BHandHLYP/6–31G(d,p)	1.432	1.419	0.957	109.3	101.0	113.4
MP2/6–31G	1.501	1.575	0.988	106.3	95.5	152.5
MP2/6–31G(d)	1.448	1.472	0.977	108.0	98.2	116.8
MP2/6–31G(d,p)	1.446	1.473	0.970	107.7	98.2	114.7
Experiment [18]	1.443	1.473	0.990	109.6	100.0	114.0

TABLE 8.4 Molecular Geometry Parameters of the *Tert*-Butyl Hydroperoxide Molecule Calculated Within MP2/6–31G(d,p)/PCM Approximation

Parameter	Solvent			
	-	CH_3CN	$CHCl_3$	DMSO
r_{O-O},Å	1.473	1.472	1.472	1.472
r_{O-H},Å	0.970	0.986	0.980	0.986
r_{C-O},Å	1.446	1.450	1.449	1.450
O-O-H,°	98.2	99.4	98.9	99.3
C-O-O, °	107.7	108.2	107.9	108.1
C-O-O-H, °	114.7	113.7	116.3	114.5

8.3.1.2 GIAO NMR ^{1}H and ^{13}C Spectra of the Tert-Butyl Hydroperoxide

With regard to the results obtained from this foregoing study, the GIAO calculations for the *tert*-butyl hydroperoxide molecule were performed at MP2/6–31G(d,p) level of theory in the approximation of the isolated molecule as well as with solvent effect accounting within PCM approach. Fig. 1 illustrates the structural model of the *tert*-butyl hydroperoxide molecule (equilibrium configuration obtained at MP2/6–31G(d,p) level

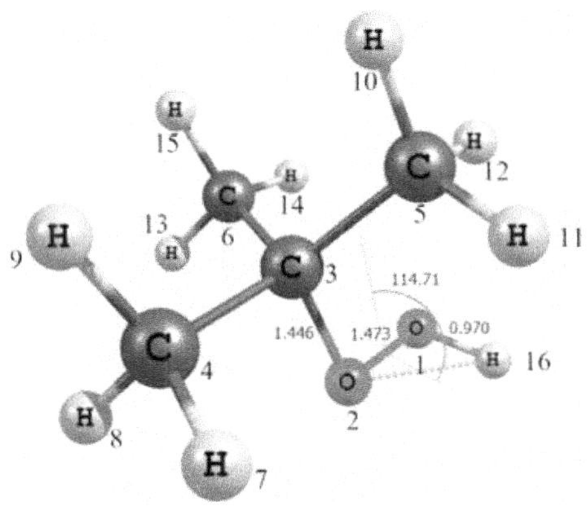

FIGURE 8.1 *Tert*-butyl hydroperoxide structural model with corresponding atom numbering (MP2/6–31G(d,p)).

of theory) with atom numbering used for GIAO NMR ^{1}H and ^{13}C spectra parameters representation.

To estimate the hydroperoxide NMR ^{1}H and ^{13}C spectra parameters the magnetic shielding constants (χ, ppm) for corresponding nuclei were calculated by GIAO method. Obtained χ values for magnetically equivalent nuclei were averaged and listed in Table 8.5. On the base of the obtained χ values the chemical shift values (δ, ppm) of the ^{1}H and ^{13}C nuclei in the hydroperoxide molecule were evaluated. TMS was used as standard, for which the molecular geometry optimization and χ calculation were performed using the same level of theory and basis set. Values of the ^{1}H and ^{13}C chemical shifts were found as the difference of the magnetic shielding tensors of the corresponding TMS and hydroperoxide nuclei.

For the MP2/6–31G(d,p) calculated ^{1}H and ^{13}C chemical shifts there was no full conformity with the experimental data. Experimental δ values for -CO-OH moiety proton in all solvents are higher than the $\delta = 6.805$ ppm calculated in the isolated molecule approximation. The labile protons signals are usually shifted to lower field in DMSO-d_6 solution. Difference between δ values of -CO-OH moiety proton for CH_3CN and $CHCl_3$ solutions may also be due to the influence of the solvent. Formation of hydroperoxides self-associates are possible in the low-polarity solvents [19] that may affect the magnitude of the proton chemical shift of the -CO-OH fragment.

Figure 8.2 presents one of many possible configurations of the *tert*-butyl hydroperoxide molecule dimer. Molecular geometry optimization of this homo-associate was performed at MP2/6–31G(d,p) theory level without solvent effect accounting. This associate is stabilized by the formation of two intermolecular hydrogen bonds: O…HO distance is 1.834 Å, bond angle O…H-O – 161.51°, O-H bond length increases up to 0.982 Å. However, the configuration of the hydroperoxide fragment is not largely changed. For this associate χ values were calculated of by GIAO method (MP2/6–31G(d,p) theory level, isolated molecule approximation) and the chemical shift values of the ^{1}H and ^{13}C nuclei were evaluated. The δ value of 10.541 ppm has been obtained for the -CO-OH moiety proton. This is significantly higher than the experimental values observed in chloroform-d and acetonitrile-d_3 solutions.

TABLE 8.5 GIAO-Magnetic Shielding Tensors for ^{1}H and ^{13}C Nuclei of the *Tert*-Butyl Hydroperoxide, Calculated At MP2/6–31G(d,p) Theory Level

Solvent	NMR ^{1}H				NMR ^{13}C			
	-COOH		$-CH_3$		-COOH		$-CH_3$	
	χ, ppm	δ, ppm	χ, ppm	δ, ppm	χ, ppm	δ, ppm	χ, ppm	δ, ppm
-	25.15	6.805	30.73	1.226	128.16	79.379	180.38	27.163
CH_3CN	23.18	8.771	30.70	1.250	127.61	80.311	180.47	27.455
$CHCl_3$	23.82	8.136	30.71	1.243	127.84	79.977	180.42	27.394
DMSO	23.16	8.778	30.70	1.238	127.64	80.219	180.47	27.391

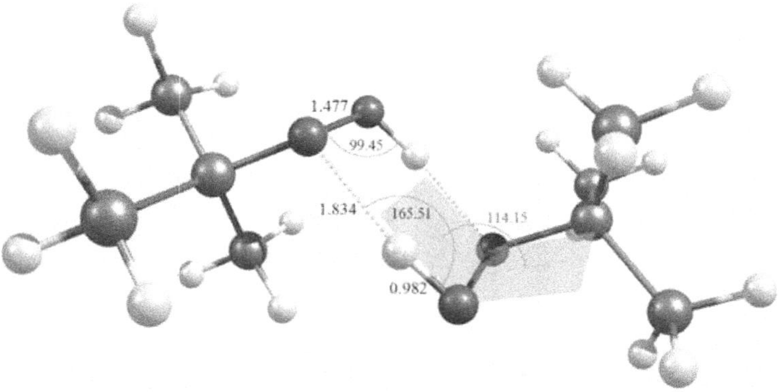

FIGURE 8.2 Structural model of the *tert*-butyl hydroperoxide dimer (MP2/6–31G(d,p)).

Study the hydroperoxide concentration effect on the signals position in the NMR ^1H spectrum in CD_3CN and $CDCl_3$ solutions has been carried out. The hydroperoxide concentration was ranged within $(2.1–500.0)\times10^{-3}$ $-(9.0–20.0)\times10^{-3}$ mol·dm^{-3} in CD_3CN and $CDCl_3$, respectively. Changing in the hydroperoxide concentration in these ranges does not lead to a change in the signal position in the spectrum. Hence, the chemical shift of the hydroperoxide group proton is independent of the hydroperoxide concentration in the system in experimental conditions. Thus, calculation and NMR ^1H spectroscopy results showed that the hydroperoxide dimers do not formed in the system. And observed chemical shift values for the -CO-OH moiety proton are due to the solvent effect. In order to account for the solvent effect in the calculation of magnetic shielding tensors the PCM approach was used. Magnetic shielding tensors for ^1H and ^{13}C nuclei of the *tert*-butyl hydroperoxide, calculated by GIAO method at MP2/6–31G(d,p) theory level with PCM solvent effect (Tables 8.5 and 8.6) have been used for the chemical shift values of the ^1H and ^{13}C nuclei estimation.

Table 8.6 illustrates the chemical shift values (δ, ppm) for ^1H and ^{13}C nuclei of the *tert*-butyl hydroperoxide, calculated at MP2/6–31G(d,p) level of theory with solvent effect accounting.

Concerning the spectral patterns of a $-CH_3$ group protons, inspection of Tables 8.5 and 8.6 reveals the following features: the pattern of NMR ^1H spectra of $(CH_3)_3C$-O-OH is rather correctly reproduced at selected

TABLE 8.6 GIAO Chemical Shifts of the ^1H and ^{13}C Nuclei of the *Tert*-Butyl Hydroperoxide (MP2/6–31G(d,p)/PCM)

Atom	CH3CN			CHCl3			DMSO		
	1	**2**	**Δ**	**1**	**2**	**Δ**	**1**	**2**	**Δ**
				***δ*, ppm NMR ^{13}C**					
C_3	80.31	80.53	0.22	79.98	80.87	0.89	80.22	76.56	3.66
C_4	27.94	26.27	1.67	27.90	25.71	2.19	27.88	26.01	1.88
C_5	27.11	26.27	0.84	27.21	25.71	1.50	27.07	26.01	1.06
C_6	27.30	26.27	1.03	27.00	25.71	1.29	27.22	26.01	1.21
σ		1.15			2.38			4.88	
				***δ*, ppm NMR ^1H**					
H_7	1.06	1.18	0.12	1.04	1.27	0.23	1.05	1.12	0.07
H_8	1.08	1.18	0.10	1.10	1.27	0.17	1.07	1.12	0.05
H_9	1.17	1.18	0.01	1.10	1.27	0.17	1.14	1.12	0.02
H_{10}	1.00	1.18	0.18	0.96	1.27	0.31	0.98	1.12	0.14
H_{11}	1.09	1.18	0.09	1.06	1.27	0.21	1.07	1.12	0.05
H_{12}	1.93	1.18	0.75	1.93	1.27	0.66	1.92	1.12	0.80
H_{13}	1.07	1.18	0.11	1.10	1.27	0.17	1.05	1.12	0.07
H_{14}	1.95	1.18	0.77	1.90	1.27	0.63	1.93	1.12	0.81
H_{15}	0.93	1.18	0.25	0.89	1.27	0.38	0.91	1.12	0.21
H_{16}	8.78	8.80	0.02	8.13	7.24	0.89	8.77	10.73	1.96
σ		0.13			0.20			0.52	

Notes: 1 – calculated, 2 – experimental; Δ – difference between the experimental and calculated hydroperoxide chemical shifts; σ – RMS errors.

computational level for all solvents; the best agreement between the experimental and calculated ^1H and ^{13}C chemical shifts of the -CO-OH moiety is observed for acetonitrile-d solution; for all cases solvent effect accounting leads to a better result compared to the isolated molecule approximation; calculated and experimental values of δ for -CO-OH group proton decrease symbiotically with the solvent polarity increasing. But in DMSO-d_6 solution a significant shift to lower field region is observed for -CO-OH group proton as compared with other solvents. Calculated δ value for this proton in DMSO is very close to the experimentally observed one in acetonitrile-d_3 solution. This shift can be explained by the formation of hydroperoxide-DMSO-d_6 hetero-associates in experimental conditions. One should note that similar values of $\delta = 10.77 \div 10.33$ ppm has a -CO-OH group proton of *tert*-butyl hydroperoxide complex-bonded with tetra-alkyl-ammonium bromides [20].

8.4 CONCLUSIONS

The influence of the solvent on the NMR spectra parameters the of *tert*-butyl hydroperoxide was investigated. It is shown that with increasing polarity of the solvent signal of the hydroperoxide group proton shifts toward weak fields. On the basis of the complex data analysis of the spectroscopic studies and molecular modeling shows that the GIAO method with the MP2/6–31G (d,p) level of theory and the PCM approximation can be used to estimate NMR ^1H and ^{13}C spectra parameters of *tert*-butyl hydroperoxide.

KEYWORDS

- **chemical shift**
- **GIAO**
- **magnetic shielding constant**
- **molecular modeling**
- **NMR spectroscopy**
- **tert-butyl hydroperoxide**

REFERENCES

1. Turovskij N. A., Raksha E. V., Gevus O. I., Opeida I. A., Zaikov G. E. Activation of 1-Hydroxycyclohexyl Hydroperoxide Decomposition in the Presence of Alk4NBr. Oxidation Communications. (2009). Vol. 32, No 1. 69–77.
2. Turovsky M. A., Raksha O. V., Opeida I. O., Turovska O. M., Zaikov G. E. Molecular modeling of aralkyl hydroperoxides homolysis. Oxidation Communications. (2007). Vol. 30, No 3. 504–512.
3. Bach R. D., Su M. D., Schlegel H. B. Oxidation of Amines and Sulfides with Hydrogen Peroxide and Alkyl Hydrogen Peroxide. The Nature of the Oxygen-Transfer Step. J. Am. Chem. Soc. (1994). Vol. 116. P.5379–5391.
4. Araújo J. Q., Carneiro J. W. M., Araujo M. T., Leite F. H. A., Taranto A. G. Interaction between artemisinin and heme. A Density Functional Theory study of structures and interaction energies. Bioorg. and Med. Chem. (2008). Vol. 16, Iss. 19. 5021–5029.
5. Siegbahn P. E. M. Modeling Aspects of Mechanisms for Reactions Catalyzed by Metalloenzymes. J. Comput. Chem. (2001). Vol. 22. 1634–1645.
6. Ryan P, Konstantinov I., Snurr R. Q., Broadbelt L. J. DFT investigation of hydroperoxide decomposition over copper and cobalt sites within metal-organic frameworks. Journal of Catalysis. (2012). Vol. 286. 95–102.
7. Litvinenko S. L., Lobachev V. L., Dyatlenko L. M., Turovskii N. A. Quantum-chemical investigation of the mechanisms of oxidation of dimethyl sulfide by hydrogen peroxide and peroxoborates. Theoretical and Experimental Chemistry. (2011). Iss. 1. 2–8.
8. Turovskij N. A., Raksha E. V., Berestneva Yu. V., Pasternak E. N., Zubritskij M. Yu., Opeida I. A., Zaikov G. E. (2013) Supramolecular decomposition of the aralkyl hydroperoxides in the presence of Et_4NBr. In: Pethrick R. A., Pearce E. M., Zaikov G. E. (ed), Polymer Products and Chemical Processes: Techniques, Analysis and Applications, Apple Academic Press, Inc., Toronto, New Jersey, p. 322.
9. Antonovskij V. L., Khursan S. L. Physical Chemistry of organic Peroxides. M.: PTC "AKADEMKNIGA," (2003). 391 p.
10. Belaykov P. A., Ananikov V. P. Modeling of NMR spectra and signal assignment using real-time DFT/GIAO calculations. Russian Chemical Bulletin. (2011). Vol. 60, Iss. 5. 783–789.
11. Vaara J. Theory and computation of nuclear magnetic resonance parameters. Phys. Chem. Chem. Phys. (2007). Vol. 9. 5399–5418.
12. H. Hock, S. Lang. Autoxydation von Kohlenwasserstoffen, IX. Mitteil.: Über Peroxyde von Benzol-Derivaten (Autoxidation of hydrocarbons IX. Msgs. About peroxides of benzene derivatives). Chem. Ber. (1944). Vol. 77 257–264.
13. Gaussian 03, Revision B.01, M. J. Frisch, G. W. Trucks, H. B. Schlegel, G. E. Scuseria, M. A. Robb, J. R. Cheeseman, J. A. Montgomery, Jr., T. Vreven, K. N. Kudin, J. C. Burant, J. M. Millam, S. S. Iyengar, J. Tomasi, V. Barone, B. Mennucci, M. Cossi, G. Scalmani, N. Rega, G. A. Petersson, H. Nakatsuji, M. Hada, M. Ehara, K. Toyota, R. Fukuda, J. Hasegawa, M. Ishida, T. Nakajima, Y. Honda, O. Kitao, H. Nakai, M. Klene, X. Li, J. E. Knox, H. P. Hratchian, J. B. Cross, C. Adamo, J. Jaramillo, R. Gomperts, R. E. Stratmann, O. Yazyev, A. J. Austin, R. Cammi, C. Pomelli, J. W. Ochterski, P. Y. Ayala, K. Morokuma, G. A. Voth, P. Salvador, J. J. Dannenberg,

V. G. Zakrzewski, S. Dapprich, A. D. Daniels, M. C. Strain, O. Farkas, D. K. Malick, A. D. Rabuck, K. Raghavachari, J. B. Foresman, J. V. Ortiz, Q. Cui, A. G. Baboul, S. Clifford, J. Cioslowski, B. B. Stefanov, G. Liu, A. Liashenko, P. Piskorz, I. Komaromi, R. L. Martin, D. J. Fox, T. Keith, M. A. Al-Laham, C. Y. Peng, A. Nanayakkara, M. Challacombe, P. M.W. Gill, B. Johnson, W. Chen, M. W. Wong, C. Gonzalez, J. A. Pople, Gaussian, Inc., Pittsburgh PA, 2003.

14. Mennucci B., Tomasi J. Continuum solvation models: A new approach to the problem of solute's charge distribution and cavity boundaries. J. Chem. Phys. (1997). Vol. 106. 5151–5158.
15. Cossi M., Scalmani G., Rega N., Barone V. New developments in the polarizable continuum model for quantum mechanical and classical calculations on molecules in solution. J. Chem. Phys. (2002). Vol. 117. 43–54.
16. Wolinski K., Hinton J. F., Pulay P. Efficient implementation of the gauge-independent atomic orbital method for NMR chemical shift calculations. *J. Am. Chem. Soc.* (1990). Vol. 112(23). 8251–8260.
17. http://www.isotope.com/uploads/File/NMR_Solvent_Data_Chart.pdf
18. Kosnikov A. Yu., Antonovskii V. L., Lindeman S. V., Antipin M. Yu., Struchkov Yu. T., Turovskii N. A., Zyat'kov I. P. X-ray crystallographic and quantum-chemical investigation of tert-butyl hydroperoxide. Theoretical and Experimental Chemistry. (1989). Vol. 25. Iss. 1. 73–77.
19. Remizov A. B., Kamalova D. I., Skochilov R. A., Batyrshin N. N., Kharlampidi Kh. E. FT-IR study of self-association of some hydropoxides. J. Mol. Structure. (2004). Vol.700. 73–79.
20. Turovskij N. A., Berestneva Yu. V., Raksha E. V., Pasternak E. N., Zubritskij M. Yu., Opeida I. A., Zaikov G. E. ^{1}H NMR study of the *tert*-butyl hydroperoxide interaction with tetraalkyl ammonium bromides. Polymers Research Journal. (2014). Vol.8, № 2. 85–92.

CHAPTER 9

CLEARING AND COOLING OF SMOKE FUMES IN PRODUCTION OF POTTERY

R. R. USMANOVA[1] and G. E. ZAIKOV[2]

[1]*Ufa State technical university of aviation, 12 Karl Marks str., Ufa 450000, Bashkortostan, Russia; E-mail: Usmanovarr@mail.ru*

[2]*N.M.Emanuel Institute of Biochemical Physics, Russian Academy of Sciences, 4 Kosygin str., Moscow 119334, Russia; E-mail: chembio@sky.chph.ras.ru*

CONTENTS

ABSTRACT

Air pollution source by manufacture of ceramic materials are emissions of a smoke from refire kilns. Designs on modernization of system of an aspiration of smoke fumes of refire kilns in manufacture ceramic and refractories are devised. Experimental researches of efficiency of clearing of gas emissions are executed. Modelling of process of a current of a gas-liquid stream is implemented in the program of computing hydrodynamics Ansys CFX. The ecological result of implementation of system consists highly clearings of a waste-heat and betterment of ecological circumstances in a zone of the factories.

9.1 INTRODUCTION

Ceramics are the foundation of many microelectronic circuits, acting as the substrate to deposit conductive, resistive, and dielectric films to form interconnections and passive components. They are formed by the bonding of a metal and a nonmetal and may exist as oxides, nitrides, carbides, or silicides. Ceramics are ideal as substrates for thick-film and thin-film circuits because they have a high electrical resistivity, are very stable chemically and thermally, and have a high melting point.

Refractory materials are the pottery work, capable to stand temperature from above 1500°C. Refractory products of a various size and the form apply in many industries at exhaustion of a steel, pig-iron, cement, to exhaust, glasses, ceramics, aluminum, copper, on petrochemical manufactures, in furnaces for incineration of rubbish, on power stations, in systems of household heating, including boiler-houses. These products are necessary at high-temperature processes and are capable to resist to any kinds of voltage (mechanical, thermal, chemical), for example, to erosive deterioration, a creep strain.

The basic source of air pollution by manufacture ceramic and refractories are emissions of a smoke from furnaces in the course of roasting. Contaminants are formed because of the maintenance of impurity in raw materials. Their composition can vary depending on raw materials source, and also depend on type of used fuel. The problem is created by emissions of fluorides (containing in mineral ore), and also sulfur oxides (containing in minerals and sulfates). If glaze it is put in the course of roasting is used. Glaze is possible to a vaporous state and is put on a surface of a finished product for formation of a glossy surface.

The bulk of steams of glaze is taken out in an aerosphere. If in the capacity of fuel for furnaces fuel oil or coal level of emissions of a dust and sulfur oxides raises is used.

9.2 SUBSTRATE MANUFACTURING

Ceramics make of various raw materials, burn in furnaces of different types, finished articles have the various form, sizes and color. The general process of manufacturing of ceramics is equal to all its aspects though by manufacture of a facing and low-ground tile, ware and ornamental products (economic-household ceramics), engineering ceramics roasting often spend to some stages. Roasting of refractory materials conduct at temperatures 2050–2850°C.

The stand-up temperature depends on composition of a product and can attain the beginning of a temperature interval of deformation. At factory the gas-cleaning installation, the cyclone separator is installed. Separation efficiency of a waste-heat under the theoretical data should make 94%. However, under the fact sheet of check separation efficiency does not exceed 70%.

The smoke fumes which are selected from the furnace at a high temperature, contain a lot of dust, resinous substances, chloride of metals and are unsuitable for swapping by gas compressors as presence of impurity in it and an at a high temperature lead rapid corrosion and to an abrasive wear of the expensive equipment – gas compressors and to formation of the adjournment consisting of resinous and other substances.

TABLE 9.1 Melting Points of Selected Ceramics

Material	Melting Point(°C)
SiC	2700
BN	2732
AlN	2232
BeO	2570
Al_2O_3	2000

For raise of efficiency of clearing of gas refire kiln redesign has been made. It has allowed to increase efficiency of process of clearing of gas emissions by 15–20%.

9.3 LABORATORY FACILITY AND TECHNIQUE OF CONDUCTING OF EXPERIMENT

Dynamic gas washer, according to Fig. 9.1, contains the vertical cylindrical case with the bunker gathering slime, branch pipes of input and an output gas streams. Inside of the case it is installed conic vortex generator, containing.

Dynamic gas washer works as follows. The Gas stream containing mechanical or gaseous impurity, acts on a tangential branch pipe in the ring space formed by the case and rotor. The liquid acts in the device by means of an axial branch pipe. at dispersion liquids the zone of contact of phases increases and, hence, the effective utilization of working volume of the device takes place more. The Invention is directed on increase of efficiency of clearing of gas from mechanical and gaseous impurity due to more effective utilization of action of centrifugal forces and increase in a surface of contact of phases. The Centrifugal forces arising at rotation of a rotor provide crushing a liquid on fine drops that causes intensive contact of gases and caught particles to a liquid. Owing to action of centrifugal forces, intensive hashing of gas and a liquid and presence of the big interphase surface of contact, there is an effective clearing of gas in a foamy layer. The water resistance of the irrigated apparatus at change of loadings on phases has been designed. Considered angular speed of twirl of a rotor and veering of twirl of guide vanes of an air swirler.

In a Fig. 9.2 results of an experimental research of efficiency of clearing of a dust are shown. For various diameter of corpuscles of a dust the increase in general efficiency of separation with decrease in concentration of corpuscles is observed.

In a Fig. 9.2 results of an experimental research of efficiency of clearing of a dust are shown. For various diameter of corpuscles d, mic of a dust the increase in general efficiency of separation with decrease in concentration of corpuscles C, % is observed.

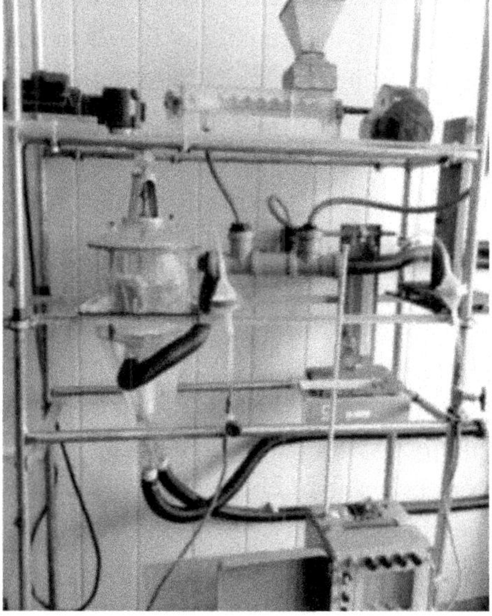

FIGURE 9.1 The laboratory facility.

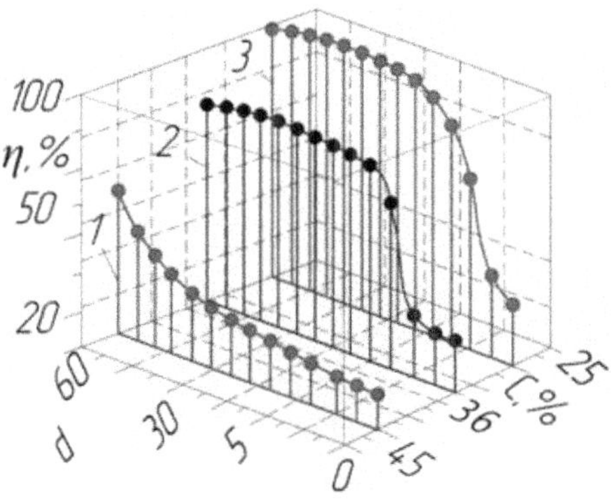

FIGURE 9.2 Dependence of efficiency of separation on diameter and concentration of corpuscles.

9.4 NUMERICAL SIMULATION AND CALCULATION OF CLEARING OF A DUST IN THE APPARATUS

The algorithm of modeling of process of separation of a dispersoid in a gas stream with irrigation by a liquid has been developed. The carried out calculations allow to define potential possibilities of a dynamic scrubber at its use in the capacity of the apparatus for clearing of gas emissions. Verification of the data gained by calculation, and also an estimation of the parameters defining possibility of separation of a dispersoid on drips of an irrigating liquid, is modeled as process of a current of a water gas stream in a packet of computing hydrodynamics Ansys CFX (Fig. 9.3). Numerical research of work of a scrubber will allow to analyze its work for the purpose of decrease of power inputs at conservation of quality of gas cleaning. The developed model helps to simulate traffic of a dusty gas stream sweepingly and visually. The model can consider modification of geometry of the apparatus. Thus, the model can be applied to optimization of a design of a dynamic scrubber.

Quality gained on the basis of conducting of computing experiment of results directly depends on quality of the built design grid. Preprocessor GAMBIT allows to create and process sweepingly geometry of investigated processes. Ansys Mesh possesses the powerful oscillator of the grids, allowing to create various types of grids: the structured hexahedral grid, automatic (not structured) hexahedral and a grid tetrahedron (Fig. 9.4). Besides, in it there is a possibility of creation of boundary layers with

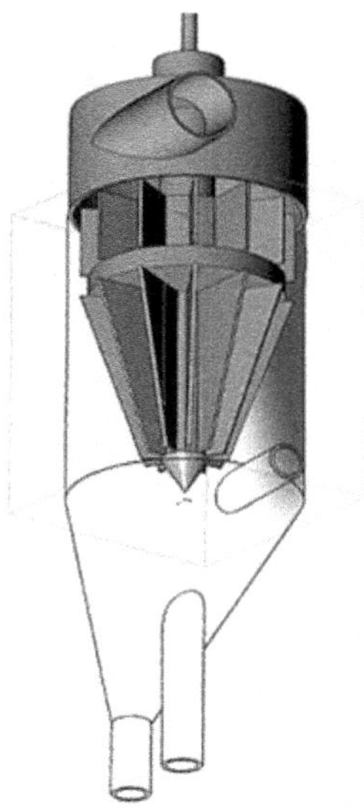

FIGURE 9.3 Geometrical model of a scrubber.

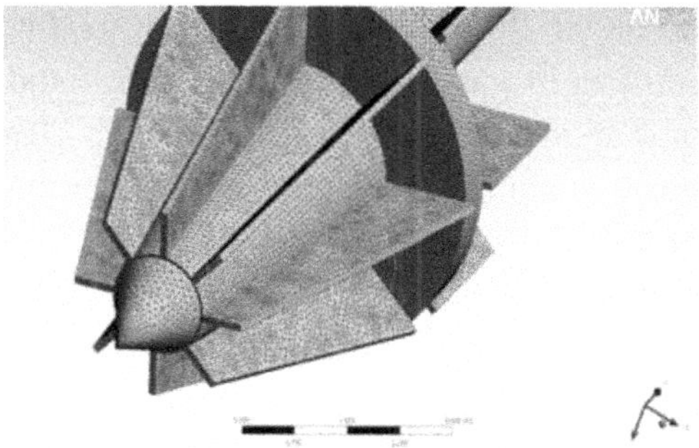

FIGURE 9.4 Typical design area, a design grid and a surface of the interface of a twirled vortex generator.

the combined grids. After construction of a grid the user has possibility to muster its quality on various parameters (displacement of elements, a relationship of sides).

In Ansys CFX possibility of reception of integrated parameters of calculation, including typical for dedusters is realized also: the hydraulic resistance, a pressure, an input, efficiency of clearing, swirling flow, and is possibility to edit the formula on which these parameters are computed.

9.5 CLEARING OF GASES OF A DUST IN THE INDUSTRY

The had results have been almost implemented in manufacture of roasting of refractory materials at conducting of redesign of system of an aspiration of smoke fumes of refire kilns. The devised scrubber is applied to clearing of smoke fumes of refire kilns of limestone in the capacity of the another echelon of clearing.

Temperature of gases of baking ovens in main flue gas breeching before a the exhaust-heat boiler 500–600°C, after exhaust-heat boiler 250°C. An average chemical compound of smoke gases (by volume): 17% $C0_2$; 16% N_2; 67% CO. Besides, in gas contains to 70 mg/m^3 SO_2; 30 mg/m^3 H_2S; 200 mg/m^3 F and 20 mg/m^3 CI. The gas dustiness on an exit from the converter reaches to 200/m^3 the Dust, as well as at a fume extraction with carbonic oxide after-burning, consists of the same components, but has the different maintenance of oxides of iron. In it than 1 micron, than in the dusty gas formed at after-burning of carbonic oxide contains less corpuscles a size less. It is possible to explain it to that at after-burning CO raises temperatures of gas and there is an additional excess in steam of oxides. Carbonic oxide before a gas heading on clearing burn in the special chamber. The dustiness of the cleared blast-furnace gas should be no more than 4 mg/m^3. The following circuit design (Fig. 9.5) is applied to clearing of the blast-furnace gas of a dust.

Gas from a furnace mouth of a baking oven 1 on gas pipes 3 and 4 is taken away in the gas-cleaning plant. In raiser and down-taking duct gas is chilled, and the largest corpuscles of a dust which in the form of sludge are trapped in the inertia sludge remover are inferred from it. In a centrifugal scrubber 5 blast-furnace gas is cleared of a coarse dust to final dust content 5–10/m^3 the Dust drained from the deduster loading pocket periodically from a feeding system of water or steam for dust moistening. The final

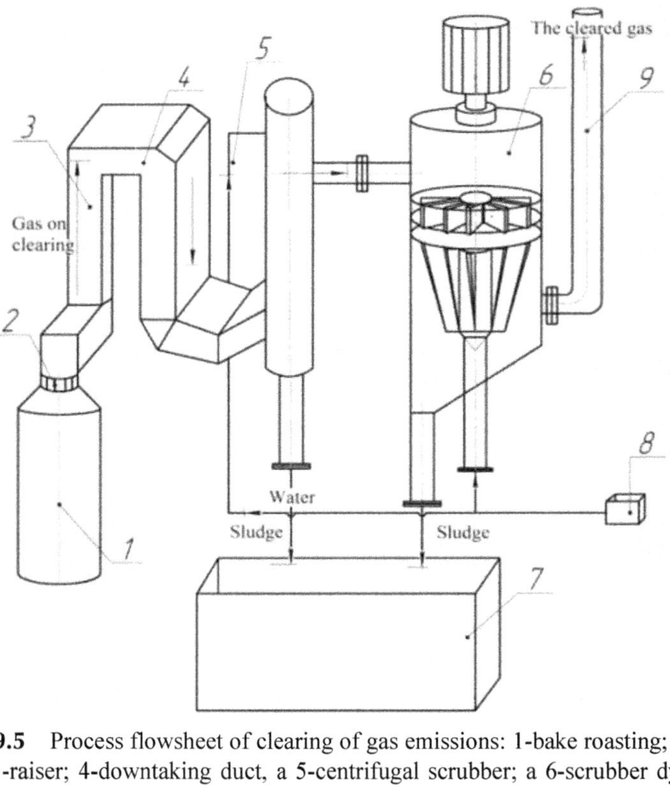

FIGURE 9.5 Process flowsheet of clearing of gas emissions: 1-bake roasting; a 2-water block; a 3-raiser; 4-downtaking duct, a 5-centrifugal scrubber; a 6-scrubber dynamic; a 7-forecastle of gathering of sludge, a 8-hydraulic hitch, a 9-chimney.

cleaning of the blast-furnace gas is carried out in a dynamic spray scrubber where there is an integration of a finely divided dust. Most the coarse dust and drops of liquid are inferred from gas in the inertia mist eliminator. The cleared gas is taken away in a collecting channel of pure gas 9, whence is fed in an aerosphere. The clarified sludge from a gravitation filter is fed again on irrigation of apparatuses. The closed cycle of supply of an irrigation water to what in the capacity of irrigations the lime milk close on the physical and chemical properties to composition of dusty gas is applied is implemented. As a result of implementation of trial installation clearings of gas emissions the maximum dustiness of the gases which are thrown out in an aerosphere, has decreased with 3950 mg/m^3 to 840 mg/m^3, and total emissions of a dust from sources of limy manufacture were scaled down about 4800 to/a to 1300 to/a.

Such method gives the chance to make gas clearing in much smaller quantity, demands smaller capital and operational expenses, reduces an atmospheric pollution and allows to use water recycling system.

9.6 CONCLUSION

1. The basic source of air pollution by manufacture ceramic and refractories are emissions of a smoke from furnaces in the course of roasting. Contaminants are formed because of the maintenance of impurity in raw materials. Their composition can vary depending on raw materials source. The special problem is called by emissions of fluorides.
2. For the first time research of hydrodynamics and dynamic spray scrubber separation in bundled software ANSYS, on the laboratory and trial installation, allowed to study character of interconnection of the basic aerohydrodynamic parameters from design features of the apparatus is conducted.
3. On the devised trial installations the results which have been had during mathematical modeling of process of motion and separation of dispersion particles from a gas stream are experimentally confirmed.
4. The ecological result of implementation of systems and recommendations consists highly clearings of a waste-heat and betterment of ecological circumstances in a zone of the factories.

KEYWORDS

- Ansys CFX
- dust
- dynamic separator
- emissions
- pottery work
- smoke fumes

REFERENCES

1. Uzhov, V. N., Valdberg, A. J., Myagkov B. I. Clearing of industrial gases of a dust, Moscow: Chemistry, 1981.
2. Pirumov, A. I. Air Dust removal, Moscow: Engineering industry, 1974.
3. Shvydky, V. S., Ladygichev, M. G. Clearing of gases. The directory, Moscow: Heat power engineering, 2002.
4. Straus, V. Industrial clearing of gases Moscow: Chemistry, 1981.
5. Kouzov, P. A., Malgin, A. D., Skryabin, G. M. Clearing of gases and air of a dust in the chemical industry. St.-Petersburg: Chemistry, 1993.
6. Vatin, N. I., Strelets, K. I. Air purification by means of apparatuses of type the cyclone separator. St.-Petersburg, 2003.
7. Aljamovskij, A. A. Solid Works 2007/2008. Computer modeling in engineering practice. SPb., 2008.
8. Kutateladze, S.S., Styrikovich, M.A. Hydrodynamics of gas-liquid systems. M: Energy, 1976.
9. The patent 2339435 Russian Federations, Dynamic spray scrubber. R.R. Usmanova, 27 November, 2008.
10. Crowe C., Sommerfield, M., Multiphase Flows with Droplets and Particles. CRC: Press, 1998.
11. Aksenov A. A., Dyadkin A. A., Gudzovsky A. V. Numerical Simulation of Car Tire Aquaplaning. Computational Fluid Dynamics '96, J.-A. Desideri, C. Hirsch, P. Le Tallec, M. Pandolfi, J. Periaux edts. John Wiley & Sons, 1996. 815–820.
12. Dukowicz J. K. A Particle-Fluid Numerical Model for Liquid Sprays. Journal of Computational Physics. Vol. 35, 1980. 229–253.
13. Harlow F. H., Welch J. E. Numerical Calculation of Time-Dependent Viscous Incompressible Flows of Fluid With Free Surface. Phys. Fluids. №. 8, 1965. 2182–2187.
14. Launder B. E., Spalding D. B. The Numerical Computation of Turbulent Flows. Comp. Meth. Appl. Mech. Eng. 1974. Vol. 3. 269–289.
15. Menter F. R., Esch T. Advanced Turbulence Modelling in CFX. CFX Update. Spring 2001. №. 20. 4–5.

CHAPTER 10

UPDATE ON CNT/POLYMER NANO-COMPOSITES: FROM THEORY TO APPLICATIONS

A. K. HAGHI

University of Guilan, Rasht, Iran

CONTENTS

ABSTRACT

In this chapter, an update on the modeling and mechanical properties of CNT/polymer nano-composites is presented. A very comprehensive references and further reading is also provided at the end of this chapter.

10.1 INTRODUCTION

Carbon nanotubes were first observed by Iijima, almost two decades ago [1], and since then, extensive work has been carried out to characterize their properties [2–4]. A wide range of characteristic parameters has been reported for carbon nanotube nanocomposites. There are contradictory reports that show the influence of carbon nanotubes on a particular property (e.g., Young's modulus) to be improving, indifferent or even deteriorating [5]. However, from the experimental point of view, it is a great challenge to characterize the structure and to manipulate the fabrication of polymer nanocomposites. The development of such materials is still largely empirical and a finer degree of control of their properties cannot be achieved so far. Therefore, computer modeling and simulation will play an ever increasing role in predicting and designing material properties, and guiding such experimental work as synthesis and characterization, For polymer nanocomposites, computer modeling and simulation are especially useful in the hierarchical characteristics of the structure and dynamics of polymer nanocomposites ranging from molecular scale, microscale to mesoscale and macroscale, in particular, the molecular structures and dynamics at the interface between nanoparticles and polymer matrix. The purpose of this review is to discuss the application of modeling and simulation techniques to polymer nanocomposites. This includes a broad subject covering methodologies at various length and time scales and many aspects of polymer nanocomposites. We organize the review as follows. In Section 10.1 we will discuss about Carbon nanotube's (CNTs) and nano composite properties. In Section 10.2, we introduce briefly the computational methods used so far for the systems of polymer nanocomposites which can be roughly divided into three types: molecular scale methods (e.g., molecular dynamics (MD), Monte Carlo (MC)), microscale methods (e.g., Brownian dynamics (BD), dissipative particle dynamics (DPD), lattice Boltzmann (LB), time dependent Ginzburg–Lanau method, dynamic density functional theory (DFT) method), and mesoscale and macroscale methods (e.g., micromechanics, equivalent-continuum and self-similar approaches, finite element method (FEM)) [6].

10.2 CARBON NANOTUBES (CNTS) AND NANO COMPOSITE PROPERTIES

10.2.1 INTRODUCTION TO CNTS

CNTs are one dimensional carbon materials with aspect ratio greater than 1000. They are cylinders composed of rolled-up graphite planes with diameters in nanometer scale [1, 7–8]. The cylindrical nanotube usually has at least one end capped with a hemisphere of fullerene structure. Depending on the process for CNT fabrication, there are two types of CNTs [7–10]: single-walled CNTs (SWCNTs) and multiwalled CNTs (MWCNTs). SWCNTs consist of a single graphene layer rolled up into a seamless cylinder whereas MWCNTs consist of two or more concentric cylindrical shells of graphene sheets coaxially arranged around a central hollow core with van der Waals forces between adjacent layers. According to the rolling angle of the graphene sheet, CNTs have three chiralities: armchair, zigzag and chiral one. The tube chirality is defined by the chiral vector, $C_h = na_1 + ma_2$ (Fig. 10.1), where the integers (n, m) are the number of steps along the unit vectors (a_1 and a_2) of the hexagonal lattice [11–12]. Using this (n, m) naming scheme, the three types of orientation of the carbon atoms around the nanotube circumference are specified. If n = m, the nanotubes are called "armchair." If m = 0, the nanotubes are called "zigzag." Otherwise, they are called "chiral." The chirality of nanotubes has significant impact on their transport properties, particularly the electronic properties. For a given (n, m) nanotube, if (2n + m) is a multiple of 3, then the nanotube is metallic, otherwise the nanotube is a semiconductor. Each MWCNT contains a multilayer of graphene, and each layer can have different chiralities, so the prediction of its physical properties is more complicated than that of SWCNT. Figure 10.1 shows the CNT with different chiralities.

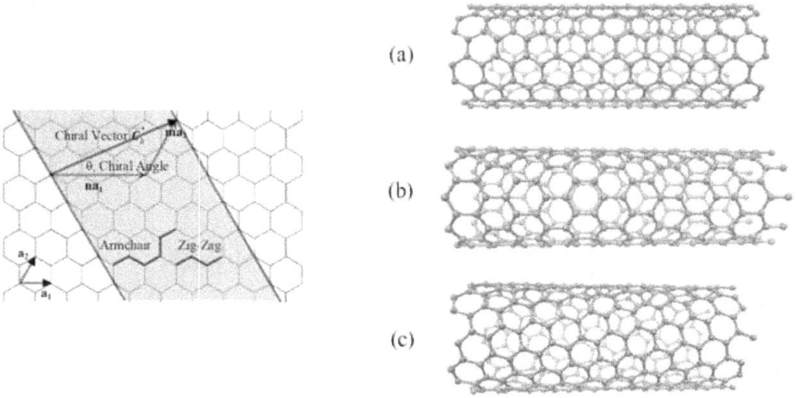

FIGURE 10.1 Schematic diagram showing how a hexagonal sheet of graphene is rolled to form a CNT with different chiralities (A: armchair; B: zigzag; C: chiral).

10.3 CLASSIFICATION OF CNT/POLYMER NANOCOMPOSITES

Polymer composites, consisting of additives and polymer matrices, including thermoplastics, thermosets and elastomers, are considered to be an important group of relatively inexpensive materials for many engineering applications. Two or more materials are combined to produce composites that possess properties that are unique and cannot be obtained each material acting alone. For example, high modulus carbon fibers or silica particles are added into a polymer to produce reinforced polymer composites that exhibit significantly enhanced mechanical properties including strength, modulus and fracture toughness. However, there are some bottlenecks in optimizing the properties of polymer composites by employing traditional micron-scale fillers. The conventional filler content in polymer composites is generally in the range of 10–70 wt. %, which in turn results in a composite with a high density and high material cost. In addition, the modulus and strength of composites are often traded for high fracture toughness [13].Unlike traditional polymer composites containing micron-scale fillers, the incorporation of nanoscale CNTs into a polymer system results in very short distance between the fillers, thus the properties of composites can be largely modified even at an extremely low content of filler. For example, the electrical conductivity of CNT/epoxy nanocomposites can be enhanced several orders of magnitude with less than 0.5 wt. %

of CNTs [14]. As described previously, CNTs are among the strongest and stiffest fibers ever known. These excellent mechanical properties combined with other physical properties of CNTs exemplify huge potential applications of CNT/polymer nanocomposites. Ongoing experimental works in this area have shown some exciting results, although the much-anticipated commercial success has yet to be realized in the years ahead. In addition, CNT/polymer nanocomposites are one of the most studied systems because of the fact that polymer matrix can be easily fabricated without damaging CNTs based on conventional manufacturing techniques, a potential advantage of reduced cost for mass production of nanocomposites in the future. Following the first report on the preparation of a CNT/polymer nanocomposite in 1994 [15], many research efforts have been made to understand their structure–property relationship and find useful applications in different fields, and these efforts have become more pronounced after the realization of CNT fabrication in industrial scale with lower costs in the beginning of the twenty-first century. According to the specific application, CNT/polymer nanocomposites can be classified as structural or functional composites [16]. For the structural composites, the unique mechanical properties of CNTs, such as the high modulus, tensile strength and strain to fracture, are explored to obtain structural materials with much improved mechanical properties. As for CNT/polymer functional composites, many other unique properties of CNTs, such as electrical, thermal, optical and damping properties along with their excellent mechanical properties, are used to develop multifunctional composites for applications in the fields of heat resistance, chemical sensing, electrical and thermal management, photoemission, electromagnetic absorbing and energy storage performances, etc.

10.4 MOLECULAR STRUCTURE OF CNTS

10.4.1 BONDING MECHANISMS

The mechanical properties of CNTs are closely related to the nature of the bonds between the carbon atoms. The bonding mechanism in a carbon nanotube system is similar to that of graphite, since a CNT can be thought of as a rolled-up graphene sheet. The atomic number for carbon is 6, and the atom electronic structure is 1 $s^2 2 s^2 2p^2$ in atomic physics notation. For a detailed description of the notation and the structure, readers may refer to basic textbooks on general chemistry or physics [197].

When carbon atoms combine to form graphite, sp^2 hybridization occurs. In this process, one *s*-orbital and two *p*-orbitals combine to form three hybrid sp^2-orbitals at 120° to each other within a plane (shown in Fig. 10.1). This in-plane bond is referred to as a σ-bond (*sigma*–bond). This is a strong covalent bond that binds the atoms in the plane, and results in the high stiffness and high strength of a CNT. The remaining *π*-orbital is perpendicular to the plane of the σ-bonds. It contributes mainly to the interlayer interaction and is called the π-bond (*pi*–bond). These out-of-planes, delocalized p-bonds interact with the p-bonds on the neighboring layer. This interlayer interaction of atom pairs on neighboring layers is much weaker than a s-bond. For instance, in the experimental study of *shellsliding* [98, 198], it was found that the shear strength between the outermost shell and the neighboring inner shell was 0.08 MPa and 0.3 MPa according to two separate measurements on two different MWCNTs. The bond structure of a graphene sheet is shown in Fig. 10.1.

10.4.2 FROM GRAPHENE SHEET TO SINGLE-WALLED NANOTUBE

There are various ways of defining a unique structure for each carbon nanotube. One way is to think of each CNT as a result of rolling a graphene sheet, by specifying the direction of rolling and the circumference of the cross-section. Shown in Fig. 10.2 is a graphene sheet with defined roll-up vector ***r***. After rolling to form a NT, the two end nodes coincide. The notation we use here is adapted from [2, 8, 199].

Note that ***r*** (bold solid line in Fig. 10.2) can be expressed as a linear combination of base vectors ***a*** and ***b*** (dashed line in Fig. 10.2),

$$r = na + mb \tag{1}$$

with n and m being integers. Different types of NT are thus uniquely defined by of the values of n and m and the ends are closed with caps for certain types of fullerenes. Three major categories of NT can also be defined based on the chiral angle u (Fig. 10.2) as follows

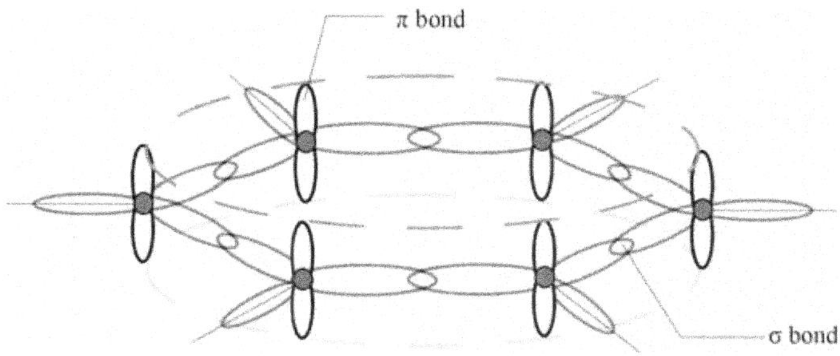

FIGURE 10.2 Basic hexagonal bonding structure for one graphite layer (the graphene sheet); carbon nuclei shown as filled circles, out-of-plane π-bonds represented as delocalized (dotted line), and σ-bonds connect the C nuclei in-plane.

$$\begin{aligned} &\theta = 0 \text{ "Zigzag"} \\ &0 < \theta < 30, \text{ "Chiral"} \\ &\theta = 30, \text{ "ArmChair"} \end{aligned} \tag{2}$$

Based on simple geometry, the diameter d and the chiral angle θ of the NT can be given as

$$d = 0.783\sqrt{(n^2 + nm + m^2)}A \tag{3}$$

$$\theta = \sin^{-1}\left[\frac{\sqrt{3}m}{2(n^2 + nm + m^2)}\right] \tag{4}$$

Most CNTs to date have been synthesized with closed ends. Fujita et al. [200] and Dresselhaus et al. [2, 201] have shown that NTs which are larger than (5, 5) and (9, 0) tubes can be capped. Based on Euler's theorem of polyhedral [202], which relates the numbers of the edges, faces and vertices, along with additional knowledge of the minimum energy structure of fullerenes, they conclude that any cap must contain 6 pentagons that are isolated from each other.

For NTs with large radius, there are different possibilities of forming caps that satisfy this requirement. The experimental results of Iijima et al. [203] and Dravid [204] indicate a number of ways that regular-shaped caps can be formed for large diameter tubes. Bill-like [205] and semitoroidal

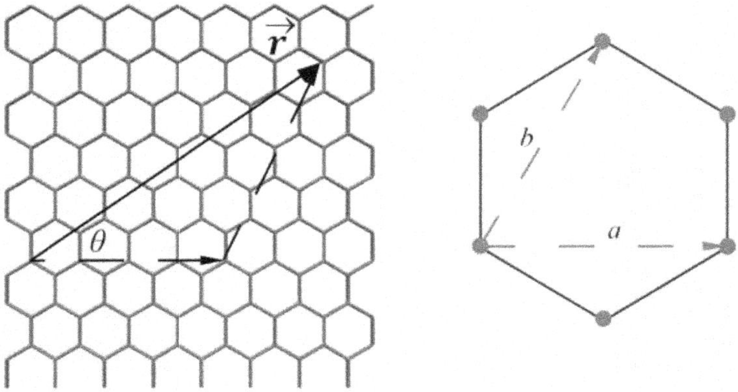

FIGURE 10.3 Definition of roll up vector as linear combinations of base vector *a* and *b*.

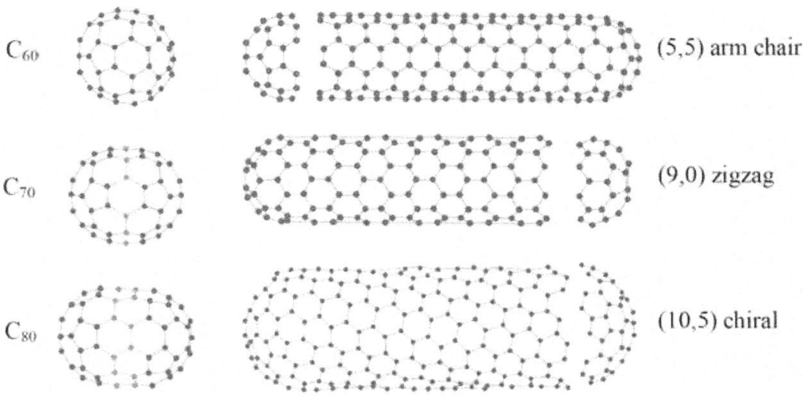

FIGURE 10.4 Examples of zigzag, chiral, and arm-chair nanotubes and their caps corresponding to different types of fullerenes.

[203] types of termination have also been reported. Experimental observation of CNTs with open ends can be found in [203].

10.4.3 MULTI-WALLED CARBON NANOTUBES AND SCROLL-LIKE STRUCTURES

The first carbon nanotubes discovered [1] were multiwalled carbon nanotubes. Transmission electron microscopy studies on MWCNTs suggest a

Russian doll-like structure (nested shells) and give interlayer spacing of approximately; 0.34 nm [206–207] close to the interlayer separation of graphite, 0.335 nm. However, Kiang et al. [208] have shown that the interlayer spacing for MWCNTs can range from 0.342 to 0.375 nm, depending on the diameter and number of nested shells in the MWCNT. The increase in intershell spacing with decreased nanotube diameter is attributed to the increased repulsive force as a result of the high curvature. The experiments by Zhou et al. [207], Amelincx et al. [209] and Lavin et al. [210] suggested an alternative "scroll" structure for some MWCNTs, like a cinnamon roll. In fact, both forms might be present along a given MWCNT and separated by certain types of defects. The energetics analysis by Lavin et al. [210] suggests the formation of a scroll, which may then convert into a stable multiwall structure composed of nested cylinders.

10.5 STRUCTURAL CHARACTERISTICS OF CARBON NANOTUBES

A single-walled carbon nanotube (SWNT) can be viewed as a graphene sheet that has been rolled into a tube. A multiwalled carbon nanotube (MWNT) is composed of concentric graphitic cylinders with closed caps at both ends and the graphitic layer spacing is about 0.34 nm. Unlike diamond, which assumes a 3-D crystal structure with each carbon atom having four nearest neighbors arranged in a tetrahedron, graphite assumes the form of a 2-D sheet of carbon atoms arranged in a hexagonal array. In this case, each carbon atom has three nearest neighbors.

The atomic structure of nanotubes can be described in terms of the tube chirality, or helicity, which is defined by the chiral vector C_h and the chiral angle θ. In Fig. 10.5, we can visualize cutting the graphite sheet along the dotted lines and rolling the tube so that the tip of the chiral vector touches its tail. The chiral vector, also known as the roll-up vector, can be described by the following equation:

$$\overrightarrow{C_h} = n\overrightarrow{a_1} + m\overrightarrow{a_2},$$

where the integers (n;m) are the number of steps along the zigzag carbon bonds of the hexagonal lattice and a1 and a2 are unit vectors. The chiral angle determines the amount of 'twist' in the tube. The chiral angles are

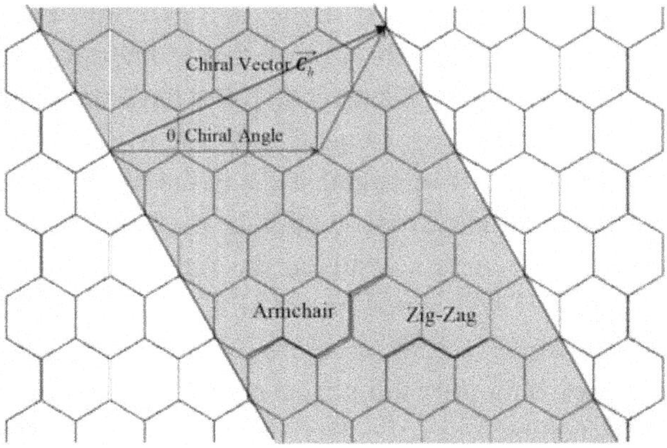

FIGURE 10.5 Schematic diagram of a hexagonal graphene sheet.

0° and 30° for the two limiting cases which are referred to as zigzag and armchair, respectively (Fig. 10.6).

In terms of the roll-up vector, the zigzag nanotube is denoted by (n, 0) and the armchair nanotube (n, n). The roll-up vector of the nanotube also defines the nanotube diameter.

The physical properties of carbon nanotubes are sensitive to their diameter, length and chirality.

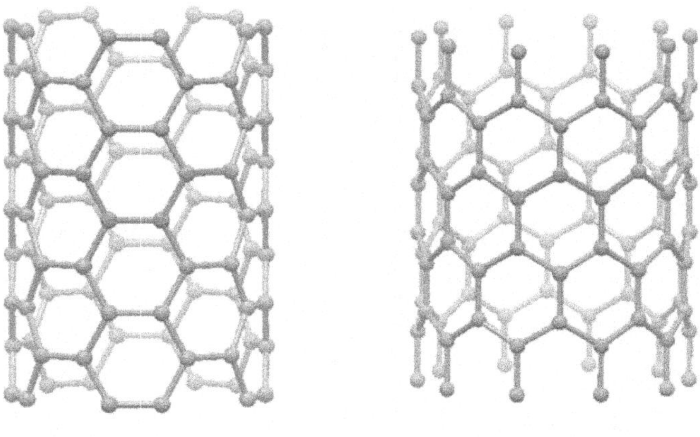

FIGURE 10.6 Schematic diagram of (a) an armchair and (b) A zigzag nanotube.

In particular, tube chirality is known to have a strong influence on the electronic properties of carbon nanotubes. Graphite is considered to be a semimetal, but it has been shown that nanotubes can be either metallic or semiconducting, depending on tube chirality [2]. The influence of chirality on the mechanical properties of carbon nanotubes has also been reported [158, 335].

10.6 CHARACTERIZATION OF CARBON NANOTUBES

Significant challenges exist in both the micromechanical characterization of nanotubes and the modeling of the elastic and fracture behavior at the nano-scale. Challenges in characterization of nanotubes and their composites include (a) complete lack of micromechanical characterization techniques for direct property measurement, (b) tremendous limitations on specimen size, (c) uncertainty in data obtained from indirect measurements, and (d) inadequacy in test specimen preparation techniques and lack of control in nanotube alignment and distribution.

In order better to understand the mechanical properties of carbon nanotubes, a number of investigators have attempted to characterize carbon nanotubes directly. Treacy et al.[138] first investigated the elastic modulus of isolated multiwalled nanotubes by measuring, in the transmission electron microscope, the amplitude of their intrinsic thermal vibration. The average value obtained over 11 samples was 1.8 TPa. Direct measurement of the stiffness and strength of individual, structurally isolated multiwall carbon nanotubes has been made with an atomic-force microscope (AFM). Wong and co-workers [139] were the first to perform direct measurement of the stiffness and strength of individual, structurally isolated multiwall carbon nanotubes using atomic force microscopy. The nanotube was pinned at one end to molybdenum disulfide surfaces and load was applied to the tube by means of the AFM tip. The bending force was measured as a function of displacement along the unpinned length, and a value of 1.26 TPa was obtained for the elastic modulus. The average bending strength measured was 14.28 GPa.

Single-walled nanotubes tend to assemble in 'ropes' of nanotubes. Salvetat and co-workers [140] measured the properties of these nanotube bundles with the AFM. As the diameter of the tube bundles increases, the axial and shear moduli decrease significantly. This suggests slipping of the nanotubes within the bundle. Walters et al. [141] further investigated the elastic strain of nanotube bundles with the AFM. On the basis of their experimental strain measurements and an assumed elastic modulus of 1.25 TPa, they calculated a yield strength of 45±7 GPa for the nanotube ropes. Indeed, their calculated value for strength would be much lower if the elastic modulus of the nanotube bundle is decreased as a consequence of slipping within the bundle, suggested by Salvetat et al. [140]. Yu and co-workers [96, 142] have investigated the tensile loading of multiwalled nanotubes and single-walled nanotube ropes. In their work, the nanotubes were attached between two opposing AFM tips and loaded under tension. For multiwalled carbon nanotubes [142] the failure of the outermost tube occurred followed by pullout of the inner nanotubes. The experimentally calculated tensile strengths of the outermost layer ranged from 11 to 63 GPa and the elastic modulus ranged from 270 to 950 GPa. In their subsequent investigation of single-walled nanotube ropes [97], they assumed that only the outermost tubes assembled in the rope carried the load during the experiment, and they calculated tensile strengths of 13 to 52 Gpa and average elastic moduli of 320–1470 GPa. Xie et al. [143] also tested ropes of multiwalled nanotubes in tension. In their experiments, the obtained tensile strength and modulus were 3.6 and 450 GPa, respectively. It was suggested that the lower values for strength and stiffness may be a consequence of defects in the CVD-grown nanotubes.

10.7 MECHANICS OF CARBON NANOTUBES

As discussed in the previous section, nanotube deformation has been examined experimentally. Recent investigations have shown that carbon nanotubes possess remarkable mechanical properties, such as exceptionally high elastic modulus [144, 145], large elastic strain and fracture strain sustaining capability [60, 146]. Similar conclusions have also been reached through some theoretical studies [147–150], although very few correlations between theoretical predictions and experimental studies have been made. In this section we examine the mechanics of both single walled and multiwalled nanotubes.

10.7.1 SINGLE-WALLED CARBON NANOTUBES

Theoretical studies concerning the mechanical properties of single-walled nanotubes have been pursued extensively. Overney et al. [147] studied the low-frequency vibrational modes and structural rigidity of long nanotubes consisting of 100, 200 and 400 atoms. The calculations were based on an empirical Keating Hamiltonian with parameters determined from first principles. A comparison of the bending stiffnesses of single-walled nanotubes and an iridium beam was presented.

The bending stiffness of the iridium beam was deduced by using the continuum Bernoulli-Euler theory of beam bending. Overney and co-workers concluded that the beam bending rigidity of a nanotube exceeds the highest values found in any other presently available materials. Besides their experimental observations, Iijima et al. [151] examined response of nanotubes under compression using molecular dynamics simulations. They simulated the deformation properties of single- and multiwalled nanotubes bent to large angles. Their experimental and theoretical results show that nanotubes are remarkably flexible. The bending is completely reversible up to angles in excess of 110°, despite the formation of complex kink shapes.

Ru [152] noticed that actual bending stiffness of single-walled nanotubes is much lower than that given by the elastic-continuum shell model if the commonly defined representative thickness is used. Ru proposed the use of an effective nanotube bending stiffness as a material parameter not related to the representative thickness. With the aid of this concept, the elastic shell equations can be readily modified and then applied to single-walled nanotubes. The computational results based on this concept show a good agreement with the results from molecular dynamics simulations.

Vaccarini et al. [153] investigated the influence of nanotube structure and chirality on the elastic properties in tension, bending, and torsion. They found that the chirality played a small influence on the nanotube tensile modulus. However, the chiral tubes exhibit asymmetric torsional behavior with respect to left and right twist, whereas the armchair and ziz-zag tubes do not exhibit this asymmetric torsional behavior. A relatively comprehensive study of the elastic properties of single-walled nanotubes was reported by Lu [148]. In this study, Lu adopted an empirical lattice dynamics model [154], which has been successfully adopted in calculating the phonon spectrum and elastic properties of graphite. In this lattice-dynamics model, atomic interactions in a single carbon layer are approximated by a sum of pair-wise harmonic potentials between atoms. The local structure of a nanotube layer is constructed from conformal mapping of a graphite sheet on to a cylindrical surface. Lu's work attempted to answer such basic questions as: (a) how do elastic properties of nanotubes depend on the structural details, such as size and chirality? And (b) how do elastic properties of nanotubes compare with those of graphite and diamond? Lu concluded that the elastic properties of nanotubes are insensitive to size and chirality. The predicted Young's modulus (~1 TPa), shear modulus (~0.45 TPa), and bulk modulus (~0.74 TPa) are comparable to those of diamond. Hernandez and co-workers [51] performed calculations similar to those of Lu and found slightly higher values (~1.24 TPa) for the Young's moduli of tubes. But unlike Lu, they found that elastic moduli are sensitive to both tube diameter and structure. Besides their unique elastic properties, the inelastic behavior of nanotubes has also received considerable attention. Yakobson and co-workers [12, 150] examined the instability behavior of carbon nanotubes beyond linear response by using a realistic many-body Tersoff-Brenner potential and molecular dynamics simulations. Their molecular-dynamics simulations show that carbon nanotubes, when subjected to large deformations, reversibly switch into different morphological patterns. Each shape change corresponds to an abrupt release of energy and a singularity in the stress/strain curve. These transformations are explained well by a continuum shell model. With properly chosen parameters, their model provided a very accurate 'roadmap' of nanotube behavior beyond the linear elastic regime. They also made molecular dynamics simulations to single- and double-walled nanotubes of different chirality and at different temperatures [149]. Their simulations show that nanotubes have an extremely large breaking strain (in the range 30–40%) and the breaking strain decreases with temperature. Yakobson [155] also applied dislocation theory to carbon nanotubes for describing their main routes of mechanical relaxation under tension. It was concluded that the yield strength of a nanotube depends on its symmetry and it was believed that there exists an intramolecular plastic flow. Under high stress, this plastic flow corresponds to a motion of dislocations along helical paths within the nanotube wall and causes a stepwise necking, a well-defined new symmetry, as the domains of different chiral symmetry are formed. As a result, both the mechanical and electronic properties of carbon nanotubes are changed.

The single walled nanotubes produced by laser ablation and arc-discharge techniques have a greater tendency to form 'ropes' or aligned bundles [156, 157]. Thus, theoretical studies have been made to investigate the mechanical properties of these nanotube bundles. Ru [114] presented a modified elastic-honeycomb model to study elastic buckling of nanotube ropes under high pressure. Ru gave a simple formula for the critical pressure as a function of nanotube Young's modulus and wall thickness-to-radius ratio. It was concluded that single- walled ropes are susceptible to elastic buckling under high pressure and elastic buckling is responsible for the pressure-induced abnormalities of vibration modes and electrical resistivity of single walled nanotubes. Popov et al. [158] studied the elastic properties of triangular crystal lattices formed by single-walled nanotubes by using analytical expressions based on a force constant lattice dynamics model [159]. They calculated various elastic constants of nanotube crystals for nanotube types, such as armchair and zigzag. It was shown that the elastic modulus, Poisson's ratio and bulk modulus clearly exhibit strong dependence on the tube radius. The bulk modulus was found to have a maximum value of 38 GPa for crystals composed of single-walled nanotubes with 0.6 nm radius.

10.7.2 MULTI-WALLED CARBON NANOTUBES

Multi-walled nanotubes are composed of a number of concentric single walled nanotubes held together with relatively weak van der Waals forces. The multilayered structure of these nanotubes further complicates the modeling of their properties.

Ruoff and Lorents [160] derived the tensile and bending stiffness constants of ideal multiwalled nanotubes in terms of the known elastic properties of graphite. It is suggested that unlike the strongly anisotropic thermal expansion in conventional carbon fibers and graphite, the thermal expansion of carbon nanotubes is essentially isotropic. However, the thermal conductivity of nanotubes is believed to be highly anisotropic and its magnitude along the axial direction is perhaps higher than that of any other material. Lu [148] also calculated the elastic properties of many multiwalled nanotubes formed by single-layer tubes by means of the empirical-lattice dynamics model. It was found that elastic properties are insensitive to different combinations of parameters, such as chirality, tube radius and numbers of layers, and the elastic properties are the same for all nanotubes with a radius larger than one nm. Interlayer van der Waals interaction has a negligible contribution to both the tensile and shear stiffness.

Govindjee and Sackman [161] were the first to examine the use of continuum mechanics to estimate the properties of multiwalled nanotubes. They investigated the validity of the continuum approach by using Bernoulli-Euler bending to infer the Young's modulus. They used a simple elastic sheet model and showed that at the nanotube scale the assumptions of continuum mechanics must be carefully respected in order to obtain reasonable results. They showed the explicit dependence of 'material properties' on system size when a continuum cross-section assumption was used. Ru [162] used the elastic-shell model to study the effect of van der Waals forces on the axial buckling of a double-walled carbon nanotube. The analysis showed that the van der Waals forces do not increase the critical axial buckling strain of a double-walled nanotube. Ru [163, 164] thereafter also proposed a multiple column model that considers the interlayer radial displacements coupled through the van der Waals forces.

This model was used to study the effect of interlayer displacements on column buckling. It was concluded that the effect of interlayer displacements could not be neglected unless the van der Waals forces are extremely strong.

Kolmogorov and Crespi [165] investigated the interlayer interaction in two-walled nanotubes. Registry-dependent two-body graphite potential was developed. It was demonstrated that the tightly constrained geometry of a multiwalled nanotube could produce an extremely smooth solid-solid interface wherein the corrugation against sliding does not grow with system size. The energetic barrier to interlayer sliding in defect-free nanotubes containing thousands of atoms can be comparable to that for a single unit cell of crystalline graphite. Although there is experimental variability in the direct characterization of carbon nanotubes, theoretical and experimental observations reveal their exceptional properties. As a consequence, there has been recent interest in the development of nanotube-based composites. Although most research has focused on the development of nanotube-based polymer composites, attempts have also been made to develop metal and ceramic-matrix composites with nanotubes as reinforcement.

10.8 NANOTUBE-BASED POLYMER COMPOSITES

The reported exceptional properties of nanotubes have motivated others to investigate experimentally the mechanics of nanotube-based composite films. Uniform dispersion within the polymer matrix and improved nanotube/matrix wetting and adhesion are critical issues in the processing of these nanocomposites. The issue of nanotube dispersion is critical to efficient reinforcement. In the work of Salvetat et al. [140] discussed earlier, slipping of nanotubes when they are assembled in ropes significantly affects the elastic properties.

In addition to slipping of tubes that are not bonded to the matrix in a composite, the aggregates of nanotube ropes effectively reduce the aspect ratio (length/diameter) of the reinforcement. It is, however, difficult to obtain a uniform dispersion of carbon nanotubes in the polymer matrix. Shaffer and Windle [166] were able to process carbon nanotube/polyvinyl-alcohol composite films for mechanical characterization. The tensile elastic modulus and damping properties of the composite films were assessed in a dynamic mechanical thermal analyzer (DMTA) as a function of nanotube loading and temperature. From the theory developed for short-fiber composites, a nanotube elastic modulus of 150 MPa was obtained from the experimental data.

This value in a microscopic composite is well below the values reported for isolated nanotubes. It is not clear whether this result is a consequence of imperfections in the graphite layers of catalytically grown nanotubes used for the investigation or whether it relates to a fundamental difficulty in stress transfer.

Qian et al. [53] characterized carbon-nanotube/polystyrene composites. With only the addition of 1% by weight (about 0.5% by volume) they achieved between 36–42% increase in the elastic stiffness and a 25% increase in the tensile strength.

Jia et al. [167] showed that the nanotubes can be initiated by a free radical initiator, AIBN (2,2'-azobisisobutyronitrile), to open their p bonds. In their study of carbon-nanotube/poly (methyl methacrylate) (PMMA) composites, the possibility exists to form a C–C bond between the nanotube and the matrix. Gong et al. [168] investigated surfactant-assisted processing of nanotube composites with a nonionic surfactant. Improved dispersion and interfacial bonding of the nanotubes in an epoxy matrix resulted in a 30% increase in elastic modulus with addition of 1 wt.% nanotubes. Lordi and Yao [102] looked at the molecular mechanics of binding in nanotube-based composites. In their work, they used force-field-based molecular-mechanics calculations to determine the binding energies and sliding frictional stresses between pristine carbon nanotubes and different polymeric matrix materials. The binding energies and frictional forces were found to play only a minor role in determining the strength of the interface. The key factor in forming a strong bond at the interface is having a helical conformation of the polymer around the nanotube. They suggested that the strength of the interface may result from molecular-level entanglement of the two phases and forced long-range ordering of the polymer.

Because the interaction at the nanotube/matrix interface is critical to understanding the mechanical behavior of nanotube-based composites, a number of researchers have investigated the efficiency of interfacial stress transfer. Wagner et al. [52] examined stress-induced fragmentation of multiwalled carbon nanotubes in polymer films. Their nanotube-containing film had a thickness of approximately 200 mm. The observed fragmentation phenomenon was attributed to either process- induced stress resulting from curing of the polymer or tensile stress generated by polymer deformation and transmitted to the nanotube. From estimated values of nanotube axial normal stress and elastic modulus, Wagner and co-workers concluded that the nanotube/ polymer interfacial shear stress is on the order of 500 MPa and higher. This value, if reliable, is an order of magnitude higher than the stress-transfer ability of current advanced composites and, therefore, such interfaces are more able than either the matrix or the nanotubes themselves to sustain shear. In further work, Lourie and Wagner [169, 170] investigated tensile and compressive fracture in nanotube-based composites.

Stress transfer has also been investigated by Raman spectroscopy. Cooper and co-workers [92] prepared composite specimens by applying an epoxy-resin/nanotube mixture to the surface of an epoxy beam. After the specimens were cured, stress transfer between the polymer and the nanotubes was detected by a shift in the G' Raman band (2610 cm^{-1}) to a lower wavenumber. The shift in the G' Raman band corresponds to strain in the graphite structure, and the shift indicates that there is stress transfer, and hence reinforcement, by the nanotubes. It was also concluded that the effective modulus of single-walled nanotubes dispersed in a composite could be over 1 TPa and that of multiwalled nanotubes was about 0.3 TPa. In their investigation of single-walled nanotube/epoxy composites, Ajayan et al. [93] suggest that their nearly constant value of the Raman peak in tension is related to tube sliding within the nanotube bundles and, hence, poor interfacial load transfer between the nanotubes. Similar results were obtained by Schadler et al. [51] Multi-walled nanotube/epoxy composites were tested in both tension and compression. The compressive modulus was found to be higher than the tensile modulus of the composites, and the Raman peak was found to shift only in compression, indicating poor interfacial load transfer in tension.

Even with improved dispersion and adhesion, micromechanical characterization of these composites is difficult because the distribution of the nanotubes is random. Thus, attempts have been made to align nanotubes in order better to elucidate the reinforcement mechanisms. Jin et al. [171] showed that aligned nanotube composites could be obtained by mechanical stretching of the composite. X-ray diffraction was used to determine the orientation and degree of alignment.

Bower et al. [60] further investigated the deformation of carbon nanotubes in these aligned films. Haggenmueller and co-workers [172] showed that melt spinning of single wall nanotubes in fiber form can also be used to create a well-aligned nanotube composite.

In addition to alignment of the carbon nanotubes, researchers have attempted to spin carbon fibers from carbon nanotubes [173–175]. Andrews et al. [173] dispersed 5 wt. % single-walled nanotubes in isotropic petroleum pitch. Compared to isotropic pitch fibers without nanotubes, the tensile strength was improved by ~90%, the elastic modulus was improved by ~150% and the electrical conductivity increased by 340%. Because the pitch matrix is isotropic, the elastic modulus is 10–20 times less than that of mesophase pitch fibers used in composite materials. Further developments in this area may potentially create a new form of carbon fiber that has exceptional flexibility as well as stiffness and strength. Vigolo and coworkers [175] technique for spinning nanotube-based fibers involves dispersing the nanotubes in surfactant solutions followed by recondensing the nanotubes in the stream of a polymer solution to form macroscopic fibers and ribbons. Their work indicates that there is preferential orientation of the nanotubes along the axis of the ribbons. Although the elastic modulus of the nanotube fibers (9–15 GPa) is far below the values for individual nanotubes or conventional carbon fibers, the demonstrated resilience of the fibers gives hope for future improvements.

10.9 MODELING AND SIMULATION TECHNIQUES

10.9.1 MODELING OF CARBON NANOTUBES BEHAVIOR

The investigations on carbon nanotubes behavior have been mainly focused on the experimental description and molecular dynamics simulations such as classical molecular dynamics, tight-binding molecular dynamics and the *ab initio* method. However, the researchers have been seeking more efficient computational methods with which it is possible to analyze the large scale of CNTs in a more general manner. Yakobson [12] found that the continuum shell model could predict all changes of buckling patterns in atomistic molecular-dynamics simulations. The analogousness of the cylindrical shell model and CNTs leads to extensive application of the shell model for CNT structural analysis. Ru [114, 115] used the CM approach and simulated the effect of van der Waals forces by applying a uniformly distributed pressure field on the wall, the pressure field was adjusted so as to give the same resultant force on each wall of the tube. It has verified that the mechanical responses of CNTs can be efficiently and reasonably predicted by the shell model provided that the parameters, such as Young's modulus and effective wall thickness, are judiciously adopted. Wang et al. [116] studied buckling of double-walled carbon nanotubes under axial loads, by modeling CNTs using solid shell elements. Han et al. [117, 118] investigated torsional buckling of a DWNT and MWNT embedded in an elastic medium. Han et al. [119] also studied bending instability of double-walled carbon nanotubes. Yao and Han [118] analyzed the thermal effect on axially compressed and torsional buckling of a multiwalled carbon nanotube. Some conclusions were drawn that at low and room temperature the critical load for infinitesimal buckling of a multiwalled carbon nanotube increases as the value of temperature change increases, while at high temperatures the critical load for infinitesimal buckling of a multiwalled carbon nanotube decreases as the value of temperature change increases. Nonlinear postbuckling behavior of carbon nanotubes under large strain is significant to which great attention is paid by some researchers [120, 121]. The torsional postbuckling behavior of single-walled or multiwalled carbon nanotubes was discussed in details by Yiao and Han [122]. The problems encountered in the numerical modeling of pristine and defective carbon nanotubes were demonstrated in details by Muc [123–125]. In the mentioned references, the linear and nonlinear (iterative) approaches were illustrated.

A successful work has been also conducted with continuum modeling such as dynamic studies. A comprehensive review of the literature dealing with the analysis of wave propagation in CNTs with the use of shell theories was presented by Liew and Wang [126], although the authors focused the attention mainly on the application of thick shell theories. They pointed out that there was growing interest in the terahertz physics of nanoscale materials and devices, which opened a new topic on phonon dispersion of CNTs, especially in the terahertz frequency range. Hu et al. [127] proposed to use nonlocal shell theories in the analysis of elastic wave propagation in single- or double-walled carbon nanotubes. However, it is worth to emphasize that the vibrational characteristics of CNTs are studied with the use of both beam theories and different variants of shell theories, for example, Natsuki et al. [128] and Ghorbanpourarani et al. [129]. Recently, thermal vibrations have attracted considerable attention, for example, Tylikowski [130], where the dynamic stability analysis was conducted with the use of stochastic methods. The cited above work demonstrates also another tendency in the dynamic analysis of multiwalled CNTs connected with the application of a multiple-elastic shell model which assumes that each of the concentric tubes of multiwall carbon nanotubes is an individual elastic shell and coupled with adjacent tubes through van der Waals interaction. The broader discussion of that class of problems was presented, for example, by Xu and Wang [131].

10.9.2 MOLECULAR SCALE METHODS

The modeling and simulation methods at molecular level usually employ atoms, molecules or their clusters as the basic units considered. The most popular methods include molecular mechanics (MM), MD and MC simulation. Modeling of polymer nanocomposites at this scale is predominantly directed toward the thermodynamics and kinetics of the formation, molecular structure and interactions.

10.9.2.1 Molecular Dynamics

MD is a computer simulation technique that allows one to predict the time evolution of a system of interacting particles (e.g., atoms, molecules, granules, etc.) and estimate the relevant physical properties [22, 23]. Specifically, it generates such information as atomic positions, velocities and forces from which the macroscopic properties (e.g., pressure, energy, heat capacities) can be derived by means of statistical mechanics. MD simulation usually consists of three constituents: (i) a set of initial conditions (e.g., initial positions and velocities of all particles in the system); (ii) the interaction potentials to represent the forces among all the particles; (iii) the evolution of the system in time by solving a set of classical Newtonian equations of motion for all particles in the system. The equation of motion is generally given by

$$\vec{F}_i(t) = m_i \frac{d^2 \vec{r}_i}{dt^2} \tag{5}$$

where is the force acting on the i-th atom or particle at time t which is obtained as the negative gradient of the interaction potential U, mi is the atomic mass and $\vec{\mathbf{r}}_i$ the atomic position. A physical simulation involves the proper selection of interaction potentials, numerical integration, periodic boundary conditions, and the controls of pressure and temperature to mimicbphysically meaningful thermodynamic ensembles. The interaction potentials together with their parameters, i.e., the so-called force field, describe in detail how the particles in a system interact with each other, i.e., how the potential energy of a system depends on the particle coordinates. Such a force field may be obtained by quantum method (e.g., ab initio), empirical method (e.g., Lennard–Jones, Mores, Born-Mayer) or quantum-empirical method (e.g., embedded atom model, glue model, bond order potential). The criteria for selecting a force field include the accuracy, transferability and computational speed. A typical interaction potential U may consist of a number of bonded and nonbonded interaction terms:

$$U\left(\vec{r}_1,\vec{r}_2,\vec{r}_3,\ldots,\vec{r}_n\right)=\sum_{i_{bond}}^{N_{bond}}U_{bond}(i_{bond},\vec{r}_a,\vec{r}_b)+\sum_{i_{angle}}^{N_{angle}}U_{angle}(i_{angle},\vec{r}_a,\vec{r}_b,\vec{r}_c)$$
$$+\sum_{i_{torsion}}^{N_{torsion}}U_{torsion}(i_{torsion},\vec{r}_a,\vec{r}_b,\vec{r}_c,\vec{r}_d)$$
$$+\sum_{i_{inversion}}^{N_{inversion}}U_{inversion}\left(i_{inversion},\vec{r}_a,\vec{r}_b,\vec{r}_c,\vec{r}_d\right)$$
$$+\sum_{i=1}^{N-1}\sum_{j>1}^{N}U_{vdw}(i,j,\vec{r}_a,\vec{r}_b)$$
$$+\sum_{i=1}^{N-1}\sum_{j>i}^{N}U_{electrostatic(i,j,\vec{r}_a,\vec{r}_b)} \quad (6)$$

The first four terms represent bonded interactions, that is, bond stretching Ubond, bond-angle bend Uangle and dihedral angle torsion Utorsion and inversion interaction Uinversion, while the last two terms are nonbonded interactions, that is, van der Waals energy Uvdw and electrostatic energy Uelectrostatic. In the equation, are the positions of the atoms or particles specifically involved in a given interaction;, and stand for the total numbers of these respective interactions in the simulated system;, and uniquely specify an individual interaction of each type; i and j in the van der Waals and electrostatic terms indicate the atoms involved in the interaction. There are many algorithms for integrating the equation of motion using finite difference methods. The algorithms of varlet, velocity varlet, leap-frog and Beeman, are commonly used in MD simulations [23]. All algorithms assume that the atomic position $\vec{r}$, velocities and accelerations can be approximated by a Taylor series expansion:

$$\vec{r}(t+\delta t)=\vec{r}(t)+\vec{v}(t)\delta t+\frac{1}{2}\vec{a}(t)\delta^2 t+\cdots \tag{7}$$

$$\vec{v}(t+\delta t)=\vec{v}(t)+\vec{a}(t)\delta t+\frac{1}{2}\vec{b}(t)\delta^2 t+\cdots \tag{8}$$

$$\vec{a}(t+\delta t)=\vec{a}(t)+\vec{b}(t)\delta t+\cdots \tag{9}$$

Generally speaking, a good integration algorithm should conserve the total energy and momentum and be time-reversible. It should also be easy to implement and computationally efficient, and permit a relatively long time step. The verlet algorithm is probably the most widely used method. It uses the positions and accelerations at time t, and the positions $\vec{r}(t-\delta t)$ from the previous step (t–δ) to calculate the new positions at (t+δt), we have:

$$\vec{r}(t+\delta t)=\vec{r}(t)+\vec{v}(t)\delta t+\frac{1}{2}\vec{a}(t)\delta t^2+\cdots \tag{10}$$

$$\vec{r}(t-\delta t)=\vec{r}(t)-\vec{v}(t)\delta t+\frac{1}{2}\vec{a}(t)\delta t^2+\cdots \tag{11}$$

$$\vec{r}(t+\delta t)=2\vec{r}(t)-\vec{r}(t-\delta t)+\vec{a}(t)\delta t^2+\cdots \tag{12}$$

The velocities at time t and can be respectively estimated

$$\vec{v}(t)=\frac{\left[\vec{r}(t+\delta t)-\vec{r}(t-\delta t)\right]}{2\delta t} \tag{13}$$

$$\vec{v}\left(t+\frac{1}{2\delta t}\right)=\frac{\left[\vec{r}\left(t+\delta t\right)-\vec{r}\left(t-\delta t\right)\right]}{\delta t} \tag{14}$$

MD simulations can be performed in many different ensembles, such as grand canonical (μVT), microcanonical (NVE), canonical (NVT) and isothermal–isobaric (NPT). The constant temperature and pressure can be controlled by adding an appropriate thermostat (e.g., Berendsen, Nose, Nose–Hoover and Nose–Poincare) and barostat (e.g., Andersen, Hoover and Berendsen), respectively. Applying MD into polymer composites allows us to investigate into the effects of fillers on polymer structure and dynamics in the vicinity of polymer–filler interface and also to probe the effects of polymer–filler interactions on the materials properties.

10.9.2.2 Monte Carlo

MC technique, also called Metropolis method [23], is a stochastic method that uses random numbers to generate a sample population of the system from which one can calculate the properties of interest. A MC simulation usually consists of three typical steps. In the first step, the physical problem under investigation is translated into an analogous probabilistic or statistical model. In the second step, the probabilistic model is solved by a numerical stochastic sampling experiment. In the third step, the obtained data are analyzed by using statistical methods. MC provides only the information on equilibrium properties (e.g., free energy, phase equilibrium), different from MD which gives nonequilibrium as well as equilibrium properties. In a NVT ensemble with N atoms, one hypothesizes a new configuration by arbitrarily or systematically moving one atom from position i→j. Due to such atomic movement, one can compute the change in the system Hamiltonian ΔH:

$$\Delta \mathrm{H} = \mathrm{H}(\mathrm{j})-\mathrm{H}(\mathrm{i}) \tag{15}$$

where H(i) and H(j) are the Hamiltonian associated with the original and new configuration, respectively.

This new configuration is then evaluated according to the following rules. If ΔH≥0, then the atomic movement would bring the system to a state of lower energy. Hence, the movement is immediately accepted and the displaced atom remains in its new position. If $\Delta H \geq 0$, the move is accepted only with a certain probability *Pi→j* which is given by

$$Pi \rightarrow j \propto \exp\left(-\frac{\Delta H}{K_B T}\right) \tag{16}$$

where is the Boltzmann constant. According to Metropolis et al. [24], one can generate a random number ζ between 0 and 1 and determine the new configuration according to the following rule:

$$\zeta \leq \exp\left(-\frac{\Delta H}{K_B T}\right) \text{; The move is accepted;} \tag{17}$$

$$\zeta > \exp\left(-\frac{\Delta H}{K_B T}\right) \text{; The move is not accepted.} \tag{18}$$

If the new configuration is rejected, one counts the original position as a new one and repeats the process by using other arbitrarily chosen atoms. In a μVT ensemble, one hypothesizes a new configuration j by arbitrarily choosing one atom and proposing that it can be exchanged by an atom of a different kind. This procedure affects the chemical composition of the system. Also, the move is accepted with a certain probability. However, one computes the energy change ΔU associated with the change in composition. The new configuration is examined according to the following rules. If $\Delta U \geq 0$, the movement of compositional change is accepted. However, if $\Delta U \geq 0$, the move is accepted with a certain probability which is given by

$$Pi \rightarrow \propto \exp\left(-\frac{\Delta H}{K_B T}\right) \tag{19}$$

where ΔU is the change in the sum of the mixing energy and the chemical potential of the mixture. If the new configuration is rejected one counts the original configuration as a new one and repeats the process by using some other arbitrarily or systematically chosen atoms. In polymer nanocomposites, MC methods have been used to investigate the molecular structure at nanoparticle surface and evaluate the effects of various factors.

10.9.3 MICROSCALE METHODS

The modeling and simulation at microscale aim to bridge molecular methods and continuum methods and avoid their shortcomings. Specifically, in nanoparticle–polymer systems, the study of structural evolution (i.e., dynamics of phase separation) involves the description of bulk flow (i.e., hydrodynamic behavior) and the interactions between nanoparticle and polymer components. Note that hydrodynamic behavior is relatively straightforward to handle by continuum methods but is very difficult and expensive to treat by atomistic methods. In contrast, the interactions between components can be examined at an atomistic level but are usually not straightforward to incorporate at the continuum level. Therefore, various simulation methods have been evaluated and extended to study the microscopic structure and phase separation of these polymer nanocomposites, including BD, DPD, LB, time-dependent Ginsburg–Landau (TDGL) theory, and dynamic DFT. In these methods, a polymer system is usually treated with a field description or microscopic particles that incorporate molecular details implicitly. Therefore, they are able to simulate the phenomena on length and time scales currently inaccessible by the classical MD methods.

10.9.3.1 Brownian Dynamics

BD simulation is similar to MD simulations [24]. However, it introduces a few new approximations that allow one to perform simulations on the microsecond timescale whereas MD simulation is known up to a few nanoseconds. In BD the explicit description of solvent molecules used in MD is replaced with an implicit continuum solvent description. Besides, the internal motions of molecules are typically ignored, allowing a much larger time step than that of MD. Therefore, BD is particularly useful for systems where there is a large gap of time scale governing the motion of different components. For example, in polymer–solvent mixture, a short time-step is required to resolve the fast motion of the solvent molecules, whereas the evolution of the slower modes of the system requires a larger time step. However, if the detailed motion of the solvent molecules is concerned, they may be removed from the simulation and their effects on the polymer are represented by dissipative (–γP) and random (σ ζ (t)) force terms. Thus, the forces in the governing Eq. (16) is replaced by a Langevin equation,

$$F_i(t) = \sum_{i \neq j} F_{ij}^c - \gamma P_i + \sigma \tau_i(t) \tag{20}$$

where F^c_{ij} is the conservative force of particle *j* acting on particle *i*, γ and σ are constants depending on the system, P_i the momentum of particle *i*, and a Gaussian random noise term. One consequence of this approximation of the fast degrees of freedom by fluctuating forces is that the energy and momentum are no longer conserved, which implies that the macroscopic behavior of the system will not be hydrodynamic. In addition, the effect of one solute molecule on another through the flow of solvent molecules is neglected. Thus, BD can only reproduce the diffusion properties but not the hydrodynamic flow properties since the simulation does not obey the Navier–Stokes equations.

10.9.3.2 Dissipative Particle Dynamics

DPD was originally developed by Hoogerbrugge and Koelman [25]. It can simulate both Newtonian and non-Newtonian fluids, including polymer melts and blends, on microscopic length and time scales. Like MD and BD, DPD is a particle-based method. However, its basic unit is not a single atom or molecule but a molecular assembly (i.e., a particle).DPD particles are defined by their mass M_i, position and momentum P_i. The interaction force between two DPD particles i and j can be described by a sum of conservative F^c_{ij}, dissipative and random forces [26–28]:

$$F_{ij} = F_{ij}^C + F_{ij}^D + F_{ij}^R \tag{21}$$

While the interaction potentials in MD are high-order polynomials of the distance between two particles, in DPD the potentials are softened so as to approximate the effective potential at microscopic length scales. The form of the conservative force in particular is chosen to decrease linearly with increasing r_{ij}. Beyond a certain cut-off separation r_c, the weight functions and thus the forces are all zero.

Because the forces are pair wise and momentum is conserved, the macroscopic behavior directly incorporates Navier–Stokes hydrodynamics. However, energy is not conserved because of the presence of the dissipative and random force terms which are similar to those of BD, but incorporate the effects of Brownian motion on larger length scales. DPD has several advantages over MD, for example, the hydrodynamic behavior is observed with far fewer particles than required in a MD simulation because of its larger particle size. Besides, its force forms allow larger time steps to be taken than those in MD.

10.9.3.3 Lattice Boltzmann

LB [29] is another microscale method that is suited for the efficient treatment of polymer solution dynamics. It has recently been used to investigate the phase separation of binary fluids in the presence of solid particles. The LB method is originated from lattice gas automaton which is constructed as a simplified, fictitious molecular dynamic in which space, time and particle velocities are all discrete. A typical lattice gas automaton consists of a regular lattice with particles residing on the nodes. The main feature of the LB method is to replace the particle occupation variables (Boolean variables), by single-particle distribution functions (real variables) and neglect individual particle motion and particle–particle correlations in the kinetic equation. There are several ways to obtain the LB equation from either the discrete velocity model or the Boltzmann kinetic equation, and to derive the macroscopic Navier–Stokes equations from the LB equation. An important advantage of the LB method is that microscopic physical interactions of the fluid particles can be conveniently incorporated into the numerical model. Compared with the Navier– Stokes equations, the LB method can handle the interactions among fluid particles and reproduce the microscale mechanism of hydrodynamic behavior. Therefore it belongs to the MD in nature and bridges the gap between the molecular level and macroscopic level. However, its main disadvantage is that it is typically not guaranteed to be numerically stable and may lead to physically unreasonable results, for instance, in the case of high forcing rate or high interparticle interaction strength.

10.9.3.4 Time-Dependent Ginzburg–Landau Method

TDGL is a microscale method for simulating the structural evolution of phase-separation in polymer blends and block copolymers. It is based on the Cahn–Hilliard–Cook (CHC) nonlinear diffusion equation for a binary blend and falls under the more general phase-field and reaction-diffusion models [30, 31]. In the TDGL method, a free-energy function is minimized to simulate a temperature quench from the miscible region of the phase diagram to the immiscible region. Thus, the resulting time-dependent structural evolution of the polymer blend can be investigated by solving the TDGL/CHC equation for the time dependence of the local blend concentration. Glotzer and co-workers have discussed and applied this method to polymer blends and particle-filled polymer systems [32]. This model reproduces the growth kinetics of the TDGL model, demonstrating that such quantities are insensitive to the precise form of the double-well potential of the bulk free-energy term. The TDGL and CDM methods have recently been used to investigate the phase-separation of polymer nanocomposites and polymer blends in the presence of nanoparticles [33–35].

10.9.3.5 Dynamic DFT Method

Dynamic DFT method is usually used to model the dynamic behavior of polymer systems and has been implemented in the software package Mesodyn™ from Accelrys [36]. The DFT models the behavior of polymer fluids by combining Gaussian mean-field statistics with a TDGL model for the time evolution of conserved order parameters. However, in contrast to traditional phenomenological free-energy expansion methods employed in the TDGL approach, the free energy is not truncated at a certain level, and instead retains the full polymer path integral numerically.

At the expense of a more challenging computation, this allows detailed information about a specific polymer system beyond simply the Flory–Huggins parameter and mobilities to be included in the simulation. In addition, viscoelasticity, which is not included in TDGL approaches, is included at the level of the Gaussian chains. A similar DFT approach has been developed by Doi and co-workers [37–38] and forms the basis for their new software tool Simulation Utilities for Soft and Hard Interfaces

(SUSHI), one of a suite of molecular and mesoscale modeling tools (called OCTA) developed for the simulation of polymer materials [38]. The essence of dynamic DFT method is that the instantaneous unique conformation distribution can be obtained from the off-equilibrium density profile by coupling a fictitious external potential to the Hamiltonian. Once such distribution is known, the free energy is then calculated by standard statistical thermodynamics. The driving force for diffusion is obtained from the spatial gradient of the first functional derivative of the free energy with respect to the density. Here, we describe briefly the equations for both polymer and particle in the diblock polymer–particle composites [39].

10.9.4 MESOSCALE AND MACROSCALE METHODS

Despite the importance of understanding the molecular structure and nature of materials, their behavior can be homogenized with respect to different aspects which can be at different scales. Typically, the observed macroscopic behavior is usually explained by ignoring the discrete atomic and molecular structure and assuming that the material is continuously distributed throughout its volume. The continuum material is thus assumed to have an average density and can be subjected to body forces such as gravity and surface forces. Generally speaking, the macroscale methods (or called continuum methods hereafter) obey the fundamental laws of: (i) continuity, derived from the conservation of mass; (ii) equilibrium, derived from momentum considerations and Newton's second law; (iii) the moment of momentum principle, based on the model that the time rate of change of angular momentum with respect to an arbitrary point is equal to the resultant moment; (iv) conservation of energy, based on the first law of thermodynamics; and (v) conservation of entropy, based on the second law of thermodynamics. These laws provide the basis for the continuum model and must be coupled with the appropriate constitutive equations and the equations of state to provide all the equations necessary for solving a continuum problem. The continuum method relates the deformation of a continuous medium to the external forces acting on the medium and the resulting internal stress and strain. Computational approaches range from simple closed-form analytical expressions to micromechanics and complex structural mechanics calculations based on beam and shell theory.

In this section, we introduce some continuum methods that have been used in polymer nanocomposites, including micromechanics models (e.g., Halpin–Tsai model, Mori–Tanaka model), equivalent-continuum model, self-consistent model and finite element analysis.

10.9.5 MICROMECHANICS

Since the assumption of uniformity in continuum mechanics may not hold at the microscale level, micromechanics methods are used to express the continuum quantities associated with an infinitesimal material element in terms of structure and properties of the micro constituents. Thus, a central theme of micromechanics models is the development of a representative volume element (RVE) to statistically represent the local continuum properties. The RVE is constructed to ensure that the length scale is consistent with the smallest constituent that has a first-order effect on the macroscopic behavior. The RVE is then used in a repeating or periodic nature in the full-scale model. The micromechanics method can account for interfaces between constituents, discontinuities, and coupled mechanical and nonmechanical properties. Our purpose is to review the micromechanics methods used for polymer nanocomposites. Thus, we only discuss here some important concepts of micromechanics as well as the Halpin–Tsai model and Mori–Tanaka model.

10.9.5.1 Basic Concepts

When applied to particle reinforced polymer composites, micromechanics models usually follow such basic assumptions as (i) linear elasticity of fillers and polymer matrix; (ii) the fillers are axisymmetric, identical in shape and size, and can be characterized by parameters such as aspect ratio; (iii) well-bonded filler–polymer interface and the ignorance of interfacial slip, filler–polymer debonding or matrix cracking. The first concept is the linear elasticity, that is, the linear relationship between the total stress and infinitesimal strain tensors for the filler and matrix as expressed by the following constitutive equations:

$$\textit{For filler}\,\sigma^f = C^f \varepsilon^f \tag{22}$$

$$For\,matrix\,\sigma^{m} = C^{m}\varepsilon^{m} \tag{23}$$

where C is the stiffness tensor. The second concept is the average stress and strain. Since the point-wise stress field α (*x*) and the corresponding strain field ε(*x*) are usually nonuniform in polymer composites, the volume–average stress $\bar{\sigma}$ and strain are $\bar{\varepsilon}$ then defined over the representative averaging volume V, respectively,

$$\bar{\sigma} = \frac{1}{v}\int \sigma(x)\,dv \tag{24}$$

$$\bar{\varepsilon} = \frac{1}{v}\int \tau(x)\,dv \tag{25}$$

Therefore, the average filler and matrix stresses are the averages over the corresponding volumes and, respectively,

$$\overline{\sigma_f} = \frac{1}{V_f}\int \sigma(x)\,dv \tag{26}$$

$$\overline{\sigma_m} = \frac{1}{V_f}\int \sigma(x)\,dv \tag{27}$$

The average strains for the fillers and matrix are defined, respectively, as

$$\overline{\varepsilon_f} = \frac{1}{V_f}\int \tau(x)\,dv \tag{28}$$

$$\overline{\varepsilon_m} = \frac{1}{V_m}\int \tau(x)\,dv \tag{29}$$

Based on the above definitions, the relationships between the filler and matrix averages and the overall averages can be derived as follows:

$$\bar{\sigma} = \overline{\sigma_f}v_f + \overline{\overline{\sigma_m}}v_m \tag{30}$$

$$\bar{\varepsilon} = \overline{\varepsilon_f}v_f + \overline{\varepsilon_m}v_m \tag{31}$$

where v_f, v_m are the volume fractions of the fillers and matrix, respectively.

The third concept is the average properties of composites which are actually the main goal of a micromechanics model. The average stiffness of the composite is the tensor C that maps the uniform strain to the average stress

$$\bar{\sigma} = \bar{\varepsilon} C \tag{32}$$

The average compliance S is defined in the same way:

$$\bar{\varepsilon} = \bar{\sigma} S \tag{33}$$

Another important concept is the strain–concentration and stress–concentration tensors A and B which are basically the ratios between the average filler strain (or stress) and the corresponding average of the composites.

$$\overline{\varepsilon_f} = \bar{\bar{\varepsilon}} A \tag{34}$$

$$\overline{\sigma_f} = \bar{\bar{\sigma}} B \tag{35}$$

Using the above concepts and equations, the average composite stiffness can be obtained from the strain concentration tensor A and the filler and matrix properties:

$$C = C_m + v_f (C_f - C_m)_A \tag{36}$$

10.9.5.2 Halpin–Tsai Model

The Halpin–Tsai model is a well-known composite theory to predict the stiffness of unidirectional composites as a functional of aspect ratio. In this model, the longitudinal E_{11} and transverse E_{22} engineering moduli are expressed in the following general form:

$$\frac{E}{E_m} = \frac{1 + \zeta \eta v_f}{1 - \eta v_f} \tag{37}$$

where E and E_m represent the Young's modulus of the composite and matrix, respectively, v_f is the volume fraction of filler, and η is given by:

$$\eta = \frac{\dfrac{E}{E_m} - 1}{\dfrac{E_f}{E_m} + \zeta_f} \tag{38}$$

where E_f represents the Young's modulus of the filler and ζ_f the shape parameter depending on the filler geometry and loading direction. When calculating longitudinal modulus E_{11}, ζ_f is equal to l/t, and when calculating transverse modulus E_{22}, ζ_f is equal to w/t. Here, the parameters of l, w and t are the length, width and thickness of the dispersed fillers, respectively. If $\zeta_f \rightarrow 0$, the Halpin–Tsai theory converges to the inverse rule of mixture (lower bound):

$$\frac{1}{E} = \frac{v_f}{E_f} + \frac{1 - v_f}{E_m} \tag{39}$$

Conversely, if $\zeta_f \rightarrow \infty$, the theory reduces to the rule of mixtures (upper bound),

$$E = E_f v_f + E_m \left(1 - v_f\right) \tag{40}$$

10.9.5.3 Mori–Tanaka Model

The Mori–Tanaka model is derived based on the principles of Eshelby's inclusion model for predicting an elastic stress field in and around ellipsoidal filler in an infinite matrix. The complete analytical solutions for longitudinal E_{11} and transverse E_{22} elastic moduli of an isotropic matrix filled with aligned spherical inclusion are [40]:

$$\frac{E_{11}}{E_m} = \frac{2A_0}{A_0 + v_f \left(A_1 + 2v_0 A_2\right)} \tag{41}$$

$$\frac{E_{22}}{E_m} = \frac{2A_0}{2A_0 + v_f \left(-2A_3 + \left(1 - v_0 A_4\right) + \left(1 + v_0\right) A_5 A_0\right)} \tag{42}$$

where E_m represents the Young's modulus of the matrix, v_f the volume fraction of filler, v_0 the Poisson's ratio of the matrix, parameters, A0, A1,…,A5 are functions of the Eshelby's tensor and the properties of the filler and the matrix, including Young's modulus, Poisson's ratio, filler concentration and filler aspect ratio [40].

10.9.5.4 Equivalent-Continuum and Self-Similar

Numerous micromechanical models have been successfully used to predict the macroscopic behavior of fiber-reinforced composites. However, the direct use of these models for nanotube-reinforced composites is doubtful due to the significant scale difference between nanotube and typical carbon fiber. Recently, two methods have been proposed for modeling the mechanical behavior of single-walled carbon nanotube (SWCN) composites: equivalent-continuum approach and self-similar approach [41]. The equivalent-continuum approach was proposed by Odegard et al. [42]. In this approach, MD was used to model the molecular interactions between SWCN–polymer and a homogeneous equivalent-continuum reinforcing element (e.g., a SWCN surrounded polymer) was constructed. Then, micromechanics are used to determine the effective bulk properties of the equivalent-continuum reinforcing element embedded in a continuous polymer. The equivalent-continuum approach consists of four major steps, as briefly described below. Step 1: MD simulation is used to generate the equilibrium structure of a SWCN–polymer composite and then to establish the RVE of the molecular model and the equivalent-continuum model. Step 2: The potential energies of deformation for the molecular model and effective fiber are derived and equated for identical loading conditions. The bonded and nonbonded interactions within a polymer molecule are quantitatively described by MM. For the SWCN/polymer system, the total potential energy U^m of the molecular model is:

$$U^m = \sum U^r(K_r) + \sum U^\theta(k_\theta) + \sum U^{vdw}(k_{vdw}) \tag{43}$$

where U^r, U^θ and U^{vdw} are the energies associated with covalent bond stretching, bond-angle bending, and van der Waals interactions, respectively. An equivalent-truss model of the RVE is used as an intermediate step to link the molecular and equivalent-continuum models. Each atom in the molecular model is represented by a pin-joint, and each truss element represents an atomic bonded or nonbonded interaction. The potential energy of the truss model is

$$U^t = \sum U^a(E^a) + \sum U^b(E^b) + \sum U^c(E^c) \tag{44}$$

where U^a, U^b and U^c are the energies associated with truss elements that represent covalent bond stretching, bond-angle bending, and van der Waals interactions, respectively. The energies of each truss element are a function of the Young's modulus, E.

Step 3: A constitutive equation for the effective fiber is established. Since the values of the elastic stiffness tensor components are not known a priori, a set of loading conditions are chosen such that each component is uniquely determined from

$$U^f = U^t = U^m \tag{45}$$

Step 4: Overall constitutive properties of the dilute and unidirectional SWCN/polymer composite are determined with Mori–Tanaka model with the mechanical properties of the effective fiber and the bulk polymer. The layer of polymer molecules that are near the polymer/nanotube interface is included in the effective fiber, and it is assumed that the matrix polymer surrounding the effective fiber has mechanical properties equal to those of the bulk polymer. The self-similar approach was proposed by Pipes and Hubert [43] which consists of three major steps:

First, a helical array of SWCNs is assembled. This array is termed as the SWCN nanoarray where 91 SWCNs make up the cross-section of the helical nanoarray. Then, the SWCN nanoarrays is surrounded by a polymer matrix and assembled into a second twisted array, termed as the SWCN nanowire Finally, the SWCN nanowires are further impregnated with a polymer matrix and assembled into the final helical array—the SWCN microfiber. The self-similar geometries described in the nanoarray, nanowire and microfiber allow the use of the same mathematical and geometric model for all three geometries [43].

10.9.5.5 Finite Element Method

FEM is a general numerical method for obtaining approximate solutions in space to initial-value and boundary-value problems including time-dependent processes. It employs preprocessed mesh generation, which enables the model to fully capture the spatial discontinuities of highly inhomogeneous materials. It also allows complex, nonlinear tensile relationships to be incorporated into the analysis. Thus, it has been widely used in mechanical, biological and geological systems. In FEM, the entire domain of interest is spatially discretized into an assembly of simply shaped subdomains (e.g., hexahedra or tetrahedral in three dimensions, and rectangles or triangles in two dimensions) without gaps and without overlaps. The subdomains are interconnected at joints (i.e., nodes). The energy in FEM is taken from the theory of linear elasticity and thus the input parameters are simply the elastic moduli and the density of the material. Since these parameters are in agreement with the values computed by MD, the simulation is consistent across the scales. More specifically, the total elastic energy in the absence of tractions and body forces within the continuum model is given by [44]:

$$U = U_v + U_k \tag{46}$$

$$U_k = \frac{1}{2}\int drp(r)\left|U_r^{\cdot}\right|^2 \tag{47}$$

$$U_v = \frac{1}{2}\int dr \sum_{\mu\nu\sigma\lambda=1}^{a} \varepsilon_{\mu\nu}(r) C_{\mu\nu\lambda\sigma^3\lambda\sigma(r)} \tag{48}$$

where U_v is the Hookian potential energy term which is quadratic in the symmetric strain tensor e, contracted with the elastic constant tensor C. The Greek indices (i.e., m, n, l, s) denote Cartesian directions. The kinetic energy U_k involves the time rate of change of the displacement field U, and the mass density ρ.

These are fields defined throughout space in the continuum theory. Thus, the total energy of the system is an integral of these quantities over the volume of the sample dυ. The FEM has been incorporated in some commercial software packages and open source codes (e.g., ABAQUS, ANSYS, Palmyra and OOF) and widely used to evaluate the mechanical properties of polymer composites. Some attempts have recently been made to apply the FEM to nanoparticle-reinforced polymer nanocomposites. In order to capture the multiscale material behaviors, efforts are also underway to combine the multiscale models spanning from molecular to macroscopic levels [45, 46].

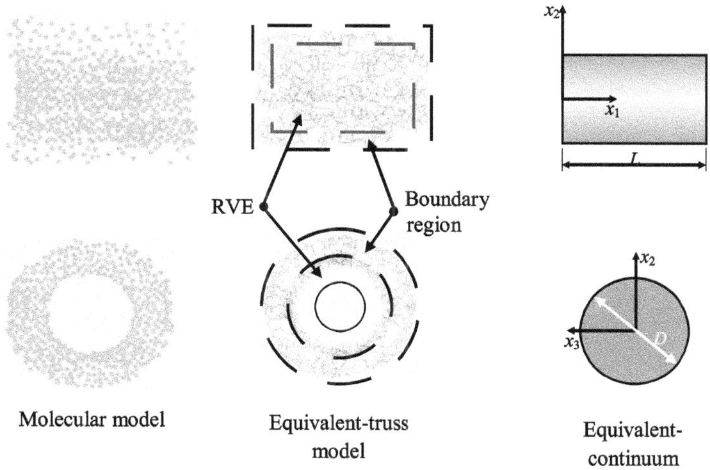

FIGURE 10.7 Equivalent-continuum modeling of effective fiber.

10.9.6 MULTI SCALE MODELING OF MECHANICAL PROPERTIES

In Odegard's study [42], a method has been presented for linking atomistic simulations of nano-structured materials to continuum models of the corresponding bulk material. For a polymer composite system reinforced with single-walled carbon nanotubes (SWNT), the method provides the steps whereby the nanotube, the local polymer near the nanotube, and the nanotube/polymer interface can be modeled as an effective continuum fiber by using an equivalent-continuum model. The effective fiber retains the local molecular structure and bonding information, as defined by molecular dynamics, and serves as a means for linking the equivalent-continuum and micromechanics models. The micromechanics method is then available for the prediction of bulk mechanical properties of SWNT/polymer composites as a function of nanotube size, orientation, and volume fraction. The utility of this method was examined by modeling tow composites that both having an interface. The elastic stiffness constants of the composites were determined for both aligned and three-dimensional randomly oriented nanotubes, as a function of nanotube length and volume fraction. They used Mori– Tanaka model [47] for random and oriented fibers position and compare their model with mechanical properties, the interface between fiber and matrix was assumed perfect. Motivated by micrographs showing that embedded nanotubes often exhibit significant curvature within the polymer, Fisher et al.[48] have developed a model combining finite element results and micromechanical methods (Mori-Tanaka) to determine the effective reinforcing modulus of a wavy embedded nanotube with perfect bonding and random fiber orientation assumption. This effective reinforcing modulus (ERM) is then used within a multiphase micromechanics model to predict the effective modulus of a polymer reinforced with a distribution of wavy nanotubes. We found that even slight nanotube curvature significantly reduces the effective reinforcement when compared to straight nanotubes. These results suggest that nanotube waviness may be an additional mechanism limiting the modulus enhancement of nanotube-reinforced polymers. Bradshaw et al. [49] investigated the degree to which the characteristic waviness of nanotubes embedded in polymers can impact the effective stiffness of these materials. A 3D finite element model of a single infinitely long sinusoidal fiber within an infinite matrix is used to numerically compute the dilute strain concentration tensor. A Mori–Tanaka model uses this tensor to predict the effective modulus of the material with aligned or randomly oriented inclusions. This hybrid finite element micromechanical modeling technique is a powerful extension of general micromechanics modeling and can be applied to any composite microstructure containing nonellipsoidal inclusions. The results demonstrate that nanotube waviness results in a reduction of the effective modulus of the composite relative to straight nanotube reinforcement. The degree of reduction is dependent on the ratio of the sinusoidal wavelength to the nanotube diameter. As this wavelength ratio increases, the effective stiffness of a composite with randomly oriented wavy nanotubes converges to the result obtained with straight nanotube inclusions.

The effective mechanical properties of carbon nanotube-based composites are evaluated by Liu and Chen [50] using a 3-D nanoscale RVE based on 3-D elasticity theory and solved by the finite element method. Formulas to extract the material constants from solutions for the RVE under three loading cases are established using the elasticity. An extended rule of mixtures, which can be used to estimate the Young's modulus in the axial direction of the RVE and to validate the numerical solutions for short CNTs, is also derived using the strength of materials theory. Numerical examples using the FEM to evaluate the effective material constants of a CNT-based composites are presented, which demonstrate that the reinforcing capabilities of the CNTs in a matrix are significant. With only about 2% and 5% volume fractions of the CNTs in a matrix, the stiffness of the composite in the CNT axial direction can increase as many as 0.7 and 9.7 times for the cases of short and long CNT fibers, respectively. These simulation results, which are believed to be the first of its kind for CNT-based composites, are consistent with the experimental results reported in the literature Schadler et al. [51], Wagner et al. [52]; Qian et al. [53]. The developed extended rule of mixtures is also found to be quite effective in evaluating the stiffness of the CNT-based composites in the CNT axial direction. Many research issues need to be addressed in the modeling and simulations of CNTs in a matrix material for the development of nanocomposites. Analytical methods and simulation models to extract the mechanical properties of the CNT-based nanocomposites need to be further developed and verified with experimental results. The analytical method and simulation approach developed in this paper are only a preliminary study. Different type of RVEs, load cases and different solution methods should be investigated. Different interface conditions, other than perfect bonding, need to be investigated using different models to more accurately account for the interactions of the CNTs in a matrix material at the nanoscale. Nanoscale interface cracks can be analyzed using simulations to investigate the failure mechanism in nanomaterials. Interactions among a large number of CNTs in a matrix can be simulated if the computing power is available. Single-walled and multiwalled CNTs as reinforcing fibers in a matrix can be studied by simulations to find out their advantages and disadvantages. Finally, large multiscale simulation models for CNT-based composites, which can link the models at the nano, micro and macro scales, need to be developed, with the help of analytical and experimental work [50]. The three RVEs proposed in Ref. [54] and shown in Fig. 10.8 are relatively simple regarding the models and scales and pictures in Fig. 10.9 are Three loading cases for the cylindrical RVE. However, this is only the first step toward more sophisticated and large scale simulations of CNT-based composites. As the computing power and confidence in simulations of CNT-based composites increase, large scale 3-D models containing hundreds or even more CNTs, behaving linearly or nonlinearly,

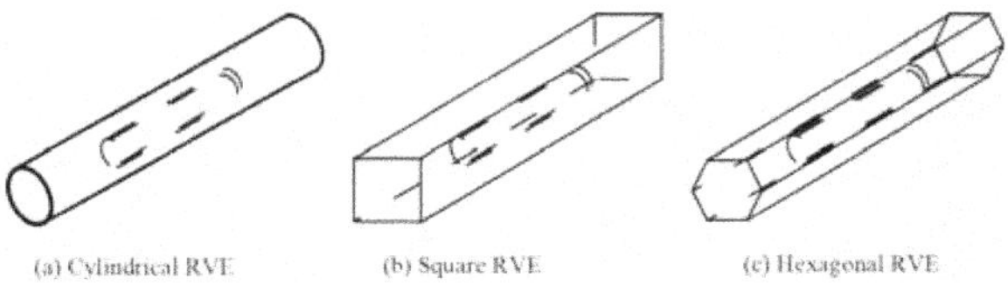

FIGURE 10.8 Three nanoscale representative volume elements for the analysis of CNT-based nanocomposites.

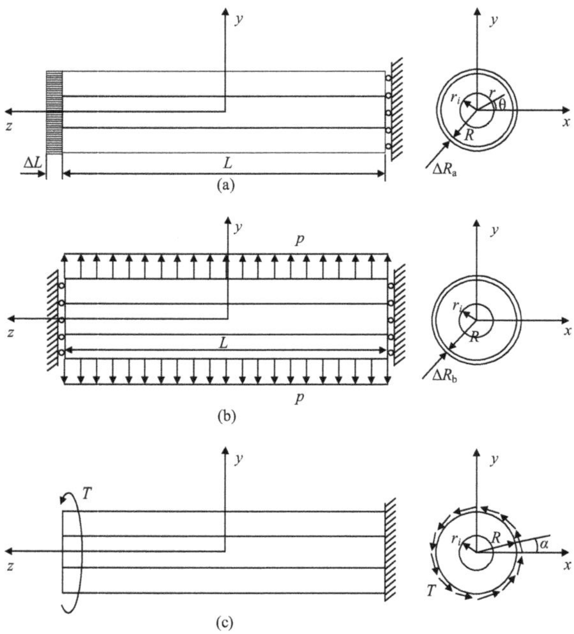

FIGURE 10.9 Three loading cases for the cylindrical RVE used to evaluate the effective material properties of the CNT-based composites. (a) Under axial stretch DL; (b) under lateral uniform load P; (c) under torsional load T.

with coatings or of different sizes, distributed evenly or randomly, can be employed to investigate the interactions among the CNTs in a matrix and to evaluate the effective material properties. Other numerical methods can also be attempted for the modeling and simulations of CNT-based composites, which may offer some advantages over the FEM approach. For example, the boundary element method, Liu et al. [54]; Chen and Liu [55], accelerated with the fast multipole techniques, Fu et al. [56]; Nishimura et al. [57], and the mesh free methods (Qian et al. [58]) may enable one to model an RVE with thousands of CNTs in a matrix on a desktop computer. Analysis of the CNT-based composites using the boundary element method is already underway and will be reported subsequently.

The effective mechanical properties of CNT based composites are evaluated using square RVEs based on 3-D elasticity theory and solved by the FEM. Formulas to extract the effective material constants from solutions for the square RVEs under two loading cases are established based on elasticity. Square RVEs with multiple CNTs are also investigated in evaluating the Young's modulus and Poisson's ratios in the transverse plane. Numerical examples using the FEM are presented, which demonstrate that the load-carrying capabilities of the CNTs in a matrix are significant. With the addition of only about 3.6% volume fraction of the CNTs in a matrix, the stiffness of the composite in the CNT axial direction can increase as much as 33% for the case of long CNT fibers [58]. These simulation results are consistent with both the experimental ones reported in the literature [50–53, 59]. It is also found that cylindrical RVEs tend to overestimate the effective Young's moduli due to the fact that they overestimate the volume fractions of the CNTs in a matrix. The square RVEs, although more demanding in modeling and computing, may be the preferred model in future simulations for estimating the effective material constants, especially when multiple CNTs need to be considered. Finally, the rules of mixtures, for both long and short CNT cases, are found to be quite accurate in estimating the effective Young's moduli in the CNT axial direction. This may suggest that 3-D FEM modeling may not be necessary in obtaining the effective material constants in the CNT direction, as in the studies of the conventional fiber reinforced composites. Efforts in comparing the results presented in this paper using the continuum approach directly with the MD simulations are underway. This is feasible now only for a smaller RVE of one CNT embedded in a matrix. In future research, the MD and continuum approach should be integrated in a multiscale modeling and simulation environment for analyzing the CNT-based composites. More efficient models of the CNTs in a matrix also need to be developed, so that a large number of CNTs, in different shapes and forms (curved or twisted), or randomly distributed in a matrix, can be modeled. The ultimate validation of the simulation results should be done with the nanoscale or microscale experiments on the CNT reinforced composites [58].

Griebel & Hamaekers [60] reviewed the basic tools used in computational nanomechanics and materials, including the relevant underlying principles and concepts. These tools range from subatomic ab initio methods to classical molecular dynamics and multiple-scale approaches. The energetic link between the quantum mechanical and classical systems has been discussed, and limitations of the standing alone molecular dynamics simulations have been shown on a series of illustrative examples. The need for multiscale simulation methods to tackle nanoscale aspects of material behavior was therefore emphasized; that was followed by a review and classification of the mainstream and emerging multiscale methods. These simulation methods include the broad areas of quantum mechanics, molecular dynamics and multiple-scale approaches, based on coupling the atomistic and continuum models. They summarize the strengths and limitations of currently available multiple-scale techniques, where the emphasis is made on the latest perspective approaches, such as the bridging scale method, multiscale boundary conditions, and multiscale fluidics. Example problems, in which multiple-scale simulation methods yield equivalent results to full atomistic simulations at fractions of the computational cost, were shown. They Compare their results with Odegard, et al. [42], the micromechanics method was BEM Halpin-Tsai Eq. [26] with aligned fiber by perfect bonding.

The solutions of the strain-energy-changes due to a SWNT embedded in an infinite matrix with imperfect fiber bonding are obtained through numerical method by Wan, et al. [61]. A “critical” SWNT fiber length is defined for full load transfer between the SWNT and the matrix, through the evaluation of the strain-energy-changes for different fiber lengths The strain-energy-change is also used to derive the effective longitudinal Young’s modulus and effective bulk modulus of the composite, using a dilute solution. The main goal of their research was investigation of strain-energy-change due to inclusion of SWNT using FEM. To achieve full load transfer between the SWNT and the matrix, the length of SWNT fibers should be longer than a ‘critical’ length if no weak interphase exists between the SWNT and the matrix [61].

A hybrid atomistic/continuum mechanics method is established in the Feng, et al.’ [62] study the deformation and fracture behaviors of carbon nanotubes (CNTs) in composites. The unit cell containing a CNT embedded in a matrix is divided in three regions, which are simulated by the atomic-potential method, the continuum method based on the modified Cauchy–Born rule, and the classical continuum mechanics, respectively. The effect of CNT interaction is taken into account via the Mori–Tanaka effective field method of micromechanics. This method not only can predict the formation of Stone–Wales (5–7–7–5) defects, but also simulate the subsequent deformation and fracture process of CNTs. It is found that the critical strain of defect nucleation in a CNT is sensitive to its chiral angle but not to its diameter. The critical strain of Stone–Wales defect formation of zigzag CNTs is nearly twice that of armchair CNTs. Due to the constraint effect of matrix, the CNTs embedded in a composite are easier to fracture in comparison with those not embedded. With the increase in the Young's modulus of the matrix, the critical breaking strain of CNTs decreases.

Estimation of effective elastic moduli of nanocomposites was performed by the version of effective field method developed in the framework of quasi-crystalline approximation when the spatial correlations of inclusion location take particular ellipsoidal forms [63]. The independent justified choice of shapes of inclusions and correlation holes provide the formulae of effective moduli which are symmetric, completely explicit and easily to use. The parametric numerical analyzes revealed the most sensitive parameters influencing the effective moduli which are defined by the axial elastic moduli of nanofibers rather than their transversal moduli as well as by the justified choice of correlation holes, concentration and prescribed random orientation of nanofibers [63].

Li and Chou [64, 65] have reported a multiscale modeling of the compressive behavior of carbon nanotube/polymer composites. The nanotube is modeled at the atomistic scale, and the matrix deformation is analyzed by the continuum finite element method. The nanotube and polymer matrix are assumed to be bonded by van der Waals interactions at the interface. The stress distributions at the nanotube/polymer interface under isostrain and isostress loading conditions have been examined. They have used beam elements for SWCNT using molecular structural mechanics, truss rod for vdW links and cubic elements for matrix. The rule of mixture was used as for comparison in this research. The buckling forces of nanotube/polymer composites for different nanotube lengths and diameters are computed. The results indicate that continuous nanotubes can most effectively enhance the composite buckling resistance.

Anumandla and Gibson [66] describes an approximate, yet comprehensive, closed form micromechanics model for estimating the effective elastic modulus of carbon nanotube-reinforced composites. The model incorporates the typically observed nanotube curvature, the nanotube length, and both 1D and 3D random arrangement of the nanotubes. The analytical results obtained from the closed form micromechanics model for nanoscale representative volume elements and results from an equivalent finite element model for effective reinforcing modulus of the nanotube reveal that the reinforcing modulus is strongly dependent on the waviness, wherein, even a slight change in the nanotube curvature can induce a prominent change in the effective reinforcement provided. The micromechanics model is also seen to produce reasonable agreement with experimental data for the effective tensile modulus of composites reinforced with multiwalled nanotubes (MWNTs) and having different MWNT volume fractions.

Effective elastic properties for carbon nanotube reinforced composites are obtained through a variety of micromechanics techniques [67]. Using the in-plane elastic properties of graphene, the effective properties of carbon nanotubes are calculated using a composite cylinders micromechanics technique as a first step in a two-step process. These effective properties are then used in the self-consistent and Mori–Tanaka methods to obtain effective elastic properties of composites consisting of aligned single or multiwalled carbon nanotubes embedded in a polymer matrix. Effective composite properties from these averaging methods are compared to a direct composite cylinders approach extended from the work of Hashin and Rosen [68] and Christensen and Lo [69]. Comparisons with finite element simulations are also performed. The effects of an interphase layer between the nanotubes and the polymer matrix as result of functionalization is also investigated using a multilayer composite cylinders approach. Finally, the modeling of the clustering of nanotubes into bundles due to interatomic forces is accomplished herein using a tessellation method in conjunction with a multiphase Mori–Tanaka technique. In addition to aligned nanotube composites, modeling of the effective elastic properties of randomly dispersed nanotubes into a matrix is performed using the Mori–Tanaka method, and comparisons with experimental data are made.

Selmi, et al. [70] deal with the prediction of the elastic properties of polymer composites reinforced with single walled carbon nanotubes. Their contribution is the investigation of several micromechanical models, while most of the papers on the subject deal with only one approach. They implemented four homogenization schemes, a sequential one and three others based on various extensions of the Mori–Tanaka (M–T) mean-field homogenization model: two-level (M–T/M–T), two-step (M–T/M–T) and two-step (M–T/Voigt). Several composite systems are studied, with various properties of the matrix and the graphene, short or long nanotubes, fully aligned or randomly oriented in 3D or 2D. Validation targets are experimental data or finite element results, either based on a 2D periodic unit cell or a 3D representative volume element. The comparative study showed that there are cases where all micromechanical models give adequate predictions, while for some composite materials and some properties, certain models fail in a rather spectacular fashion. It was found that the two-level (M–T/M–T) homogenization model gives the best predictions in most cases. After the characterization of the discrete nanotube structure using a homogenization method based on energy equivalence, the sequential, the two-step (M–T/M–T), the two-step (M–T/Voigt), the two-level (M–T/M–T) and finite element models were used to predict the elastic properties of SWNT/polymer composites. The data delivered by the micromechanical models are compared against those obtained by finite element analyzes or experiments. For fully aligned, long nanotube polymer composite, it is the sequential and the two-level (M–T/M–T) models which delivered good predictions. For all composite morphologies (fully aligned, two-dimensional in-plane random orientation, and three-dimensional random orientation), it is the two-level (M–T/M–T) model which gave good predictions compared to finite element and experimental results in most situations. There are cases where other micromechanical models failed in a spectacular way.

Luo, et al. [71] have used multiscale homogenization (MH) and FEM for wavy and straight SWCNTs, they have compare their results with Mori-Tanaka, Cox, Halpin-Tsai, Fu, et al. [72], Lauke [72], Tserpes, et al. [73] used 3D elastic beam for C-C bond and, 3D space frame for CNT and progressive fracture model for prediction of elastic modulus, they used rule of mixture for compression of their results. Their assumption was embedded a single SWCNT in polymer with Perfect bonding. The multiscale modeling, Monte Carlo, FEM and using equivalent continuum method was used by Spanos and Kontsos [74] and compared with Zhu, et al. [75] and Paiva, et al. [76]'s results.

Md. A. Bhuiyan et al. [77] studied the effective modulus of CNT/PP composites using FEA of a 3D RVE which includes the PP matrix, multiple CNTs and CNT/PP interphase and accounts for poor dispersion and non homogeneous distribution of CNTs within the polymer matrix, weak CNT/polymer interactions, CNT agglomerates of various sizes and CNTs orientation and waviness. Currently, there is no other model, theoretical or numerical, that accounts for all these experimentally observed phenomena and captures their individual and combined effect on the effective modulus of nanocomposites. The model is developed using input obtained from experiments and validated against experimental data. CNT reinforced PP composites manufactured by extrusion and injection molding are characterized in terms of tensile modulus, thickness and stiffness of CNT/PP interphase, size of CNT agglomerates and CNT distribution using tensile testing, AFM and SEM, respectively. It is concluded that CNT agglomeration and waviness are the two dominant factors that hinder the great potential of CNTs as polymer reinforcement. The proposed model provides the upper and lower limit of the modulus of the CNT/PP composites and can be used to guide the manufacturing of composites with engineered properties for targeted applications. CNT agglomeration can be avoided by employing processing techniques such as sonication of CNTs, stirring, calendaring, etc., whereas CNT waviness can be eliminated by increasing the injection pressure during molding and mainly by using CNTs with smaller aspect ratio. Increased pressure during molding can also promote the alignment of CNTs along the applied load direction. The 3D modeling capability presented in this study gives an insight on the upper and lower bound of the CNT/PP composites modulus quantitatively by accurately capturing the effect of various processing parameters. It is observed that when all the experimentally observed factors are considered together in the FEA the modulus prediction is in good agreement with the modulus obtained from the experiment. Therefore, it can be concluded that the FEM models proposed in this study by systematically incorporating experimentally observed characteristics can be effectively used for the determination of mechanical properties of nanocomposite materials. Their result is in agreement with the results reported in Ref. [78]. The theoretical micromechanical models, shown in Fig. 10.10, are used to confirm that our FEM model predictions follow the same trend with the one predicted by the models as expected.

For reasons of simplicity and in order to minimize the mesh dependency on the results the hollow CNTs are considered as solid cylinders of circular cross-sectional area with an equivalent average diameter, shown in Fig. 10.11, calculated by equating the volume of the hollow CNT to the solid one [78].

The micromechanical models used for the comparison were Halpin–Tsai (H–T) [79] and Tandon–Weng (T–W) [40] model and the comparison

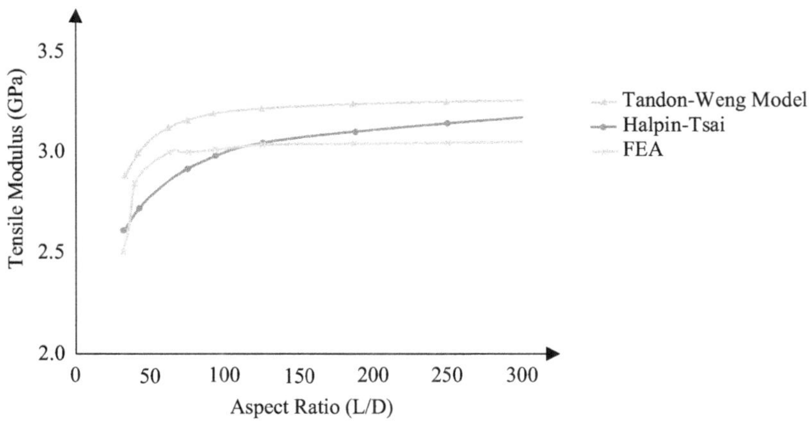

FIGURE 10.10 Effective modulus of 5 wt.% CNT/PP composites: theoretical models vs. FEA.

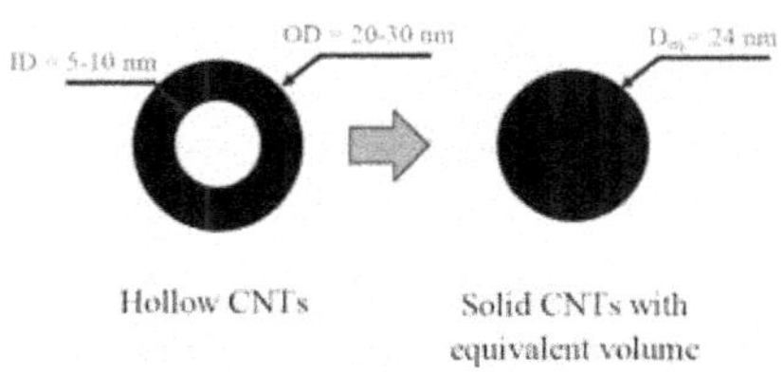

FIGURE 10.11 Schematic of the CNTs considered for the FEA.

was performed for 5 wt.% CNT/PP. It was noted that the H–T model results to lower modulus compared to FEA because H–T equation does not account for maximum packing fraction and the arrangement of the reinforcement in the composite. A modified H–T model that account for this has been proposed in the literature [80]. The effect of maximum packing fraction and the arrangement of the reinforcement within the composite become less significant at higher aspect ratios [81].

A finite element model of carbon nanotube, interphase and its surrounding polymer is constructed to study the tensile behavior of embedded short carbon nanotubes in polymer matrix in presence of vdW interactions in interphase region by Shokrieh and Rafiee [82].The interphase is modeled using nonlinear spring elements capturing the force-distance curve of vdW interactions. The constructed model is subjected to tensile loading to extract longitudinal Young's modulus. The obtained results of this work have been compared with the results of previous research of the same authors [83] on long embedded carbon nanotube in polymer matrix. It shows that the capped short carbon nanotubes reinforce polymer matrix less efficient than long CNTs.

Despite the fact that researches have succeeded to grow the length of CNTs up to 4 cm as a world record in US Department of Energy Los Alamos National Laboratory [84] and also there are some evidences on producing CNTs with lengths up to millimeters [85], CNTs are commercially available in different lengths ranging from 100 nm to approximately 30 lm in the market based on employed process of growth. Chemists at Rice University have identified a chemical process to cut CNTs into short segments. As a consequent, it can be concluded that the SWCNTs with lengths smaller than 1000 nm do not contribute significantly in reinforcing polymer matrix. On the other hand, the efficient length of reinforcement for a CNT with (10, 10) index is about 1.2 lm and short CNT with length of 10.8 lm can play the same role as long CNT reflecting the uppermost value reported in our previous research. Finally, it is shown that the direct use of Halpin–Tsai equation to predict the modulus of SWCNT/composites overestimates the results. It is also observed that application of previously developed long equivalent fiber stiffness is a good candidate to be used in Halpin–Tsai equations instead of Young's modulus of CNT. Halpin–Tsai equation is not an appropriate model for smaller lengths, since there is not any reinforcement at all for very small lengths [82].

Earlier, a nano-mechanical model has been developed by Chowdhury et al. [86] to calculate the tensile modulus and the tensile strength of randomly oriented short carbon nanotubes (CNTs) reinforced nanocomposites, considering the statistical variations of diameter and length of the CNTs. According to this model, the entire composite is divided into several composite segments which contain CNTs of almost the same diameter and length. The tensile modulus and tensile strength of the composite are then calculated by the weighted sum of the corresponding modulus and strength of each composite segment. The existing micromechanical approach for modeling the short fiber composites is modified to account for the structure of the CNTs, to calculate the modulus and the strength of each segmented CNT reinforced composites. Multi-walled CNTs with and without intertube bridging (see Fig. 10.12) have been considered. Statistical variations of the diameter and length of the CNTs are modeled by a normal distribution. Simulation results show that CNTs intertube bridging; length and diameter affect the nanocomposites modulus and strength. Simulation results have been compared with the available experimental results and the comparison concludes that the developed model can be effectively used to predict tensile modulus and tensile strength of CNTs reinforced composites.

The effective elastic properties of carbon nanotube-reinforced polymers have been evaluated by Tserpes and Chanteli [104] as functions of material and geometrical parameters using a homogenized RVE. The RVE

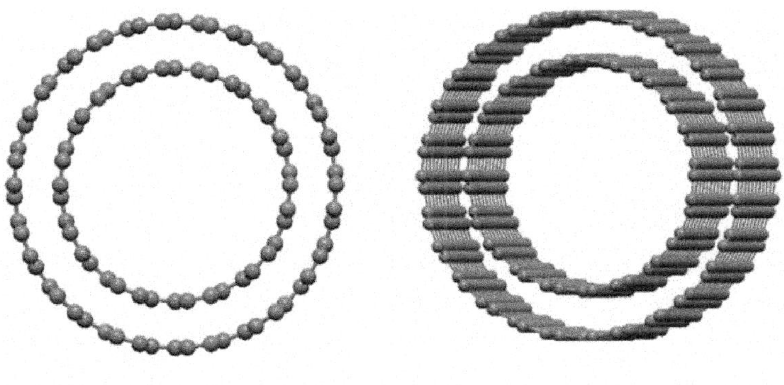

(a) (b)

FIGURE 10.12 Schematic of MWNT with intertube bridging. (a) Top view and (b) oblique view.

consists of the polymer matrix, a multiwalled carbon nanotube (MWCNT) embedded into the matrix and the interface between them. The parameters considered are the nanotube aspect ratio, the nanotube volume fraction as well as the interface stiffness and thickness. For the MWCNT, both isotropic and orthotropic material properties have been considered. Analyzes have been performed by means of a 3D FE model of the RVE. The results indicate a significant effect of nanotube volume fraction. The effect of nanotube aspect ratio appears mainly at low values and diminishes after the value of 20. The interface mostly affects the effective elastic properties at the transverse direction. Having evaluated the effective elastic properties of the MWCNT-polymer at the microscale, the RVE has been used to predict the tensile modulus of a polystyrene specimen reinforced by randomly aligned MWCNTs for which experimental data exist in the literature. A very good agreement is obtained between the predicted and experimental tensile moduli of the specimen. The effect of nanotube alignment on the specimen's tensile modulus has been also examined and found to be significant since as misalignment increases the effective tensile modulus decreases radically. The proposed model can be used for the virtual design and optimization of CNT-polymer composites since it has proven capable of assessing the effects of different material and geometrical parameters on the elastic properties of the composite and predicting the tensile modulus of CNT-reinforced polymer specimens.

10.9.7 MODELING OF THE INTERFACE

10.9.7.1 Introduction

The superior mechanical properties of the nanotubes alone do not ensure mechanically superior composites because the composite properties are strongly influenced by the mechanics that govern the nanotube–polymer interface. Typically in composites, the constituents do not dissolve or merge completely and therefore, normally, exhibit an interface between one another, which can be considered as a different material with different mechanical properties. The structural strength characteristics of composites greatly depend on the nature of bonding at the interface, the mechanical load transfer from the matrix (polymer) to the nanotube and the yielding of the interface. As an example, if the composite is subjected to tensile loading and there exists perfect bonding between the nanotube and polymer and/or a strong interface then the load (stress) is transferred to the nanotube; since the tensile strength of the nanotube (or the interface) is very high the composite can withstand high loads. However, if the interface is weak or the bonding is poor, on application of high loading either the interface fails or the load is not transferred to the nanotube and the polymer fails due to their lower tensile strengths. Consider another example of transverse crack propagation. When the crack reaches the interface, it will tend to propagate along the interface, since the interface is relatively weaker (generally) than the nanotube (with respect to resistance to crack propagation). If the interface is weak, the crack will cause the interface to fracture and result in failure of the composite. In this aspect, carbon nanotubes are better than traditional fibers (glass, carbon) due to their ability to inhibit nano and micro cracks. Hence, the knowledge and understanding of the nature and mechanics of load (stress) transfer between the nanotube and polymer and properties of the interface is critical for manufacturing of mechanically enhanced CNT-polymer composites and will enable in tailoring of the interface for specific applications or superior mechanical properties. Broadly, the interfacial mechanics of CNT-polymer composites is appealing from three aspects: mechanics, chemistry, and physics. From a mechanics point of view, the important questions are:

1. the relationship between the mechanical properties of individual constituents, that is, nanotube and polymer, and the properties of the interface and the composite overall.
2. the effect of the unique length scale and structure of the nanotube on the property and behavior of the interface.
3. ability of the mechanics modeling to estimate the properties of the composites for the design process for structural applications.

From a chemistry point of view, the interesting issues are:

1. The chemistry of the bonding between polymer and nanotubes, especially the nature of bonding (e.g., covalent or noncovalent and electrostatic).
2. The relationship between the composite processing and fabrication conditions and the resulting chemistry of the interface.

3. The effect of functionalization (treatment of the polymer with special molecular groups like hydroxyl or halogens) on the nature and strength of the bonding at the interface. From the physics point of view, researchers are interested in:
   a. The CNT-polymer interface serves as a model nano-mechanical or a lower dimensional system (1D) and physicists are interested in the nature of forces dominating at the nano-scale and the effect of surface forces (which are expected to be significant due to the large surface to volume ratio).
   b. The length scale effects on the interface and the differences between the phenomena of mechanics at the macro (or meso) and the nano-scale.

10.9.7.2 Some Methods in Interface Modeling

Computational techniques have extensively been used to study the interfacial mechanics and nature of bonding in CNT-polymer composites. The computational studies can be broadly classified as atomistic simulations and continuum methods. The atomistic simulations are primarily based on molecular dynamic simulations (MD) and density functional theory (DFT) [53, 87, 88]. The main focus of these techniques was to understand and study the effect of bonding between the polymer and nanotube (covalent, electrostatic or Van Der Waals forces) and the effect of friction on the interface. The continuum methods extend the continuum theories of micromechanics modeling and fiber-reinforced composites (elaborated in the next section) to CNT-polymer composites [50, 89] and explain the behavior of the composite from a mechanics point of view.

On the experimental side, the main types of studies that can be found in literature are as follows:

1. Researchers have performed experiments on CNT-polymer bulk composites at the macroscale and observed the enhancements in mechanical properties (like elastic modulus and tensile strength) and tried to correlate the experimental results and phenomena with continuum theories like micromechanics of composites or Kelly Tyson shear lag model [51, 90–92].

2. Raman spectroscopy has been used to study the reinforcement provided by carbon nanotubes to the polymer, by straining the CNT-polymer composite and observing the shifts in Raman peaks [93–95].
3. In situ TEM straining has also been used to understand the mechanics, fracture and failure processes of the interface. In these techniques, the CNT-polymer composite (an electron transparent thin specimen) is strained inside a TEM and simultaneously imaged to get real-time and spatially resolved (1 nm) information [96].

10.9.7.3 Numerical Approach

A MD model may serve as a useful guide, but its relevance for a covalent-bonded system of only a few atoms in diameter is far from obvious. Because of this, the phenomenological multiple column models that considers the interlayer radial displacements coupled through the van der Waals forces is used. It should also be mentioned the special features of load transfer, in tension and in compression, in MWNT-epoxy composites studied by Schadler et al. [51] who detected that load transfer in tension was poor in comparison to load transfer in compression, implying that during load transfer to MWNTs, only the outer layers are stressed in tension due to the telescopic innerwall sliding (reaching at the shear stress 0.5 MPa [97, 98]), whereas all the layers respond in compression. It should be mentioned that NTCMs usually contain not individual, separated SWCNTs, but rather bundles of closest-packed SWCNTs [53] where the twisting of the CNTs produces the radial force component giving the rope structure more stable than wires in parallel. Without strong chemically bonding, load transfer between the CNTs and the polymer matrix mainly comes from weak electrostatic and van der Waals interactions, as well as stress/deformation arising from mismatch in the coefficients of thermal expansion [99]. Numerous researchers [100] have attributed lower than- predicted CNT-polymer composite properties to the availability of only a weak interfacial bonding. So Frankland et al. [101] demonstrated by MD simulation that the shear strength of a polymer/nanotube interface with only van der Waals interactions could be increased by over an order of magnitude at the occurrence of covalent bonding for only 1%

of the nanotubes carbon atoms to the polymer matrix. The recent force-field-based molecular-mechanics calculations [102] demonstrated that the binding energies and frictional forces play only a minor role in determining the strength of the interface. The key factor in forming a strong bond at the interface is having a helical conformation of the polymer around the nanotube; polymer wrapping around nanotube improves the polymer-nanotube interfacial strength, although configurationally thermodynamic considerations do not necessarily support these architectures for all polymer chains [5]. Thus, the strength of the interface may result from molecular-level entanglement of the two phases and forced long-range ordering of the polymer. To ensure the robustness of data reduction schemes that are based on continuum mechanics, a careful analysis of continuum approximations used in macromolecular models and possible limitations of these approaches at the nanoscale are additionally required that can be done by the fitting of the results obtained by the use of the proposed phenomenological interface model with the experimental data of measurement of the stress distribution in the vicinity of a nanotube.

Meguid et.al [103] investigated the interfacial properties of carbon nanotube (CNT) reinforced polymer composites by simulating a nanotube pullout experiment. An atomistic description of the problem was achieved by implementing constitutive relations that are derived solely from interatomic potentials. Specifically, they adopt the Lennard-Jones (LJ) interatomic potential to simulate a nonbonded interface, where only the van der Waals (vdW) interactions between the CNT and surrounding polymer matrix were assumed to exist. The effects of such parameters as the CNT embedded length, the number of vdW interactions, the thickness of the interface, the CNT diameter and the cut-off distance of the LJ potential on the interfacial shear strength (ISS) are investigated and discussed. The problem is formulated for both a generic thermoset polymer and a specific two-component epoxy based on diglycidyl ether of bisphenol A (DGEBA) and triethylene tetramine (TETA) formulation. The study further illustrated that by accounting for different CNT capping scenarios and polymer morphologies around the embedded end of the CNT, the qualitative correlation between simulation and experimental pullout profiles can be improved. Only vdW interactions were considered between the atoms in the CNT and the polymer implying a nonbonded system. The vdW interactions were simulated using the LJ potential, while the CNT was described using the Modified Morse potential. The results reveal that the ISS shows a linear dependence on the vdW interaction density and decays significantly with increasing nanotube embedded length. The thickness of the interface was also varied and our results reveal that lower interfacial thicknesses favor higher ISS. When incorporating a 2.5 cut-off distance to the LJ potential, the predicted ISS shows an error of approximately 25.7% relative to a solution incorporating an infinite cut-off distance. Increasing the diameter of the CNT was found to increase the peak pullout force approximately linearly. Finally, an examination of polymeric and CNT capping conditions showed that incorporating an end cap in the simulation yielded high initial pullout peaks that better correlate with experimental findings. These findings have a direct bearing on the design and fabrication of carbon nanotube reinforced epoxy composites.

Fiber pullout tests have been well recognized as the standard method for evaluating the interfacial bonding properties of composite materials. The output of these tests is the force required to pullout the nanotube from the surrounding polymer matrix and the corresponding interfacial shear stresses involved. The problem is formulated using a representative volume element (RVE) which consists of the reinforcing CNT, the surrounding polymer matrix, and the CNT/polymer interface as depicted in Figure 10.13 (a,b) shows a schematic of the pullout process, where x is the pullout distance and L is the embedded length of the nanotube. The atomistic-based continuum (ABC) multiscale modeling technique is used

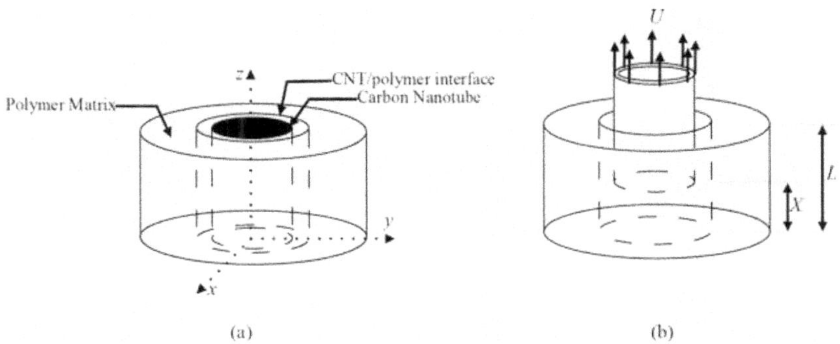

FIGURE 10.13 Schematic depictions of (a) the representative volume element and (b) the pullout process.

to model the RVE. The approach adopted here extends the earlier work of Wernik and Meguid [104].

The new features of the current work relate to the approach adopted in the modeling of the polymer matrix and the investigation of the CNT polymer interfacial properties as appose to the effective mechanical properties of the RVE. The idea behind the ABC technique is to incorporate atomistic interatomic potentials into a continuum framework. In this way, the interatomic potentials introduced in the model capture the underlying atomistic behavior of the different phases considered. Thus, the influence of the nanophase is taken into account via appropriate atomistic constitutive formulations. Consequently, these measures are fundamentally different from those in the classical continuum theory. For the sake of completeness, Wernik and Meguid provided a brief outline of the method detailed in their earlier work [103, 104].

The cumulative effect of the vdW interactions acting on each CNT atom is applied as a resultant force on the respective node which is then resolved into its three Cartesian components. This process is depicted in Fig. 10.14 during each iteration of the pullout process, the above expression is reevaluated for each vdW interaction and the cumulative resultant force and its three Cartesian components are updated to correspond to the latest pullout configuration. Figure 15 shows a segment of the CNT with the cumulative resultant vdW force vectors as they are applied to the CNT atoms.

Yang et al. [105], investigated the CNT size effect and weakened bonding effect between an embedded CNT and surrounding matrix

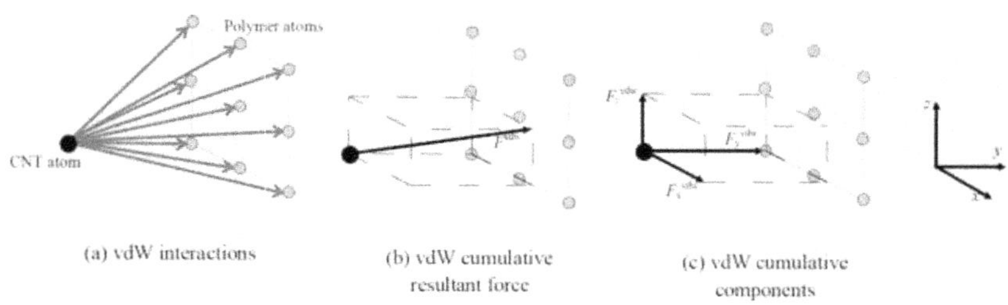

FIGURE 10.14 The process of nodal vdW force application. (a) vdW interactions on an individual CNT atom, (b) the cumulative resultant vdW force, and (c) the cumulative vdW Cartesian components.

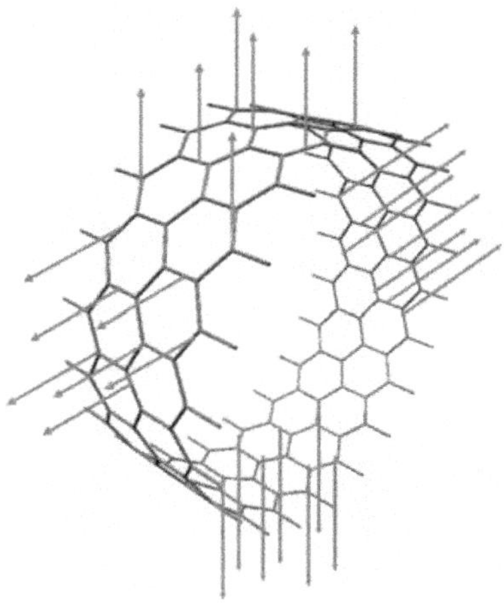

FIGURE 10.15 Segment of CNT with cumulative resultant vdW force vectors.

were characterized using MD simulations. Assuming that the equivalent continuum model of the CNT atomistic structure is a solid cylinder, the transversely isotropic elastic constants of the CNT decreased as the CNT radius increased. Regarding the elastic stiffness of the nanocomposite unit cell, the same CNT size dependency was observed in all independent components, and only the longitudinal Young's modulus showed a positive reinforcing effect whereas other elastic moduli demonstrated negative reinforcing effects as a result of poor load transfer at the interface. To describe the size effect and weakened bonding effect at the interface, a modified multiinclusion model was derived using the concepts of an effective CNT and effective matrix. During the scale bridging process incorporating the MD simulation results and modified multiinclusion model, we found that both the elastic modulus of the CNT and the adsorption layer near the CNT contributed to the size-dependent elastic modulus of the nanocomposites. Using the proposed multiscale bridging model, the elastic modulus for nanocomposites at various volume fractions and CNT sizes could be estimated. Among three major factors (CNT waviness, the dispersion state, and adhesion between the CNT and matrix), the proposed model considered only the weakened bonding effect. However, the present multiscale framework can be easily applied in considering the aforementioned factors and describing the real nanocomposite microstructures. In addition, by considering chemically grafted molecules (covalent or noncovalent bonds) to enhance the interfacial load transfer mechanism in MD simulations, the proposed multiscale approach can offer a deeper understanding of the reinforcing mechanism, and a more practical analytical tool with which to analyze and design functional nanocomposites. The analytical estimation reproduced from the proposed multiscale model can also provides useful information in modeling finite element-based representative volume elements of nanocomposite microstructures for use in multifunctional design.

The effects of the interphase and RVE configuration on the tensile, bending and torsional properties of the suggested nanocomposite were investigated by Ayatollahi et al. [106]. It was found that the stiffness of the nanocomposite could be affected by a strong interphase much more than by a weaker interphase. In addition, the stiffness of the interphase had the maximum effect on the stiffness of the nanocomposite in the bending loading conditions. Furthermore, it was revealed that the ratio of Le/Ln in RVE can dramatically affect the stiffness of the nanocomposite especially in the axial loading conditions.

For carbon nanotubes not well bonded to polymers, Jiang et al. [107] established a cohesive law for carbon nanotube/polymer interfaces. The cohesive law and its properties (e.g., cohesive strength, cohesive energy) are obtained directly from the Lennard–Jones potential from the van der Waals interactions. Such a cohesive law is incorporated in the micromechanics model to study the mechanical behavior of carbon nanotube-reinforced composite materials. Carbon nanotubes indeed improve the mechanical behavior of composite at the small strain. However, such improvement disappears at relatively large strain because the completely debonded nanotubes behave like voids in the matrix and may even weaken the composite. The increase of interface adhesion between carbon nanotubes and polymer matrix may significantly improve the composite behavior at the large strain [108].

Zalamea et al. [109] employed the shear transfer model as well as the shear lag model to explore the stress transfer from the outermost layer to the interior layers in MWCNTs. Basically, the interlayer properties between graphene layers were designated by scaling the parameter of shear transfer efficiency with respect to the perfect bonding. Zalamea et al. pointed out that as the number of layers in MWCNTs increases, the stress transfer efficiency decreases correspondingly. Shen et al. [110] examined load transfer between adjacent walls of DWCNTs using MD simulation, indicating that the tensile loading on the outermost wall of MWCNTs cannot be effectively transferred into the inner walls. However, when chemical bonding between the walls is established, the effectiveness can be dramatically enhanced. It is noted that in the above investigations, the loadings were applied directly on the outermost layers of MWCNTs; the stresses in the inner layers were then calculated either from the continuum mechanics approach [109] or MD simulation [110]. Shokrieh and Rafiee [82, 83] examined the mechanical properties of nanocomposites with capped single-walled carbon nanotubes (SWCNTs) embedded in a polymer matrix. The load transfer efficiency in terms of different CNTs' lengths was the main concern in their examination. By introducing an interphase to represent the vdW interactions between SWCNTs and the surrounding matrix, Shokrieh and Rafiee [82, 83] converted the atomistic SWCNTs into an equivalent continuum fiber in finite element analysis. The idea of an equivalent solid fiber was also proposed by Gao and Li [111]to replace the atomistic structure of capped SWCNTs in the nanocomposites' cylindrical unit cell. The modulus of the equivalent solid was determined based on the atomistic structure of SWCNTs through molecular structure mechanics [112]. Subsequently, the continuum-based shear lag analysis was carried out to evaluate the axial stress distribution in CNTs. In addition, the influence of end caps in SWCNTs on the stress distribution of nanocomposites was also taken into account in their analysis. Tsai and Lu [113] characterized the effects of the layer number, intergraphic layers interaction, and aspect ratio of MWCNTs on the load transfer efficiency using the conventional shear lag model and finite element analysis. However, in their analysis, the interatomistic characteristics of the adjacent graphene layers associated with different degrees of interactions were simplified by a thin interphase with different moduli. The atomistic interaction between the grapheme layers was not taken into account in their modeling of MWCNTs. In light of the forgoing investigations, the equivalent solid of SWCNTs was developed by several researchers and then implemented as reinforcement in continuum-based nanocomposite models.

Nevertheless, for MWCNTs, the subjects concerning the development of equivalent continuum solid are seldom explored in the literature. In fact, how to introduce the atomistic characteristics, i.e., the interfacial properties of neighboring graphene layers in MWCNTs, into the equivalent continuum solid is a challenging task as the length scales used to describe the physical phenomenon are distinct. Thus, a multiscale based simulation is required to account for the atomistic attribute of MWCNTs into an equivalent continuum solid. In T.C. Lu, J.L. Tsai's study [113], the multiscale approach was used to investigate the load transfer efficiency from surrounding matrix to DWCNTs. The analysis consisted of two stages. First, a cylindrical DWCNTs equivalent continuum was proposed based on MD simulation where the pullout extension on the outer layer was performed in an attempt to characterize the atomistic behaviors between neighboring graphite layers. Subsequently, the cylindrical continuum (denoting the DWCNTs) was embedded in a unit cell of nanocomposites, and the axial stress distribution as well as the load transfer efficiency of the DWCNTs was evaluated from finite element analysis. Both single-walled carbon nanotubes (SWCNTs) and DWCNTs were considered in the simulation and the results were compared with each other.

An equivalent cylindrical solid to represent the atomistic attributes of DWCNTs was proposed in this study. The atomistic interaction of adjacent graphite layers in DWCNTs was characterized using MD simulation based on which a spring element was introduced in the continuum equivalent solid to demonstrate the interfacial properties of DWCNTs. Subsequently, the proposed continuum solid (denotes DWCNTs) was embedded in the matrix to form DWCNTs nanocomposites (continuum model), and the load transfer efficiency within the DWCNTs was determined from FEM analysis. For the demonstration purpose, the DWCNTs with four different lengths were considered in the investigation. Analysis results illustrate that the increment of CNTs' length can effectively improve the load transfer efficiency in the outermost layers, nevertheless, for the inner layers, the enhancement is miniature. On the other hand, when the covalent bonds between the adjacent graphene layers are crafted, the load carrying capacity in the inner layer increases as so does the load transfer efficiency of DWCNTs. As compared to SWCNTs, the DWCNTs still possess the less capacity of load transfer efficiency even though there are covalent bonds generated in the DWCNTs.

10.9.8 CONCLUDING REMARKS

Many traditional simulation techniques (e.g., MC, MD, BD, LB, Ginzburg–Landau theory, micromechanics and FEM) have been employed, and some novel simulation techniques (e.g., DPD, equivalent-continuum and self-similar approaches) have been developed to study polymer nanocomposites. These techniques indeed represent approaches at various time and length scales from molecular scale (e.g., atoms), to microscale (e.g., coarse-grains, particles, monomers) and then to macroscale (e.g., domains), and have shown success to various degrees in addressing many aspects of polymer nanocomposites. The simulation techniques developed thus far have different strengths and weaknesses, depending on the need of research. For example, molecular simulations can be used to investigate molecular interactions and structure on the scale of 0.1–10 nm. The resulting information is very useful to understanding the interaction strength at nanoparticle–polymer interfaces and the molecular origin of mechanical improvement. However, molecular simulations are computationally very demanding, thus not so applicable to the prediction of mesoscopic structure and properties defined on the scale of 0.1–10 mm, for example, the dispersion of nanoparticles in polymer matrix and the morphology of polymer nanocomposites. To explore the morphology on these scales, mesoscopic simulations such as coarse-grained methods, DPD and dynamic mean field theory are more effective. On the other hand, the macroscopic properties of materials are usually studied by the use of mesoscale or macroscale techniques such as micromechanics and FEM. But these techniques may have limitations when applied to polymer nanocomposites because of the difficulty to deal with the interfacial nanoparticle–polymer interaction and the morphology, which are considered crucial to the mechanical improvement of nanoparticle-filled polymer nanocomposites. Therefore, despite the progress over the past years, there are a number of challenges in computer modeling and simulation. In general, these challenges represent the work in two directions. First, there is a need to develop new and improved simulation techniques at individual time and length scales. Secondly, it is important to integrate the developed methods at wider range of time and length scales, spanning from quantum mechanical domain (a few atoms) to molecular domain (many atoms), to mesoscopic domain (many monomers or chains), and finally to macroscopic domain (many domains or structures), to form a useful tool for exploring the structural, dynamic, and mechanical properties, as well as optimizing design and processing control of polymer nanocomposites. The need for the second development is obvious. For example, the morphology is usually determined from the mesoscale techniques whose implementation requires information about the interactions between various components (e.g., nanoparticle–nanoparticle and nanoparticle–polymer) that should be derived from molecular simulations. Developing such a multiscale method is very challenging but indeed represents the future of computer simulation and modeling, not only in polymer nanocomposites but also other fields. New concepts, theories and computational tools should be developed in the future to make truly seamless multiscale modeling a reality. Such development is crucial in order to achieve the longstanding goal of predicting particle–structure property relationships in material design and optimization.

The strength of the interface and the nature of interaction between the polymer and carbon nanotube are the most important factors governing the ability of nanotubes to improve the performance of the composite. Extensive research has been performed on studying and understanding CNT-polymer composites from chemistry, mechanics and physics aspects. However, there exist various issues like processing of composites and experimental challenges, which need to be addressed to gain further insights into the interfacial processes.

KEYWORDS

- **carbon nanotubes**
- **mechanical properties**
- **modeling**
- **simulation**
- **updates**

REFERENCES

1. Iijima, S., Helical microtubules of graphitic carbon. Nature, 1991, 354(6348), 56–58.
2. Dresselhaus, M., G. Dresselhaus, P. Eklund, Science of fullerenes and carbon nanotubes: their properties and applications. 1996: Academic press.
3. Saito, R., G. Dresselhaus, M. Dresselhaus, Physical properties of carbon nanotubes. Vol. 4. 1998: World Scientific. 517.
4. Harris, P., P. Harris, Carbon nanotubes and related structures: new materials for the 20-first century. 2001: Cambridge University Press. 279.
5. Wagner, H., R. Vaia, Nanocomposites: issues at the interface. Materials Today, 2004, 7(11), 38–42.
6. Zeng, Q., A. Yu, G. Lu, Multiscale modeling and simulation of polymer nanocomposites. Progress in Polymer Science, 2008, 33(2), 191–269.
7. Bethune, D., et al., Cobalt-catalyzed growth of carbon nanotubes with single-atomic-layer walls. 1993.
8. Dresselhaus, M., G. Dresselhaus, R. Saito, Physics of carbon nanotubes. Carbon, 1995, 33(7), 883–891.
9. Thostenson, E., Z. Ren, T. Chou, Advances in the science and technology of carbon nanotubes and their composites: a review. Composites Science and Technology, 2001, 61(13), 1899–1912.
10. Yakobson, B., P. Avouris, Mechanical properties of carbon nanotubes, in Carbon nanotubes. 2001, Springer. 287–327.
11. Dresselhaus, M., P. Avouris, Introduction to carbon materials research, in Carbon nanotubes. 2001, Springer. 90.
12. Yakobson, B., C. Brabec, J. Bernholc, Nanomechanics of carbon tubes: instabilities beyond linear response. Physical review letters, 1996, 76(14), 2511–2514.
13. Ajayan, P., L. Schadler, P. Braun, Nanocomposite science and technology. 2006, John Wiley & Sons. 550.
14. Li, J., et al., Correlations between percolation threshold, dispersion state, and aspect ratio of carbon nanotubes. Advanced Functional Materials, 2007, 17(16), 3207–3215.
15. Ajayan, P., et al., Aligned carbon nanotube arrays formed by cutting a polymer resin—nanotube composite. Science, 1994, 265(5176), 1212–1214.
16. Du, J., J. Bai, H. Cheng, The present status and key problems of carbon nanotube based polymer composites. Express Polymer Letters, 2007, 1(5), 253–273.
17. Lee, et al., Simulation of polymer melt intercalation in layered nanocomposites. The Journal of chemical physics, 1998, 109(23), 10321–10330.
18. Smith, G., et al., A molecular dynamics simulation study of the viscoelastic properties of polymer nanocomposites. The Journal of chemical physics, 2002, 117(20), 9478–9489.
19. Zeng, Q., et al., Molecular dynamics simulation of organic-inorganic nanocomposites: Layering behavior and interlayer structure of organoclays. Chemistry of Materials, 2003, 15(25), 4732–4738.
20. Zeng, Q., et al., Interfacial Structure and Interactions in Clay-Polymer Nanocomposites. 18–33.
21. Vacatello, M., Predicting the Molecular Arrangements in Polymer-Based Nanocomposites. Macromolecular theory and simulations, 2003, 12(1), 86–91.

22. Allen, M., D. Tildesley, Computer simulation of liquids, 1987, New York: Oxford, 1989, 385: 88–98.
23. Frenkel, D., B. Smit, Understanding molecular simulation: from algorithms to applications. Vol. 1. 2001: Academic press. 430.
24. Metropolis, N., et al., Equation of state calculations by fast computing machines. The Journal of chemical physics, 2004, 21(6), 1087–1092.
25. Hoogerbrugge, P., J. Koelman, Simulating microscopic hydrodynamic phenomena with dissipative particle dynamics. EPL (Europhysics Letters), 1992, 19(3), 155.
26. Gibson, J., K. Chen, S. Chynoweth, Simulation of particle adsorption onto a polymer-coated surface using the dissipative particle dynamics method. Journal of colloid and interface science, 1998, 206(2), 464–474.
27. Dzwinel, W., D. Yuen, A two-level, discrete particle approach for large-scale simulation of colloidal aggregates. International Journal of Modern Physics C, 2000, 11(05), 1037–1061.
28. Dzwinel, W., D. Yuen, K. Boryczko, Mesoscopic dynamics of colloids simulated with dissipative particle dynamics and fluid particle model. Molecular modeling annual, 2002, 8(1), 33–43.
29. Chen, S., G. Doolen, Lattice Boltzmann method for fluid flows. Annual review of fluid mechanics, 1998, 30(1), 329–364.
30. Cahn, J., On spinodal decomposition. Acta metallurgica, 1961, 9(9), 795–801.
31. Cahn, J., Free energy of a nonuniform system. II. Thermodynamic basis. The Journal of chemical physics, 2004, 30(5), 1121–1124.
32. Lee, B., J. Douglas, S. Glotzer, Filler-induced composition waves in phase-separating polymer blends. Physical Review E, 1999, 60(5), 5812–5830.
33. Ginzburg, V., et al., Simulation of hard particles in a phase-separating binary mixture. arXiv preprint cond-mat/9905284, 1999: 18–28.
34. Qiu, F., et al., Phase separation under shear of binary mixtures containing hard particles. Langmuir, 1999, 15(15), 4952–4956.
35. He, G., A. Balazs, Modeling the dynamic behavior of mixtures of diblock copolymers and dipolar nanoparticles. Journal of Computational and Theoretical Nanoscience, 2005, 2(1), 99–107.
36. Altevogt, P., et al., The MesoDyn project: software for mesoscale chemical engineering. Journal of Molecular Structure: THEOCHEM, 1999, 463(1), 139–143.
37. Morita, H., T. Kawakatsu, M. Doi, Dynamic density functional study on the structure of thin polymer blend films with a free surface. Macromolecules, 2001, 34(25), 8777–8783.
38. Powell IV, A., R. Arroyave, Open source software for materials and process modeling. JOM, 2008, 60(5), 32–39.
39. Ginzburg, V., et al., Modeling the dynamic behavior of diblock copolymer/particle composites. Macromolecules, 2000, 33(16), 6140–6147.
40. Tandon, G., G. Weng, The effect of aspect ratio of inclusions on the elastic properties of unidirectionally aligned composites. Polymer composites, 1984, 5(4), 327–333.
41. Odegard, G., R. Pipes, P. Hubert, Comparison of two models of SWCN polymer composites. Composites Science and Technology, 2004, 64(7), 1011–1020.
42. Odegard, G., et al., Constitutive modeling of nanotube–reinforced polymer composites. Composites Science and Technology, 2003, 63(11), 1671–1687.

43. Pipes, R., P. Hubert, Helical carbon nanotube arrays: mechanical properties. Composites Science and Technology, 2002, 62(3), 419–428.
44. Rudd, R., J. Broughton, Concurrent coupling of length scales in solid state systems. physica status solidi (b), 2000, 217(1), 251–291.
45. Starr, F., S. Glotzer. Simulations of filled polymers on multiple length scales. in MRS Proceedings. 2000: Cambridge University Press.
46. Glotzer, S., F. Starr. Towards multiscale simulations of filled and nanofilled polymers. in AIChE Symposium Series. 2001: New York; American Institute of Chemical Engineers; 1998.
47. Mori, T., K. Tanaka, Average stress in matrix and average elastic energy of materials with misfitting inclusions. Acta metallurgica, 1973, 21(5), 571–574.
48. Fisher, F., R. Bradshaw, L. Brinson, Fiber waviness in nanotube-reinforced polymer composites—I: Modulus predictions using effective nanotube properties. Composites Science and Technology, 2003, 63(11), 1689–1703.
49. Bradshaw, R., F. Fisher, L. Brinson, Fiber waviness in nanotube-reinforced polymer composites—II: modeling via numerical approximation of the dilute strain concentration tensor. Composites Science and Technology, 2003, 63(11), 1705–1722.
50. Liu, Y., X. Chen, Evaluations of the effective material properties of carbon nanotube-based composites using a nanoscale representative volume element. Mechanics of materials, 2003, 35(1), 69–81.
51. Schadler, L., S. Giannaris, P. Ajayan, Load transfer in carbon nanotube epoxy composites. Applied Physics Letters, 1998, 73(26), 3842–3844.
52. Wagner, H., et al., Stress-induced fragmentation of multiwall carbon nanotubes in a polymer matrix. Applied Physics Letters, 1998, 72(2), 188–190.
53. Qian, D., et al., Load transfer and deformation mechanisms in carbon nanotube-polystyrene composites. Applied Physics Letters, 2000, 76(20), 2868–2870.
54. Liu, Y., X. Chen, Modeling and analysis of carbon nanotube-based composites using the FEM and BEM. Submitted to CMES: Computer Modeling in Engineering and Science, 2002: 88–97.
55. Liu, Y., J. Luo, N. Xu, Modeling of interphases in fiber-reinforced composites under transverse loading using the boundary element method. Journal of applied mechanics, 2000, 67(1), 41–49.
56. Fu, Y., et al., A fast solution method for three-dimensional many-particle problems of linear elasticity. International Journal for Numerical Methods in Engineering, 1998, 42(7), 1215–1229.
57. Nishimura, N., K. Yoshida, S. Kobayashi, A fast multipole boundary integral equation method for crack problems in 3D. Engineering Analysis with Boundary Elements, 1999, 23(1), 97–105.
58. Qian, D., W. Liu, R. Ruoff, Mechanics of C60 in nanotubes. The journal of physical chemistry B, 2001, 105(44), 10753–10758.
59. Chen, X., Y. Liu, Square representative volume elements for evaluating the effective material properties of carbon nanotube-based composites. Computational Materials Science, 2004, 29(1), 1–11.
60. Bower, C., et al., Deformation of carbon nanotubes in nanotube–polymer composites. Applied Physics Letters, 1999, 74(22), 3317–3319.

61. Wan, H., F. Delale, L. Shen, Effect of CNT length and CNT-matrix interphase in carbon nanotube (CNT) reinforced composites. Mechanics research communications, 2005, 32(5), 481–489.
62. Shi, D., et al., Multiscale analysis of fracture of carbon nanotubes embedded in composites. International journal of fracture, 2005, 134(3–4), 369–386.
63. Buryachenko, V., et al., Multi-scale mechanics of nanocomposites including interface: experimental and numerical investigation. Composites Science and Technology, 2005, 65(15), 2435–2465.
64. Li, C., T. Chou, Multiscale modeling of carbon nanotube reinforced polymer composites. Journal of nanoscience and nanotechnology, 2003, 3(5), 423–430.
65. Li, C., T. Chou, Multiscale modeling of compressive behavior of carbon nanotube/polymer composites. Composites Science and Technology, 2006, 66(14), 2409–2414.
66. Anumandla, V., R. Gibson, A comprehensive closed form micromechanics model for estimating the elastic modulus of nanotube-reinforced composites. Composites Part A: Applied Science and Manufacturing, 2006, 37(12), 2178–2185.
67. Seidel, G., D. Lagoudas, Micromechanical analysis of the effective elastic properties of carbon nanotube reinforced composites. Mechanics of materials, 2006, 38(8), 884–907.
68. Hashin, Z., B. Rosen, The elastic moduli of fiber-reinforced materials. Journal of applied mechanics, 1964, 31(2), 223–232.
69. Christensen, R., K. Lo, Solutions for effective shear properties in three phase sphere and cylinder models. Journal of the Mechanics and Physics of Solids, 1979, 27(4), 315–330.
70. Selmi, A., et al., Prediction of the elastic properties of single walled carbon nanotube reinforced polymers: a comparative study of several micromechanical models. Composites Science and Technology, 2007, 67(10), 2071–2084.
71. Luo, D., W. Wang, Y. Takao, Effects of the distribution and geometry of carbon nanotubes on the macroscopic stiffness and microscopic stresses of nanocomposites. Composites Science and Technology, 2007, 67(14), 2947–2958.
72. Fu, S., et al., On the elastic stress transfer and longitudinal modulus of unidirectional multishort-fiber composites. Composites Science and Technology, 2000, 60(16), 3001–3012.
73. Tserpes, K., et al., Multi-scale modeling of tensile behavior of carbon nanotube-reinforced composites. Theoretical and Applied Fracture Mechanics, 2008, 49(1), 51–60.
74. Spanos, P., A. Kontsos, A multiscale Monte Carlo finite element method for determining mechanical properties of polymer nanocomposites. Probabilistic Engineering Mechanics, 2008, 23(4), 456–470.
75. Zhu, J., et al., Reinforcing epoxy polymer composites through covalent integration of functionalized nanotubes. Advanced Functional Materials, 2004, 14(7), 643–648.
76. Paiva, M., et al., Mechanical and morphological characterization of polymer–carbon nanocomposites from functionalized carbon nanotubes. Carbon, 2004, 42(14), 2849–2854.
77. Bhuiyan, M., et al., Defining the lower and upper limit of the effective modulus of CNT/polypropylene composites through integration of modeling and experiments. Composite Structures, 2013, 95: 80–87.

78. Papanikos, P., D. Nikolopoulos, K. Tserpes, Equivalent beams for carbon nanotubes. Computational Materials Science, 2008, 43(2), 345–352.
79. Affdl, J., J. Kardos, The Halpin-Tsai equations: a review. Polymer Engineering & Science, 1976, 16(5), 344–352.
80. Landel, R., L. Nielsen, Mechanical properties of polymers and composites. 1993: CRC Press. 553.
81. Tucker III, C., E. Liang, Stiffness predictions for unidirectional short-fiber composites: review and evaluation. Composites Science and Technology, 1999, 59(5), 655–671.
82. Shokrieh, M., R. Rafiee, Investigation of nanotube length effect on the reinforcement efficiency in carbon nanotube based composites. Composite Structures, 2010, 92(10), 2415–2420.
83. Shokrieh, M., R. Rafiee, On the tensile behavior of an embedded carbon nanotube in polymer matrix with nonbonded interphase region. Composite Structures, 2010, 92(3), 647–652.
84. Haghi, A., G. Zaikov, Carbon Nanotubes and Related Structures, in Handbook of Research on Functional Materials: Principles, Capabilities and Limitations. 2014, CRC Press. 147–159.
85. Esawi, A., M. Farag, Carbon nanotube reinforced composites: potential and current challenges. Materials & Design, 2007, 28(9), 2394–2401.
86. Chowdhury, S., et al., Modeling the effect of statistical variations in length and diameter of randomly oriented CNTs on the properties of CNT reinforced nanocomposites. Composites Part B: Engineering, 2012, 43(4), 1756–1762.
87. Gou, J., et al., Computational and experimental study of interfacial bonding of single-walled nanotube reinforced composites. Computational Materials Science, 2004, 31(3), 225–236.
88. Li, S., et al., Electrical properties of soluble carbon nanotube/polymer composite films. Chemistry of Materials, 2005, 17(1), 130–135.
89. Liu, Y., X. Chen, Continuum models of carbon nanotube-based composites using the boundary element method. Electronic Journal of Boundary Elements, 2007, 1(2), 18–33.
90. Thostenson, E., T. Chou, Aligned multiwalled carbon nanotube-reinforced composites: processing and mechanical characterization. Journal of Physics D: Applied Physics, 2002, 35(16), 77–90.
91. Daniel Wagner, H., Nanotube–polymer adhesion: a mechanics approach. Chemical Physics Letters, 2002, 361(1), 57–61.
92. Cooper, C., R. Young, M. Halsall, Investigation into the deformation of carbon nanotubes and their composites through the use of Raman spectroscopy. Composites Part A: Applied Science and Manufacturing, 2001, 32(3), 401–411.
93. Ajayan, P., et al., Single-walled carbon nanotube–polymer composites: strength and weakness. Advanced Materials, 2000, 12(10), 750–753.
94. Valentini, L., et al., Physical and mechanical behavior of single-walled carbon nanotube/polypropylene/ethylene–propylene–diene rubber nanocomposites. Journal of Applied Polymer Science, 2003, 89(10), 2657–2663.
95. Paipetis, A., et al., Stress transfer from the matrix to the fiber in a fragmentation test: Raman experiments and analytical modeling. Journal of composite materials, 1999, 33(4), 377–399.

96. Qian, D., et al., Mechanics of carbon nanotubes. Applied mechanics reviews, 2002, 55(6), 495–533.
97. Yu, M., et al., Strength and breaking mechanism of multiwalled carbon nanotubes under tensile load. Science, 2000, 287(5453), 637–640.
98. Yu, M., B. Yakobson, R. Ruoff, Controlled sliding and pullout of nested shells in individual multiwalled carbon nanotubes. The journal of physical chemistry B, 2000, 104(37), 8764–8767.
99. Liao, K., S. Li, Interfacial characteristics of a carbon nanotube–polystyrene composite system. Applied Physics Letters, 2001, 79(25), 4225–4227.
100. Andrews, R., M. Weisenberger, Carbon nanotube polymer composites. Current Opinion in Solid State and Materials Science, 2004, 8(1), 31–37.
101. Frankland, S., V. Harik, Analysis of carbon nanotube pull-out from a polymer matrix. Surface Science, 2003, 525(1), p. L103-L108.
102. Lordi, V., N. Yao, Molecular mechanics of binding in carbon-nanotube–polymer composites. Journal of Materials Research, 2000, 15(12), 2770–2779.
103. Wernik, J., B. Cornwell-Mott, S. Meguid, Determination of the interfacial properties of carbon nanotube reinforced polymer composites using atomistic-based continuum model. International Journal of Solids and Structures, 2012, 49(13), 1852–1863.
104. Wernik, J., S. Meguid, Multiscale modeling of the nonlinear response of nano-reinforced polymers. Acta Mechanica, 2011, 217(1–2), 1–16.
105. Yang, S., et al., Multiscale modeling of size-dependent elastic properties of carbon nanotube/polymer nanocomposites with interfacial imperfections. Polymer, 2012, 53(2), 623–633.
106. Ayatollahi, M., S. Shadlou, M. Shokrieh, Multiscale modeling for mechanical properties of carbon nanotube reinforced nanocomposites subjected to different types of loading. Composite Structures, 2011, 93(9), 2250–2259.
107. Jiang, L., et al., A cohesive law for carbon nanotube/polymer interfaces based on the van der Waals force. Journal of the Mechanics and Physics of Solids, 2006, 54(11), 2436–2452.
108. Tan, H., et al., The effect of van der Waals-based interface cohesive law on carbon nanotube-reinforced composite materials. Composites Science and Technology, 2007, 67(14), 2941–2946.
109. Zalamea, L., H. Kim, R. Pipes, Stress transfer in multiwalled carbon nanotubes. Composites Science and Technology, 2007, 67(15), 3425–3433.
110. Shen, G., S. Namilae, N. Chandra, Load transfer issues in the tensile and compressive behavior of multiwall carbon nanotubes. Materials Science and Engineering: A, 2006, 429(1), 66–73.
111. Gao, X., K. Li, A shear-lag model for carbon nanotube-reinforced polymer composites. International Journal of Solids and Structures, 2005, 42(5), 1649–1667.
112. Li, C., T. Chou, A structural mechanics approach for the analysis of carbon nanotubes. International Journal of Solids and Structures, 2003, 40(10), 2487–2499.
113. Tsai, Y., et al., Production of carbon nanotubes by single-pulse discharge in air. Journal of Materials Processing Technology, 2009, 209(9), 4413–4416.
114. Ru, C., Elastic buckling of single-walled carbon nanotube ropes under high pressure. Physical Review B, 2000, 62(15), 10405.

115. Ru, C., Column buckling of multiwalled carbon nanotubes with interlayer radial displacements. Physical Review B, 2000, 62(24), 16962.
116. Wang, C., et al., Buckling of double-walled carbon nanotubes modeled by solid shell elements. Journal of applied physics, 2006, 99(11), 114317.
117. Han, Q., G. Lu, Torsional buckling of a double-walled carbon nanotube embedded in an elastic medium. European Journal of Mechanics-A/Solids, 2003, 22(6), 875–883.
118. Zhou, W., et al., Copper catalyzing growth of single-walled carbon nanotubes on substrates. Nano Letters, 2006, 6(12), 2987–2990.
119. Han, S., X. Liu, C. Zhou, Template-free directional growth of single-walled carbon nanotubes on a-and r-plane sapphire. Journal of the American Chemical Society, 2005, 127(15), 5294–5295.
120. Wang, X., H. Yang, X. Yin, Axially critical load of multiwall carbon nanotubes under thermal environment. Journal of Thermal Stresses, 2005, 28(2), 185–196.
121. Leung, A., et al., Postbuckling of carbon nanotubes by atomic-scale finite element. Journal of applied physics, 2006, 99(12), 124308.
122. Yao, X., Q. Han, Torsional buckling and postbuckling equilibrium path of double-walled carbon nanotubes. Composites Science and Technology, 2008, 68(1), 113–120.
123. Muc, A., Modelling of carbon nanotubes behavior with the use of a thin shell theory. Journal of Theoretical and Applied Mechanics, 2011, 49: 531–540.
124. Muc, A., Design and identification methods of effective mechanical properties for carbon nanotubes. Materials and Design, 2010, 31(4), 1671–1675.
125. Muc, A., A. Banas, M. lgorzata Chwa, Free Vibrations of Carbon Nanotubes with Defects. Mechanics and Mechanical Engineering, 2013, 17(2), 157–166.
126. Liew, K., Q. Wang, Analysis of wave propagation in carbon nanotubes via elastic shell theories. International Journal of Engineering Science, 2007, 45(2), 227–241.
127. Hu, Y., et al., Nonlocal shell model for elastic wave propagation in single-and double-walled carbon nanotubes. Journal of the Mechanics and Physics of Solids, 2008, 56(12), 3475–3485.
128. Natsuki, T., Q. Ni, M. Endo, Analysis of the vibration characteristics of double-walled carbon nanotubes. Carbon, 2008, 46(12), 1570–1573.
129. Ghorbanpourarani, A., et al., Transverse vibration of short carbon nanotubes using cylindrical shell and beam models. Proceedings of the Institution of Mechanical Engineers, Part C: Journal of Mechanical Engineering Science, 2010, 224(3), 745–756.
130. Tylikowski, A., Instability of thermally induced vibrations of carbon nanotubes. Archive of Applied Mechanics, 2008, 78(1), 49–60.
131. Wang, J., et al., Capacitance properties of single wall carbon nanotube/polypyrrole composite films. Composites Science and Technology, 2007, 67(14), 2981–2985.
132. Belytschko, T., et al., Atomistic simulations of nanotube fracture. Physical Review B, 2002, 65(23), 235430.
133. Yakobson, B., C. Brabec, J. Bernholc, Structural mechanics of carbon nanotubes: From continuum elasticity to atomistic fracture. Journal of Computer-Aided Materials Design, 1996, 3(1–3), 173–182.
134. Huang, Y., J. Wu, K. Hwang, Thickness of graphene and single-wall carbon nanotubes. Physical Review B, 2006, 74(24), 245413.

135. Wu, J., K. Hwang, Y. Huang, An atomistic-based finite-deformation shell theory for single-wall carbon nanotubes. Journal of the Mechanics and Physics of Solids, 2008, 56(1), 279–292.
136. Peng, J., et al., Can a single-wall carbon nanotube be modeled as a thin shell? Journal of the Mechanics and Physics of Solids, 2008, 56(6), 2213–2224.
137. Kalamkarov, A., et al., Analytical and numerical techniques to predict carbon nanotubes properties. International Journal of Solids and Structures, 2006, 43(22), 6832–6854.
138. Treacy, M., T. Ebbesen, J. Gibson, Exceptionally high Young's modulus observed for individual carbon nanotubes. 1996: 76–83.
139. Wong, E., P. Sheehan, C. Lieber, Nanobeam mechanics: elasticity, strength, and toughness of nanorods and nanotubes. Science, 1997, 277(5334), 1971–1975.
140. Salvetat, J., et al., Elastic and shear moduli of single-walled carbon nanotube ropes. Physical review letters, 1999, 82(5), 944.
141. Walters, D., et al., Elastic strain of freely suspended single-wall carbon nanotube ropes. Applied Physics Letters, 1999, 74(25), 3803–3805.
142. Yu, M., et al., Tensile loading of ropes of single wall carbon nanotubes and their mechanical properties. Physical review letters, 2000, 84(24), 5552.
143. Xie, S., et al., Mechanical and physical properties on carbon nanotube. Journal of Physics and Chemistry of Solids, 2000, 61(7), 1153–1158.
144. Wong, S., et al., Carbon nanotube tips: high-resolution probes for imaging biological systems. Journal of the American Chemical Society, 1998, 120(3), 603–604.
145. Treacy, M., T. Ebbesen, J. Gibson, Exceptionally high Young's modulus observed for individual carbon nanotubes. 1996.
146. Falvo, M., et al., Bending and buckling of carbon nanotubes under large strain. Nature, 1997, 389(6651), 582–584.
147. Overney, G., W. Zhong, D. Tomanek, Structural rigidity and low frequency vibrational modes of long carbon tubules. Zeitschrift für Physik D Atoms, Molecules and Clusters, 1993, 27(1), 93–96.
148. Lu, J., Elastic properties of single and multilayered nanotubes. Journal of Physics and Chemistry of Solids, 1997, 58(11), 1649–1652.
149. Yakobson, B., et al., High strain rate fracture and C-chain unraveling in carbon nanotubes. Computational Materials Science, 1997, 8(4), 341–348.
150. Bernholc, J., et al., Theory of growth and mechanical properties of nanotubes. Applied Physics A: Materials Science & Processing, 1998, 67(1), 39–46.
151. Iijima, S., et al., Structural flexibility of carbon nanotubes. The Journal of chemical physics, 1996, 104(5), 2089–2092.
152. Ru, C., Effective bending stiffness of carbon nanotubes. Physical Review B, 2000, 62(15), 9973.
153. Vaccarini, L., et al., Mechanical and electronic properties of carbon and boron–nitride nanotubes. Carbon, 2000, 38(11), 1681–1690.
154. Al-Jishi, R., G. Dresselhaus, Lattice-dynamical model for graphite. Physical Review B, 1982, 26: 4514–4522.
155. Yakobson, B., G. Samsonidze, G. Samsonidze, Atomistic theory of mechanical relaxation in fullerene nanotubes. Carbon, 2000, 38(11), 1675–1680.
156. Journet, C., et al., Large-scale production of single-walled carbon nanotubes by the electric-arc technique. Nature, 1997, 388(6644), 756–758.

157. Thess, A., et al., Crystalline ropes of metallic carbon nanotubes. Science-AAAS-Weekly Paper Edition, 1996, 273(5274), 483–487.
158. Popov, V., V. Van Doren, M. Balkanski, Elastic properties of crystals of single-walled carbon nanotubes. Solid state communications, 2000, 114(7), 395–399.
159. Popov, V., V. Van Doren, M. Balkanski, Lattice dynamics of single-walled carbon nanotubes. Physical Review B, 1999, 59(13), 8355.
160. Ruoff, R., D. Lorents, Mechanical and thermal properties of carbon nanotubes. Carbon, 1995, 33(7), 925–930.
161. Govindjee, S., J. Sackman, On the use of continuum mechanics to estimate the properties of nanotubes. Solid state communications, 1999, 110(4), 227–230.
162. Ru, C., Effect of van der Waals forces on axial buckling of a double-walled carbon nanotube. Journal of applied physics, 2000, 87(10), 7227–7231.
163. Ru, C., Axially compressed buckling of a double-walled carbon nanotube embedded in an elastic medium. Journal of the Mechanics and Physics of Solids, 2001, 49(6), 1265–1279.
164. Ru, C., Degraded axial buckling strain of multiwalled carbon nanotubes due to interlayer slips. Journal of applied physics, 2001, 89(6), 3426–3433.
165. Kolmogorov, A., V. Crespi, Smoothest bearings: interlayer sliding in multiwalled carbon nanotubes. Physical review letters, 2000, 85(22), 4727–4730.
166. Shaffer, M., A. Windle, Fabrication and characterization of carbon nanotube/poly (vinyl alcohol) composites. Advanced Materials, 1999, 11(11), 937–941.
167. Jia, Z., et al., Study on poly (methyl methacrylate)/carbon nanotube composites. Materials Science and Engineering: A, 1999, 271(1), 395–400.
168. Gong, X., et al., Surfactant-assisted processing of carbon nanotube/polymer composites. Chemistry of Materials, 2000, 12(4), 1049–1052.
169. Lourie, O., H. Wagner, Transmission electron microscopy observations of fracture of single-wall carbon nanotubes under axial tension. Applied Physics Letters, 1998, 73(24), 3527–3529.
170. Lourie, O., D. Cox, H. Wagner, Buckling and collapse of embedded carbon nanotubes. Physical review letters, 1998, 81: 1638–1641.
171. Jin, L., C. Bower, O. Zhou, Alignment of carbon nanotubes in a polymer matrix by mechanical stretching. Applied Physics Letters, 1998, 73(9), 1197–1199.
172. Haggenmueller, R., et al., Aligned single-wall carbon nanotubes in composites by melt processing methods. Chemical Physics Letters, 2000, 330(3), 219–225.
173. Andrews, R., et al., Nanotube composite carbon fibers. Applied Physics Letters, 1999, 75(9), 1329–1331.
174. Gommans, H., et al., Fibers of aligned single-walled carbon nanotubes: Polarized Raman spectroscopy. Journal of applied physics, 2000, 88(5), 2509–2514.
175. Vigolo, B., et al., Macroscopic fibers and ribbons of oriented carbon nanotubes. Science, 2000, 290(5495), 1331–1334.
176. Li, Y., et al., Adsorption thermodynamic, kinetic and desorption studies of Pb2+ on carbon nanotubes. Water research, 2005, 39(4), 605–609.
177. Kalamkarov, A., On the determination of effective characteristics of cellular plates and shells of periodic structure. Mechanics of solids, 1987.
178. Kalamkarov, A., Composite and reinforced elements of construction. 1992: Wiley Chichester. 340.

179. Kalamkarov, A., A. Kolpakov, Analysis, design, and optimization of composite structures. 1997: J. Wiley & Sons.
180. Kalamkarov, A., A. Georgiades, Asymptotic homogenization models for smart composite plates with rapidly varying thickness: Part I—theory. International Journal for Multiscale Computational Engineering, 2004, 2(1).
181. Reddy, J., Mechanics of laminated composite plates and shells: theory and analysis. 2003: CRC press. 430.
182. Kalamkarov, A., V. Veedu, M. Ghasemi Nejhad, Mechanical properties modeling of carbon single-walled nanotubes: an asymptotic homogenization method. Journal of Computational and Theoretical Nanoscience, 2005, 2(1), 124–131.
183. Machida, T., et al., Coherent control of nuclear-spin system in a quantum-Hall device. Applied Physics Letters, 2003, 82(3), 409–411.
184. Hart, J., A. Rappe, van der Waals functional forms for molecular simulations. The Journal of chemical physics, 1992, 97(2), 1109–1115.
185. Tersoff, J., R. Ruoff, Structural properties of a carbon-nanotube crystal. Physical review letters, 1994, 73(5), 676.
186. Brenner, D., Empirical potential for hydrocarbons for use in simulating the chemical vapor deposition of diamond films. Physical Review B, 1990, 42(15), 9458.
187. Cornell, W., et al., A second generation force field for the simulation of proteins, nucleic acids, and organic molecules. Journal of the American Chemical Society, 1995, 117(19), 5179–5197.
188. Gelin, B., Molecular modeling of polymer structures and properties. 1994: Hanser/Gardner Publications.
189. Walther, M., B. Fischer, P. Uhd Jepsen, Noncovalent intermolecular forces in polycrystalline and amorphous saccharides in the far infrared. Chemical Physics, 2003, 288(2), 261–268.
190. László, I., A. Rassat, The geometric structure of deformed nanotubes and the topological coordinates. Journal of chemical information and computer sciences, 2003, 43(2), 519–524.
191. Jorgensen, W., D. Severance, Chemical chameleons: hydrogen bonding with imides and lactams in chloroform. Journal of the American Chemical Society, 1991, 113(1), 209–216.
192. Allinger, N., Y. Yuh, J. Lii, Molecular mechanics. The MM3 force field for hydrocarbons. 1. Journal of the American Chemical Society, 1989, 111(23), 8551–8566.
193. Walther, J., et al., Carbon nanotubes in water: structural characteristics and energetics. The journal of physical chemistry B, 2001, 105(41), 9980–9987.
194. Robertson, D., D. Brenner, J. Mintmire, Energetics of nanoscale graphitic tubules. Physical Review B, 1992, 45(21), 12592.
195. Odegard, G., et al., Equivalent-continuum modeling of nano-structured materials. Composites Science and Technology, 2002, 62(14), 1869–1880.
196. Odegard. G, et al., Equivalent-Continuum Modeling of Nano-Structured Materials. 2001.
197. Brown, T., Chemistry: the central science. 2009: Pearson Education. 543.
198. Fennimore, A., et al., Rotational actuators based on carbon nanotubes. Nature, 2003, 424(6947), 408–410.

199. Saito, R., et al., Electronic structure of chiral graphene tubules. Applied Physics Letters, 1992, 60(18), 2204–2206.
200. Fujita, M., et al., Formation of general fullerenes by their projection on a honeycomb lattice. Physical Review B, 1992, 45(23), 13834.
201. Dresselhaus, M., G. Dresselhaus, P. Eklund, Fullerenes. Journal of Materials Research, 1993, 8(08), 2054–2097.
202. Yuklyosi, K., Encyclopedic Dictionary of Mathematics. 1977, MIT Press, Cambridge.
203. Iijima, S., Growth of carbon nanotubes. Materials Science and Engineering: B, 1993, 19(1), 172–180.
204. Dravid, V., et al., Buckytubes and derivatives: their growth and implications for buckyball formation. Science, 1993, 259(5101), 1601–1604.
205. Ichihashi, T., Y. Ando, Pentagons, heptagons and negative curvature in graphite microtubule growth. Nature, 1992, 356(6372), 776–778.
206. Saito, Y., et al., Interlayer spacings in carbon nanotubes. Physical Review B, 1993, 48: 1907–1909.
207. Zhou, O., et al., Defects in carbon nanostructures. Science, 1994, 263(5154), 1744–1747.
208. Kiang, C., et al., Size effects in carbon nanotubes. Physical review letters, 1998, 81(9), 1869.
209. Amelinckx, S., et al., A structure model and growth mechanism for multishell carbon nanotubes. Science, 1995, 267(5202), 1334–1338.
210. Gerard Lavin, J., et al., Scrolls and nested tubes in multiwall carbon nanotubes. Carbon, 2002, 40(7), 1123–1130.
211. He, Z., et al., Etchant-induced shaping of nanoparticle catalysts during chemical vapor growth of carbon nanofibers. Carbon, 2011, 49(2), 435–444.
212. Bárdos, L., H. Baránková, Cold atmospheric plasma: Sources, processes, and applications. Thin Solid Films, 2010, 518(23), 6705–6713.
213. Unrau, C., R. Axelbaum, C. Lo, High-Yield Growth of Carbon Nanotubes on Composite Fe/Si/O Nanoparticle Catalysts: A Car– Parrinello Molecular Dynamics and Experimental Study. The Journal of Physical Chemistry C, 2010, 114(23), 10430–10435.
214. Kruusenberg, I., et al., Effect of purification of carbon nanotubes on their electrocatalytic properties for oxygen reduction in acid solution. Carbon, 2011, 49(12), 4031–4039.
215. Ebbesen, T., P. Ajayan, Large-scale synthesis of carbon nanotubes. Nature, 1992, 358(6383), 220–222.
216. Sakurai, T., et al., Scanning tunneling microscopy study of fullerenes. Progress in surface science, 1996, 51(4), 263–408.
217. Zhao, X., et al., Morphology of carbon nanotubes prepared by carbon arc. Japanese journal of applied physics, 1996, 35(part 1), 4451–4456.
218. Zhao, X., et al., Preparation of high-grade carbon nanotubes by hydrogen arc discharge. Carbon, 1997, 35(6), 775–781.
219. Zhao, X., et al., Morphology of carbon allotropes prepared by hydrogen arc discharge. Journal of crystal growth, 1999, 198: 934–938.
220. Anazawa, K., et al., High-purity carbon nanotubes synthesis method by an arc discharging in magnetic field. Applied Physics Letters, 2002, 81(4), 739–741.

221. Jiang, Y., et al., Influence of NH3 atmosphere on the growth and structures of carbon nanotubes synthesized by the arc-discharge method. Inorganic Materials, 2009, 45(11), 1237–1239.
222. Parkansky, N., et al., Single-pulse arc production of carbon nanotubes in ambient air. Journal of Physics D: Applied Physics, 2004, 37(19), 2715.
223. Jung, S., et al., High-yield synthesis of multiwalled carbon nanotubes by arc discharge in liquid nitrogen. Applied Physics A, 2003, 76(2), 285–286.
224. Prasek, J., et al., Methods for carbon nanotubes synthesis—review. Journal of Materials Chemistry, 2011, 21(40), 15872–15884.
225. Montoro, L., R. Lofrano, J. Rosolen, Synthesis of single-walled and multiwalled carbon nanotubes by arc-water method. Carbon, 2005, 43(1), 200–203.
226. Guo, J., et al., Structure of nanocarbons prepared by arc discharge in water. Materials Chemistry and physics, 2007, 105(2), 175–178.
227. Xing, G., S. Jia, Z. Shi, The production of carbon nano-materials by arc discharge under water or liquid nitrogen. New Carbon Materials, 2007, 22(4), 337–341.
228. Iijima, S., T. Ichihashi, Single-shell carbon nanotubes of 1-nm diameter. 1993: 8–15.
229. Ajayan, P., Nanotubes from carbon. Chemical reviews, 1999, 99(7), 1787–1800.
230. Seraphin, S., Single-Walled Tubes and Encapsulation of Nanocrystals into Carbon Clusters. Journal of the Electrochemical Society, 1995, 142(1), 290–297.
231. Tomita, M., T. Hayashi, Single-wall carbon nanotubes growing radially from Ni fine particles formed by arc evaporation. Jpn J Appl Phys, 1994(2), 4–11.
232. Saito, Y., et al., Carbon nanocapsules and single–layered nanotubes produced with platinum-group metals (Ru, Rh, Pd, Os, Ir, Pt) by arc discharge. Journal of applied physics, 1996, 80(5), 3062–3067.
233. Chen, B., et al., Fabrication and Dispersion Evaluation of Single-Wall Carbon Nanotubes Produced by FH-Arc Discharge Method. Journal of nanoscience and nanotechnology, 2010, 10(6), 3973–3977.
234. Chen, B., S. Inoue, Y. Ando, Raman spectroscopic and thermogravimetric studies of high-crystallinity SWNTs synthesized by FH-arc discharge method. Diamond and related materials, 2009, 18(5), 975–978.
235. Zhao, J., W. Bao, X. Liu, Synthesis of SWNTs from charcoal by arc-discharging. Journal of Wuhan University of Technology-Mater. Sci. Ed., 2010, 25(2), 194–196.
236. Wang, H., et al., Influence of Mo on the growth of single-walled carbon nanotubes in arc discharge. Journal of nanoscience and nanotechnology, 2010, 10(6), 3988–3993.
237. Liang, C., Z. Li, S. Dai, Mesoporous carbon materials: synthesis and modification. Angewandte Chemie International Edition, 2008, 47(20), 3696–3717.
238. Hutchison, J., et al., Double-walled carbon nanotubes fabricated by a hydrogen arc discharge method. Carbon, 2001, 39(5), 761–770.
239. Sugai, T., et al., New synthesis of high-quality double-walled carbon nanotubes by high-temperature pulsed arc discharge. Nano Letters, 2003, 3(6), 769–773.
240. Huang, H., et al., High-quality double-walled carbon nanotube super bundles grown in a hydrogen-free atmosphere. The journal of physical chemistry B, 2003, 107(34), 8794–8798.
241. Qiu, J., et al., Synthesis of double-walled carbon nanotubes from coal in hydrogen-free atmosphere. Fuel, 2007, 86(1), 282–286.

242. Qiu, H., et al., Synthesis and Raman scattering study of double-walled carbon nanotube peapods. Solid state communications, 2006, 137(12), 654–657.
243. Liu, Q., et al., Semiconducting properties of cup-stacked carbon nanotubes. Carbon, 2009, 47(3), 731–736.
244. Liu, C., H. Cheng, Carbon nanotubes for clean energy applications. Journal of Physics D: Applied Physics, 2005, 38(14), p. R231.
245. Ando, Y., et al., Multiwalled carbon nanotubes prepared by hydrogen arc. Diamond and related materials, 2000, 9(3), 847–851.
246. Pillai, S., et al., The Effect of Calcination on Multi-Walled Carbon Nanotubes Produced by Dc-Arc Discharge. Journal of nanoscience and nanotechnology, 2008, 8(7), 3539–3544.
247. Dunens, O., K. MacKenzie, A. Harris, Large-scale synthesis of double-walled carbon nanotubes in fluidized beds. Industrial and Engineering Chemistry Research, 2010, 49(9), 4031–4035.
248. Ikegami, T., et al., Optical measurement in carbon nanotubes formation by pulsed laser ablation. Thin Solid Films, 2004, 457(1), 7–11.
249. Bolshakov, A., et al., A novel CW laser–powder method of carbon single-wall nanotubes production. Diamond and related materials, 2002, 11(3), 927–930.
250. Zhang, M., et al., Strong, transparent, multifunctional, carbon nanotube sheets. Science, 2005, 309(5738), 1215–1219.
251. Lebel, L., et al., Preparation and mechanical characterization of laser ablated single-walled carbon-nanotubes/polyurethane nanocomposite microbeams. Composites Science and Technology, 2010, 70(3), 518–524.
252. Kusaba, M., Y. Tsunawaki, Production of single-wall carbon nanotubes by a XeCl excimer laser ablation. Thin Solid Films, 2006, 506: 255–258.
253. Stramel, A., et al., Pulsed laser deposition of carbon nanotube and polystyrene–carbon nanotube composite thin films. Optics and lasers in Engineering, 2010, 48(12), 1291–1295.
254. Bonaccorso, F., et al., Pulsed laser deposition of multiwalled carbon nanotubes thin films. Applied Surface Science, 2007, 254(4), 1260–1263.
255. Steiner III, S., et al., Nanoscale zirconia as a nonmetallic catalyst for graphitization of carbon and growth of single-and multiwall carbon nanotubes. Journal of the American Chemical Society, 2009, 131(34), 12144–12154.
256. Tempel, H., R. Joshi, J. Schneider, Ink jet printing of ferritin as method for selective catalyst patterning and growth of multiwalled carbon nanotubes. Materials Chemistry and physics, 2010, 121(1), 178–183.
257. Smajda, R., et al., Synthesis and mechanical properties of carbon nanotubes produced by the water assisted CVD process. Physica status solidi (b), 2009, 246(11–12), 2457–2460.
258. Byon, H., et al., A synthesis of high purity single-walled carbon nanotubes from small diameters of cobalt nanoparticles by using oxygen-assisted chemical vapor deposition process. BULLETIN-KOREAN CHEMICAL SOCIETY, 2007, 28(11), 2056.
259. Varshney, D., B.R. Weiner, G. Morell, Growth and field emission study of a monolithic carbon nanotube/diamond composite. Carbon, 2010, 48(12), 3353–3358.

260. Chatrchyan, S., et al., Search for supersymmetry at the LHC in events with jets and missing transverse energy. Physical review letters, 2011, 107(22), 221804.
261. Brown, B., et al., Growth of vertically aligned bamboo-like carbon nanotubes from ammonia/methane precursors using a platinum catalyst. Carbon, 2011, 49(1), 266–274.
262. Xu, Y., et al., Chirality-enriched semiconducting carbon nanotubes synthesized on high surface area MgO-supported catalyst. Materials Letters, 2011, 65(12), 1878–1881.
263. Wang, W., Y. Zhu, L. Yang, ZnO–SnO2 hollow spheres and hierarchical nanosheets: hydrothermal preparation, formation mechanism, and photocatalytic properties. Advanced Functional Materials, 2007, 17(1), 59–64.
264. Fotopoulos, N., J. Xanthakis, A molecular level model for the nucleation of a single-wall carbon nanotube cap over a transition metal catalytic particle. Diamond and related materials, 2010, 19(5), 557–561.
265. Sharma, R., et al., Evaluation of the role of Au in improving catalytic activity of Ni nanoparticles for the formation of one-dimensional carbon nanostructures. Nano Letters, 2011, 11(6), 2464–2471.
266. Palizdar, M., et al., Investigation of Fe/MgO Catalyst Support Precursors for the Chemical Vapour Deposition Growth of Carbon Nanotubes. Journal of nanoscience and nanotechnology, 2011, 11(6), 5345–5351.
267. Tomie, T., et al., Prospective growth region for chemical vapor deposition synthesis of carbon nanotube on C–H–O ternary diagram. Diamond and related materials, 2010, 19(11), 1401–1404.
268. Narkiewicz, U., et al., Catalytic decomposition of hydrocarbons on cobalt, nickel and iron catalysts to obtain carbon nanomaterials. Applied Catalysis A: General, 2010, 384(1), 27–35.
269. He, D., et al., Growth of carbon nanotubes in six orthogonal directions on spherical alumina microparticles. Carbon, 2011, 49(7), 2273–2286.
270. Shukla, B., et al., Interdependency of gas phase intermediates and chemical vapor deposition growth of single wall carbon nanotubes. Chemistry of Materials, 2010, 22(22), 6035–6043.
271. Afolabi, A., et al., Synthesis and purification of bimetallic catalyzed carbon nanotubes in a horizontal CVD reactor. Journal of Experimental Nanoscience, 2011, 6(3), 248–262.
272. Zhu, J., M. Yudasaka, S. Iijima, A catalytic chemical vapor deposition synthesis of double-walled carbon nanotubes over metal catalysts supported on a mesoporous material. Chemical Physics Letters, 2003, 380(5), 496–502.
273. Ramesh, P., et al., Selective chemical vapor deposition synthesis of double-wall carbon nanotubes on mesoporous silica. The journal of physical chemistry B, 2005, 109(3), 1141–1147.
274. Hiraoka, T., et al., Selective synthesis of double-wall carbon nanotubes by CCVD of acetylene using zeolite supports. Chemical Physics Letters, 2003, 382(5), 679–685.
275. Flahaut, E., C. Laurent, A. Peigney, Catalytic CVD synthesis of double and triple-walled carbon nanotubes by the control of the catalyst preparation. Carbon, 2005, 43(2), 375–383.

276. Xiang, X., et al., Co-based catalysts from Co/Fe/Al layered double hydroxides for preparation of carbon nanotubes. Applied Clay Science, 2009, 42(3), 405–409.
277. Lyu, S., et al., High-quality double-walled carbon nanotubes produced by catalytic decomposition of benzene. Chemistry of Materials, 2003, 15(20), 3951–3954.
278. Zhang, D., et al., Preparation and desalination performance of multiwall carbon nanotubes. Materials Chemistry and physics, 2006, 97(2), 415–419.
279. Sano, N., S. Ishimaru, H. Tamaon, Synthesis of carbon nanotubes in graphite microchannels in gas-flow and submerged-in-liquid reactors. Materials Chemistry and physics, 2010, 122(2), 474–479.
280. Jiang, Q., et al., Preparation and characterization on the carbon nanotube chemically modified electrode grown in situ. Electrochemistry Communications, 2008, 10(3), 424–427.
281. Scheibe, B., E. Borowiak-Palen, R. Kalenczuk, Enhancement of thermal stability of multiwalled carbon nanotubes via different silanization routes. Journal of Alloys and Compounds, 2010, 500(1), 117–124.
282. Feng, J., et al., One-step fabrication of high quality double-walled carbon nanotube thin films by a chemical vapor deposition process. Carbon, 2010, 48(13), 3817–3824.
283. Li, G., Synthesis of well-aligned carbon nanotubes on the NH_3 pretreatment Ni catalyst films. Russian Journal of Physical Chemistry A, 2010, 84(9), 1560–1565.
284. Liu, Z., et al., Aligned, Ultralong Single-Walled Carbon Nanotubes: From Synthesis, Sorting, to Electronic Devices. Advanced Materials, 2010, 22(21), 2285–2310.
285. Kim, H., et al., Synthesis of carbon nanotubes with catalytic iron-containing proteins. Carbon, 2011, 49(12), 3717–3722.
286. Cui, T., et al., Temperature effect on synthesis of different carbon nanostructures by sulfur-assisted chemical vapor deposition. Materials Letters, 2011, 65(3), 587–590.
287. Grazhulene, S., et al., Adsorption properties of carbon nanotubes depending on the temperature of their synthesis and subsequent treatment. Journal of Analytical Chemistry, 2010, 65(7), 682–689.
288. Du, G., Y. Zhou, B. Xu, Preparation of carbon nanotubes by pyrolysis of dimethyl sulfide. Materials Characterization, 2010, 61(4), 427–432.
289. Kim, S., L. Gangloff, Growth of carbon nanotubes (CNTs) on metallic underlayers by diffusion plasma-enhanced chemical vapor deposition (DPECVD). Physica E: Low-dimensional Systems and Nanostructures, 2009, 41(10), 1763–1766.
290. Wang, H., J. Moore, Different growth mechanisms of vertical carbon nanotubes by rf-or dc-plasma enhanced chemical vapor deposition at low temperature. Journal of Vacuum Science & Technology B, 2010, 28(6), 1081–1085.
291. Luais, E., et al., Preparation and modification of carbon nanotubes electrodes by cold plasmas processes toward the preparation of amperometric biosensors. Electrochimica Acta, 2010, 55(27), 7916–7922.
292. Sun, X., et al., The effect of catalysts and underlayer metals on the properties of PECVD-grown carbon nanostructures. Nanotechnology, 2010, 21(4), 045201.
293. Häffner, M., et al., Plasma enhanced chemical vapor deposition grown carbon nanotubes from ferritin catalyst for neural stimulation microelectrodes. Microelectronic Engineering, 2010, 87(5), 734–737.

294. Vollebregt, S., et al., Growth of High-Density Self-Aligned Carbon Nanotubes and Nanofibers Using Palladium Catalyst. Journal of electronic materials, 2010, 39(4), 371–375.
295. Duy, D., et al., Growth of carbon nanotubes on stainless steel substrates by DC-PECVD. Applied Surface Science, 2009, 256(4), 1065–1068.
296. Jang, S., et al., Flexible, transparent single-walled carbon nanotube transistors with graphene electrodes. Nanotechnology, 2010, 21(42), 425201.
297. Ono, Y., et al., Thin film transistors using PECVD-grown carbon nanotubes. Nanotechnology, 2010, 21(20), 202–205.
298. Yang, G., et al., Enhancement mechanism of field electron emission properties in hybrid carbon nanotubes with tree-and wing-like features. Journal of Solid State Chemistry, 2009, 182(12), 3393–3398.
299. Seo, J., et al., Metal-free CNTs grown on glass substrate by microwave PECVD. Current Applied Physics, 2010, 10(3), p. S447-S450.
300. Bu, I., S. Oei, Hydrophobic vertically aligned carbon nanotubes on Corning glass for self cleaning applications. Applied Surface Science, 2010, 256(22), 6699–6704.
301. Bu, I., K. Hou, D. Engstrom, Industrial compatible regrowth of vertically aligned multiwall carbon nanotubes by ultrafast pure oxygen purification process. Diamond and related materials, 2011, 20(5), 746–751.
302. Vinten, P., et al., Origin of periodic rippling during chemical vapor deposition growth of carbon nanotube forests. Carbon, 2011, 49(15), 4972–4981.
303. Kim, D., et al., Growth of vertically aligned arrays of carbon nanotubes for high field emission. Thin Solid Films, 2008, 516(5), 706–709.
304. Yamada, T., et al., Size-selective growth of double-walled carbon nanotube forests from engineered iron catalysts. Nature Nanotechnology, 2006, 1(2), 131–136.
305. Kim, M., et al., Growth characteristics of carbon nanotubes via aluminum nanopore template on Si substrate using PECVD. Thin Solid Films, 2003, 435(1), 312–317.
306. Sui, Y., et al., Growth of carbon nanotubes and nanofibers in porous anodic alumina film. Carbon, 2002, 40(7), 1011–1016.
307. Lee, O., et al., Synthesis of carbon nanotubes with identical dimensions using an anodic aluminum oxide template on a silicon wafer. Synthetic metals, 2005, 148(3), 263–266.
308. Chang, W., et al., Fabrication of free-standing carbon nanotube electrode arrays on a quartz wafer. Thin Solid Films, 2010, 518(22), 6624–6629.
309. Byeon, H., et al., Growth of ultra long multiwall carbon nanotube arrays by aerosol-assisted chemical vapor deposition. Journal of nanoscience and nanotechnology, 2010, 10(9), 6116–6119.
310. Jeong, N., et al., Microscopic and Spectroscopic Analyzes of Pt-Decorated Carbon Nanowires Formed on Carbon Fiber Paper. Microscopy and Microanalysis, 2013, 19(S5), 198–201.
311. Liu, J., et al., Nitrogen-doped carbon nanotubes with tunable structure and high yield produced by ultrasonic spray pyrolysis. Applied Surface Science, 2011, 257(17), 7837–7844.
312. Khatri, I., et al., Synthesis of single walled carbon nanotubes by ultrasonic spray pyrolysis method. Diamond and related materials, 2009, 18(2), 319–323.

313. Camarena, J., et al., Molecular Assembly of Multi-Wall Carbon Nanotubes with Amino Crown Ether: Synthesis and Characterization. Journal of nanoscience and nanotechnology, 2011, 11(6), 5539–5545.
314. Sadeghian, Z., Large-scale production of multiwalled carbon nanotubes by low-cost spray pyrolysis of hexane. New Carbon Materials, 2009, 24(1), 33–38.
315. Pinault, M., et al., Carbon nanotubes produced by aerosol pyrolysis: growth mechanisms and post-annealing effects. Diamond and related materials, 2004, 13(4), 1266–1269.
316. Nebol'sin, V., A. Vorob'ev, Role of surface energy in the growth of carbon nanotubes via catalytic pyrolysis of hydrocarbons. Inorganic Materials, 2011, 47(2), 128–132.
317. Lara-Romero, J., et al., Temperature effect on the synthesis of multiwalled carbon nanotubes by spray pyrolysis of botanical carbon feedstocks: turpentine, α-pinene and β-pinene. Fullerenes, Nanotubes, and Carbon Nanostructures, 2011, 19(6), 483–496.
318. Kumar, R., R. Tiwari, O. Srivastava, Scalable synthesis of aligned carbon nanotubes bundles using green natural precursor: neem oil. Nanoscale research letters, 2011, 6(1), 1–6.
319. Paul, S., S. Samdarshi, A precursor for carbon nanotube synthesis. New Carbon Materials, 2011, 26(2), 85–88.
320. Duan, W., Q. Wang, Water transport with a carbon nanotube pump. ACS nano, 2010, 4(4), 2338–2344.
321. Ionescu, M., et al., Nitrogen-doping effects on the growth, structure and electrical performance of carbon nanotubes obtained by spray pyrolysis method. Applied Surface Science, 2012, 258(10), 4563–4568.
322. Kucukayan, G., et al., An experimental and theoretical examination of the effect of sulfur on the pyrolytically grown carbon nanotubes from sucrose-based solid state precursors. Carbon, 2011, 49(2), 508–517.
323. Clauss, C., M. Schwarz, E. Kroke, Microwave-induced decomposition of nitrogen-rich iron salts and CNT formation from iron (III)–melonate Fe Carbon, 2010, 48(4), 1137–1145.
324. Kuang, Z., et al., Biomimetic chemosensor: designing peptide recognition elements for surface functionalization of carbon nanotube field effect transistors. ACS nano, 2009, 4(1), 452–458.
325. Du, G., et al., Solid-phase transformation of glass-like carbon nanoparticles into nanotubes and the related mechanism. Carbon, 2008, 46(1), 92–98.
326. El Hamaoui, B., et al., Solid-State Pyrolysis of Polyphenylene–Metal Complexes: A Facile Approach Toward Carbon Nanoparticles. Advanced Functional Materials, 2007, 17(7), 1179–1187.
327. Rudin, A., P. Choi, The Elements of Polymer Science & Engineering. 2012: Academic Press.
328. Li, J., et al., Bottom-up approach for carbon nanotube interconnects. Applied Physics Letters, 2003, 82(15), 2491–2493.
329. Jasti, R., C. Bertozzi, Progress and challenges for the bottom-up synthesis of carbon nanotubes with discrete chirality. Chemical Physics Letters, 2010, 494(1), 1–7.
330. Omachi, H., et al., A Modular and Size-Selective Synthesis of [n] Cycloparaphenylenes: A Step toward the Selective Synthesis of [n, n] Single-Walled Carbon Nanotubes. Angewandte Chemie International Edition, 2010, 49(52), 10202–10205.

331. Segawa, Y., et al., Concise synthesis and crystal structure of [12] cycloparaphenylene. Angewandte Chemie International Edition, 2011, 50(14), 3244–3248.
332. Omachi, H., Y. Segawa, K. Itami, Synthesis and racemization process of chiral carbon nanorings: a step toward the chemical synthesis of chiral carbon nanotubes. Organic letters, 2011, 13(9), 2480–2483.
333. Iwamoto, T., et al., Selective and random syntheses of [n] cycloparaphenylenes (n = 8–13) and size dependence of their electronic properties. Journal of the American Chemical Society, 2011, 133(21), 8354–8361.
334. Fort, E., L. Scott, Gas-phase Diels–Alder cycloaddition of benzyne to an aromatic hydrocarbon bay region: Groundwork for the selective solvent-free growth of armchair carbon nanotubes. Tetrahedron Letters, 2011, 52(17), 2051–2053.
335. Hernandez, E., et al., Elastic properties of C and B *x* C *y* N *z* composite nanotubes. Physical review letters, 1998, 80(20), 4502.
336. Machida, K., Principles of molecular mechanics. 1999: Wiley. 544.
337. Rappé, A., et al., UFF, a full periodic table force field for molecular mechanics and molecular dynamics simulations. Journal of the American Chemical Society, 1992, 114(25), 10024–10035.
338. Mayo, S., B. Olafson, W. Goddard, DREIDING: a generic force field for molecular simulations. Journal of Physical Chemistry, 1990, 94(26), 8897–8909.
339. Kelly, B., Physics of graphite. Vol. 3. 1981: Applied Science London. 237.
340. Kudin, K., G. Scuseria, B. Yakobson, C_2 F, BN, and C nanoshell elasticity from ab initio computations. Physical Review B, 2001, 64(23), 235406.
341. Jorgensen, W., D. Severance, Aromatic-aromatic interactions: free energy profiles for the benzene dimer in water, chloroform, and liquid benzene. Journal of the American Chemical Society, 1990, 112(12), 4768–4774.
342. Heermann, D., Computer-Simulation Methods. 1990: Springer.
343. Haile, J., Molecular dynamics simulation: elementary methods. 1992, John Wiley & Sons, Inc.
344. Sohlberg, K., et al., Continuum methods of mechanics as a simplified approach to structural engineering of nanostructures. Nanotechnology, 1998, 9(1), 30.
345. Kresin, V., A. Aharony, Fully collapsed carbon nanotubes. Nature, 1995, 377, 135.

CHAPTER 11

3D RECONSTRUCTION FROM TWO VIEWS OF SINGLE 2D IMAGE AND ITS APPLICATIONS IN PORE ANALYSIS OF NANOFIBROUS MEMBRANE

B. HADAVI MOGHADAM and A. K. HAGHI

University of Guilan, Rasht, Iran

CONTENTS

ABSTRACT

This chapter provides a detailed review on relevant approach of 3D reconstruction from two views of single 2D image and it potential applications in pore analysis of electrospun nanofibrous membrane. The review has concisely demonstrated that 3D reconstruction consists of three steps which is equivalent to the estimation of a specific geometry group. These steps include: estimation of the epipolar geometry existing between the stereo image pair, estimation of the affine geometry, and also camera calibration. The advantage of this system is that the 2D images do not need to be calibrated in order to obtain a reconstruction. Results for both the camera calibration and reconstruction are presented to verify that it is possible to obtain a 3D model directly from features in the images. Finally, the applications of 3D reconstruction in pore structure characterization of electrospun nanofibrous membrane are discussed.

11.1 INTRODUCTION

3D reconstruction is a challenging task in computer vision and has received considerable attention recently due to the loss of a dimension (the depth of the image) in the process of photographing image and the usefulness of the recovered 3D model for a variety of applications, such as city planning, cartography, architectural design, *y*-through simulations. The key task in 3D reconstruction is to recover high-quality and detailed 3D models from two or more view of images, which may be taken from widely separated viewpoints. Due to the complexity of the images, conventional modeling techniques are very time-consuming and recreating detailed geometry become very difficult. In order to overcome these difficulties, some works have been inclined towards image-based modeling techniques, using images to drive the 3D reconstruction. However, in many image-based modeling techniques, the image are reconstructed using camera calibrated images or, when this is not the case, it is nontrivial to establish correspondences between different views of image. Image-based modeling relies on a set of techniques for creating 3D representation of a 2D image from one or more views of image [1–7].

Generally, the 3D reconstruction consists of three steps: (i) estimation of the epipolar geometry existing between the two view of image pair, which involves feature matching in both images, (ii) estimation of the affine geometry which considered as a process of finding a special plane in projective space by means of vanishing points, and (iii) camera calibration by which it is possible to obtain a 3D model of the 2D image (Fig. 11.1) [8].

3D reconstruction from a number of perspective images is one of the fundamental problems of computer vision, while reconstruction from two views is the simplest one. To the best of our knowledge, earliest work in this field concentrated on calibrated cameras, from which it is possible to obtain a Euclidean (sometimes called metric) reconstruction of the image [9].

Finding the 2D point matches between images, which is known as the correspondence, is the first problem encountered to the process. There

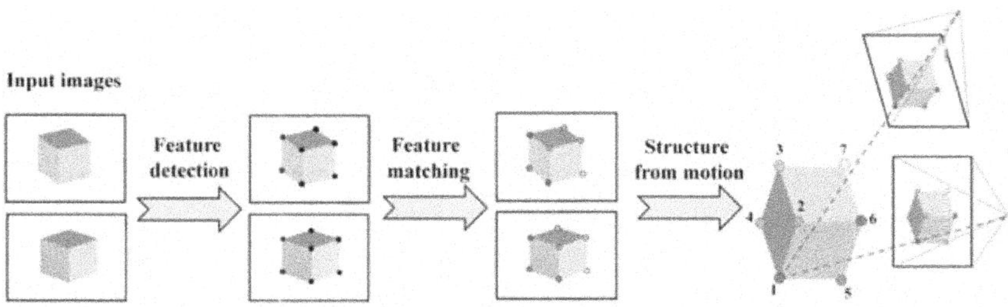

FIGURE 11.1 3D camera poses and positions.

are many automated techniques for finding correspondences between two images, but most of them work on the basis that the same 3D point in the world (e.g., the window corner) will have a similar appearance in different images, particularly if those images are taken close together. The second challenge is how to determine where each photo was taken and the 3D location of each image point, given just a set of corresponding 2D points among the photos. If one knew the camera poses but not the point positions, one could find the points through triangulation; conversely, if one knew the 3D points, one could find the camera poses through a process similar to triangulation called resectioning. Unfortunately, one knows neither the camera poses nor the points [10–13].

For a 2D image pair, the individual steps of the 3D reconstruction algorithm are as follows:

1. Features are detected in each image independently.
2. A set of initial feature matches is calculated.
3. The fundamental matrix is calculated using the set of initial matches.
4. False matches are discarded and the fundamental matrix is refined.
5. Projective camera matrices are established from the fundamental matrix.
6. Vanishing points on three different planes and in three different directions are calculated from parallel lines in the images.
7. The plane at infinity is calculated from the vanishing points in both images.
8. The projective camera matrices are upgraded to affine camera matrices using the plane at infinity.
9. The camera calibration matrix (established separately to the reconstruction process) is used to upgrade the affine camera matrices to metric camera matrices.
10. Triangulation methods are used to obtain a full 3D reconstruction with the help of the metric camera matrices.
11. If needed, dense matching techniques are employed to obtain a 3D texture map of the model to be reconstructed.

Figure 11.2 shows the algorithm for individual steps of the 3D reconstruction.

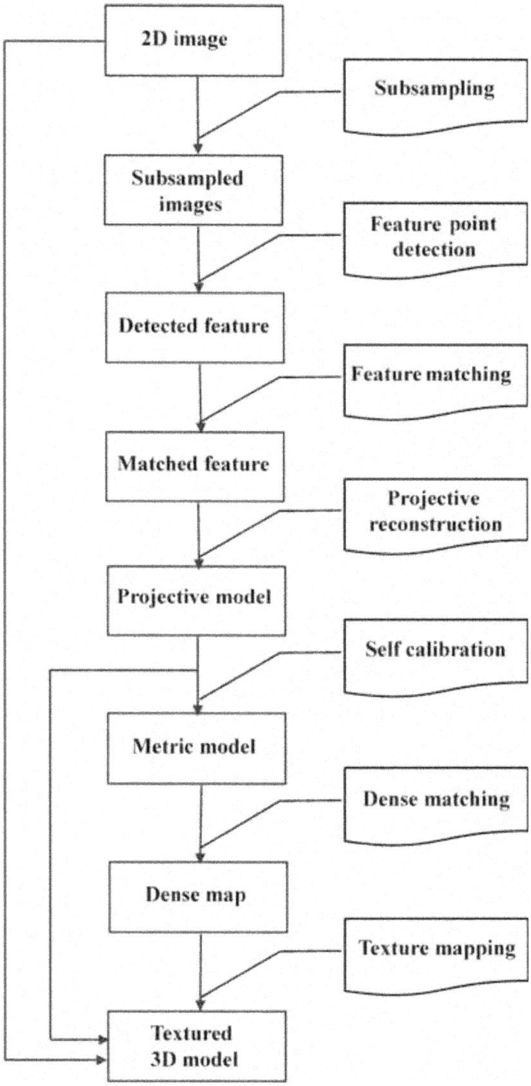

FIGURE 11.2 3D modeling framework.

11.2 CLASSIFICATION OF 3D VISION

Stratification of 3D vision makes it easier to perform a reconstruction. Up to this point it is possible to obtain the 3D geometry of the image, but as only a restricted number of features are extracted, it is not possible to obtain a very complete textured 3D model [8].

3D vision can be divided into four geometry groups or strata, of which Euclidean geometry is one. The simplest group is projective geometry, which forms the basis of all other groups. The other groups include affine geometry, metric geometry and then Euclidean geometry. These geometries are subgroups of each other, metric being a subgroup of affine geometry, and both these being subgroups of projective geometry [8].

These categories are described in Table 11.1 and some characteristics of these categories are summarized.

11.2.1 PROJECTIVE GEOMETRY

The basis of most computer vision tasks, especially in the fields of 3D reconstruction from images and camera self-calibration are Algebraic and projective geometry forms [8].

In mathematics, projective geometry is an elementary nonmetrical form of geometry, that is invariant under projective transformations and it is not based on a concept of distance. This means that projective geometry has a different setting, projective space, and a selective set of basic geometric concepts in comparison to elementary geometry. In a given

TABLE 11.1 Characteristics of 3D Geometric Transformation

Transformation	DoF	Trans. Matrix	Distortion	Invariants
Projective	15	$[\tilde{H}]_{4\times4}$	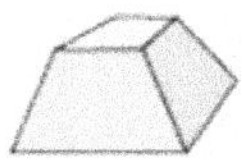	straight lines
Affine	12	$[A]_{3\times4}$	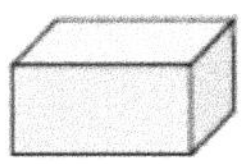	parallelism
Metric (similarity)	7	$[sR\|t]_{3\times4}$	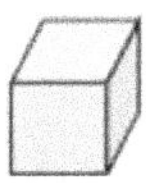	angles
Euclidean (rigid)	6	$[R\|t]_{3\times4}$	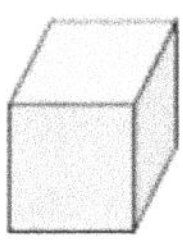	lengths

dimension, projective space has more points than Euclidean space and geometric transformations are allowed that move the extra points (called "points at infinity") to traditional points, and vice versa. In two dimensions it begins with the study of configurations of points and lines. In higher dimensional spaces there are considered hyperplanes (that always meet), and other linear subspaces, which exhibit the principle of duality. The simplest illustration of duality is in the projective plane, where the statements "two distinct points determine a unique line" (i.e., the line through them) and "two distinct lines determine a unique point" (i.e., their point of intersection) show the same structure as propositions. Properties meaningful in projective geometry are respected by this new idea of transformation, which is more radical in its effects than expressible by a transformation matrix and translations (the affine transformations).

Theory of perspective is one source for projective geometry. Another difference from elementary geometry is the way in which parallel lines can be said to meet in a point at infinity, once the concept is translated into projective geometry's terms. Again this notion has an intuitive basis, such as railway tracks meeting at the horizon in a perspective drawing. See projective plane for the basics of projective geometry in two dimensions (Fig. 11.3).

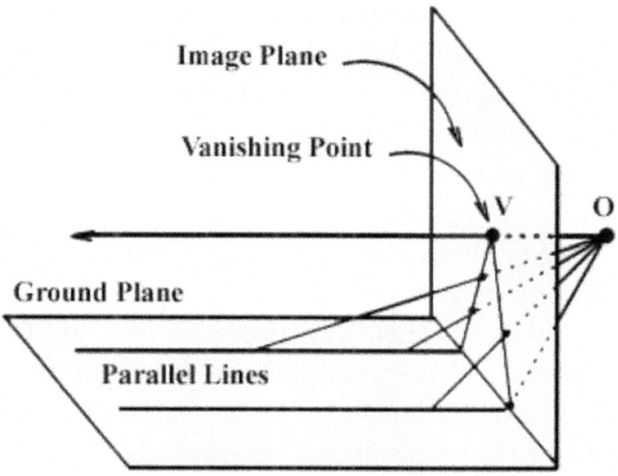

FIGURE 11.3 A two dimensional construction of perspective viewing which illustrates the formation of a vanishing point.

The projective algebraic geometry (the study of projective varieties) and projective differential geometry (the study of differential invariants of the projective transformations) are two research subfields of projective geometry.

11.2.1.1 Projective Space and Its Homogeneous Coordinates

An n-dimensional projective space P^n is the set of one-dimensional subspaces (i.e., lines through the origin) of the vector space R^{n+1}. A point P in P^n can then be assigned homogeneous coordinates $X = [X_1, X_2, \ldots, X_{n+1}]^T$ among which at least one X_i is nonzero. For any nonzero $\lambda \in R$ the coordinates $Y = [\lambda X_1, \lambda X_2, \ldots, \lambda X_{n+1}]^T$ represent the same point P in P^n. We say that X and Y are equivalent, denoted by $X \sim Y$.

11.2.1.2 Topological Models for the Projective Space P^2

In projective geometry the linear group of transformations of the line into itself contains three parameters, if we may use this convenient analytic term to stand for the geometric conception of degrees of freedom. The establishment, upon the line, of a definite number- system dependent on three assumed points 0, 1, and ∞ determines completely the theory of segments and distance – the difference between two numbers giving the distance between the points to which they are affixed. The postulate of congruence, namely, that from a given point of the line and in either direction there may be laid off upon the line one and only one segment equal to a given segment of the line, has an evident interpretation.

Figure 11.4 demonstrates two equivalent geometric interpretations of the 2D projective space P^2.

According to the definition, it is simply a family of 1D lines $\{L\}$ in R^3 through a point o (typically chosen to be the origin of the coordinate frame). Hence, P^2 can be viewed as a 2D sphere S^2 with any pair of antipodal points (e.g., p and p' in the figure) identified as one point in P^2. On the right-hand side of Figure 4, lines through the center o in general intersect with the plane $\{z = 1\}$ at a unique point except when they lie on the plane

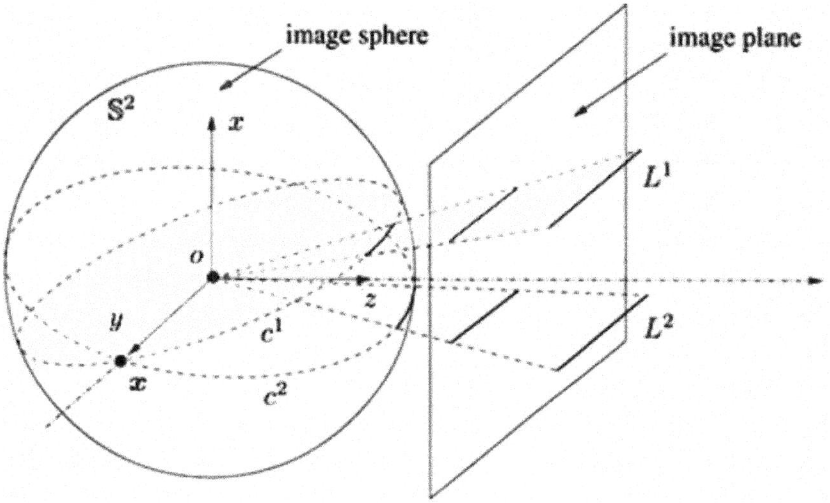

FIGURE 11.4 Topological models for projective space P^2 [16].

$\{z = 0\}$. Lines in the plane $\{z = 0\}$ simply form the 1D projective space P^1 (which is in fact a circle).

Hence, P^2 can be viewed as a 2D plane R^2 (i.e., $\{z = 1\}$) with a circle P^1 attached. If we adopt the view that lines in the plane $\{z = 0\}$ intersect the plane $\{z = 1\}$ infinitely far, this circle P^1 then represents a line at infinity. Homogeneous coordinates for a point on this circle then take the form $[x, y, 0]$; on the other hand, all regular points in R^2 have coordinates $[x, y, 1]^T$. In general, any projective space P^n can be visualized in a similar way: P^3 is then R^3 with a plane P^2 attached at infinity; and P^n is R^n with P^{n-1} attached at infinity.

Using this definition, R^n with its homogeneous representation can then be identified as a subset of P^n that includes exactly those points with coordinates $X = [x_1, x_2, \ldots, x_{n+1}]^T$ where $x_{n+1} \neq 0$. Therefore, we can always normalize the last entry to 1 by dividing X by x_{n+1} [16, 17].

11.2.2 AFFINE GEOMETRY

Affine geometry is not concerned with the notions of circle, angle and distance. It's a known dictum that in affine geometry all triangles are the same. In this context, the word affine was first used by Euler (affinis).

In modern parlance, Affine Geometry is a study of properties of geometric objects that remain invariant under affine transformations (mappings).

Affine geometry is established by finding the plane at infinity in projective space for both images. The usual method of finding the plane is by determining vanishing points in both images and then projecting them into space to obtain points at infinity. Vanishing points are the intersections of two or more imaged parallel lines. This process is unfortunately very difficult to automate, as the user generally has to select the parallel lines in the images. Some automatic algorithms try to find dominant line orientations in histograms [18].

A rigid motion is an affine map, but not a linear map in general. Also, given an $m \times n$ matrix A and a vector $b \in R^m$, the set $U = \{x \in R^n \mid Ax = b\}$ of solutions of the system $Ax = b$ is an affine space, but not a vector space (linear space) in general [19].

Analytically, affine transformations are represented in the matrix form $f(x) = Ax + b$, where the determinant $\det(A)$ of a square matrix A is not 0. In a plane, the matrix is 2×2; in the space, it is 3×3. One way to arrive at the matrix representation is to select two points (two origins) and associate with each an appropriate number of independent vectors (2 in the plane, 3 in the space), to form an affine basis. b is then the translation that maps one of the selected points onto another.

A three-dimensional incidence space is a triple $(S;L;P)$ consisting of a nonempty set S (whose elements are called points) and two nonempty disjoint families of proper subsets of S denoted by L (lines) and P (planes) respectively, which satisfy the following conditions:

1. Every line (element of L) contains at least two points, and every plane (element of P) contains at least three points.
2. If x and y are distinct points of S, then there is a unique line L such that $x, y \in L$. Notation. The line given by 2 is called xy.
3. If x, y and z are distinct points of S and $z \notin xy$, then there is a unique plane P such that $x, y, z \in P$.
4. If a plane P contains the distinct points x and y, then it also contains the line xy.
5. If P and Q are planes with a nonempty intersection, then $P \cap Q$ contains at least two points.

Affine transformations preserve collinearity of points: if three points belong to the same straight line, their images under affine transformations also belong to the same line and, in addition, the middle point remains between the other two points. As further examples, under affine transformations parallel lines remain parallel, concurrent lines remain concurrent (images of intersecting lines intersect), the ratio of length of line segments of a given line remains constant, the ratio of areas of two triangles remains constant (and hence the ratio of any areas remain constant), ellipses remain ellipses and the same is true for parabolas and hyperbolas (Fig. 11.5) [20].

11.2.3 METRIC GEOMETRY

This stratum corresponds to the group of similarities. The transformations in this group are Euclidean transformations such as rotation and

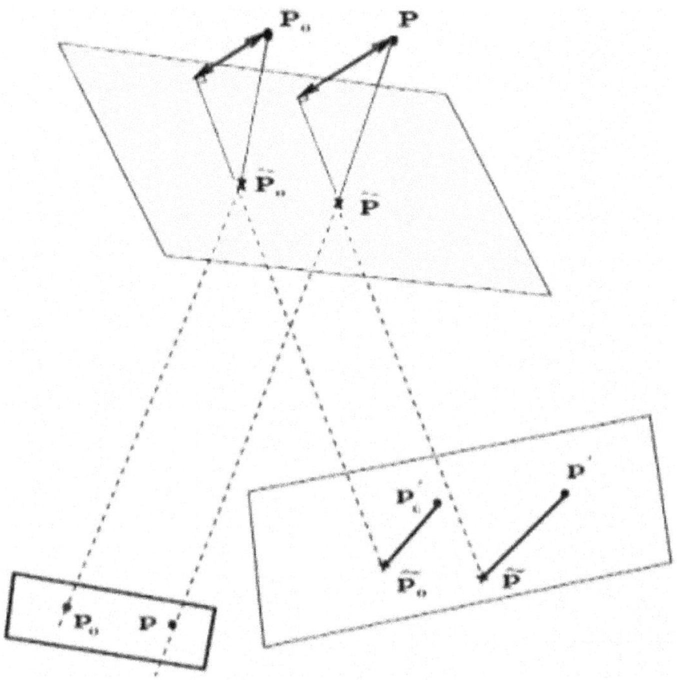

FIGURE 11.5 Affine structure under parallel projection is d_p/d_o [21].

translation. The metric stratum allows for a complete reconstruction up to an unknown scale [8].

Let X be an arbitrary set. A function $d: X \times X \rightarrow R \cup \{\infty\} f_1 g$ is a metric on X if the following conditions are satisfied for all $x, y, z \in$ X.

1. Positiveness: $d(x, y) \rangle 0$ if $x \neq y$, and $d(x, x) = 0$.
2. Symmetry: $d(x, y) = d(y, x)$.
3. Triangle inequality: $d(x, z) \leq d(x, y) + d(y, z)$.

A metric space is a set with a metric on it. In a formal language, a metric space is a pair (X, d) where d is a metric on X. Elements of X are called points of the metric space; $d(x, y)$ is referred to as the distance between points x and y. When the metric in question is clear from the context, we also denote the distance between x and y by $|x\,y|$. Unless different metrics on the same set x are considered, we will omit an explicit reference to the metric and write "a metric space X instead of "a metric space $(X; d)$." In most textbooks, the notion of a metric space is slightly narrower than our definition: traditionally one consider metrics with finite distance between points. If it is important for a particular consideration that d takes only finite values, this will be specified by saying that d is a finite metric [22, 23].

Let X and Y be metric spaces. Then

1. A sequence in X cannot have more than one limit.
2. A point $x \in X$ is an accumulation point of a set $S \subset X$ (i.e., belongs to the closure of S) if and only if there exists a sequence $\{x_n\}_{n=1}^{\infty}$ such that $x_n \in S$ for all n and $x_n \rightarrow x$. In particular, S is closed if and only if it contains all limits of sequences contained within S.
3. A map $f: X \rightarrow Y$ is continuous at a point $x \in X$ if and only if $f(x_n) \rightarrow f(x)$ for any sequence $\{x_n\}$ converging to x [22].

11.2.4 EUCLIDEAN GEOMETRY

Geometry (from the Greek "geo" = earth and "metria" = measure) arose as the field of knowledge dealing with spatial relationships. Geometry can be split into Euclidean geometry and analytical geometry. Analytical geometry deals with space and shape using algebra and a coordinate system. Euclidean geometry deals with space and shape using a system of logical deductions. A geometry in which Euclid's fifth postulate holds, sometimes also called parabolic geometry. Two-dimensional Euclidean geometry is called plane geometry, and three-dimensional Euclidean geometry is called solid geometry. Hilbert proved the consistency of Euclidean geometry.

Euclidean geometry describes a 3D world very well (Fig. 11.6). As an example, the sides of objects have known or calculable lengths, intersecting lines determine angles between them, and lines that are parallel on a plane will never meet. But when it comes to describing the imaging process of a camera, the Euclidean geometry is not sufficient, as it is not possible to determine lengths and angles anymore, and parallel lines may intersect.

Euclidean geometry is the same as metric geometry, the only difference being that the relative lengths are upgraded to absolute lengths.

Euclidean geometry is simply metric geometry, but incorporates the correct scale of the image.

The scale can be fixed by knowing the dimensions of a certain object in the image [6].

In Euclidean geometry, second-order curves such as ellipses, parabolas and hyperbolas are easily defined.

In affine geometry it is possible to deal with ratios of vectors and barycenter's of points, but there is no way to express the notion of length of a line segment or to talk about orthogonality of vectors. A Euclidean

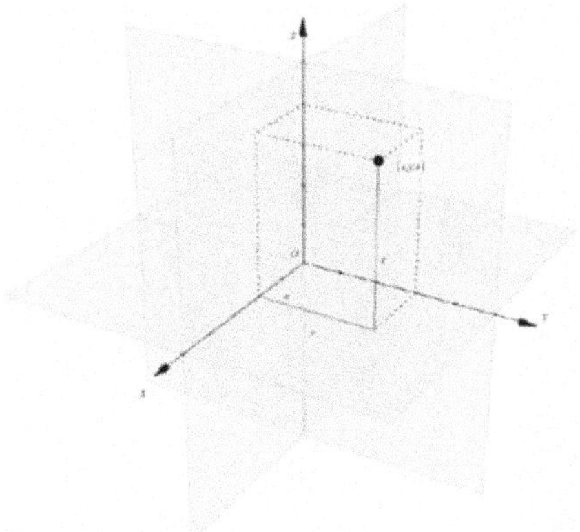

FIGURE 11.6 3D Euclidean space.

structure allows us to deal with metric notions such as orthogonality and length (or distance) [24–26].

11.3 VANISHING POINTS AND LINES

A vanishing point (VP) is defined as Parallel lines of the world are projected into perspective images as intersecting lines and their point of intersection (Fig. 11.7). Each set of parallel lines is thus associated to a VP. Popular applications of VPs in computer vision are calibration, rotation estimation and 3D reconstruction [27, 28].

The VP is the image of the point at infinity where the intersections of two or more parallel lines. It is computed as the least squares solution for the intersection of sets of images of parallel 3D line segments.

Vanishing lines are determined from vanishing points and parallelism constraints. Two or more vanishing points parallel to a plane define its vanishing line. A vanishing point belongs to the 3D direction perpendicular to a plane, completely defines the vanishing line [2].

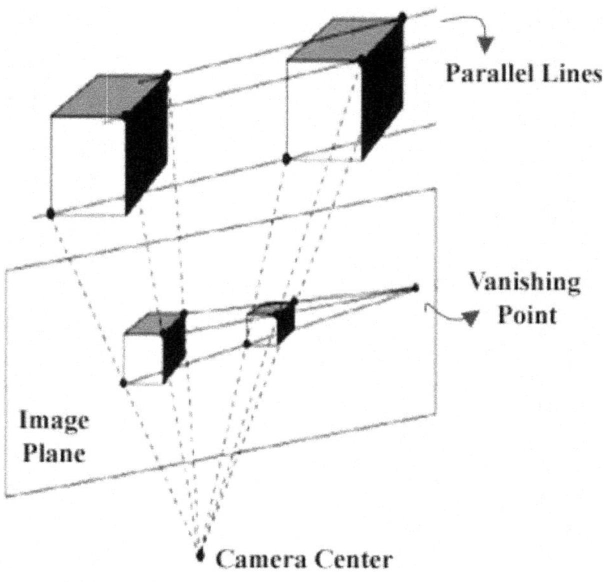

FIGURE 11.7 Vanishing point.

Each vanishing point corresponds to an orientation in the 3D image and when the camera geometry is known, these orientations can be recovered. Even without this information, vanishing points can be used to group segments on the image with the same 3D orientation. Because of its important role in 3D reconstruction, the detection of the vanishing points in a image has to be effective, especially when no human intervention is required [12].

Estimating vanishing points in an image provides strong cues to make inferences about the 3D structures of a 2D image, such as depth and object dimension, because they are invariant features and they have many applications ranging from autonomous navigation to single-view reconstruction [28].

11.4 IMAGE FEATURES

In computer vision and image processing, a feature is a piece of information which is relevant for solving the computational task related to a certain application. Features may be specific structures in the image such as points, edges or objects. Features may also be the result of a general neighborhood operation or feature detection applied to the image. Other examples of features are related to motion in image sequences, to shapes defined in terms of curves or boundaries between different image regions, or to properties of such a region. The feature concept is very general and the choice of features in a particular computer vision system may be highly dependent on the specific problem at hand [29–32].

The first kind of feature that you may notice are specific locations in the images, such as mountain peaks, building corners, doorways, or interestingly shaped patches of snow. These kinds of localized features are often called key point features or interest points (or even corners) and are often described by the appearance of patches of pixels surrounding the point location. Another class of important features are edges, for example, the profile of the mountains against the sky. These kinds of features can be matched based on their orientation and local appearance (edge profiles) and can also be good indicators of object boundaries and occlusion events in image sequences. Edges can be grouped into longer curves and straight line segments, which can be directly matched, or analyzed to find vanishing points and hence internal and external camera parameters.

11.4.1 EDGES

Edges are points where there is a boundary (or an edge) between two image regions. In general, an edge can be of almost arbitrary shape, and may include junctions. In practice, edges are usually defined as sets of points in the image which have a strong gradient magnitude. Furthermore, some common algorithms will then chain high gradient points together to form a more complete description of an edge. These algorithms usually place some constraints on the properties of an edge, such as shape, smoothness, and gradient value. Locally, edges have a one-dimensional structure [33–35].

11.4.2 CORNERS/INTEREST POINTS

The terms corners and interest points are used somewhat interchangeably and refer to point-like features in an image, which have a local two dimensional structure. The name "Corner" arose since early algorithms first performed edge detection, and then analyzed the edges to find rapid changes in direction (corners). These algorithms were then developed so that explicit edge detection was no longer required, for instance by looking for high levels of curvature in the image gradient. It was then noticed that the so-called corners were also being detected on parts of the image which were not corners in the traditional sense (for instance a small bright spot on a dark background may be detected). These points are frequently known as interest points, but the term "corner" is used by tradition [34, 36].

11.4.3 BLOBS/REGIONS OF INTEREST OR INTEREST POINTS

A blob (alternately known as a binary large object, basic large object) is a collection of binary data stored as a single entity in a database management system. Blobs are typically images, audio or other multimedia objects, though sometimes binary executable code is stored as a blob. Database support for blobs is not universal. Blobs provide a complementary description of image structures in terms of regions, as opposed to corners that are more point-like. Nevertheless, blob descriptors often contain a preferred point (a local maximum of an operator response or a center of gravity) which means that many blob detectors may also be regarded as interest point operators. Blob detectors can detect areas in an image which are too smooth to be detected by a corner detector [37, 38].

11.5 FEATURE DETECTION

Feature detection is an important early vision problem. Feature detection includes the use of gray level statistics and the detection of edges and corners. Methods based on detecting edges and corners are particularly useful in applications such as analysis of aerial images of urban images, airport facilities, and image to map matching, etc. Algorithms based on gray level statistics are applicable to a wider variety of images such as desert images and vegetation, which may or may not contain any man-made structures. Features, by definition, are locations in the image that are perceptually interesting. One can characterize an image feature detection algorithm by two attributes: (a) Generality, and (b) Robustness. Given that the nature of salient features vary from application to application, it is desirable that a feature selection algorithm be as general as possible. In case of structured objects such features could be corners and locations with significant curvature changes. When analyzing human faces, features of interest could be the eyes, nose, mouth, etc. [39].

The aim of key point detection is to determine points on a surface (a 3D face in our case) which can be identified with high repeatability in different range images of the same surface in the presence of noise and pose variations. In addition to repeatability, the features extracted from these key points should be sufficiently distinctive in order to facilitate accurate matching. Methods based on image features: Such techniques extract features such as edges, corners, contours, and centroid of specific region from the images and use the correlation between these features to determine the optimal alignment between the images. However, robustness cannot be guaranteed using only some sparse features. Human assistance is often needed in these algorithms, or the correct-match rate (CMT) will be relatively low [40, 41]. Table 2 are summarized some Common feature detectors and their classification.

TABLE 11.2 Common Feature Detectors and Their Classification

Feature detector	Edge	Corner	Blob
Canny	X		
Sobel	X		
Harris & Stephens / Plessey	X	X	
Difference of Gaussians		X	X
Determinant of Hessian		X	X
MSER			X

11.6 FEATURE MATCHING

The matching strategy is based on a similarity measure that reflects the positioning of features extracted through the first-order derivative operators, and which also quantifies the contribution of any additional attribute which can be associated to these features [39, 42].

The goal of correspondence estimation is to take a raw set of images and to find sets of matching 2D pixels across all the images. Each set of matching pixels ideally represents a single point in 3D [10].

The above features are sufficiently well characterized to use a very simple one-shot matching scheme. Each feature is represented by a set of affine invariants which are combined into a feature vector. A Mahalanobis distance metric is used to measure similarity between any two feature vectors. Hence for two images we have a set of feature vectors $v^{(i)}$ and $w^{(i)}$ and a distance measure $d(v, w)$. The matching scheme proceeds as follows:

1. Calculate distance matrix m_{ij} between pairs of features across the 2 images. $m_{ij} = d(v^{(i)}, w^{(i)})$
2. Identify potential matches (i, j) such that feature i is the closest feature in the first image to feature j in the second image and vice versa.
3. Score matches using an ambiguity measure.
4. Select unambiguous matches (or best "n" matches) [43].

An alternative to tracking features from one frame to another is to detect features independently in each image, and then search for corresponding features on the basis of a "matching score" that measure how likely it is for two features to correspond to the same point in space. The generally accepted technique consists in first establishing putative correspondences among a small number of features, and then trying to extend the matching to additional features whose score falls within a given threshold by applying robust statistical techniques [16].

Feature extractors such as SIFT take an image and return a set of pixel locations that are highly distinctive. For each of these features, the detector also computes a "signature" for the neighborhood of that feature, also known as a feature descriptor: a vector describing the local image appearance around the location of that feature. Once features have been extracted from each image, then features match by finding similar features in other images [10].

For example, in the study of 3D object recognition and reconstruction by Brown et al. [44], SIFT (Scale Invariant Feature Transform) has been used for feature matching. These locate interest points at maxima/minima of a difference of Gaussian function in scale-space. Each interest point has an associated orientation, which is the peak of a histogram of local orientations. This gives a similarity invariant frame in which a descriptor vector is sampled. Though a simple pixel resampling would be similarity invariant, the descriptor vector actually consists of spatially accumulated gradient measurements. This spatial accumulation is important for shift invariance, since the interest point locations are typically accurate in the 1–3 pixel range [6]. Illumination invariance is achieved by using gradients (which eliminates bias) and normalizing the descriptor vector (which eliminates gain). Once features have been extracted from all n images (linear time), they must be matched. Since multiple images may view the same point in the world, each feature is matched to k nearest neighbors (typically k = 4). This can be done in $O(n\log n)$ time by using a k-d tree to find approximate nearest neighbors [44].

However, in the past ten years, more powerful feature extractors have been developed that achieve invariance to a wide class of image transformations, including rotations, scales, changes in brightness or contrast, and, to some extent, changes in viewpoint (Fig. 11.8). These techniques allow us to match features between images taken with different cameras, with different zoom and exposure settings, from different angles, and V in some cases V at completely different times of day. Thus, recent feature

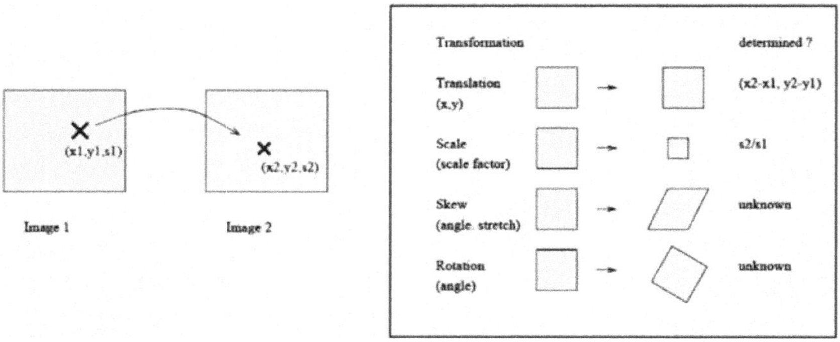

FIGURE 11.8 Hypothesized correspondence [43].

extractors open up the possibility of matching features in Internet collections, which vary along all of these dimensions (and more) [10].

The performance of matching algorithm at a particular threshold can be quantified by first counting the number of true and false matches and match failures, using the following definitions:

- *TP*: true positives, i.e., number of correct matches;
- *FN*: false negatives, matches that were not correctly detected;
- *FP*: false positives, estimated matches that is incorrect;
- *TN*: true negatives, nonmatches that were correctly rejected.
- True positive rate *FPR*,

$$TRP = \frac{TP}{TP + FN} = \frac{TP}{P} \tag{1}$$

- False positive rate *FPR*,

$$FPR = \frac{FP}{FP + TN} = \frac{FP}{N} \tag{2}$$

Any particular matching strategy (at a particular threshold or parameter setting) can be rated by the TPR and FPR numbers: ideally, the true positive rate will be close to 1, and the false positive rate close to 0. Figure 11.9 shows how we can plot the number of matches and nonmatches as a function of interfeature distance *d*.

The problem with using a fixed threshold is that it is difficult to set; the useful range of thresholds can vary a lot as we move to different parts

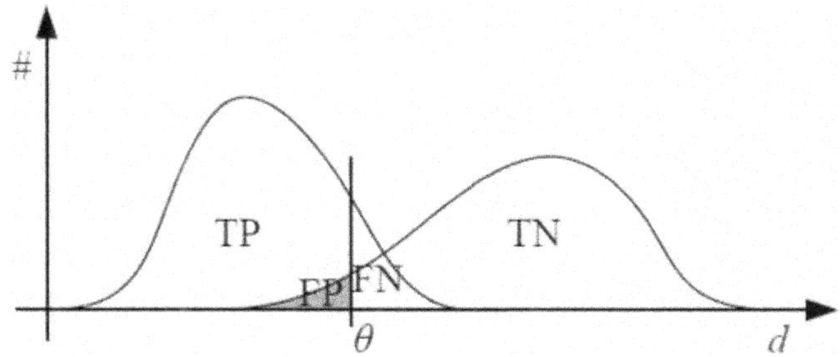

FIGURE 11.9 The distribution of positives (matches) and negatives (nonmatches) as a function of interfeature distance *d*. As the threshold is increased, the number of true positives (TP) and false positives (FP) increases.

of the feature space a better strategy in such cases is to simply match the nearest neighbor in feature space. Since some features may have no matches (e.g., they may be part of background clutter in object recognition, or they may be occluded in the other image), a threshold is still used to reduce the number of false positives.

11.7 IMAGE MATCHING

Image matching is a fundamental aspect of many problems in computer vision, including object or image recognition, solving for 3D structure from multiple images, stereo correspondence, and motion tracking [45].

During this stage, the objective is to find all matching images, that is, those that view a common subset of 3D points. Connected sets of image matches will later become 3D models.

From the feature matching step, we have identified images with a large number of matches between them. Since each image could potentially match every other one, this problem appears at first to be quadratic in the number of images. However, we have found it necessary to match each image only to a small number of neighboring images in order to get good solutions for the camera positions. We consider a constant number m images, that have the greatest number of (unconstrained) feature matches to the current image, as potential image matches (we use m = 6).

11.8 RECONSTRUCTION FROM TWO CALIBRATED VIEWS

The interesting feature of the constraint is that although it is nonlinear in the unknown camera poses, it can be solved by two linear steps in closed form. Therefore, in the absence of any noise or uncertainty, given two images taken from calibrated cameras, one can in principle recover camera pose and position of the points in space with a few steps of simple linear algebra.

11.8.1 CAMERA MODEL

A camera is usually described using the pinhole model. The pinhole camera model describes the mathematical relationship between the coordinates of a 3D point and its projection onto the image plane of an ideal pinhole camera, where the camera aperture is described as a point and no lenses are used to focus light. The model does not include, for example, geometric distortions or blurring of unfocused objects caused by lenses and finite sized apertures. It also does not take into account that most practical cameras have only discrete image coordinates. This means that the pinhole camera model can only be used as a first order approximation of the mapping from a 3D scene to a 2D image. Its validity depends on the quality of the camera and, in general, decreases from the center of the image to the edges as lens distortion effects increase.

Some of the effects that the pinhole camera model does not take into account can be compensated for, for example by applying suitable coordinate transformations on the image coordinates, and other effects are sufficiently small to be neglected if a high quality camera is used. This means that the pinhole camera model often can be used as a reasonable description of how a camera depicts a 3D scene, for example in computer vision and computer graphics.

The perspective pinhole camera model described by Eqs. (1) or (2) has retained the physical meaning of all parameters involved. In particular, the last entry of both x' and X is normalized to 1 so that the other entries may correspond to actual 2D and 3D coordinates (with respect to the metric unit chosen for respective coordinate frames). However, such normalization is not always necessary as long as we know that it is the direction of those homogeneous vectors that matters. For instance, the two vectors $[X, Y, Z, I]^T$, $[XW, YW, ZW, w]^T$ R^4 can be used to represent the same point in R^3. Similarly, we can use to represent a point $[x, y, I]^T$ on the 2D image plane as long as $x'/z' = x$ and $y'/z' = y$. However, we may run into trouble if the last entry W or z' happens to be 0.

The following equation expresses the relation between image points and world points. A camera is usually described using the pinhole model. A 3D point $M = [X, Y, Z]^T$ in a Euclidean world coordinate system and the retinal image coordinates $m = [u, v]^T$ are related by the following Eq. (3):

$$s\tilde{m} = P\tilde{M} \tag{3}$$

where s is a scale factor, $\tilde{m} = [u, v, I]^T$ and $\tilde{M} = [X, Y, Z, I]^T$ are the homogeneous coordinates of vector m and M, and P is a 3×4 matrix representing the collineation: $P^3 \rightarrow P^2$. P is called the perspective projection matrix.

Figure 10 illustrates this process. The figure shows the case where the projection center is placed at the origin of the world coordinate frame and the retinal plane is at $Z = f = 1$. Then

$$u = \frac{fx}{z}, v = \frac{FY}{Z}$$

$$P = [I_{3\times3} \quad 0_3] \tag{4}$$

The optical axis passes through the center of projection (camera) C and is orthogonal to the retinal plane. The point c is called the principal point, which is the intersection of the optical axis with the retinal plane. The focal length f of the camera is also shown, which is the distance between the center of projection and the retinal plane (Fig. 11.10).

If only the perspective projection matrix P is available, it is possible to recover the coordinates of the optical center or camera.

The world coordinate system is usually defined as follows: the positive Y-direction is pointing upwards, the positive X-direction is pointing to the right and the positive Z-direction is pointing into the page [46, 47].

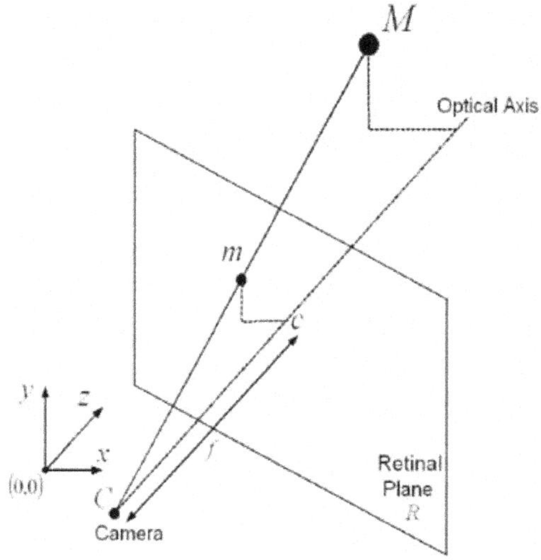

FIGURE 11.10 Perspective projection.

The camera calibration matrix is specified by a five parameter upper triangular matrix

$$K = \begin{pmatrix} f & k & u_0 \\ 0 & rf & v_0 \\ 0 & 0 & 1 \end{pmatrix} \tag{5}$$

An image point x is related to a point in the camera's coordinate system x_c as $x = Kx_c$. Parameter f is the focal length of the camera. The aspect ratio of the camera r depends on the relative scaling of the vertical and horizontal camera axes. The line from the camera center perpendicular to the image intersects the image at the principal point with coordinates $(u_0, v_0)^T$

The skew, k, is a factor dependent on the physical angle θ between the u and v axes in the sensor array, given by $k = f\cot(\theta)$. Note that radial lens distortion is ignored, but can be corrected in cases where it is significant. In many cases a simplified camera model may be used. A CCD camera, for example, has zero skew ($k = 0$) and unit aspect ratio ($r = 1$). The resultant simplified or natural camera is

$$K = \begin{pmatrix} f & 0 & u_0 \\ 0 & f & v_0 \\ 0 & 0 & 1 \end{pmatrix} \tag{6}$$

The more general camera model (5) does apply in certain situations however. In many situations the principal point is located near the center of the image, and often can be approximated by the image center. However, it cannot be assumed that this is always the case because photographs (and images) are sometimes cropped before display [18].

11.8.2 EPIPOLAR GEOMETRY

Consider two images of the same image taken from two distinct vantage points. If we assume that the camera is calibrated, the homogeneous image coordinates x and the spatial coordinates X of a point p, with respect to the camera frame, are related by

$$\lambda X = \Pi_0 X \tag{7}$$

where $\Pi_0 = [I, 0]$. That is, the image x differs from the actual 3D coordinates of the point by' an unknown (depth) scale $\lambda \in R^+$. For simplicity, we will assume that the image is static (that is, there are no moving objects) and that the position of corresponding feature points across images is available. If we call x_1, x_2 the corresponding points in two views, they will then be related by a precise geometric relationship.

Assume we have taken two images upon a same object from two different perspectives. Its epipolar geometry is showed in Fig. 11.11.

In the Fig. 11.11, C_1 and C_2 are the centers of two cameras, with m_1 and m_2 as their projection matrices respectively. M is a point in 3D space. m_1 and m_2 are the projections of M in the two image planes respectively. The line connecting C_1 and C_2 is called the baseline, and it intersects the two image planes at points e and e', respectively, called epipoles. The plane π containing the baseline is called the epipolar plane, which intersects the two image planes at two epipolar lines l_1 and l_2, respectively [8, 16, 48–53].

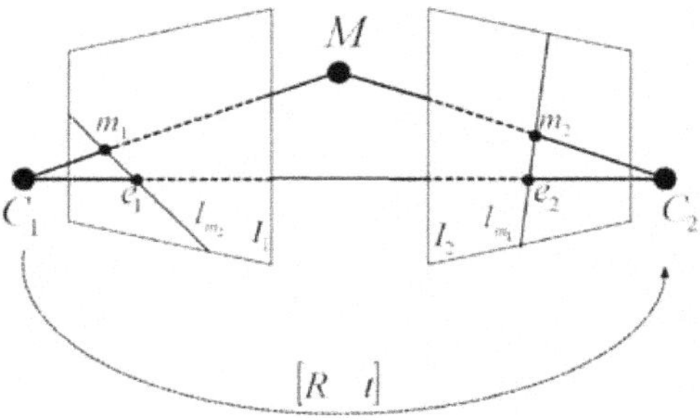

FIGURE 11.11 Epipolar geometry between two views [16].

11.8.2.1 The Epipolar Constraint and the Essential Matrix

An orthonormal reference frame is associated with each camera, with its origin 0 at the optical center and the *z*-axis aligned with the optical axis. The relationship between the 3D coordinates of a point in the inertial "world" coordinate frame and the camera frame can be expressed by a rigid-body transformation. Without loss of generality, we can assume the world frame to be one of the cameras, while the other is positioned and oriented according to a Euclidean transformation $g = (R, T) \in SE$. If we call the 3D coordinates of a point p relative to the two camera frames $X_1 \in R^3$ and $X_2 \in R^3$, they are related by a rigid-body transformation in the following way:

$$X_2 = RX_1 + T \tag{8}$$

Now let $x_1, x_2 \in R^3$ be the homogeneous coordinates of the projection of the same point p in the two image planes. Since $X_i = \lambda_i x_i$, $i = 1, 2$, this equation can be written in terms of the image coordinates X_1 and the depths λ_i as

$$\lambda_2 x_2 = R\lambda_1 x_1 + T \tag{9}$$

In order to eliminate the depths λ_i in the preceding equation, premultiply both sides by $\hat{T}$ to obtain

$$\lambda_2 \hat{T} X_2 = \hat{T} R \lambda_1 X_1 \tag{10}$$

Since the vector $\hat{T} x_2 = T \times x_2$ is perpendicular to the vector x_2, the inner product $\langle x_2, \hat{T} x_2 \rangle = x_2^T \hat{T} x_2$ is zero. Premultiplying the previous equation by

x_2^T yields that the quantity $x_2^T \hat{T} R \lambda_1 x_1$ is zero. Since $\lambda_1 \rangle 0$, we have proven the following result:
Consider two images x_1, x_2 of the same point p from two camera positions with relative pose (R, T), where $R \in SO(3)$ is the relative orientation and $T \in R^3$ is the relative position. Then x_1, x_2 satisfy

$$\left\langle x_2, \hat{T} \times R x_1 \right\rangle = 0, \text{ or } x_2{}^T \hat{T} R x_1 = 0 \tag{11}$$

The matrix $E = \hat{T}R \quad R \in R^3$ in the epipolar constraint Eq. (11) is called the essential matrix.

In computer vision, the essential matrix is a 3 × 3 matrix, E, with some additional properties described below, which relates corresponding points in 2D images assuming that the cameras satisfy the pinhole camera model.

Longuet-Higgins' [27] paper includes an algorithm for estimating E from a set of corresponding normalized image coordinates as well as an algorithm for determining the relative position and orientation of the two cameras given that E is known. Finally, it shows how the 3D coordinates of the image points can be determined with the aid of the essential matrix.

Given an essential matrix $E = TR$ that defines an epipolar relation between two images x_1, x_2, we have:

1. The two epipoles $e_1, e_2 \in R^3$, with respect to the first and second camera frames, respectively, are the left and right null spaces of E, respectively:

$$e_2{}^T E = 0, Ee_1 = 0 \tag{12}$$

That is, $e_2 \sim T$ and $e_1 \sim R^T T$. We recall that ~ indicates equality up to a scalar factor.

2. The (coinages of) epipolar lines l_1, $l_2 \in R^3$ associated with the two image points x_1, x_2 can be expressed as

$$l_2 \sim Ex_1, l_1 \sim E^T x_2 \in R^3 \tag{13}$$

where l_1, l_2 are in fact the normal vectors to the epipolar plane expressed with respect to the two camera frames, respectively.

3. In each image, both the image point and the epipole lie on the epipolar line

$$l_i^T e_i = 0, \quad l_i^T x_i = 0, \quad i = 1,2 \tag{14}$$

11.8.2.2 Estimation of the Essential Matrix

Given a set of corresponding image points it is possible to estimate an essential matrix which satisfies the defining epipolar constraint for all the points in the set. However, if the image points are subject to noise, which is the common case in any practical situation, it is not possible to find an essential matrix which satisfies all constraints exactly.

Depending on how the error related to each constraint is measured, it is possible to determine or estimate an essential matrix which optimally satisfies the constraints for a given set of corresponding image points. The most straightforward approach is to set up a total least squares problem, commonly known as the eight-point algorithm [8, 16, 48–53].

11.8.2.3 Determining *R* and *t* from *E*

Given that the essential matrix has been determined for a stereo camera pair, for example, using the estimation method above this information can be used for determining also the rotation and translation (up to a scaling) between the two camera's coordinate systems. In these derivations E is seen as a projective element rather than having a well-determined scaling.

The following method for determining R and t is based on performing a *SVD* of E. It is also possible to determine R and t without an *SVD*, for example, following Longuet-Higgins' paper.

An *SVD* of E gives

$$E = U\Sigma V^T \quad (15)$$

where U and V are orthogonal 3×3 matrices and Σ is a 3×3 diagonal matrix with

$$\Sigma = \begin{pmatrix} s & 0 & 0 \\ 0 & s & 0 \\ 0 & 0 & 0 \end{pmatrix} \quad (16)$$

The diagonal entries of Σ are the singular values of E which, according to the internal constraints of the essential matrix, must consist of two identical and one zero value.

To summarize, given E there are two opposite directions which are possible for t and two different rotations which are compatible with this essential matrix. In total this gives four classes of solutions for the rotation and translation between the two camera coordinate systems. On top of that, there is also an unknown scaling $s>0$ for the chosen translation direction.

It should be noted that the above determination of R and t assumes that E satisfy the internal constraints of the essential matrix. If this is not the case which, for example, typically is the case if E has been estimated from real (and noisy) image data, it has to be assumed that it approximately satisfy the internal constraints. The vector $\hat{t}$ is then chosen as right singular vector of E corresponding to the smallest singular value.

11.8.2.4 3D Points From Corresponding Image Points

The problem to be solved there is how to compute (x_1, x_2, x_3) given corresponding normalized image coordinates (y_1, y_2) and $\left(y_1', y_2'\right)$. If the essential matrix is known and the corresponding rotation and translation transformations have been determined, this algorithm (described in Longuet-Higgins' paper) provides a solution.

Let r_k denote row k of the rotation matrix R:

$$R = \begin{pmatrix} r_1 \\ r_2 \\ r_3 \end{pmatrix} \tag{17}$$

Combining the above relations between 3D coordinates in the two coordinate systems and the mapping between 3D and 2D points described earlier gives

$$y_1' = \frac{x_1'}{x_3'} = \frac{r_1.(\tilde{x}-t)}{r_3.(\tilde{x}-t)} = \frac{r_1.\left(y - {}^{t}\!/_{x_3}\right)}{r_3.\left(\tilde{x} - {}^{t}\!/_{x_3}\right)} \tag{18}$$

or

$$x_3 = \frac{\left(r_1 - y_1' r_3\right).t}{\left(r_1 - y_1' r_3\right).y} \tag{19}$$

Once x_3 is determined, the other two coordinates can be computed as

$$\begin{pmatrix} x_1 \\ x_2 \end{pmatrix} = x_3 \begin{pmatrix} y_1 \\ y_2 \end{pmatrix} \tag{20}$$

The above derivation is not unique. It is also possible to start with an expression for y_2' and derive an expression for x_3 according to

$$x_3 = \frac{\left(r_1 - y_2' r_3\right).t}{\left(r_1 - y_2' r_3\right).y} \tag{21}$$

In the ideal case, when the camera maps the 3D points according to a perfect pinhole camera and the resulting 2D points can be detected without any noise, the two expressions for x_3 are equal. In practice, however, they are not and it may be advantageous to combine the two estimates of x_3, for example, in terms of some sort of average.

There are also other types of extensions of the above computations which are possible. They started with an expression of the primed image coordinates and derived 3D coordinates in the unprimed system. It is also possible to start with unprimed image coordinates and obtain primed 3D coordinates, which finally can be transformed into unprimed 3D coordinates. Again, in the ideal case the result should be equal to the above expressions, but in practice they may deviate.

A final remark relates to the fact that if the essential matrix is determined from corresponding image coordinate, which often is the case when 3D points are determined in this way, the translation vector t is known only up to an unknown positive scaling. As a consequence, the reconstructed 3D points, too, are undetermined with respect to a positive scaling [8, 16, 48–53].

11.8.3 PLANAR HOMOGRAPHY

Historically, homographies (and projective spaces) have been introduced to study perspective and projections in Euclidean geometry, and the term "homography," which, etymologically, roughly means "similar drawing" date from this time. At the end of nineteenth century, formal definitions of projective spaces were introduced, which differed from extending Euclidean or affine spaces by adding points at infinity. The term "projective transformation" originated in these abstract constructions. These constructions divide into two classes that have been shown to be equivalent. A projective space may be constructed as the set of the lines of a vector space over a given field (the above definition is based on this version); this construction facilitates the definition of projective coordinates and allows using the tools of linear algebra for the study of homographies. The alternative approach consists in defining the projective space through a set of axioms, which do not involve explicitly any field (incidence geometry, see also synthetic geometry); in this context, collineations are easier to define than homographies, and homographies are defined as specific collineations, thus called "projective collineations" [8, 16, 54, 55].

A 2D point (x, y) in an image can be represented as a 3D vector $x = (x_1, x_2, x_3)$ where $x = \frac{x_1}{x_3}$ and $y = \frac{x_2}{x_3}$. This is called the homogeneous representation of a point and it lies on the projective plane P^2. A homography is an invertible mapping of points and lines on the projective plane P^2.

Other terms for this transformation include collineation, projectivity, and planar projective transformation. Hartley and Zisserman [11] provide the specific definition that a homography is an invertible mapping from P^2 to itself such that three points lie on the same line if and only if their mapped points are also collinear. They also give an algebraic definition by proving the following theorem: A mapping from $P^2 \rightarrow P^2$ is a projectivity if and only if there exists a nonsingular 3×3 matrix H such that for any point in P^2 represented by vector x it is true that its mapped point equals Hx. This tells us that in order to calculate the homography that maps each x_i to its corresponding x_1' it is sufficient to calculate the 3×3 homography matrix, H. It should be noted that H can be changed by multiplying by an arbitrary nonzero constant without altering the projective transformation. Thus H is considered a homogeneous matrix and only has 8 degrees of freedom even though it contains 9 elements. This means there are 8 unknowns that need to be solved for. Typically, homographies are estimated between images by finding feature correspondences in those images. The most commonly used algorithms make use of point feature correspondences, though other features can be used as well, such as lines or conics.

$$\underset{x'}{\begin{bmatrix} wx' \\ wy' \\ w \end{bmatrix}} = \underset{H}{\begin{bmatrix} * & * & * \\ * & * & * \\ * & * & * \end{bmatrix}} \underset{x}{\begin{bmatrix} x \\ y \\ 1 \end{bmatrix}} \tag{22}$$

$$\lambda_2 x_2' = H\lambda_1 x_1' \quad \lambda_2 x_2' = HX \quad X = \begin{bmatrix} X & Y & 1 \end{bmatrix}^T \tag{23}$$

$$\widehat{x_2'} H_1 x_1' = 0 \tag{24}$$

The homography transformation has 8 degrees of freedom and there are other simpler transformations that still use the 3 × 3 matrix but contain specific constraints to reduce the number of degrees of freedom (Fig. 11.12) [54].

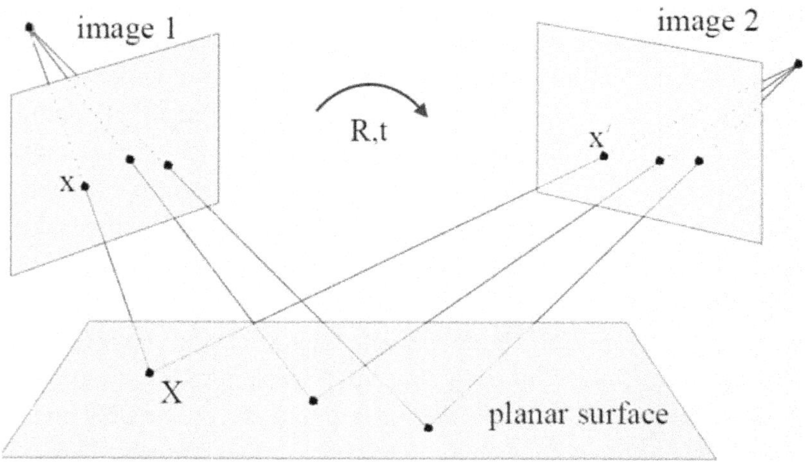

FIGURE 11.12 The geometry of a homography mapping [55].

11.8.3.1 Relationships Between the Homography and the Essential Matrix

In practice, especially when the image is piecewise planar, we often need to compute the essential matrix E with a given homography H computed from some four points known to be planar; or in the opposite situation, the essential matrix E may have been already estimated using points in general position, and we then want to compute the homography for a particular (usually smaller) set of coplanar points. We hence need to understand the relationship between the essential matrix E and the homography H [54–56].

11.8.4 FUNDAMENTAL MATRIX

Structure from uncalibrated images only leads to a projective reconstruction. This image data is usually in the form of corners (high curvature points), as they can be easily represented and manipulated in projective geometry [8].

The essential matrix can be seen as a precursor to the fundamental matrix. Both matrices can be used for establishing constraints between matching image points, but the essential matrix can only be used in relation to calibrated cameras since the inner camera parameters must be known in order to achieve the normalization. If, however, the cameras are calibrated the essential matrix can be useful for determining both the relative position and orientation between the cameras and the 3D position of corresponding image points (Fig. 11.13).

11.8.4.1 Fundamental Matrix Estimation

Let be the fundamental matrix. Some important equations are listed here: (40)

$$x_2'^T F x_1' = 0 \tag{25}$$

$$\begin{matrix} l_1 \sim F^T x_2' & l_i^T x_i' = 0 & l_2 \sim F^T x_1' \\ Fe_1 = 0 & l_i^T e_i = 0 & e_2^T F = 0 \end{matrix} \tag{26}$$

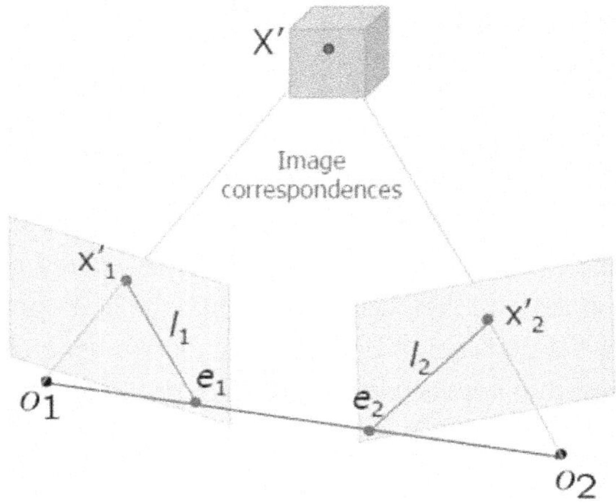

FIGURE 11.13 Project of 2D Point to 3D Point With Uncalibrated Camera.

In the case of calibrated cameras, becomes the essential matrix, which relates the normalized points in two images.

After finding a set of matching points from two images using key point descriptors, we find the fundamental matrix using normalized 8-point algorithm [8, 16].

11.9 ROBUST ESTIMATION

Robust methods for determining the fundamental matrix are especially important when dealing with real image data, which is able to detect outliers in the correspondences. As the fundamental matrix has only seven degrees of freedom, it is possible to estimate F directly using only 7 point matches. In general more than 7 point matches are available and a method for solving the fundamental matrix using 8 point matches. The points in both images are usually subject to noise and therefore a minimization technique is implemented. A robust method is very useful as the technique will ignore these false matches in the estimation of the fundamental matrix [54]. The two most commonly used approaches of robust estimation are RANSAC and LMS [16].

11.9.1 RANSAC

RANSAC (Random Sample Consensus) is the most commonly used robust estimation method for homographies and this method makes decisions based on the number of data points within a distance threshold [14]. The idea of the algorithm is pretty simple; for a number of iterations, a random sample of four correspondences is selected and a homography H is computed from those four correspondences. Each other correspondence is then classified as an inlier or outlier depending on its concurrence with H. After all of the iterations are done, the iteration that contained the largest number of inliers is selected. H can then be recomputed from all of the correspondences that were considered as inliers in that iteration.

One important issue when applying the RANSAC algorithm described above is to decide how to classify correspondences as inliers or outliers. Statistically speaking, the goal is to assign a distance threshold, t, (e.g., between x' and Hx), such that with a probability α the point is an inlier.

Another issue is to decide how many iterations to run the algorithm. It will likely be infeasible to try every combination of four correspondences, and thus the goal becomes to determine the number of iterations, N, that ensures with a probability p that at least one of the random samples will be free from outliers. Hartley and Zisserman [54] show that $N = \log(1-p)\big/\log\left(1-(1-\epsilon)^s\right)$, where ϵ is the probability that a sample correspondence is an outlier and s is the number of correspondences used in each iteration, four in this case. If ϵ is unknown, the data can be probed to adaptively determine ϵ and N.

11.9.2 LEAST MEDIAN OF SQUARES REGRESSION

This is one way to deal with the fact that sum of squared difference algorithms such as the Algebraic distance version of DLT is not very robust with respect to outliers. There is a lot of research on the topic of ways to improve the robustness of regression methods. One example would be to replace the squared distance with the absolute value of the distance. This improves the robustness since outliers aren't penalized as severely as when they are squared. A popular approach with respect to homography estimation is Least Median of Squares or (LMS) estimation. As described in Ref. [57], this method replaces the sum with the median of the squared residuals. LMS works very well if there are less than 50% outliers and has the advantage over RANSAC that it requires no setting of thresholds or a priori knowledge of how much error to expect (unlike the setting of the t and $\in$ parameters in RANSAC).

The major disadvantage of LMS is that it would be unable to cope with more than half the data being outliers. In this case, the median distance would be to an outlier correspondence [54].

11.9.3 FUTURE WORK

It is interesting to note that RANSAC itself is not a great solution when there is a high percentage of outliers, as its computational cost can blow up with the need for too many iterations. While RANSAC and LMS are the most commonly used methods for robust homography estimation, there may be an opening for research into whether there are other methods aside from those two that would do well in the presence of a very high number of outlier data. One potentially successful approach could be to try to fit a lower order transformation from a set of correspondences, such as a similarity transform. While this would be unlikely to perfectly segment outliers from inliers, it could be useful for removing the obvious outliers. By disproportionally removing more outliers than inliers, a situation where RANSAC would have previously failed can be brought into a realm where it could work [54].

Also Brown et al. [44] parameterize each camera using seven parameters. These are a rotation vector $\Theta_i = \Theta_{i1} \quad \Theta_{i2} \quad \Theta_{i3}$ translation $t_i = [t_{i1} \quad t_{i2} \quad t_{i3}]$ and focal length f_i. The calibration matrix is then

$$K_i = \begin{bmatrix} f_i & 0 & 0 \\ 0 & f_i & 0 \\ 0 & 0 & 1 \end{bmatrix} \tag{27}$$

and the rotation matrix (using exponential representation)

$$R_i = e^{[\theta_i]_x}, \quad [\theta_i]_x = \begin{bmatrix} 0 & -\theta_{i3} & \theta_{i2} \\ \theta_{i3} & 0 & -\theta_{i1} \\ -\theta_{i2} & \theta_{i1} & 0 \end{bmatrix} \tag{28}$$

Each pairwise image match adds four constraints on the camera parameters while adding three unknown structure parameters $X = \begin{bmatrix} X_1 & X_2 & X_3 \end{bmatrix}$

$$\tilde{u}_i = K_i X_{ci} \tag{29}$$

$$\tilde{u}_j = K_i X_{cj} \tag{30}$$

$$X_{ci} = R_i X + t_i \tag{31}$$

$$X_{cj} = R_j X + t_j \tag{32}$$

where $\tilde{u}_i, \tilde{u}_j$ are the homogeneous image positions in camera *i* and *j*, respectively. The single remaining constraint (4 equations minus 3 unknowns = 1 constraint) expresses the fact that the two camera rays $\tilde{p}_i, \tilde{p}_j$ and the translation vector between camera centers t_{ij} are coplanar, and hence their scalar triple product is equal to zero

$$\tilde{p}_i^T \left[t_{ij} \right] \times \tilde{p}_j = 0 \tag{33}$$

Writing $\tilde{p}_i, \tilde{p}_j$ and t_{ij} in terms of camera parameters

$$\tilde{p}_i = R_i^T K_i^{-1} \tilde{u}_i \tag{34}$$

$$\tilde{p}_j = R_j^T K_j^{-1} \tilde{u}_j \tag{35}$$

$$t_{ij} = R_j^T t_j - R_i^T t_i \tag{36}$$

And substituting in Eq. (1) gives,

$$\tilde{u}_i^T F_{ij} \tilde{u}_j = 0 \tag{37}$$

where

$$F_{ij} = K_i^{-T} R_i \left[R_j^T t_j - R_i^T t_i \right] \times R_j^T K_j^{-1} \tag{38}$$

This is the well-known epipolar constraint. Image matching entails robust estimation of the fundamental matrix F_{ij}. Since Eq. (37) is nonlinear in the camera parameters, it is commonplace to relax the nonlinear constraints and estimate a general 3×3 matrix F_{ij}. This enables a closed form solution via SVD.

In this work are used RANSAC to robustly estimate F and hence find a set of inliers that have consistent epipolar geometry. An image match is declared if the number of RANSAC inliers $n_{inliers} \rangle n_{match}$, where the minimum number of matches n_{match} is a constant (typically around 20). Then 3D objects/images are identified as connected components of image matches [44].

11.10 BUNDLE ADJUSTMENT

Bundle adjustment is the problem of refining a visual reconstruction to produce jointly optimal 3D structure and viewing parameter (camera pose and/or calibration) estimates. Optimal means that the parameter estimates are found by minimizing some cost function that quantifies the model fitting error, and jointly that the solution is simultaneously optimal with respect to both structure and camera variations. The name refers to the 'bundles' of light rays leaving each 3D feature and converging on each camera center, which are 'adjusted' optimally with respect to both feature and camera positions. Equivalently unlike independent model methods, which merge partial reconstructions without updating their internal structure all of the structure and camera parameters are adjusted together 'in one bundle.' Bundle adjustment is really just a large sparse geometric parameter estimation problem, the parameters being the combined 3D feature coordinates, camera poses and calibrations. Almost everything that we will say can be applied to many similar estimation problems in vision, photogrammetry, industrial metrology, surveying and geodesy (Fig. 11.14). [44, 58–60]

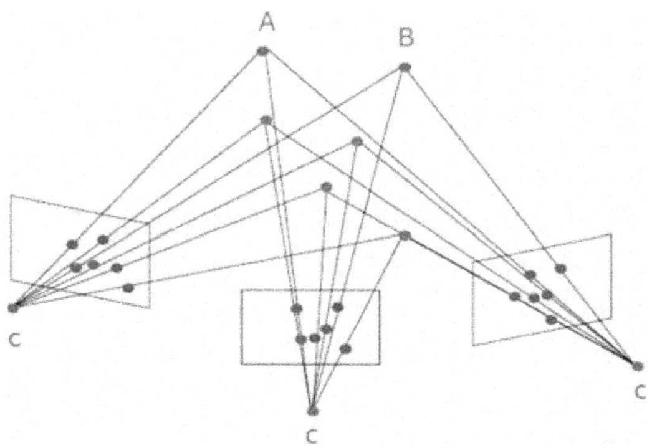

FIGURE 11.14 Jointly optimal 3D structure bundle adjustment.

11.11 VISUALIZATION

Visualization of the model during image acquisition allows the operator to interactively verify that an adequate set of input images has been collected for the modeling task, while automatic image selection keeps storage requirements to a minimum [16, 61].

Single 3D points cannot provide a global illustration about the structure of the object. Thereby, the creation of the 3D object model is a requirement in order to depict the formation and the real conditions of the object. Furthermore, the 3D reconstruction is a requirement for further processes such as the application of visualization and graphics techniques. In any case the 3D object model can be developed under standard or special CAD software [62].

There are many different visualization methods from simple hand drawings, to CAD designs, GIS systems, 3D representations, animations and walkthroughs or even stereoscopic representations in virtual reality applications.

The 3D representation has to provide adequate geometry characteristics and detailed enough textures for the archeologists to be able to work on them. While the process of creating the virtual models can be complex, there are various techniques that try to automate the whole process as much as possible. During the visualization process there are two options, to visualize the actual objects or to visualize the reconstructed objects [62].

When all pairs of images have been matched, we can construct an image connectivity graph to represent the connections between the images in the collection. An image connectivity graph contains a node for each image, and an edge between any pair of images that have matching features. To create this visualization, the graph was embedded in the plane using the neato tool in the Graphviz graph visualization toolkit. Neato works by modeling the graph as a mass-spring system and solving for an embedding whose energy is a local minimum. The image connectivity graph for this collection has several notable properties. There is a large, dense cluster in the center of the graph that consists of photos that are mostly wide-angle, frontal, well-lit shots of the fountain. Other images, including the Bleaf (nodes) corresponding to tightly cropped details, and nighttime images, are more loosely connected to this core set [10, 64, 66].

11.12 EPIPOLAR RECTIFICATION

Image rectification is an important component of stereo computer vision algorithms. Rectification is a process used to facilitate the analysis of a stereo pair of images by making it simple to enforce the two view geometric constraint. We assume that a pair of 2D images of a 3D object or environment is taken from two distinct viewpoints and their epipolar geometry has been determined. Corresponding points between the two images must satisfy the so-called epipolar constraint. For a given point in one image, we have to search for its correspondence in the other image along an epipolar line. In general, epipolar lines are not aligned with coordinate axis and are not parallel. Such searches are time consuming since we must compare pixels on skew lines in image space. These types of algorithms can be simplified and made more efficient if epipolar lines are axis aligned and parallel. This can be realized by applying 2D projective transforms, or homographies, to each image. This process is known as image rectification. Generally, the rectification process expects to be provided with two rectangular images as well as a fundamental matrix and a set of point matches such as those used to calculate the fundamental matrix and also rectification can be used to recover 3D structure from an image pair without appealing to 3D geometry notions like cameras (Fig. 11.15) [49, 51].

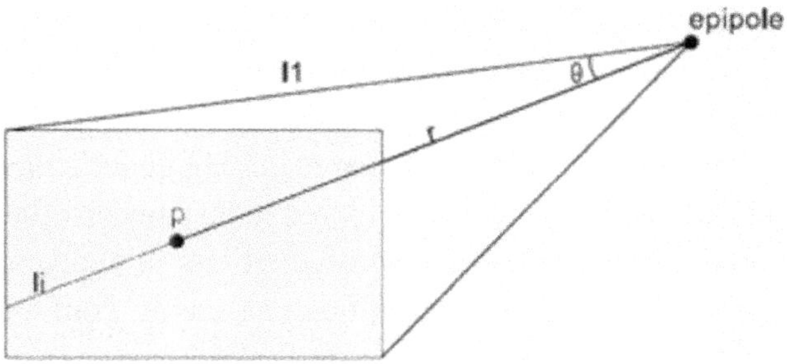

FIGURE 11.15 Epipolar rectification: a point p is encoded by a pair of (r; θ) [14].

Some previous techniques for finding image rectification homographies involve 3D constructions. These methods find the 3D line of intersection between image planes and project the two images onto the plane containing this line that is parallel to the line joining the optical centers. While this approach is easily stated as a 3D geometric construction, its realization in practice is somewhat more involved and no consideration is given to other more optimal choices. A strictly 2D approach that does attempt to optimize the distorting effects of image rectification can be found in. Their distortion minimization criterion is based on a simple geometric heuristic which may not lead to optimal solutions. The important advantage of rectification is that computing correspondences is made simpler, because search is done along the horizontal lines of the rectified images [18, 49–53].

11.13 DENSE MATCHING

A dense matching algorithm is integrated as preprocessing. It computes the correspondence information for only those sufficiently textured areas. Matching is propagated from the most reliably matched pixels to their neighbors. Propagation is stopped when texture cue is not sufficient [67–69].

Traditional dense matching problems such as stereo or optical flow deal with the "instance matching" scenario, in which the two input images contain different viewpoints of the same image or object. More recently, researchers have pushed the boundaries of dense matching to estimate correspondences between images with different images or objects. This advance beyond instance matching leads to many interesting new applications, such as semantic image segmentation, image completion, image classification, and video depth estimation [45].

There are two major challenges when matching generic images: image variation and computational cost. Compared to instances, different images and objects undergo much more severe variations in appearance, shape, and background clutter. These variations can easily confuse low level matching functions. At the same time, the search space is much larger, since generic image matching permits no clean geometric constraints. Without any prior knowledge on the images' spatial layout, in principle we must search every pixel to find the correct match [70–72].

Recent innovations in matching algorithms considerably improved the quality of elevation data, generated automatically from aerial images. Traditional matching, originally introduced more than two decades ago, usually applies feature based algorithms. These algorithms first extract feature points and then search the corresponding features in the overlapping images. The restriction to matches of selected points usually provides correspondences at high certainty. However, feature based matching was also introduced to avoid problems due to limited computational resources. In contrast, recent algorithms aim on dense, pixel-wise matches. By these means 3D point clouds and Digital Surface Models (DSM) are generated at a resolution, which corresponds to the ground sampling distance GSD of the original images. To compute pixel matches even for regions with very limited texture, additional constraints are required [65, 66].

11.14 TRIANGULATION

Triangulation is the process of determining the location of a point by measuring angles to it from known points at either end of a fixed baseline, rather than measuring distances to the point directly (trilateration). The point can then be fixed as the third point of a triangle with one known side and two known angles. The 3D position of the points can be obtained easily by triangulation. The process of triangulation is needed to find the intersection of two known rays in space. Due to measurement noise in images and some inaccuracies in the calibration matrices, these two rays will not generally meet in a unique point [8, 73–78].

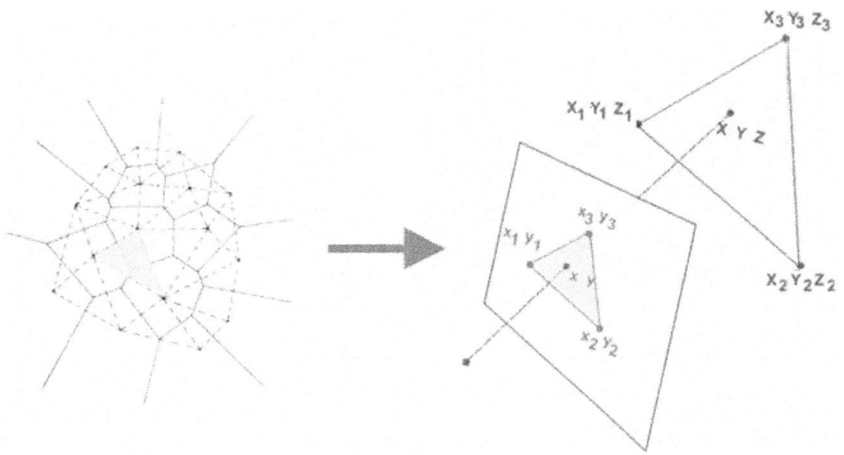

FIGURE 11.16 Corresponding triangles in 2D image point & 3D space point (for affine coordinates).

The triangulation problem is a small cog in the machinery of computer vision, but in many applications of scene reconstruction it is a critical one, on which ultimate accuracy depends [76].

By investigation of Voronoi diagrams goes the investigation of related constructs. Among them, the Delaunay triangulation is most prominent. It contains a (straight-line) edge connecting two sites in the plane if and only if their Voronoi regions share a common edge. The structure was introduced by Voronoi [1908] for sites that form a lattice and was extended by Delaunay [1934] to irregularly placed sites by means of the empty-circle method: Consider all triangles formed by the sites such that the circumcircle of each triangle is empty of other sites. The set of edges of these triangles gives the Delaunay triangulation of the sites (Fig. 11.16) [60, 78].

11.15 TEXTURE MAPPING

Texture mapping has traditionally been used to add realism to computer graphics images. In its basic form, texture mapping lays an image (the texture) onto an object in a image. In recent years, this technique has moved from the domain of software rendering systems to that of high performance graphics hardware, also Texture mapping ensures that "all the right things" happen as a textured polygon is transformed and rendered.

In order to render the image from novel viewpoints, we need a model of its, surfaces, so that we can texture-map images onto it. Clearly, the handful of point features we have reconstructed so far is not sufficient [16].

More general forms of texture mapping generalize the image to other information; an "image" of altitudes, for instance, can be used to control shading across a surface to achieve such effects as bump-mapping [79,80].

The texture mapping can be done as follows:

1. Back project the 3D mesh, a set of wire-frame patches, into each frame.
2. Extract texture patches (photometric information) of each wire-frame patches.
3. Use photometric and geometric information, that is, the angles between the line of sight of views and the normal of the wire-frame patch), to create the mapping texture patch.
4. Map the texture patches to the corresponding wire-frame patches.

One simple use of texture mapping is to draw antialiased points of any width. In this case the texture image is of a filled circle with a smooth (anti-aliased) boundary. When a point is specified, its coordinates indicate the center of a square whose width is determined by the point size. The texture coordinates at the square's corners are those corresponding to the corners of the texture image. This method has the advantage that any point shape may be accommodated simply by varying the texture image [14, 80].

11.16 APPLICATIONS OF 3D RECONSTRUCTION IN PORE STRUCTURE CHARACTERIZATION

Nanofibrous media have low basis weight, high permeability and small pore size that make them appropriate for a wide range of filtration applications. In addition, nanofiber membrane offers unique properties like high specific surface area (ranging from 1 to 35 m^2/g depending on the diameter of fibers), good interconnectivity of pores and potential to incorporate active chemistry or functionality on nanoscale. Therefore, nanofibrous membranes are extensively being studied for air and liquid filtration. The performance of filtration media in all of these industries is determined by the pore structure characteristics of the media. There are several well documented approaches in literature to evaluate these pore structure properties including scanning electron microscopy (SEM) analysis, gas pycnometry and adsorption, flow and mercury porosimetry, theoretical modeling, and most recently micro computed tomography (Micro-CT). Because of advantages and disadvantages of Correspondence current approaches, virtues and pitfalls of each technique should be scrutinized. Sometimes a combination of techniques is required. However, a single nondestructive and capable of providing a comprehensive set of data is the most attractive option. These techniques are also used for pore structure characterization of nanofibrous membranes, but the low stiffness and high pressure sensitivity of nanofiber mats limit application of these techniques, because these methods cannot be used for porosity measurement of various surface layers and can only measure the total porosity of nanofiber mat (Fig. 11.17). Therefore, an accurate estimation of porosity in these grades of materials (Nanofiber mat) is a difficult task [81–86].

Porous media typically contain an interconnected three-dimensional network of channels of nonuniform size and shape. Usually porosity is determined for materials with a three-dimensional structure, for example, relatively thick nonwoven fabrics. Nevertheless, for two-dimensional

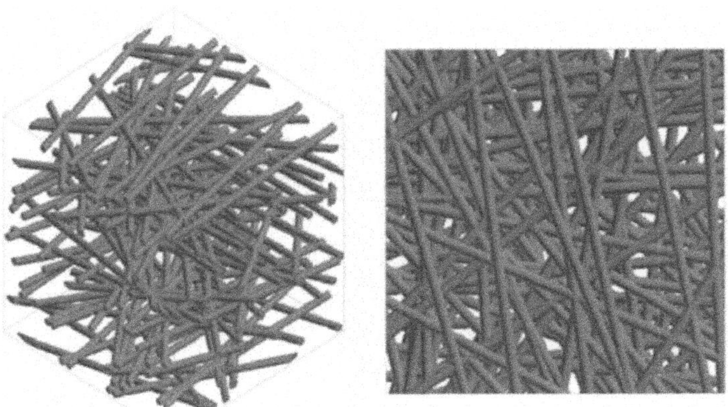

(a) (b)

FIGURE 11.17 SEM image of electrospun nanofibrous membrane: (a) 3D (b) 2D.

textiles such as woven fabrics and relatively thin nonwovens it is often assumed that porosity and POA are equal [87, 88].

SEM analysis of electrospun fibrous membrane by incorporating different image analysis methods is one of the renowned methods for researchers to measure porosity parameters of woven fabric, nonwoven and membranes, and nanofiber membrane. Although image analysis of SEM micrographs for geometrical characterization is useful for measuring total porosity, pore shape, pore size and pore size distribution of relatively thin nonwovens, it cannot be applied to multilayer electrospun fibrous analysis. Another problem encounter to this method is that it is not possible to measure 3D pore characteristic of the membrane and it is limited on relatively small fields of view [84].

Three-dimensional analysis is possible with image analysis techniques. Image-based techniques can be prohibitively tedious because enough pores must be analyzed to give an adequate statistical representation [89].

3D pore structures of nanofibrous membrane have been evaluated by nondestructive three-dimensional (3D) laser scanning confocal microscope (LSCM) and micro computed tomography (Micro-CT), 3D electron back-scatter diffraction (EBSD), nuclear magnetic resonance imaging [84–99].

Several instrumental characterization techniques have been suggested to obtain 3D volume images of pore space, such as X-ray computed micro tomography and magnetic resonance computed micro tomography. However, these techniques may be limited by their resolution. So, the 3D stochastic reconstruction of porous media from statistical information (produced by analysis of 2D photomicrographs) has been suggested. Although pore network models can be two- or three-dimensional, 2D image analysis, due to their restricted information about the whole microstructure, was unable to predict morphological characteristics of porous membrane. Therefore, 3D reconstruction of porous structure will lead to significant improvement in predicting the pore characteristics. Recently research work has focused on the 3D image analysis of porous membranes.

Wiederkehr et al. [100] in their study of three-dimensional reconstruction of pore network used an image morphing technique to construct a three-dimensional multiphase model of the coating from a number of such cross section images. They show that the technique can be successfully applied to light microscopy images to reconstruct 3D pore networks. The reconstructed volume was converted into a tetrahedron-based mesh representation suited for the use in finite element applications using a marching cubes approach. Comparison of the results for three-dimensional data and two-dimensional cross-section data suggested that the 3D-simulation should be more realistic due to the more exacter representation of the real microstructure.

Delerue et al. [101] used skeletization method to obtain a reconstructed image of the spatialized pore sizes distribution, that is, a map of pore sizes, in soil or any porous media. The Voronoi diagram, as an important step towards the calculation of pore size distribution both in2D and 3D media, was employed to determine the pore space skeleton. Each voxel has been assigned a local pore size and a reconstructed image of a spatialized local pore size distribution was created. The reconstructed image not only provides a means for calculating the global volume versus size pore distribution, but also performs fluid invasion simulation which take into account the connectivity of and constrictions in the pore network. In this case, mercury intrusion in a 3D soil image was simulated.

Al-Raoush et al. [102] employed a series of algorithms, based on the three-dimensional skeletonization of the pore space in the form of nodes connected to paths, to extract pore network structure from high-resolution three-dimensional synchrotron microtomography images of unconsolidated porous media systems. They used dilation algorithms to generate inscribed spheres on the nodes and paths of the medial axis to represent pore-bodies and pore-throats of the network, respectively. The authors have also determined the pore network structure, that is, three-dimensional spatial distribution (x-, y-, and z-coordinates) of pore-bodies and pore-throats, pore-body and pore-throat sizes, and the connectivity, as well as the porosity, specific surface area, and representative elementary volume analysis on the porosity. They show that X-ray microtomography is an effective tool to nondestructively extract the structure of porous media. They concluded that spatial correlation between pore-bodies in the network is important and controls many processes and phenomena in single and multiphase flow and transport problems. Furthermore, the impact of resolution on the properties of the network structure was also investigated and the results showed that it has a significant impact and can be controlled by two factors: the grain size/resolution ratio and the uniformity of the system.

In another study, Liang et al. [103] proposed a truncated Gaussian method based on Fourier transform to generate 3D pore structure from 2D images of the sample. The major advantage of this method is that the Gaussian field is directly generated from its autocorrelation function and also the use of a linear filter transform is avoided. Moreover, it is not required to solve a set of nonlinear equations associated with this transform. They show that the porosity and autocorrelation function of the reconstructed porous media, which are measured from a 2D binarized image of a thin section of the sample, agree with measured values. By truncating the Gaussian distribution, 3D porous media can be generated. The results for a Berea sandstone sample showed that the mean pore size distribution, taken as the result of averaging between several serial cross-sections of the reconstructed 3D representation, is in good agreement with the original thresholded 2D image. It is believed that by 3D reconstruction of porous media, the macroscopic properties of porous structure such as permeability, capillary pressure, and relative permeability curves can be determined.

Diógenes et al. [104] reported the reconstruction of porous bodies from 2D photomicrographic images by using simulated annealing techniques. They proposed the following methods to reconstruct a well-connected pore space: (i) Pixel-based Simulated Annealing (PSA), and (ii) Objective-based Simulated Annealing (OSA). The difference between the present methods and other research studies, which tried to reconstruct porous media using pixel-movement based simulated techniques, is that this method is based in moving the microstructure grains (spheres) instead of the pixels. They applied both methods to reconstruct reservoir rocks microstructures, and compared the 2D and 3D results with microstructures reconstructed by truncated Gaussian methods. They found that PSA method is not able to reconstruct connected porous media in 3D, while the OSA reconstructed microstructures with good pore space connectivity. The OSA method also tended to have better permeability determination results than the other methods. These results indicated that the OSA method can reconstruct better microstructures than the present methods.

In another study, a 3D theoretical model of random fibrous materials was employed by Faessel et al. [105]. They used X-ray tomography to find 3D information on real networks. Statistical distributions of fibers morphology properties (observed at microscopic scale) and topological characteristics of networks (derived from mesoscopic observation), is built using mathematical morphology tools. The 3D model of network is assembled to simulate fibrous networks. They used a number of parameter describing a fiber, such as length, thickness, parameters of position, orientation and curvature, which derived from the morphological properties of the real network.

11.17 CONCLUSIONS

The topic of obtaining 3D models from images is a new research field in Pore analysis of nanofibrous membrane. A set of techniques for creating a 3D representation of a view from one or more 2D images can be used to image-based modeling. A solution to this lack of 3D content is to convert existing 2D material to 3D. A detailed review on 3D image reconstruction from two views of single 2D image has been done in this contribution. The paper has concisely demonstrated that there are three steps for 3D reconstruction, comprising estimation of the epipolar geometry existing between the 2D image pair, estimation of the affine geometry, and also camera calibration. The obtained results for both the camera calibration and reconstruction showed that it is possible to obtain a 3D model directly from features in the images.

There are a number of extensions to this work:

1. If two or more views of an object are available, but under conditions which prevent computation of corresponding points matches, single view reconstructions can be created and joined. This might be necessary, for example, when two views of an object have very limited overlap.
2. Once the camera is calibrated using orthogonality of vanishing points and rectification of planes, it may then be used in the reconstruction of other surfaces.

3. For an optimal reconstruction the final 3D model should be 'polished' by minimizing reprojection errors in the image subject to the parallelism and orthogonality constraints obtained from the 2D image.

Although image analysis of SEM micrographs for geometrical characterization is useful for measuring total porosity, pore shape, pore size and pore size distribution of relatively thin nonwovens, it cannot be applied to multilayer electrospun fibrous analysis. Another problem encounter to this method is that it is not possible to measure 3D pore characteristic of the membrane and it is limited on relatively small fields of view. It is believed that the three-dimensional reconstruction of porous media, from the information obtained from a two-dimensional analysis of photomicrographs, will bring a promising future to nanoporous membranes.

NOMENCLATURE

2D	Two-Dimensional
3D	Three-Dimensional
SVD	Singular Value Decomposition of a Matrix
LMS	Least Median of Squares
CAD	Computer-Aided Design
GIS	Geographic Information System
CCD	Charged Coupled Device
SO (3)	Special Orthogonal Group, 3D Rotation Group
SE (3)	Special Euclidean Group, the group of rigid-body motions in 3D
DLT	Direct Linear Transform
VP	Vanishing Point
SIFT	Scale Invariant Feature Transform
CMT	Correct Match Rate
TP	True Positives
FN	False Negatives
FP	False Positives
TN	True Negatives
TPR	True Positive Rate

Micro-CT	Micro Computed Tomography
SEM	Scanning Electron Microscopy
LSCM	Laser Scanning Confocal Microscope
EBSD	Electron Back-Scatter Diffraction
POA	Pore Open Area
PSA	Pixel-based Simulated Annealing
OSA	Objective-based Simulated Annealing

KEYWORDS

- **2D image**
- **3D reconstruction**
- **epipolar geometry**
- **geometry groups**
- **nanofibrous membrane**

REFERENCES

1. Akash M. Kushal, Vikas Bansal Subhashis Banerjee, "A simple method for interactive 3D reconstruction and camera calibration from a single view," ICVGIP, 2002.
2. Peter F. Sturm, Stephen J. Maybank, "A Method for Interactive 3D Reconstruction of Piecewise Planar Objects from Single Images," The 10th British Machine Vision Conference (BMVC '99) 265–274, 1999.
3. Criminisi, A., Reid, I., Zisserman, A. "Single View Metrology," International Journal of Computer Vision 40(2), 123–148, 2000.
4. Paul Ernest Debevec, "Modeling and Rendering Architecture from Photographs," PhD thesis, University Of California At Berkeley, 1996.
5. Etienne Grossmann, Diego Ortin and José Santos-Victor, "Single and Multi-View Reconstruction of Structured Images," The 5th Asian Conference on Computer Vision, 23–25, 2002.
6. David Liebowitz, Antonio Criminisi and Andrew Zisserman, "Creating Architectural Models from Images," The Eurographics Association and Blackwell Publishers, Volume 18, Number 3, 1999.
7. Marta Wilczkowiak, Edmond Boyer, Peter Sturm, "Camera Calibration and 3D Reconstruction from Single Images Using Parallelepipeds," 8th International Conference on Computer Vision (ICCV '01) 1, 142–148, 2001.

8. Arne Henrichsen, "3D Reconstruction and Camera Calibration from 2D Images," MSc Thesis, University Of Cape Town, 2000.
9. Richard Hartley and Chanop Silpa-Anan, "Reconstruction from two views using approximate calibration," Proc. 5th Asian Conf. Comput. Vision, 2002.
10. Noah Snavely, Member IEEE, Ian Simon, Michael Goesele, Member IEEE, Richard Szeliski, Fellow IEEE, Steven M. Seitz, Senior Member IEEE, "Image Reconstruction and Visualization From Community Photo Collections," Proceedings of the IEEE, Vol. 98, No. 8, 2010.
11. Liang Zhang, Senior Member, IEEE, Carlos Vázquez, Member, IEEE, and Sebastian Knorr, "3D-TV Content Creation: Automatic 2D-to-3D Video Conversion," IEEE TRANSACTIONS ON BROADCASTING, VOL. 57, NO. 2, 2011.
12. Fernanda A. Andal, Gabriel Taubin, and Siome Goldenstein, "Vanishing Point Detection by Segment Clustering on the Projective Space," Trends and Topics in Computer Vision, Lecture Notes in Computer Science Volume 6554, pp. 324–33, 2012.
13. S. Ourselin, A. Roche, G. Subsol, X. Pennec, N. Ayache, "Reconstructing a 3D structure from serial histological sections," Image and Vision Computing, Vol. 19, 25–31, 2000.
14. Dang Trung Kien, "A Review of 3D Reconstruction from Video Sequences," University of Amsterdam ISIS Technical Report Series, 2005.
15. Ying Sun, Eric Bergerson, "Automated 3d Reconstruction Of Tree-Like Structures From Two Orthogonal Views," Browse Conference Publications, IEEE Xplore Digital library, 1988.
16. Yi Ma, Stefano Soatto, Jana Kosecka, S. Shankar Sastry, "An Invitation to 3D Vision," Springer Science and Business Media, LLC, Volume 26, 2003.
17. Edwin Bidwell Wilson, "Projective and Metric Geometry," Annals of Mathematics, Second Series, Vol. 5, No. 3, pp. 145–150, 1904.
18. David Liebowitz, Antonio Criminisi and Andrew Zisserman, "Creating Architectural Models from Images," EUROGRAPHICS'99, Volume 18, Number 3, 1999.
19. Sophie Germain, "Basics of Affine Geometry," Geometric Methods and Applications, 2011.
20. Leichtweiß, Kurt, "Affine geometry of convex bodies," Heidelberg: Barth, 1998.
21. Amnon shashua, nassir navab, "relative affine structure: theory and application to 3D reconstruction from perspective view," IEEE Computer Society, 1994.
22. Dmitri Burago, Yuri Burago, Sergei Ivanov, "A Course in Metric Geometry," Department of Mathematics, Pennsylvania State University, 2001.
23. Ilkka Holopainen, "Metric Geometry," 2009.
24. Havel, Timothy F. "Some examples of the use of distances as coordinates for Euclidean geometry" Journal of Symbolic Computation, Vol. 11, No. 5, pp. 579–593, 1991.
25. Hermann Minkowski, "Basics of Euclidean Geometry," Geometric Methods and Applications, 2011.
26. F. Devernay and O. Faugeras, "From Projective to Euclidean Reconstruction," International Conference on Computer Vision and Pattern Recognition – 1996.
27. Bazin, J.C., Seo, Y., Pollefeys, M. "Globally optimal line clustering and vanishing point estimation in Manhattan world." In: CVPR, 2012.
28. Lilian Zhang, Reinhard Koch, "Vanishing Points Estimation and Line Classification in a Manhattan World," Institute of Computer Science, University of Kiel, Germany, 2012.

29. Thaddeus Beier, "Feature-Based Image Metamorphosis," Computer Graphics, Vol. 26, No. 2, 1992.
30. Piotr Doll'ar, Zhuowen Tu, Hai Tao, Serge Belongie, "Feature Mining for Image Classification," University of California, 2007.
31. Xiang Sean Zhou, Thomas S. Huang, "Image Retrieval: Feature Primitives, Feature Representation, and Relevance Feedback," Beckman Institute for Advanced Science and Technology, University of Illinois at Urbana Champaign, Urbana, USA, 2000.
32. Robert M. Haralick, K. Shanmugam, Itshak Dinstein, "Textural features for Image Classification," IEEE Transactions on systems, Man and cybernetics, Vol. SMC-3, No. 6, pp. 610–621,1973.
33. McGuire, Morgan, John F. Hughes. "Hardware-determined feature edges" In Proceedings of the 3rd international symposium on Non-photorealistic animation and rendering, pp. 35–47. ACM, 2004.
34. Harris, Chris, and Mike Stephens. "A combined corner and edge detector" In Alvey vision conference, vol. 15, p. 50. 1988.
35. Yamakawa, Soji, Kenji Shimada. "Polygon crawling: Feature-edge extraction from a general polygonal surface for mesh generation" In proceedings of the 14th International Meshing Roundtable, pp. 257–274. Springer Berlin Heidelberg, 2005.
36. Azad, Pedram, Tamim Asfour, and Rüdiger Dillmann. "Combining Harris interest points and the SIFT descriptor for fast scale-invariant object recognition" In Intelligent Robots and Systems, IROS 2009. IEEE/RSJ International Conference on, pp. 4275–4280. IEEE, 2009.
37. Yu, Tsz-Ho, Oliver J. Woodford, Roberto Cipolla. "An evaluation of volumetric interest points" In 3D Imaging, Modeling, Processing, Visualization and Transmission (3DIMPVT), 2011 International Conference on, pp. 282–289. IEEE, 2011.
38. Luis Ferraz, Xavier Binefa, "A scale invariant interest point detector for discriminative blob detection," Universitat Autonoma de Barcelona, Department of Computing Science, Barcelona, Spain, 2009.
39. Manjunath, B. S. "A New Approach to Image Feature Detection with Applications," Department of Electrical and Computer Engineering, University of California, Santa Barbara, 1994.
40. Qiaoliang Li, Guoyou Wang, Jianguo Liu, Member, IEEE, and Shaobo Chen, "Robust Scale-Invariant Feature Matching for Remote Sensing Image Registration," IEEE Geoscience And Remote Sensing Letters, VOL. 6, NO. 2, 2009.
41. Ali Cafer Gurbuz, "Feature Detection Algorithms In Computed Images," PhD Thesis, Georgia Institute of Technology, 2008.
42. Frank Candocia and Malek Adjouadi, "A Similarity Measure for Stereo Feature Matching," IEEE Transactions On Image Processing, VOL. 6, NO. 10, 1997.
43. Adam Baumberg, "Reliable Feature Matching Across Widely Separated Views," Computer Vision and Pattern Recognition, 2000.
44. M. Brown, D. G. Lowe, "Unsupervised 3D Object Recognition and Reconstruction in Unordered Datasets," Department of Computer Science, University of British Columbia,2006.
45. David G. Lowe, "Distinctive Image Features from Scale-Invariant Keypoints," Computer Science Department University of British Columbia, 2004.

46. Janne Heikkila, "Geometric camera calibration Using circular control points," IEEE Transactions on pattern analysis and machine intelligence, VOL. 22, NO. 10, 2000.
47. Branislav Micusık, Tomas Pajdla, "Estimation of omnidirectional camera model from epipolar geometry," In Proceedings of Conference on Computer Vision and Pattern Recognition, IEEE Computer Society, Madison, Wisconsin, USA, 2003.
48. Marc Pollefeys, Luc Van Gool, Marc Proesmans, "Euclidean 3D reconstruction from image sequences with variable focal lengths," Computer Vision—ECCV'96, 1996.
49. Charles Loop, Zhengyou Zhang, "Computing Rectifying Homographies for Stereo Vision," IEEE Conference on Computer Vision and Pattern Recognition, Vol.1, pages 125–131, 1999.
50. Andrea Fusiello, Emanuele Trucco, Alessandro Verri, "A compact algorithm for rectification of stereo pairs," Machine Vision and Applications, Vol. 12, pp. 16–22, 2000.
51. Daniel Oram, "Rectification for Any Epipolar Geometry," BMVC, comp.leeds.ac.uk, 2001.
52. Marc Pollefeys, Reinhard Koch, Luc Van Gool, "A simple and efficient rectification method for general motion," Computer Vision, 1999.
53. Jan Sandr, "Epipolar Rectification for Stereovision," Research Reports of CMP, Czech Technical University in Prague, No. 4, 2009.
54. Elan Dubrofsky, "Homography Estimation," MSC thesis, The University Of British Columbia, 2009.
55. Zhang, Z., Hanson, A. R. "Scaled Euclidean 3D reconstruction based on externally uncalibrated cameras," Browse Conference Publications, Computer Vision, 1995.
56. Zhang, Z., Hanson, A. R. "3D reconstruction based on homography mapping," Proc. ARPA96, 1007–1012, 1996.
57. Barinova, O., Konushin, V., Yakubenko, A., Lee, K., Lim, H., Konushin, A. "Fast automatic single-view 3D reconstruction of urban images" Computer Vision – ECCV 2008, Lecture Notes in Computer Science Volume 5303, pp. 100–113, 2008.
58. Mouragnon, Etienne, Maxime Lhuillier, Michel Dhome, Fabien Dekeyzer, and Patrick Sayd. "Real time localization and 3d reconstruction," In Computer Vision and Pattern Recognition, IEEE Computer Society Conference on, vol. 1, pp. 363–370, 2006.
59. Triggs, B., McLauchlan, P. F., Hartley, R. I., Fitzgibbon, A. W. "Bundle adjustment—a modern synthesis," In Vision algorithms: theory and practice, Springer Berlin Heidelberg, pp. 298–372, 2000.
60. Engels. C, Stewénius. H, Nistér. D, "Bundle adjustment rules," Photogrammetric computer vision, 2, 2006.
61. Rachmielowski, A., Birkbeck, N., Jagersand, M., Cobzas, D, "Real-time visualization of monocular data for 3D reconstruction," In Computer and Robot Vision, CRV'08. Canadian Conference on IEEE, pp. 196–202, 2008.
62. Ioannides, M., Stylianidis, E., Stylianou, S, "3D reconstruction and visualization in cultural heritage." In Proceedings of the 19th International CIPA Symposium, 258–262, 2003.
63. Snavely, Keith N. "Image reconstruction and visualization from internet photo collections," PhD thesis, University of Washington, 2008.

64. Mayer, H., Neubiberg, "3D reconstruction and visualization of urban images from uncalibrated wide-baseline image sequences," Photogrammetric Fernerkundung Geoinformation, Vol. 167, No. 3, 2007.
65. Haala, N., "The Landscape of Dense Image Matching Algorithms," 2013.
66. Haala, Norbert, "Multiray photogrammetry and dense image matching" Photogramtrische Woche, 2011.
67. Wei, Yichen, Maxime Lhuillier, Long Quan, "Fast segmentation-based dense stereo from quasi-dense matching," Asian Conference on Computer Vision. 2004.
68. Braux-Zin. J., Dupont. R., Bartoli. A, "A General Dense Image Matching Framework Combining Direct and Feature-based Costs," 2013.
69. Rothermel, Mathias, Norbert Haala. "Potential of Dense Matching for the Generation of High Quality Digital Elevation models" In Proceedings of ISPRS Hannover Workshop High-Resolution Earth Imaging for Geospatial Information, pp. 331–343. 2011.
70. Kim, Jaechul, Ce Liu, Fei Sha, and Kristen Grauman. "Deformable spatial pyramid matching for fast dense correspondences" In Computer Vision and Pattern Recognition (CVPR), 2013 IEEE Conference on, pp. 2307–2314. IEEE, 2013.
71. Leordeanu, Marius, Andrei Zanfir, and Cristian Sminchisescu. "Locally affine sparse-to-dense matching for motion and occlusion estimation" ICCV, 2013.
72. Megyesi, Zoltán. "Dense Matching Methods for 3D Image Reconstruction from Wide Baseline Images" PhD dissertation, France, Eotvos Lorand University, 2009.
73. Bartoli, Adrien, and Peter Sturm, "Structure-from-motion using lines: Representation, triangulation, and bundle adjustment" Computer Vision and Image Understanding, Vol. 100, No. 3, pp. 416–441, 2005.
74. Kim, Deok-Soo, Donguk Kim, Youngsong Cho, Kokichi Sugihara. "Quasi-triangulation and interworld data structure in three dimensions" Computer-Aided Design, Vol. 38, No. 7, pp. 808–819, 2006.
75. Lee, Der-Tsai, and Bruce J. Schachter. "Two algorithms for constructing a Delaunay triangulation," International Journal of Computer and Information Sciences, Vol. 9, No. 3, pp. 219–242, 1980.
76. Richard I. Hartle, "Triangulation," Computer Vision and Image Understanding, Vol. 68, No. 2, pp. 146–157, 1997.
77. Kirby, Robion C., and Lawrence C. Siebenmann, "On the triangulation of manifolds and the Hauptvermutung" Bull. Amer. Math. Soc, Vol. 75, pp. 742–749, 1969.
78. Aurenhammer, Franz. "Voronoi diagrams—a survey of a fundamental geometric data structure," ACM Computing Surveys (CSUR), Vol. 23. No. 3, pp. 345–405, 1991.
79. Heckbert, Paul S., "Fundamentals of texture mapping and image warping," MSc thesis, University of California, Berkeley, 1989.
80. Cohen, Michael, Claude Puech, and Francois Sillion, "Texture Mapping as a Fundamental Drawing Primitive," 1993.
81. Barhate, R. S., Seeram Ramakrishna. "Nanofibrous filtering media: filtration problems and solutions from tiny materials" Journal of membrane science, Vol. 296, No. 1, pp.1–8, 2007.
82. Cho, D., Naydich, A., Frey, M. W., Joo, Y. L. "Further improvement of air filtration efficiency of cellulose filters coated with nanofibers via inclusion of electrostatically active nanoparticles," Polymer, Vol. 54, No. 9, pp. 2364–2372, 2013.

83. Jena, Akshaya, and Krishna Gupta. "Characterization of pore structure of filtration media" Fluid/Particle Separation Journal, Vol. 14, No. 3, pp. 227–241, 2002.
84. Bagherzadeh, R., Latifi, M., Najar, S. S., Kong, L. "Three dimensional pore structure analysis of Nano/Microfibrous scaffolds using confocal laser scanning microscopy," Journal of Biomedical Materials Research Part A, Vol. 101, No. 3, pp. 765–774, 2013.
85. Ghasemi Mobarakeh, L., Semnani, D., Morshed, M., "A Novel Method for Porosity Measurement of Various Surface Layers of Nanofibers Mat Using Image Analysis for Tissue Engineering Applications," Journal of Applied Polymer Science, Vol. 106, pp. 2536–2542, 2007.
86. Sreedhara, Sudhakara Sarma, and Narasinga Rao Tata. "A Novel Method for Measurement of Porosity in Nanofiber Mat using Pycnometer in Filtration," Journal of Engineered Fibers and Fabrics, Vol. 8, Issue 4, 2013.
87. Strange, J. H., M. Rahman, and E. G. Smith. "Characterization of porous solids by NMR," Physical review letters, Vol. 71, No. 21, pp. 3589, 1993.
88. Ziabari, Mohammad, Vahid Mottaghitalab, and Akbar Khodaparast Haghi. "Evaluation of electrospun nanofiber pore structure parameters" Korean Journal of Chemical Engineering, Vol. 25, No. 4, pp. 923–932, 2008.
89. Nimmo, J. R. "Porosity and pore size distribution" Encyclopedia of Soils in the Environment, Vol. 3, pp. 295–303, 2004.
90. He, Mei, Yong Zeng, Xuejun Sun, and D. Jed Harrison. "Confinement effects on the morphology of photopatterned porous polymer monoliths for capillary and microchip electrophoresis of proteins" Electrophoresis, Vol. 29, No. 14, pp.2980–2986, 2008.
91. Al-Kharusi, A. S., Martin J. B., "Network extraction from sandstone and carbonate pore space images" Journal of Petroleum Science and Engineering, Vol. 56, No. 4, pp. 219–231, 2007.
92. Al-Raoush, R. I., C. S. Willson., "Extraction of physically realistic pore network properties from three-dimensional synchrotron X-ray microtomography images of unconsolidated porous media systems" Journal of hydrology, Vol. 300. No. 1, pp. 44–64, 2005.
93. Al-Raoush, Riyadh., "Change in microstructure parameters of porous media over representative elementary volume for porosity" Particulate Science and Technology, Vol. 30, No. 1, pp. 1–16, 2012.
94. Miao, J., Ishikawa, T., Johnson, B., Anderson, E. H., Lai, B., Hodgson, K. O., "High resolution 3D x-ray diffraction microscopy," Physical review letters, Vol. 89, No. 8, pp. 088303, 2002.
95. Rollett, A. D., Lee, S. B., Campman, R., Rohrer, G. S. "Three-dimensional characterization of microstructure by electron back-scatter diffraction," Annu. Rev. Mater. Res., Vol. 37, pp. 627–658, 2007.
96. Prior, D. J., Boyle, A. P., Brenker, F., Cheadle, M. C., Day, A., Lopez, G., Zetterström, L, "The application of electron backscatter diffraction and orientation contrast imaging in the SEM to textural problems in rocks," American Mineralogist, Vol. 84, pp. 1741–1759, 1999.
97. Davies, P. A., Randle, V., "Combined application of electron backscatter diffraction and stereophotogramtry in fractography studies," Journal of microscopy, Vol. 204, No. 1, pp. 29–38, 2001.

98. Uchic, M. D., Groeber, M. A., Dimiduk, D. M., Simmons, J. P., "3D microstructural characterization of nickel super alloys via serial-sectioning using a dual beam FIB-SEM," Scripta Materialia, Vol. 55, No. 1, pp. 23–28, 2006.
99. Khokhlov, A. G., Valiullin, R. R., Stepovich, M. A., Kärger, J., "Characterization of pore size distribution in porous silicon by NMR cryoporosimetry and adsorption methods," Colloid Journal, Vol. 70, No. 4, pp. 507–514, 2008.
100. Wiederkehr, T., B. Klusemann, D. Gies, H. Müller, B. Svendsen, "An image morphing method for 3D reconstruction and FE-analysis of pore networks in thermal spray coatings," Computational Materials Science, Vol. 47, pp. 881–889, 2010.
101. Delerue, J. F., E. Perrie, Z. Y. Yu and B. Velde, "New Algorithms in 3D Image Analysis and their Application to the Measurement of a Spatialized Pore Size Distribution in Soils," Phys. Chem. Earth(A), Vol. 24, No. 7, pp. 639–644, 1999.
102. Al-Raoush, R. I., C. S. Willson, "Extraction of physically realistic pore network properties from three-dimensional synchrotron X-ray microtomography images of unconsolidated porous media systems," Journal of Hydrology, Vol. 300, pp. 44–64, 2005.
103. Liang, Z. R., C.P. Fernandes, F.S. Magnani, P.C. Philippi, " A reconstruction technique for three-dimensional porous media using image analysis and Fourier transforms," Journal of Petroleum Science and Engineering Vol. 21, 273–283,1998.
104. Diógenes, A. N., L. O. E. dos Santos, C. P. Fernandes, A. C. Moreira, C. R. Apolloni, "Porous Media Microstructure Reconstruction Using Pixel-Based And Object-Based Simulated Annealing – Comparison With Other Reconstruction Methods," Engenharia Térmica (Thermal Engineering), Vol. 8, No. 02, pp. 35–41, 2009.
105. Faessel, M., C. Delisee, F. Bos, P. Castera, "3D Modelling of random cellulosic fibrous networks based on X-ray tomography and image analysis," Composites Science and Technology 65, pp. 1931–1940, 2005.

CHAPTER 12

SOLAR POWER HARVESTING BY PHOTOVOLTAIC MATERIALS: A COMPREHENSIVE REVIEW

M. KANAFCHIAN

University of Guilan, Rasht, Iran

CONTENTS

ABSTRACT

Today, energy is an important requirement for both industrial and daily life, as well as political, economical, and military issues between countries. While the energy demand is constantly increasing every day, existing energy resources are limited and slowly coming to an end. Due to all of these conditions, researchers are directed to develop new energy sources which are abundant, inexpensive, and environmentally friendly. The solar cells, which directly convert sunlight into electrical energy, can meet these needs of mankind. This study reviews the efforts in incorporating of solar cells into textile materials.

12.1 INTRODUCTION

12.1.1 THE SIGNIFICANCE OF SOLAR ENERGY

There is increasing concern worldwide about the huge dependence on oil as a source of energy. Although coal resources are immense, their extraction and use tend to potentially damaging environmental problems. There are concerns too about nuclear fission as a source of energy, and controlled fusion has yet to be proven feasible. Consequently, there is growing interest in harnessing energy by other means that do not rely on the consumption of reserves, as with oil and coal, nor involve dangers (real and perceived) with utilizing nuclear fission. In recent years alternative renewable energies including that obtained by solar cells have attracted much attention due to exhaustion of other conventional energy resources especially fossil-based fuels. There are numerous examples of attempts to utilize solar energy for storage of chemical energy (mimicking natural photosynthesis) and for the storage of electrical energy in batteries. Nevertheless, greatest success has been achieved in the direct conversion of solar power into electrical energy, with the aid of photovoltaic devices (PV). Apart from using an endless source of energy, the application of photovoltaic devices offers a number of other advantages: the technology is clean and noiseless, most applicable and promising alternative energy using limitless sun light as raw material, maintenance costs are very small, and the technology is attractive for remote areas, which are difficult to supply by conventional means from a grid [1]. The only drawback is the initial price of commercially available solar panels. However, the decrease to 20% of the price in 1980 [2] already makes the solar energy a suitable solution for the remote areas with no electric network. This currently refers to almost 1.6 billion people, mainly in the developing countries. The total electric power-generation capacity (based on all available sources, including fossil fuels) installed worldwide was about 3500 GW in 2003 and it is expected to increase to almost 7000 GW in next 30 years [3]. Half of the new capacity will be installed in the developing countries, so the need for cheap and preferably renewable energy sources is obvious. Moreover, these countries typically have very good insolation conditions, which favors the solar energetics. According to the optimistic PV market scenario, almost 656 GWp of total solar power can be expected by 2030 if very large scale PV systems in the deserts are used [4].

12.1.2 HOW SOLAR CELL WORKS

Sunlight may be converted into electricity by heating water until it vaporizes, and then passing the steam through a conventional turbine set. This triple conversion of energy has losses at each step. PV cells are more efficient device to convert optical energy directly to electrical energy. To see how this works, we need a little semiconductor science. A typical PV cell is made up of special materials called semiconductors. These are typically made of silicon, which is doped to create a more powerful current. Electrons within the doped silicon have an uneven number making them unstable and looking to move around. When sunlight hits the cell, the semiconductor absorbs some of the energy of the rays. The energy then knocks electrons loose, allowing them to flow freely. PV cells also all have one or more electric field that acts to force electrons freed by light absorption to flow in a certain direction. This flow of electrons is a current, and by placing metal contacts on the top and bottom of the PV cell, we can draw that current off for external use to power a calculator. This current, together with the cell's voltage (which is a result of its built-in electric field or fields), defines the power (or wattage) that the solar cell can produce [5].

It is important to note that thermodynamics limits the efficiency for conversion of solar radiation by a simple photovoltaic cell to 32%, and today's crystalline cells can approach this; multiple junction cells have a thermodynamic limiting efficiency of 66% and triple junction cells have achieved 32%. As mentioned above, in a semiconductor, valence electrons are freed into the solid, to travel to the contacts, when light below a wavelength specific to that solid is absorbed. This threshold wavelength must be selected to provide an optimum match to the spectrum of the light source; too short a wavelength will allow much of the radiation to pass through unabsorbed, too long a wavelength will extract only part of the energy of the shorter-wavelength photons. The typical solar spectrum requires a threshold absorption wavelength of 800 nm, in the near infra-red, to provide the maximum electrical conversion, but this still loses some 40% of the solar radiation. Further losses occur through incomplete absorption or reflection of even this higher energy radiation. The current that is delivered by a solar cell depends directly on the number of photons that are absorbed: the closer the match to the solar spectrum and the larger the cell area the better. As well as a current, electrical energy has a potential associated with it and this is determined by the internal construction of the cell, although it is ultimately limited by the threshold for electron release. In conventional silicon solar cells, there is an electrical field built into the cell by the addition of minute amounts of intentional impurities to the two cell halves; this produces what is known as a PN junction (Fig. 12.1). The PN junction is a built-in asymmetry in the solar cell, where an electric field ensures that the electron will travel in one direction, while the hole travels in the opposite direction. The junction separates the photon-generated electrons from the effective positive charges that pull them back and in so doing, generates a potential of 0.5 volts, approximately. This potential depends on the amounts of dopants, and on the actual semiconductor itself. Multiple junction cells have integrated layers stacked together

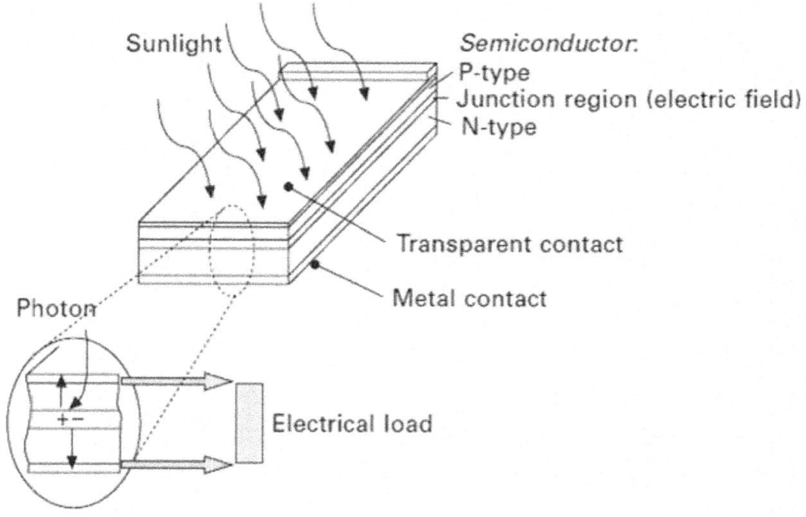

FIGURE 12.1 Schematic of a PV cell [6].

during manufacture, producing greater voltages than single junctions. The power developed by a solar cell depends on the product of current and voltage. To operate any load will require a certain voltage and current, and this is tailored by adding cells in series and/or parallel, just as with conventional dry-cell batteries. Some power is lost before it reaches the intended load, by too high a resistance in the cells, their contacts and leads. Note that at least one of the electrical contacts must be semitransparent to allow light through, without interfering with the electrical current. Thus optimizing the design of a solar cell array involves optics, materials science, and electronics [6].

12.2 SOLAR CELL SUBSTANCES

12.2.1 SILICON

Silicon is the most known and the 2nd most widespread element in the Earth crust. Although it is one of the most studied and used elements, there exist many open questions and there even may be some interesting properties or new forms of silicon we do not know about yet. Thanks to the electronic industry and its rapid development in the second half of the twentieth century, we can make our lives easier by enjoying a rich variety of electronic devices based on silicon components. One of the critical properties that make silicon suitable as a solar cell material is that it is a semiconductor, possessing a band-gap. This band-gap is a range of energies that the electrons in the materials are not allowed to have. The electron can either have an energy placing it in its ground energy state in the valence band, or it can be in an excited state in the conduction band. The electron can transition from valence band to conduction band and back through excitation and recombination processes. The energy required for an excitation may come from a photon, being the smallest package of energy one can divide light into. The sunlight consists of photons with a wide range of energies. The energy of the photon corresponds to what we observe as the color of the light, where the blue light consists of photons with a higher energy, and the red light consists of photons with lower energy. The energy of the photon also corresponds to a wavelength of the light, where the blue light has a shorter wavelength, and the red light has a longer wavelength. When a photon hits the silicon, it may be absorbed by an electron in the silicon, providing enough energy for the electron to be excited from its ground energy state in the valence band to an excited state in the conduction band. Such an absorption process may only take place if the photon carries an energy corresponding to at least the band gap energy. The electron being excited will leave behind a hole in the valence band; an electron-hole pair is created. In a solar cell, the electron-hole pair moves by diffusion until it reaches the PN junction. As such, the electron may reach one of the electrical contacts, while the hole reaches the other contact, as a result of a combination of random diffusion and directional drift in an electric field. This is the principal mechanism for current generation in a solar cell. Only photons with high enough energy may be absorbed by the electrons. A photon with energy lower than the band-gap energy will not carry sufficient energy to lift the electron to the conduction band, and will as such not be absorbed in the semiconductor. Hence, its energy will not be converted into electricity. On the other hand, photons with high energy can create an electron-hole pair, lifting the electron high above the conduction band edge. However, all the excess energy that is put into the electron will be rapidly lost, as the electron will collide with other electrons or atoms, losing energy until it reaches the conduction band edge. This loss process is called thermalization. In a real solar cell, not all generated electron-hole pairs will contribute to current generation. There is always the chance that an electron finds a hole on its way to the contacts and relaxes back across the band gap, in a process called recombination. Recombination may happen slowly in the bulk of a high-quality silicon wafer, but it will always take place even in a perfect material. These unavoidable recombination mechanisms are termed intrinsic recombination mechanisms. In a more realistic material, recombination happens faster. Examples of recombination-active areas are crystal defects or impurities in the silicon, highly doped silicon, silicon crystal boundaries or wafer surfaces and metal-silicon interfaces, such as contacts. By combining intrinsic recombination mechanisms with spectrum losses, we reach a maximum efficiency of a solar cell, known as the Shockley-Queisser limit [7], which for silicon under a special irradiance is around 29%

[8]. Currently, the record efficiency of a silicon solar cell is 25% [9], which is actually quite close to the theoretical maximum of 29% given by the Shockley-Queisser limit. Hence, for a further successful progress of the solar energetics we still need a lot of innovations leading, above all, to the competitive price of the solar cells and solar panels. Comparison of energy payback times shows that the thin film solar cells have potential for achieving low costs. The significant reduction of the raw material needs and faster production time make them an attractive alternative to the crystalline Si solar cells, although the efficiencies are still much lower. The theoretical limit for the efficiency of the single junction crystalline silicon solar cell given by the energy of its band gap is around 27% [10]. The disordered structure of the material in thin film solar cells leads to much lower efficiencies, however, a significant improvements can be achieved by the application of a multi layered structure of stacked cells. So, the efficiency increase still remains to be an important task for the solar cell research.

12.2.2 CONDUCTIVE POLYMERS

Among the many literature of conducting materials, a tremendous amount of publications can be found in the field of conducting polymers [11]. In 1862, at the College of London Hospital, Letheby attained a partly conductive material by oxidizing aniline in a sulfuric acid solution. Better conductivity results were gained after the successful synthesis of polyacetylene [12]. Polyacetylene was first synthesized in 1958 as a highly crystalline high molecular weight conjugated polymer by Natta et al. after the polymerization of acetylene in hexane using an initiator. However, this received little attention due to the air sensitive, infusible and insoluble properties aside from the preparation method [13]. In the early 1970s, even though the inorganic explosive polymer polysulfurnitride was found to be very conductive at low temperatures, interest shifted to organic conductors. In 1974, by using a Ziegler-Natta catalyst, a silvery film of polyacetylene was prepared. In spite of its similar appearance to metals, this film is not conductive. In 1977, MacDiarmid, Shirakawa and Heeger modified the polyacetylene film through a partial oxidation treatment with oxidizing agents, such as halogens or AsF_5, making the film significantly conductive [14]. Since this breakthrough, many other conducting polymers have been synthesized and characterized, such as polyaniline (PANI), polythiophenes (PTh) and polypyrrole (PPy). In the beginning of the 1980s, as the interest in conductive materials significantly increased, the problems with early conducting polymers became clear. The lack of processability in comparison to plastic materials was a big disadvantage. In addition, insolubility, an infusible and brittle nature, and the lack of air stability of the conductive polymers prevented the integration of these conductive materials into new application areas. In the late 1980s, some of the problems, such as solubility and air durability, were overcome with modifications of the polymers [15]. In 2000, Heeger, MacDiarmid and Shirakawa were awarded a Nobel prize in chemistry in recognition of the significance of their research in the development of conducting polymers.

12.2.2.1 Molecular Structure

Organic chemistry states that double bonds can be isolated, conjugated or cumulated [11]. Conjugated bonds require strict alternation of double and single bonds which is also seen in the structure of conducting polymers [15]. In the ground state, a carbon atom has an electron configuration of 1 $s^2$2 $s^2$2p^2. It is expected for C to form only 2 bonds with the neighboring atoms since the 2s shell is completely occupied. However, it forms 4 bonds due to hybridization. In conjugated polymers, carbon makes sp^2 hybridization in which the 2s orbital pairs with two 2p orbitals to form sp^2 hybrid orbitals. At the end of hybridization, three sp^2 and one p orbitals are formed. Two of the three sp^2 orbitals on each carbon atom are bonded to another carbon atom next to it and the last sp^2 orbital is bonded either to hydrogen or any of the side groups. The bonding between these atoms is covalent bonding and in this case, it is called 'σ bond' which has a cylindrical symmetry around the internuclear axis. The unhybridized p orbitals of neighboring carbon atoms overlap and form a 'π bond' [16]. The π electrons are not strongly bound. As a result of the weak interaction between them, they can be readily delocalized, which in this situation, contribute electrical conductivity to the polymer [17]. According to the simple free electron molecular orbital model in the theories of Hünkel and Bloch, when the molecular chain length is long enough (showing metallic transport properties), π electrons are delocalized over the entire chain, which form a very small band gap. As a result, a conjugated polymer which has an alternation of double and single bonds can be conductive when conditions are right. In this case, the electrons delocalized over the conjugated chain are predicted to be evenly spaced out. Thus, all bonds are assumed to be equivalent [11]. However, under simplified conditions, the σ and π bonds have different bonding lengths [15]. Thus, the conductivity of conjugated polymers cannot be explained with conventional theories used in semiconductor physics. In Fig. 12.2, the commonly studied conducting polymer structures are shown.

12.2.2.2 Conduction Mechanism

Materials can be generalized into three groups in terms of electrical conductivity: insulators, semiconductors and conductors. Conductive polymers are in the group of semiconductors. Electronic energy is important in determining the electric conductivity of materials. In metals, free electrons and electrons, which have similarly low binding energy, can move easily between atoms. The degenerated atomic orbitals overlap in all directions and form molecular orbitals in metals. In a solid state structure, the number of resulting molecular orbitals formed by neighboring

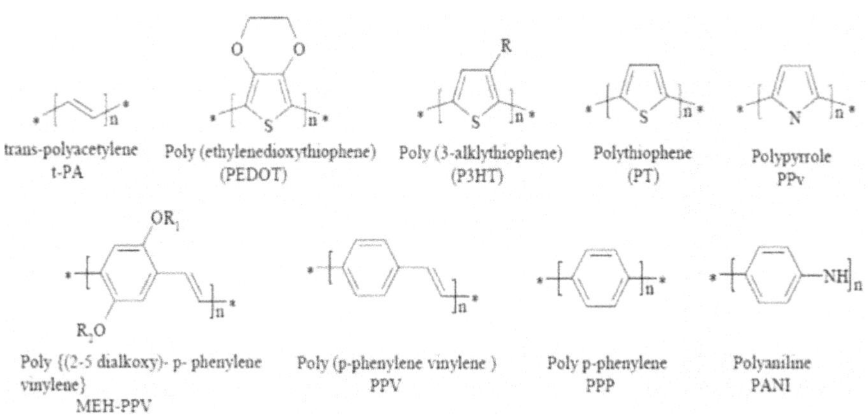

FIGURE 12.2 Molecular structure of some conducting polymers.

atoms whose energy levels are equivalent is very high, up to 1022 for a 1 cm^3 metal piece. This amount of molecular orbitals spaced together in a similar energy range leads to a formation of continuous band energies. In a metal atom, excluding an inert gas form, the valence orbitals are not fully filled, and give birth to molecular orbitals which are not fully filled either. On the other hand, above a certain energy level, which can be zero in the case of metals, all molecular orbitals appear to be empty. The energy states which are forbidden for electron occupation between the highest occupied (valence) and the lowest unoccupied (conduction) orbitals, is called the band gap [15].

Electrical conductivity depends highly on the band gap. If the gap is small, the electrons can be excited even with a small amount of thermal energy. In metals, there is either no gap or the bands are overlapped [18]. Materials with a band gap of 1–5 eV or below are considered semiconductors and generally, they have a conductivity range from 10–9 S/cm to 103 S/cm, which is lower than the metallic conductivity range of 106 S/cm. Most of the semiconductors are solid crystalline inorganic materials, but some conjugated polymers can also show semiconducting properties with appropriate doping. In 1955, Peierls described theoretically that for a one-dimensional metallic molecule with a partly filled band, the regular chain structure will never be stable. Peierls instability provides a better interpretation of the conduction mechanism in polymers so that even a small conformational distortion can change the band gap which directly affects the conductivity of the polymer [19]. In conductive polymers, there are no holes or free electron formation in the conduction bands, unlike crystalline semiconductors. The difference from inorganic semiconductors lies in the structure of the polymer backbone, which is deformed after the polymer is oxidized or reduced by the dopants. For example, after an electron is removed from the valence band of a conducting polymer, a cation is formed. The hole left in the valence band is not completely empty. In other words, it cannot delocalize completely as it is expected to do by the classical band theory. Instead, partial delocalization occurs which also causes structural deformation while spreading along the surrounding monomer units. The energy that was created by the electron is attempted to be balanced and this effort results in a new equilibrium condition [20]. A polaron has two defects:

charge and defect. In this concept, conducting polymers can be classified into two main groups: degenerate ground state and nondegenerate ground state. The main difference between these polymers is apparent when they are subsequently charged. When conducting polymers with degenerate ground states are charged subsequently, a second polaron is formed. In this case, the polarons are independent and not bound to each other. As a result, they can be split into two phases of opposite orientation with equal energy. These are called solitons. Solitons could also be neutral, caused by isomerization. Although neutral solitons do not carry any charge, they can move along the chain and hop between localized states of adjacent polymer chains. For conducting polymers with nondegenerate ground states (such as PPy, PTh, polyphenylene, PANI, etc.) there is no soliton formation, but pairs of defects are created. These are called bipolarons. In this case, PPy can be given as an example for bipolaron formation due to its highly disordered structure. Bipolaron formation takes place when two polarons are formed on the same chain [20].

12.2.2.3 Conducting Polymer Applications in PV Technology

Conductive polymers have been expected to yield several applications due to their electrical properties and wide color variation despite their poor mechanical properties, they have the potential to replace metallic materials in many applications, which is attributable to their lightweight and semiconducting nature [21]. An application area for conducting polymers is definitely solar cell technology. It has been known for a long time that solar radiation has a great potential as an energy source which can be used in various ways [22]. Photovoltaic technology is based on conversion of sunlight into electricity. PV cells generate direct current electric power from semiconductors after they are illuminated by photons. Inorganic materials, silicon and other semiconductors are used in PV cells because by their nature, electrons in their structure can be excited from valence band into the conduction band when they are impinged upon by photons, and that generation leads the creation of electron-hole pairs. As a result, electrons move towards the P-type material and holes move towards the N-type material, leading to an electric current with an external circuit [23]. To lower the manufacturing cost, organic materials which can easily be processed are seen as an alternative way for PV cell production [24]. Organic solar cells differ from inorganic PV cells in their production technique, character of the materials used in the cell structure and the device design [25]. In a simple organic solar cell system, an organic semiconductor, which consists of donor and acceptor layers, is sealed between two metallic electrodes, indium tin oxide (ITO) and Al. MDMO-PPV: *poly (2-methoxy-5-(3',7'-dimethyloctyloxy)-1,4-phenyl-ene-vinylene)*, RRP3HT: *regioregular poly(3-hexylthiophene)*, PCPDTBT: *poly[2,6-(4,4-bis-(2-ethylhexyl)-4Hcyclopenta[2,1-b;3,4-b_]-di-thiophene)-alt-4,7(2,1,3benzothiadiazole)* and PCBM: *(6,6)-phenyl-C61-butyric acid methyl ester,* which are some of the polymers used in the organic PV cell system. The mismatch between the electronic band structures of donor and acceptor is considered as the driving force of the electron transfer. Subsequent to sun illumination, an electron is excited from highest occupied molecular orbital (HOMO) to the lowest unoccupied molecular orbital (LUMO) of the donor. This photoinduced electron is first transferred from the excited state of the donor to the LUMO of the acceptor and then carried out to the Al electrode. Similarly, holes left in the HOMO of the donor are moved through the working electrode, ITO. Thus, with an external circuit, direct current is generated [26].

Among the wide range of conjugated polymers already developed, perhaps the diethoxy substituted thiophene poly(3,4-ethylenedioxythiophene) or PEDOT, also known under the trade name Baytron® or Clevios, is one of the most promising conducting polymer to date [27, 28]. It was synthesized by researchers at Bayer AG in Germany in 1988 with excellent conductivity (300 S/cm) [27, 29]. Since then, it has shown extraordinary scope for smart textiles, electronics, and optoelectronics applications [30]. PEDOT is highly conductive in its oxidized (doped) state, where the molecular backbone contains mobile carriers (holes). The chemical structures of 3,4-ethylenedioxythiophene (EDOT) monomer and PEDOT polymer are shown in Fig. 12.3. The presence of only two reactive hydrogen atoms at the 2 and 5 positions on the EDOT monomer ring gives PEDOT a very regular molecular structure, which makes it a stable polymer [31, 32].

PEDOT has a stiff conjugated aromatic backbone structure, which makes it insoluble and infusible in most organic and inorganic solvents [33–35].

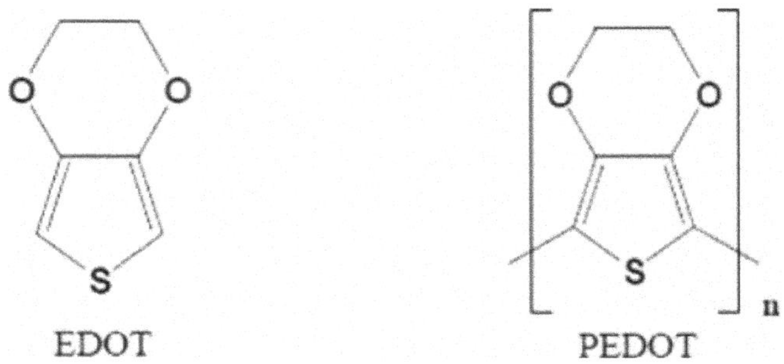

FIGURE 12.3 Molecular structure of EDOT monomer and PEDOT polymer.

Bayer resolved this problem by introducing polystyrene sulfonic acid (PSS), a water-soluble polyanion, during the polymerization of PEDOT as a charge balancing dopant [36]. This water-soluble PEDOT:PSS complex, as a commercial product known as BAYTRON PTM, has electrical and film-forming properties that somehow allow fairly stable, transparent conductive polymer coatings on a variety of substrates. However, PSS itself is a nonconducting material, which limits the conductivity of the PEDOT:PSS complex to the ~1–10 S/cm range [37]. In order to use the conductive polymer PEDOT for electronics and optoelectronics applications, it should be applied or deposited as a thin film on the surface of different substrates. However, due to the lower solubility of PEDOT in most organic and inorganic solvents, the simplest coating techniques such as spray coating, spin casting, and solution casting are difficult to use with it. Deposition of PEDOT polymer directly on the surface of desired substrates could therefore be an efficient way of obtaining highly conductive, flexible, and thin uniform films.

12.3 SMART TEXTILES

12.3.1 DEFINITION

Today, people want to make their lives more comfortable and talk a lot about intelligent devices. An intelligent device means a system that can evaluate the environment and respond to the results of the evaluation [38, 39]. These systems can give us information we need in any particular situation and help us to plan our everyday lives [40]. In order to make human life healthier, more comfortable, and safer, intelligent devices are being brought one stage closer to the individual by means of easy-to-use wearable interfaces [41–43]. In this regard, the multifunctional fabrics commonly known as *smart textiles*, *electronic textiles (E-textiles)*, or *textronics* are good candidates for making our daily lives healthier, safer, and more comfortable [41]. The term *smart textiles* or *textronics* refers to interdisciplinary approaches in the process of producing and designing textile materials, which started about the year 2000 [44]. It is an emerging field of research that connects the textile industry with other disciplines such as information technology, microsystems, and materials science with elements of automation and metrology [44, 45].

Smart textiles are defined as "*the synergic combination of electronic and conventionally used textile materials. They are able to sense the physical stimuli from the external environment and then perform some reactions against them*" [46]. These external stimuli and the reactions of smart textiles can be in the form of electrical, thermal, chemical or magnetic signals. Smart textiles not only protect the human body from extreme environments, but can also monitor, do local computing, and communicate wirelessly [47]. E-textiles because of their multifunctional interactivity in wearable devices that are adaptable, flexible, and conform to the human body are successful promoters of better quality of life and progress in the field of biomedicine, as well as in several other health-based disciplines such as rehabilitation, telemedicine, ergonomics, tele-assistance, and sports medicine [41, 48–51].

During the last decade, apart from biomedical applications, smart and interactive textiles have been widely used in technical clothing [52], flexible solar cell panels [14, 53], protective garments for electromagnetic shielding and static charge dissipation [54], heating elements [55], fabrics for dust and germ-free clothes [56, 57], pressure sensors [50, 58], chemical sensors [59], power sources [60], and wireless devices [61].

12.3.2 PV CELL IN TEXTILE

Conventional solar cells are mostly encased between glass substrates, giving the structure fragility, heaviness and rigidness that cause problems in their storage and transportations. Therefore, attention has turned to organic solar cells which are more flexible and lighter. The idea of manufacturing flexible solar cells is not novel. Several companies and research groups have been working on flexible solar cells that can generate enough electricity to run daily electronics. Konarka Technologies Inc. has recently introduced "power plastic" which is capable of running portable electronic devices and formed with a primary electrode, an organic polymer blend and a transparent electrode encased between the plastic substrate and the transparent packaging material. Another flexible solar cell design has been developed by Sky Station International Inc. In this research, a flexible sheet material is embedded with a solar cell for stratospheric vehicles. There are some other flexible solar cell devices designed for different applications, such as a solar tensile pavilion designed by Nicholas Goldsmith and a solar awning designed by Zebetakis et al. [62]. However, attaching conventional solar panels onto textiles or forming them within/on plastics does not meet expectations in terms of flexibility and wearability. To contribute flexibility and wearability to the system, textile materials can be directly used as structural elements that replace plastic based materials. Recently, Winther-Jensen et al., Krebs et al. and Biancardo et al. developed a strategy that involves thermal lamination of a thin layer of polyethylene (PE) onto textile followed by a plasma treatment which contributed adhesion to the surface and lastly, PEDOT formation on the top [63]. The main aspect of the textile based solar cells is the level of electrical conductivity on the surface, which enables them to reach an appropriate level of current collection efficiency.

As mentioned above, despite the many attractions of solar cells as vehicles for providing energy, the way in which they are constructed provides problems in application. Typically, solar cells are either encased between glass plates, which are rigid and heavy, or the cells are covered by glass. Glass plates are fragile, and so care has to be taken with their storage and transport. The rigid nature of solar cells requires their attachment to flat surfaces and, since they are used outdoors, they have to be protected from any atmospheric pollution and adverse weather. It is, therefore, not surprising that increasing attention has been turned to the construction of lighter, flexible cells, which can still withstand unfavorable environments, yet nevertheless maintain the durability required. Several examples have recently appeared of solar cells in plastic films. Examples are the 'power plastic' developed by Konarka Technologies Inc. [64] in which the cell is apparently coated or printed onto the plastic surface, and an integrated flexible solar cell developed at Sky Station International Inc. for high-altitude and stratospheric applications [65]. Iowa Thin Technologies Inc. have claimed the development of a roll-to-roll manufacturing method for integrating solar modules on plastic [66].

However, while the successful incorporation of solar cells into plastics represents an important step in the expansion of solar cell technology, still greater expansion of the technology would be achieved through their incorporation into textile fabrics, especially if the cells are fully integrated into the fabrics. Textiles are materials with a huge range of applications and markets, and they can be produced by a number of fabrication processes, all of which offer enormous versatility for tailoring fabric shape and properties. The different types of woven, knitted and nonwoven constructions that can nowadays be achieved seem almost infinite, and indeed technical uses are beginning to be found for crocheted and embroidered constructions [67, 69].

Some examples of the application of solar cells in textiles have now been reported. Architect, Nicholas Goldsmith, has designed a solar tensile pavilion which, while providing shade and shelter, can also capture sunlight, and transform it into electrical power [70]. The skin of the tent consists of amorphous silicon cells, encapsulated and laminated to contoured panels of the woven fabric. Zabetakis et al. have recently described the design of an awning, which provides protection from the sun and at the same time converts the incident sunlight into electricity [71]. A type of sail has been designed in which attached to the sailcloth are strips of flexible solar cells [72]. The incorporation of the solar cells has been designed to allow for expansion of the sailcloth when wind blows on it. Textiles offer further advantages too. Fabrics can be wound and rewound. Thus, Warema Renkhoff GmbH in Germany has developed a retractable woven canopy [73], in which energy is collected from the solar cells integrated into the canopy, and then stored. The stored energy is applied to the operation of the winding drive motor. In addition, textile fabrics can be readily installed into structures with complex geometries, a feature which opens up a number of potential applications.

12.3.3 PV TEXTILES BY PV FIBERS

12.3.3.1 Conductive Fibers

Generally, conductive fibers can be divided into two categories: Naturally and Treated conductive fibers.

I) Naturally Conductive Fibers

The fibers that can be produced purely from inherently conductive materials, such as metals, metal alloys, carbon sources, and conjugated polymers (ICPs), are associated with the class of naturally conductive fibers. In their pure form, these fibers have high conductivity values. these fibers can be divided into three categories:

a) Metallic Fibers

Metallic fibers, as suggested by the name, are the first man-made fibers to be developed from metals or metal alloys. These fibers have very thin metal filaments with diameter ranging from 1 to 80 μm, which can be produced by the bundle-drawing or shaving process [74]. Metallic fibers have very high conductivity (106 S/cm) with a wide range of mechanical properties [75]. Despite the fact that they have extraordinary electro-mechanical properties, metallic fibers have limited textile applications because of their low flexibility, stiffness, high weight, high cost, low compatibility with other materials, and poor weaving properties [74, 76].

b) Carbon Fibers

Carbon fibers, which were invented by Edison in 1879, are used as the most demanding materials in high-tech industrial applications such as structural composites in aerospace, transportation, and defense-related products. Carbon fibers are petroleum-based products and they can be produced from petroleum pitch and polyacrylonitrile (PAN) [77]. The heat treatment of PAN, also called graphitization, strongly influences the electrical and mechanical properties of carbon fibers. Carbon fibers have graphite structure, which means that they have conductivity values similar to those of metals, that is, 104–106 S/cm [78]. Carbon fiber composites are usually appropriate for structural applications when high strength, stiffness, lower weight, and extraordinary fatigue characteristics are required [79, 80]. For smart and interactive textile applications, carbon fibers cannot be easily integrated into knitted or weaved structures because of their high stiffness and brittleness. In addition, esthetic considerations and health-related issues are also strong reasons for the use of carbon fibers being limited in the clothing industry.

c) Conjugated Polymer Fibers

The conjugated polymers or ICPs are organic materials that conduct electricity. Due to their high conductivity, lower weight, and environmental stability, they have a very important place in the field of smart and interactive textiles. Several attempts have been made to produce conductive fibers. Pomfret et al. [81] and Okuzaki et al. [82] produced polyaniline and PEDOT:PSS fibers, respectively, by using a one-step wet spinning process. However, it is very difficult to use these fibers for interactive applications. The major reasons are poor mechanical strength, brittleness, a lower production rate, and difficult processing. The production of pure PEDOT fibers with high conductivity values, from 150 to 250 S/cm, by a chemical polymerization method has also been reported, but due to their microscale size and brittle nature, useful applications could not be found [83].

II) Treated Conductive Fibers

The conductive fibers that can be produced by the combination of two or more materials, such as nonconductive and conductive materials, are known as treated conductive fibers or extrinsically conductive fibers. Special treatments involve the mixing, blending, or coating of inherently insulating materials such as polyethylene (PE), polypropylene (PP), polystyrene (PS), or textile fibers with highly conductive materials such as metals, carbon black, or ICPs. The conductive fibers obtained, also known as conductive polymer composites (CPCs), can have a combination of the electrical and mechanical properties of the treated materials. These fibers can be classified further as:

a) Conductive Filled Fibers

Conductive filled fibers are a class of conductive fibers that can be produced by adding conductive fillers such as metallic powder, carbon black, carbon nanotubes, or conjugated polymers to nonconductive polymers such as PP, PS, or PE [78, 84–86]. Usually, melt spinning and solution spinning techniques are used to produce filled conductive fibers. However, in order to get homogenous distribution of conductive particles or ICPs in polymers, the parent materials are well mixed before the spinning process. However, the conductive fibers produced by the solution spinning process have better electrical properties than those produced by the melt spinning process, but the need for large quantities of solvents, their separation, and possible health hazards has made this process obsolete [87]. The electrical conductivity values of melt-spun conductive fibers strongly depend on two parameters: volume loading of filler and filler shape. The fibrous conductive filler can give higher conductivity in melt-spun fibers than irregular and spherical particulates [88]. The melt spinning process is the most economical and least complex process for the production of filled conductive fibers, but the lower electrical and mechanical properties of conductive fibers limit their use in smart and interactive textile applications.

b) Conductive Coated Fibers

The conductive fibers that can be produced by coating insulating materials with highly conductive materials, such as metals, metal alloys, carbon black, carbon nanotubes, and ICPs, are known as conductive coated fibers [59, 89–91]. With coating processes, not only can the highest conductivities be achieved but also the mechanical properties of conductive fibers can be enhanced. Depending on the type of conductive materials, different coating techniques can be used to transform polymeric or textile fibers into electro-active fibers. To apply metallic coatings, sputtering, vacuum deposition, electroless plating, carbonizing, and filling or loading fibers are the most extensively used methods [74, 92]. High conductivities similar to those of metals (106 S/cm) can be achieved with these methods, but high cost, stiffness, brittleness, high weight, and lower levels of comfort limit their application in the textile field.

Carbon black, carbon nanotubes, and ICPs can be applied as a thin layer to a fiber surface by the solution casting, inkjet printing, in-situ polymerization, VPP, CVD, or PACVD methods [91, 93–96]. The thin layers of coating offer reasonably high conductivity and also preserve the flexibility and elasticity of substrate fibers [96]. However, the use of carbon black and carbon nanotubes for wearable applications is very limited because of health-hazard issues. On the other hand, due to their high conductivity, exceptional environmental stability, lower weight, and high compliance, ICPs are reasonable candidates for functionalization of different polymeric and textile materials. However, difficulty in processing, poor adhesion to substrate, and durability issues are some disadvantages of ICPs. By using suitable coating techniques and compatible substrates, these problems can be minimized [97].

12.3.3.2 PV FIBERS

A basic technique to manufacture PV textiles is based on the development of photovoltaic fibers using Si-based /organic semiconducting coating or incorporation of dye-sensitized cells (DSC). Availability of PV fiber offer more freedom in the selection of structure for various type of applications [98–102]. The development of photovoltaic fibers offers advantages to manufacture large area active surfaces and higher flexibility to weave or knit etc40. Although, the problem of manufacturing textile structure by using dye-sensitized cells (DSC-PV) fibers into textile structure is still alive and require a optimization with respect to textile manufacturing operations. In a typical research work the working electrode of DSCPV fiber is prepared by coating Ti wire with a porous layer of TiO_2. This working electrode is embedded in an electrolyte with titanium counter electrode. The composite structure is coated with a transparent cladding to ensure protection and structural integrity.

The electrons from dye molecules are excited by photo energy and penetrated into the conduction band of TiO_2 and move to the counter electrode through external circuit and regenerate the electrolyte by happening of redox reaction. Ultimately the electrolyte regenerates the dye by means of reduction reaction.

The performance of DSC fiber is majorly depends on the grade of TiO_2 coating and its integration to Ti substrate. The integrity of Ti with TiO_2 will depend on the surface cleanliness and roughness of Ti, affinity between Ti and TiO_2 and other defects. The deposition of TiO_2 dye on Ti wire surface is performed by strategy. The integrity of coating on Ti substrate is tested by using peel test, tensile test, four point bending test and scratch test. The amount of discontinuities is measured by optional microscopy and SEM [103–109]. The photovoltaic potential of dye-sensitized solar cells (DSSC) of Poly(vinyl alcohol) (PVA) was improved by spun it into nanofibers by electrospinning technique using PVA solution containing silver nitrate ($AgNO_3$). The silver nanoparticles were generated in electrospun PVA nanofibers after irradiation with UV light of 310~380 nm wavelength. Electrospun PVA/Ag nanofibers have exhibited Isc, FF, Voc, and η showed the values of 11.9~12.5 mA/cm^2, 0.55~10.59, 0.70~10.71 V, and 4.73~14.99%, respectively. When the silver was loaded upto 1% as dope additives in PVA solution, the resultant electrospun PVA/Ag nanofibers exhibited power conversion efficiency 4.99%, which is higher than that of DSSC using electrospun PVA nanofibers without Ag nanoparticles [110]. Ramier et al., concluded that the feasibility of producing textile structure from DSCPV fiber is quite good. The deposition of TiO_2 on flexible fiber is expected to be quite fruitful in order to maintain the structural integrity without comparing with PV performance [111]. Fiber based organic PV devices inroads their applications in electronics, lighting, sensing and thermoelectric harvesting. By successful patch up between commodity fiber and photovoltaic concept, a very useful and cost effective way of power harvesting is matured [112–114]. Coner et al. [115], have developed a photovoltaic fiber by deposition of small Molecular weight organic compound in the form of concentric layer on long fibers. They manufactured the optical photovoltaic (OPV) fiber by vacuum thermal evaporation (VTE) of concentric thin films upto 0.48 mm thickness on polyamide coated silica fiber. Different control devices are based on OPV cells containing identical layer structures deposited on polyimide substrates. The OPV based fiber cells were defined by the shape of the substrate and 1 mm long cathodes. All fiber surfaces were cleaned well prior to deposition. Lastly, they concluded that performance of OPV fiber cells from Indium tin oxide (ITO) is inferior in terms of changes in illumination angle, enabling OPV fiber containing devices to outperform its planar analog under favorable operating conditions. Light emitting devices are designed in such a way that becomes friendly to weave it. The light trapping on fiber surface can be improved by using external dielectric coating which is coupled with protective coating to enhance its service time. Successful PV fiber can be manufactured by opting appropriate material with more improved fabrication potential [116].

DSCs are low cost, applicable in wide range of application and simple to manufacture. These merits of DSCPV fiber make it a potential alternative to the conventional silicon and thin film PV devices55. DSC works on the principle of optoelectronically active cladding on an optical fiber. This group was manufactured two types of PV fiber using polymethylmethacrylate (PMMA) baltronic quality diameter 1.3 to 2.0 mm and photonium quality glass fiber with diameter 1.0 to 1.5 mm. Both virgin fiber were made electronically conductive by deposition of 130 nm thick layer of ZnO:Al by atomic layer deposition technique with the help of P400 equipment. The high surface area photoelectric film for DSC was prepared in two steps. In first step TiO_2 in the form of solution or paste having TiO_2 nanoparticle is deposited on electronically conductive surface. In the second step dry layer of TiO_2 is sintered at 450–500°C for 30 min to ensure proper adhesion to the fiber surface. PMMA fiber is suitable to survive upto 85°C. Hence mechanical compression is alternate technique to ensure the fixation.

Glass fiber is capable to withstand with sintering temperatures which inroads the possibilities of preparation of porous photoelectrodes on them. Commercially available TiO_2 paste was diluted to achieve appropriate viscosity with tarpin oil to make suitable for dip-coating. TiO_2 film was formed by dip coating and dried at room temperature upto 30 min before proper sintering between 475 to 500°C. Appropriately sintered fibers were then immersed in dye solution consisting of 0.32 mL of the cis-bis(isothiocyanate)bis(2,2-bipyridyl-4–4' dicarboxylato)-ruthenium(II) bis-tetrabutyl ammonium, Solaronix SA with trade name N719 dye in absolute ethanol for 48h. The dyeing of nonporous Polymethylmethacrylate (PMMA) fiber coated with nonporous TiO_2 layer was performed in the same dye bath. After complete sanitization the excess dye was rinsed away with ethanol. A electrolyte solution was prepared with 0.5 M 4-tert butylpyridine and 0.5 M Lil, 0.05 M I_2 in methoxypropionitrile (MePRN) with 5 wt% polyvinylidene fluoride-hexafluoro-propylene (PVDE-HFP) added as gelatinizing agent as used by Wang et al. [117]. Finally gelatinized iodine electrolyte was added next with dip coating from hot solution. Lastly the carbon based counter electrode was coated by means of a gel prepared by exhaustive grinding of 1.4 g graphite powder and 0.49 grade carbon black simultaneously.

12.3.4 CONDUCTIVE LAYERS FOR PVS

Solar cells require two contacts in a sandwich configuration, with the active semiconductor between them. At least one of the contacts must be transparent to solar radiation, over the waveband which the semiconductor absorbs. Conventional crystalline silicon cells have a thick metal contact on the back surface, and a gridded metal contact on the front surface. They do not require complete coverage of the front surface because the top layer of silicon is sufficiently conducting ('heavily doped') to deliver the photocurrent without significant resistance losses. Amorphous silicon cells, like most thin-film cells, have a more insulating top semiconductor layer and so require the whole surface to be contacted. This is achieved without blocking the incoming light, by a layer of transparent conducting oxide (TCO) such as ZnO or Indium tin oxide (ITO), often with other elements added to enhance the conductivity without reducing the transparency. In addition, a fine gridded contact is often superimposed to further improve the current collection efficiency.

It is also possible to build the cell with the light entering through the substrate: this is known as the superstrate configuration. Typically a glass sheet, coated with TCO is then covered with thin-film semiconductor, which in turn is covered by an opaque metal layer. Textile substrates are unlikely to be sufficiently transparent that this construction may be used, although it is not necessary for the material to be transparent enough to show an image through it, only to be translucent. Hence the textile material is likely to be coated with a conducting layer that may be opaque. Alternatively, the textile may itself form part of the active cell structure, either enabling the conduction of photo-current generated within an imposed semiconductor, or in future materials, enabling photo-currents to be generated within the fabric (probably by semiconducting polymers). Organic solar cells or polymer solar cells still have a long process of development before they can be considered for this application, perhaps being a more intractable problem than the inverse application of light-emitting polymers. Unless the textile can be made from conducting fibers, an additional material is needed, that should not degrade the other desirable textile properties. It is possible, of course, to incorporate metal fibers within a fabric, but it is questionable whether these can make sufficiently good contact to the whole thin-film cell area. They may be helpful in addition to a thinner, continuous conductor. Although polycrystalline silicon can be sufficiently conducting to enable only partial coverage by the contacting layer, it generally requires much higher processing temperatures, before or after its deposition, than most textiles can withstand. We therefore require a continuous conducting layer to be placed over the whole textile surface before adding the semiconductor. The choice of materials (probably metals) is determined by electrical conductance (lower conductance requires thicker layers), compatibility with the substrate (chemically, physically, and during processing), and compatibility with the semiconductor (chemically and work function value). In addition, the mechanical behavior of the conductor in a cell that can flex or twist will depend on its composition, thickness and adhesion. There are several options to be considered, using both physical and chemical coating techniques, to obtain the correct composition and conformity with a non-planar surface. These will deposit metals or TCOs at the limited temperatures allowed by the preferred textiles. Physical coating methods include sputtering from a solid target in an argon atmosphere and simple evaporation from a solid source (either thermally or with an electron beam), and both are widely used in the optics, electronics, and engineering industries. Chemical methods are based either on the decomposition of a volatile compound of the conductor by heat or electrical plasma, or on the deposition of a coating from a solution, usually driven by an electrical current. These methods are also in widespread use within industries. There are a few additional methods that are applicable for certain materials, such as dip- and spray- and spin-coating, that would be appropriate for the conducting polymers that are being developed now. These do not conduct well enough for photovoltaic cells, because they generate low voltage electrons that cannot overcome the resistance of a long path through a poor conductor. The compromise solution to retaining substrate flexibility with high electrical conductivity may be to use a conducting polymer layer on the textile, or incorporated within its structure, superimposed by a thin, more-conducting metallic layer. If this layer breaks during flexure then the minute gaps will be bridged by the underlying organic conductor, whose limited conductivity will not be a problem over such short paths. This also allows the semiconductor to be in contact with its preferred metal (say aluminum for silicon), avoiding the inclusion of resistive barriers that can exist at other semiconductor/conductor interfaces.

12.3.5 MANUFACTURING OF PV CELLS

ITO was used as a common transparent electrode in polymer-based solar cells due to its remarkable efficiency and ability of light transmission. However, it is quite expensive and generally too brittle to be used with flexible textile substrates. Therefore, highly conductive poly (3,4-ethylenedioxythiophene) doped with poly(styrene sulfonate) PEDOT:PSS, carbon nano-tube or metal layers are used to substitute ITO electrode. This can be a promising way to develop PV textiles for smart application due to its low cost and easy application features for future photovoltaic textile applications. A group of scientists has demonstrated the fabrication of an organic photovoltaic device with improved power conversion efficiency by reducing lateral contribution of series resistance between subcells through active area partitioning by introducing a patterned structure of insulating partitioning walls inside the device. Thus, the method of the present invention can be effectively used in the fabrication and development of a next-generation large area organic thin layer photovoltaic cell device [118]. The manufacturing of organic photovoltaic cells can be possible at reasonable cost by two techniques:

12.3.5.1 Roll-to-Roll Coating Technique

A continuous roll-to-roll nanoimprint lithography (R2RNIL) technique can provide a solution for high-speed large-area nanoscale patterning with greatly improved throughput. In a typical process, four inch wide area was printed by continuous imprinting of nanogratings by using a newly developed apparatus capable of roll-to- roll imprinting (R2RNIL) on flexible web base. The 300 nm line width grating patterns are continuously transferred on flexible plastic substrate with greatly enhanced throughput by roll-to- roll coating technique. European Union has launched an European research project "HIFLEX" under the collaboration with Energy research Centre of the Netherland (ECN) to commercialize the roll to roll technique. Highly flexible Organic Photovoltaics (OPV) modules will allow the cost-effective production of large-area optical photovoltaic modules with commercially viable Roll-to-Roll compatible printing and coating techniques. Coatema, Germany with Renewable Technologies and Konarka Technologies has started a joint project to manufacture commercial coating machine. Coatema, Germany along with US Company Solar Integrated Technologies (SIT) has developed a process of hot-melt lamination of flexible photovoltaic films by continuous roll-to-roll technique [119]. Roll-to-roll (R2R) processing technology is still in neonatal stage. The novel innovative aspect of R2R technology is related to the roll to roll deposition of thin films on textile surfaces at very high speed to make photovoltaic process cost effective. This technique is able to produce direct pattern of the materials [120, 121].

12.3.5.2 Thin-Film Deposition Techniques

Various companies of the world have claimed the manufacturing of various photovoltaic thin films of amorphous silicon (a-Si), copper indium selenide (CIGS), cadmium telluride (CdTe) and dye-sensitized solar cell (DSSC) successfully. Thin film photovoltaics became cost effective after the invention of highly efficient deposition techniques. These deposition techniques offer more engineering flexibilities to increase cell efficiencies, reflectance and dielectric strength, as well as act as a barrier to ensure a long life of the thin film photovoltaics and create high vapor barrier to save the chemistry of these types of photovoltaics [122, 123]. A fiber shaped organic photovoltaic cell was produced by using concentric thin layer of small molecular organic compounds. Thin metal electrode are exhibited 0.5% efficiency of solar power conversion to electricity which is lower than 0.76% that of the planner control device of fiber shape organic PV cells. Results are encouraged to the researchers to explore the possibility of weaving these fibers into fabric form.

The thin film deposition of photovoltaic materials takes place by electron beam, resistance heating and sputtering techniques. These technologies differ from each other in terms of degree of sophistication and quality of film produced. A resistance-heated evaporation technology is relatively simple and inexpensive, but the material capacity is very small which restricts its use for commercial production line. Sputtering technique can be used to deposit on large areas and complex surfaces. Electron beam evaporation is the most versatile technique of vacuum evaporation and deposition of pure elements, including most metals, numerous alloys and compounds.

12.3.5.3 Dye-Sensitized PV

An exhaustive research on photovoltaic fibers based on dye-sensitized TiO_2-coated fibers has opened up various gateways for novel PV applications of textiles. The cohesion and adhesion of the TiO_2 layer are identified as crucial factors in maintaining PV efficiency after weaving operation. By proper control of tension on warp and weft fibers, high PV efficiency of woven fabrics is feasible. The deposition of thin porous films of ZnO on metalized textiles or textile-compatible metal wires by template assisted electro-deposition technique is possible. A sensitizer was adsorbed and the performance as photoelectrodes in dye-sensitized photovoltaic cells was investigated. The thermal instability of textiles restricts its use as photovoltaic material because process temperatures are needed to keep below 150°C. Therefore, the electrode position of semiconductor films from low-temperature aqueous solutions has become a most reliable technique to develop textile based photovoltaics. Among low-temperature solution based photovoltaic technologies; dye sensitized solar cell technology appears most feasible. If textile materials are behaved as active textiles, the maximum electrode distance in the range of 100 μm has to be considered. Loewenstein et al., and Lincot et al., have used Ag coated polyamide threads and fibers to deposit porous ZnO as semiconductor material. The crystalline ZnO films were prepared in a cathodic electrode position reaction induced by oxygen reduction in an aqueous electrolyte in presence of Zn^{2+} and eosin Y as structure-directing agent [124, 125].

Bedeloglu et al. [126], were used nontransparent nonconductive flexible polypropylene (PP) tapes as substrate without use of ITO layer. PP tapes were gently cleaned in methanol, isopropanol, and distilled water respectively and then dried in presence of nitrogen. 100 nm thick Ag layer was deposited by thermal evaporation technique. In next step, a thin layer of poly(3,4-ethylenedioxythiophene) doped: poly(styrene sulfonate) PEDOT: PSS mixture solution was dip coated on PP tapes. Subsequently, poly [2-methoxy-5-(3, 7- dimethyloctyloxy)-1–4-phenylene vinylene] and 1-(3-methoxycarbonyl)-propyl-1- phenyl(6,6)C61, MDMO: PPV: PCBM or poly(3-hexylthiophene) and 1-(3-methoxycarbonyl)- propyl-1-phenyl(6,6)C61, P3HT: PCBM blend were dip coated onto PP tapes. Finally, a thin layer of LiF (7 nm) and Al (10 nm) were deposited by thermal evaporation technique.

The enhanced conductivity will always useful to improve the photovoltaic potential of poly(3,4-ethylene dioxythiophene):poly(styrene sulfonate) (PEDOT: PSS). Photovoltaic scientific community found that the conductivity of poly(3,4-ethylene dioxythiophene): poly(styrene sulfonate) (PEDOT: PSS), film is enhanced by over 100-folds if a liquid or solid organic compound, such as methyl sulfoxide (DMSO), N,Ndimethylformamide (DMF), glycerol, or sorbitol, is added to the PEDOT: PSS aqueous solution. The conductivity enhancement is strongly dependent on the chemical structure of the organic compounds. The aqueous PEDOT: PSS can be easily converted into film form on various substrates by conventional solution processing techniques and these films have excellent thermal stability and high transparency in the visible range [127–130]. Some organic solvents such as ethylene glycol (EG), 2-nitroethanol, methyl sulfoxide or 1-methyl-2-pyrrolidinone are tried to enhance the conductivity of PEDOT: PSS. The PEDOT: PSS film which is soluble in water becomes insoluble after treatment with EG. Raman spectroscopy indicates that interchain interaction increases in EG treated PEDOT: PSS by conformational changes of the PEDOT chains, which change from a coil to linear or expanded-coil structure. The electron spin resonance (ESR) was also used to confirm the increased interchain interaction and conformation changes as a function of temperature. It was found that EG treatment of PEDOT: PSS lowers the energy barrier for charge among the PEDOT chains, lowers the polaron concentration in the PEDOT: PSS film by w 50%, and increases the electrochemical activity of the PEDOT: PSS film in NaCl aqueous solution by w100%. Atomic force microscopy (AFM) and contact angle measurements were used to confirm the change in surface morphology of the PEDOT: PSS film. The presence of organic compounds was helpful to increase the conductivity which was strongly dependent on the chemical structure of the organic compounds, and observed only with organic compound with two or more polar groups. Experimental data were enough to make a statement that the conductivity enhancement is due to the conformational change of the PEDOT chains and the driving force is the interaction between the dipoles of the organic compound and dipoles on the PEDOT chains [131].

12.3.6 CHARACTERIZATION OF PV TEXTILES

Characterization of various photovoltaic textiles is essential to prove its performance before send to the market. Various characterization techniques collectively ensure the perfect achievement of the targets to manufacture the desired product.

12.3.6.1 Thickness and Morphology of PV Textiles

Scanning electron microscope is used to investigate the thickness and morphology of various donor, acceptor layers. Scanning electron microscopes from LEO Supra 35 and others can be used to measure the existence and thickness of various coated layers on various textile surfaces at nanometer level. Various layers on photovoltaic fibers become clearly visible with 50000X magnification. The thickness of the layers can be seen from SEM photographs by bright interface line between the polymer anode and the photoactive layer.

12.3.6.2 Current and Voltage

In order to characterize the Photovoltaic fibers open circuit voltage, short circuit current density, current and voltage at the maximum power point under an illumination of 100 mW/cm^2 are carried out. In order to calculate the Photovoltaic efficiency of Photovoltaic textiles, current verses voltage study is essential. To achieve this target a computer controlled source meter equipped with a solar simulator under a range of illumination power is required with proper calibration. All photoelectrical characterizations are advised to conduct under nitrogen or argon atmosphere inside a glove box to maintain the preciseness of observations. The overall efficiency of the PV devices can be represented by following equation:

$$\eta = \frac{V_{OC} \times I_{SC} \times FF}{P_{in}} = \frac{P_{out}}{P_{in}} \tag{1}$$

where, V_{oc} is the open circuit voltage (for I=0) typically measured in volt (V); I_{sc} is the short circuit current density (for V=0) in ampere/square meter (A/m^2); P_{out} is the output electrical power of the device under illumination; P_{in} the incident solar radiation in (watt/square meter) W/m^2; FF is the fill factor and can be explained by the following relationship:

$$FF = \frac{I_{mpp} \times V_{mpp}}{I_{SC} \times V_{OC}} \tag{2}$$

where, V_{mpp} voltage at the maximum power point (MPP); I_{mpp} is the current at the maximum power point (MPP); where the product of the voltage and current is maximized.

To assure an objective measurement for precise comparison of various photovoltaic devices, characterization has to be performed under identical conditions. An European research group has used Keithley 236 source measure unit in dark simulated AM 1.5 global solar conditions at an intensity of 100 mW/cm^2. The solar simulator unit made by K.H. Steuernagel Lichttechnik GmbH was calibrated with the help of standard crystalline silicon diode. PV fibers were illuminated through the cathode side and I-V characteristics were measured. The semilogarithmic I-V curves demonstrate the current density versus voltage behavior of photovoltaic fibers under various conditions. It gives a comparative picture of voltage Vs current density as a function of various light intensities. Durisch et al., has developed a computer based testing instrument to measure the performance of solar cells under actual outdoor conditions. This testing system consist a sun tracked specimen holder, digital multimeters, devices to apply different electronic loads and a computer based laser printer. Pyranometers, pyrheliometers and a reference cell is used to measure and record the insulation. This instrument is able to test wide dimensions of photovoltaic articles ranging from 3 mm × 3 mm to 1 m × 1.5 m. The major part of world's energy scientist community predicts that photovoltaic energy will play a decisive role in any sustainable energy future [132].

12.3.6.3 Mechanical Characterization

Textile substrates are subjected to different stresses under various situations. Hence usual tensile characterization is essential for photovoltaic textiles. For tensile testing of PV fibers, the constant rate of extension (CRE) based tensile testing machines are used at 1 mm per minute deformation rate using Linear Variable Differential Transformers (LVDT) displacement sensor. Fracture phenomenon is recorded by means of high resolution video camera integrated with tensile testers. To study about the adhesion and crack formation in coating on textile structures, generally 30 mm gauge length is used in case of photovoltaic fibers. Fiber strength measuring tensile tester, integrated with an appropriate optical microscope to record the images of specimen at an acquisition rate of about one frame per second is used to record the dynamic fracture of PV fibers. Different softwares are available to analyze the image data like PAXit, Clemex, and Digimizer etc.

12.3.6.4 Absorption Spectra of Solid Films

Various spectrophotometers like Varian Carry 3G UV-Visible were used to observe the ultraviolet visible absorption spectra of photovoltaic films. The thin films are prepared to study the absorption spectrum of solid films. In a typical study, a thin film was prepared by spin coating of solution containing 10 mg of P3HT and 8 mg of PCBM and 4.5 mg of MDMO-PPV and 18 mg of PCBM (in case of 1:4)/ml with chlorobenzene as solvent. A typical absorption spectra of MDMO-PPV:PCBM and P3HT.

12.3.6.5 X-Ray Diffraction of Photovoltaic Structures

Crystallization process is very common phenomenon that takes place during photovoltaic structure development. The content of crystalline and amorphous regions in photovoltaic structures influences the photoactivity of photovoltaic structures. X-ray diffraction technique is capable to characterize the amount of total crystallinity, crystal size and crystalline orientation in photovoltaic structures. Presently, thin film photovoltaics are highly efficient devices being developed in different crystallographic forms: epitaxial, microcrystalline, polycrystalline, or amorphous. Critical structural and microstructural parameters of these thin film photovoltaics are directly related to the photovoltaic performance. Various X-ray techniques like X-ray diffraction for phase identification, texture analysis, high-resolution *x*-ray diffraction, diffuse scattering, X-ray reflectivity are used to study the fine structure of photovoltaic devices [133].

12.3.6.6 Raman Spectroscopy

The Raman Effect takes place when light rays incidents upon a molecule and interact with the electron cloud and the bonds of that molecule. Spontaneous Raman effect is a form of scattering when a photon excites the molecule from the ground state to a virtual energy state. When the molecule relaxes it emits a photon and it returns to a different rotational or vibrational state. Raman spectroscopy is majorly used to confirm the chemical bonds and symmetry of molecules. It provides a fingerprint to identify the molecules. The fingerprint region for organic molecules remains in the (wavenumber) range of 500–2000 cm^{-1}. Spontaneous Raman spectroscopy is used to characterize superconducting gap excitations and low frequency excitations of the solids. Raman scattering by anisotropic crystal offers information related to crystal orientation. The polarization of the Raman scattered light with respect to the crystal and the polarization of the laser light can be used to explore the degree of orientation of crystals [134]. The in situ morphological and optoelectronic changes in various photovoltaic materials can be observed by observing the changes in the Raman and photoluminescence (PL) feature with the help of a spectrometer. Various spectrums can be recorded at a definite integration time after avoiding any possibility of laser soaking of the sample67.

12.3.7 CHALLENGES TO BE MET

The active cell components, the semiconducting layers, therefore have a great part to play in determining the cell performance, not only from an efficiency point of view but also from the load matching perspective. The separate improvement of both output current and voltage demands effective optical absorption and efficient photogenerated charge separation. There may be unavoidable trade-offs in meeting both of these simultaneously. Textile substrates offer new ways of improving the optical absorption in thin layers by scattering light that passes through the layer back into the layer for a second chance of absorption. Rough surfaces in conventional silicon cells are sometimes a source of electrical problems but substrate texturing has always been seen as a possible tool for advanced thin-film cell design. Transferring some of the best features of amorphous silicon cell technology to a flexible textile substrate device [135] would provide more rapid evolution of the new cells than a separate development based on single crystal cell technology, given the processing temperature restraints of the preferred textiles. There are still plenty of challenges in ensuring efficient charge collection from a stack of thin layers on a non-planar base. Finally the word textile still often suggests applications in clothing, but users would expect to have the same performance from a solar cloth as from a passive cloth, with respect to folding, wear, washing, abrasion, etc. Few of these properties have yet been tested with coatings of the type used in solar cells, nor have solar cells been assessed in such conditions, even if hermetically sealed with a thin polymer layer. Low cost production and easy processing of organic solar cells comparing to conventional silicon-based solar cells make them interesting and worth employing for personal use and large scale applications. Today, the smart textiles as the part of technical textiles using smart materials including PV materials, conductive polymers, shape memory materials, etc. are developed to mimic the nature in order to form novel materials with a variety of functionalities. The solar cell-based textiles have found its application in various novel field and promising development obtaining new features. These PV textiles have found its application in military applications, where the soldiers need electricity for the portable devices in very remote areas. The PV textile materials can be used to manufacture power wearable, mobile and stationary electronic devices to communicate, lighten, cool and heat, etc. by converting sun light into electrical energy. The PV materials can be integrated onto the textile structures especially on clothes, however, the best promising results from an efficient PV fiber has to be come which can constitute a variety of smart textile structures and related products [136]. Fossil fuels lead to the emission of CO_2 and other pollutants and consequently human health is under pressure due to adverse environmental conditions. In consequence of that renewable energy options have been explored widely in last decades [137, 138].

Unprecedented characteristics of PV cells attract maximum attention in comparison of other renewable energy options which has been proved by remarkable growth in global photovoltaic market4. Organic solar cells made of organic electronic materials based on liquid crystals, polymers, dyes, pigment etc. attracted maximum attention of scientific and industrial community due to low weight, graded transparency, low cost, low bending rigidity and environmental friendly processing potential [139, 140]. Various PV materials and devices similar to solar cells integrated with textile fabrics can harvest power by translating photon energy into electrical energy. The successful integration of solar cells into textiles has to take into account the type of fiber used and the method of textile fabrication. The selection of a type of fiber is strongly influenced by its ability to withstand prolonged irradiation by ultraviolet (UV) light. Fiber selection is also governed by the temperatures required to lay down the thin films comprising the solar cell, although it has been shown that nanocrystalline silicon thin films, an improved form of amorphous silicon, can be successfully deposited at temperatures as low as 200°C, and under the appropriate conditions even single crystal silicon may be grown, albeit epitaxially on silicon wafers [141]. These two factors of UV resistance and maximum temperature restrict the choice of commodity fibers.

Commercial polyolefin fibers melt below 200°C. Cotton, wool, silk and acrylic fibers start to decompose below this temperature. Polyamide fibers are likely to be too susceptible to UV radiation. Polyethylene terephthalate (PET) fibers, however, are viable substrates. They melt at 260–270°C and exhibit good stability to UV light [142]. They are commercially attractive too because of their existing widespread use. Thus, fabrics composed of a large variety of PET grades are currently available. PET fibers possess good mechanical properties and are resistant to most forms of chemical attack. They are less resistant to alkalis, but a solar cell would not normally find use in an alkaline environment. Fabrics constructed from PET fibers should, therefore, be suitable as substrates for solar cells, while also possessing flexibility and conformability to any desired shape. Fabrics composed of glass fibers produced from E-glass formulations16 could also be used. One advantage could lie in their transparency, as with plate glass in conventional solar cells. Moreover, the price of E-glass is similar to that of PET [143]. E-glass fibers, however, suffer from poor resistance to acids and alkalis and are prone to flexural rupture. Other glasses are more stable to environments of extreme pH. S-glass is used for specialist sports equipment and aerospace components, but its price is approximately five times that of E-glass [144].

Many high-performance fibers, which readily withstand temperatures up to 300–400°C, could also be considered, although aramids would not be sufficiently stable to UV radiation. Examples of suitable high-performance fibers could be polybenzimidazole (PBI) fibers, polyimide (PI) fibers and polyetheretherketone (PEEK) fibers. However, these fibers are expensive. There are clear commercial attractions, therefore, in adopting PET fibers. The type of textile fabric construction is also important to the performance of the solar cell. The type of construction affects the physical and mechanical properties of a fabric, and also its effectiveness as an electrical conductor. Where conduction in textile fabrics is required, woven fabrics are generally considered to be best in that they possess good dimensional stability and can be constructed to give desired flexibilities and conformations. Moreover, the yarn paths in woven structures are well ordered, which allows the design of complex woven fabric-based electrical circuits. Knitted structures, on the other hand, do not retain their shapes so well, and the rupture of a yarn may cause laddering. These problems are heightened if the shape of the fabric is continually changing, as in apparel usage. Nonwoven fabrics do not, as yet, generally possess the strength and dimensional stability of woven fabrics. More significantly, the construction of electrical circuits in them is limited, because their yarn paths are highly unoriented. Embroidery, however, may offer an opportunity for circuit design [145, 146].

12.4 CONCLUSION

In these times of uncertain energy sources and with the impacts that they are having on the atmosphere, clean, renewable solar energy is a good alternative. However, it is still more expensive than methods used currently by consumers, and inn order to be a viable option, it has to be attainable and affordable. With the use of different materials and thin film technologies, the use of photovoltaic solar energy for the average person is becoming more of a reality. The incorporation of photovoltaics into textiles has been explored new inroads for potential use in intelligent clothing in more smart ways. Incorporation of organic solar cells into textiles has been realized encouraging performances. The functionality of the photovoltaic textiles does not limited by mechanical stability of photovoltaics. Polymer-based solar cell materials and manufacturing techniques are suitable and applicable for flexible. The manufactured photovoltaic fibers may also be used to manufacture functional yarns by spinning and then fabric by weaving and knitting. Photovoltaic tents, curtains, tarpaulins and roofing are available to use the solar power to generate electricity in more green and clean fashion.

KEYWORDS

- **photovoltaic**
- **polybenzimidazole**
- **polyetheretherketone**
- **polyimide**
- **solar cell**
- **textile**

REFERENCES

1. Grätzel M., 'Photoelectrochemical cells,' *Nature*, 414, 2001, 338–344.
2. Hassett, R. J. *Future visions of U.S. photovoltaics industry*. In *Proceedings of the 4th IEA–PVPS international conference*, Osaka, Japan (2003). IEA-PVPS.
3. Sellers, R. *PV: Progress and Promise*. In *Proceedings of the 4th IEA–PVPS international conference*, Osaka, Japan (2003). IEA-PVPS.
4. Kurokawa, K. *PV Industries: Future Visions*. In *Proceedings of the 4th IEA–PVPS international conference*, Osaka, Japan (2003). IEA-PVPS.
5. Boutrois, J. P., Jolly, R., Petrescu, C. 1997, 'Process of Polypyrrole Deposit on Textile. Product Characteristics and Applications,' *Synthetic Metals*, vol. 85, pp. 1405–1406.
6. Goetzberger A., Hebling C., Schock H.-W., 'Photovoltaic materials, history, status and outlook,' *Materials Science and Engineering: R: Reports*, 40, 2003, 1–46.
7. Shockley, W., Queisser, H. J. "Detailed Balance Limit of Efficiency of p-n Junction Solar Cells," *Journal of Applied Physics*, vol. 32, no. 3, pp. 510–519, 1961.
8. Kerr, M. J., Campbell, P., Cuevas, A. "Lifetime and efficiency limits of crystalline silicon solar cells," in *Proceedings of the 29th IEEE Photovoltaic Specialists Conference*, 2002, pp. 438–441.
9. Zhao, J., Wang, A., Green, M. A. "24.5% Efficiency silicon PERT cells on MCZ substrates and 24.7% efficiency PERL cells on FZ substrates," *Progress in Photovoltaics: Research and Applications*, vol. 7, no. 6, pp. 471–474, Nov. 1999
10. Shah, A.V., Platz R., Keppner, H. *Thin–film silicon solar cells: A review and selected trends*. Sol. Energy Mater. Sol. Cells, 38, 501 (1995).
11. Roth, S., Carroll, D. 2004, *One-Dimensional Metals: Conjugated Polymers, Organic Crystals, Carbon Nanotubes*, 2nd edition.
12. Dinh, H.N. 1998, *'Electrochemical and structural studies of polyaniline film growth and degradation at different substrate surfaces,'* Ph.D. Thesis, University of Calgary.
13. Kuran, W. 2001, '*Principles of Coordination Polymerization, Coordination Polymerization of Alkynes*,' Wiley, pp. 379.
14. Chiang, C.K, Fincher, C.R., Park, Y.W., Heeger, A.J., Shirakawa, H., Louis, E.J., Gau, S.C. & MacDiarmid, A.G. 1977, 'Electrical Conductivity in Doped Polyacetylene,' *Physical Review Letters*, vol. 39, no. 17, pp. 1098.
15. Neuendorf, A.J. 2003, '*High Pressure Synthesis of Conductive Polymers*,' Ph.D. Thesis, Faculty of Science, Griffith University.
16. Nardes, A.M. 2007, '*On the conductivity of PEDOT: PSS thin films*,' Ph.D. Thesis, Technische Universiteit Eindhoven, the Netherlands.
17. Kumar, D. & Sharma, R.C. 1998, 'Advances in Conductive Polymers,' *European polymer journal*, vol. 34, pp. 1053–1060.
18. Streetman, B. & Banerjee, S. 2006, *Solid State Electronic Devices*, Prentice Hall.
19. Heeger, A.J. 2000, 'Semiconducting and Metallic Polymers: The Fourth Generation of Polymeric Materials,' *Nobel Lecture, The Nobel Prize in Chemistry 2000.*
20. George, P.M. 2005, '*Novel Polypyrrole Derivatives to Enhance Conductive Polymer-Tissue Interactions*,' Ph.D. Thesis, Massachusetts Institute of Technology, USA.
21. Kim, B., Koncar, V. & Devaux, E. 2004, 'Electrical Properties of Conductive Polymers: PET –Nanocomposites' Fibres,' *AUTEX Research Journal*, vol. 4, no. 1.
22. McDaniels, D.K. 1991, *The Sun.*, Krieger Pub Co.

23. Kasap, S.O. 2005, *Principles of Electronic Materials and Devices*, McGraw Hill Higher Education.
24. Spanggaard, H. & Krebs, F.C. 2004, 'A brief history of the development of organic and polymeric photovoltaics,' *Sol. Energy Mater. Sol. Cells*, vol. 83, pp. 125.
25. Hoppe, H. 2004, '*Nanomorphology – Efficiency Relationship in Organic Bulk Heterojunction Plastic Solar Cells*,' Ph.D. Thesis, Kepler University Linz.
26. Mayer, A.C., Scully, S.R., Hardin, B.E., Rowell, M.W. & McGehee, M.D. 2007, 'Polymer Based Solar Cells,' *Materials Today*, vol. 10, no. 11.
27. Groenendaal, L., et al., *Poly(3,4-ethylenedioxythiophene) and Its Derivatives: Past, Present, and Future.* Advanced Materials, 2000. 12(7): p. 481–494.
28. Kirchmeyer, S. and K. Reuter, *Scientific importance, properties and growing applications of poly(3,4-ethylenedioxythiophene).* Journal of Materials Chemistry, 2005. 15(21): p. 2077–2088.
29. Jonas, F. and L. Schrader, *Conductive modifications of polymers with polypyrroles and polythiophenes.* Synthetic Metals, 1991. 41(3): p. 831–836.
30. Carpi, F. and D. Rossi, *Colours from electroactive polymers: electrochromic, electroluminescent and laser devices based on organic materials.* Optics Laser Technology, 2006. 38: p. 292–305
31. Dietrich, M., et al., *Electrochemical and spectroscopic characterization of polyalkylenedioxythiophenes.* Journal of Electroanalytical Chemistry, 1994. 369(1–2): p. 87–92.
32. Cui, X. and D.C. Martin, *Electrochemical deposition and characterization of poly(3,4-ethylenedioxythiophene) on neural microelectrode arrays.* Sensors and Actuators B: Chemical, 2003. 89(1–2): p. 92–102.
33. Bhattacharya, A. and A. De, *Conducting polymers in solution – Progress toward processibility.* Journal of Macromolecular Science – Reviews in Macromolecular Chemistry and Physics, 1999. 39(1): p. 17–56.
34. Laforgue, A. and L. Robitaille, *Production of Conductive PEDOT Nanofibers by the Combination of Electrospinning and Vapor-Phase Polymerization.* Macromolecules, 2010. 43(9): p. 4194–4200.
35. Pei, Q., et al., *Electrochromic and highly stable poly(3,4-ethylenedioxythiophene) switches between opaque blue-black and transparent sky blue.* Polymer, 1994. 35(7): p. 1347–1351.
36. Asplund, M., H. Holst, and O. Inganäs, *Composite biomolecule/PEDOT materials for neural electrodes.* Biointerphases, 2008. 3(3): p. 83–93.
37. Talo, A., et al., *Polyaniline/epoxy coatings with good anticorrosion properties.* Synthetic Metals, 1997. 85(1–3): p. 1333–1334.
38. Norstebo, C.A., *Intelligent Textiles, Soft Products.* Journal of Future Materials, 2003 p. 1–14.
39. Das, S., *Mobility and Resource Management in Smart Home Environments*, in *Embedded and Ubiquitous Computing*, L. Yang, et al., Editors. 2004, Springer Berlin Heidelberg. p. 1109–1111.
40. Mattmann, C., F. Clemens, and G. Tröster, *Sensor for Measuring Strain in Textile.* Sensors, 2008. 8(6): p. 3719–3732.
41. Carpi, F. and D.D. Rossi, *Electroactive Polymer-Based Devices for e-Textiles in Biomedicine.* IEEE Transactions on Information Technology in Biomedicine, 2005. 9(3): p. 295–318.

42. De Rossi, D., A. Della Santa, and A. Mazzoldi, *Dressware: wearable hardware.* Materials Science and Engineering: C, 1999. 7(1): p. 31–35.
43. Billinghurst, M. and T. Starner, *Wearable devices: new ways to manage information.* Computer, 1999. 32(1): p. 57–64.
44. Zieba, J. and M. Frydrysiak, *Textronics- Electrical and Electronic Textiles. Sensors for Breathing Frequency Measurement* Fibers & Textiles in Eastern Europe, 2006. 14(5, 59): p. 43–48.
45. Marculescu, D., et al., *Electronic textiles: A platform for pervasive computing.* Proceedings of the IEEE, 2003. 91(12): p. 1995–2018.
46. Singh, M.K., *The state-of-art Smart Textiles*, 2004, http://www.ptj.com.pk/Web%20 2004/08–2004/Smart%20Textiles.html.
47. Shahhaidar, E., *Decreasing Power Consumption of a Medical E-textile*, in *World Academy of Science, Engineering and Technology* 2008.
48. Hung, K., Y.T. Zhang, and B. Tai. *Wearable medical devices for tele-home healthcare.* in *Engineering in Medicine and Biology Society, 2004. IEMBS '04. 26th Annual International Conference of the IEEE.* 2004.
49. Akita, J., et al., *Wearable electromyography measurement system using cable-free network system on conductive fabric.* Artificial Intelligence in Medicine, 2008. 42(2): p. 99–108.
50. Huang, C.-T., et al., *A wearable yarn-based piezo-resistive sensor.* Sensors and Actuators A: Physical, 2008. 141(2): p. 396–403.
51. McCann, J., R. Hurford, and A. Martin. *A design process for the development of innovative smart clothing that addresses end-user needs from technical, functional, esthetic and cultural view points.* in *Wearable Computers, 2005. Proceedings. Ninth IEEE International Symposium on.* 2005.
52. Mondal, S., *Phase change materials for smart textiles – An overview.* Applied Thermal Engineering, 2008. 28(11–12): p. 1536–1550.
53. Fan, X., et al., *Wire-Shaped Flexible Dye-sensitized Solar Cells.* Advanced Materials, 2008. 20(3): p. 592–595.
54. Yajima, T., M.K. Yamada, and M.S. Tanaka, *Protection effects of a silver fiber textile against electromagnetic interference in patients with pacemakers.* Journal of Artifical Organics, 2002. 5: p. 175–178.
55. Yang, X., et al., *Vapor phase polymerization of 3,4-ethylenedioxythiophene on flexible substrate and its application on heat generation.* Polymers for Advanced Technologies, 2011. 22(6): p. 1049.
56. Xue, P., et al., *Electrically conductive yarns based on PVA/carbon nanotubes.* Composite Structures, 2007. 78(2): p. 271–277.
57. Dastjerdi, R., M. Montazer, *A review on the application of inorganic nanostructured materials in the modification of textiles: Focus on antimicrobial properties.* Colloids and Surfaces B: Biointerfaces, 2010. 79(1): p. 5–18.
58. Rothmaier, M., M.P. Luong, F. Clemens, *Textile Pressure Sensor Made of Flexible Plastic Optical Fibers.* Sensors, 2008. 8: p. 4318–4329.
59. Kincal, D., et al., *Conductivity switching in polypyrrole-coated textile fabrics as gas sensors.* Synthetic Metals, 1998. 92: p. 53–56.
60. Bessette, R.R., et al., *Development and characterization of a novel carbon fiber based cathode for semifuel cell applications.* Journal of Power Sources, 2001. 96(1): p.40- 244.

61. Meoli, D. and T. May-Plumlee, *Interactive Electronic Textile Development: A review of Technologies*. Journal of Textile and Apparel, Technology and Management, 2002. 2(2).
62. Mattila, H.R. 2006, *Intelligent Textiles and Clothing*, CRC Press.
63. Krebs, F.C., Biancardo, M., Winther-Jensen, B., Spanggrad, H. & Alstrup, J. 2005, 'Strategies for incorporation of polymer photovoltaics into garments and textiles,' *Solar Energy Materials & Solar Cells*, vol. 90, pp. 1058–1067.
64. Konarka Technologies, Inc., *Konarka Builds Power Plastic*, http://www.konarka.com/ technology/photovoltaics.php
65. Lee Y.-C., Chen S. M.-S., Lin Y.-L., Mason B.G., Novakovskaia E.A., Connell, V.R., *Integrated Flexible Solar Cell Material and Method of Production* (Sky Station International Inc.), US Patent 6,224,016, May 2001.
66. Iowa Thin Film Technologies, Inc., *Revolutionary Package of Proprietary Solar/Semiconductor Technologies*, http://www.iowathinfilm.com/technology/index.html
67. Mowbray J.L., 'Wider choice for narrow fabrics,' *Knitting International*, 2002, 109, 36–38.
68. Ellis J.G., 'Embroidery for engineering and surgery' in *Proceedings of the Textile Institute World Conference*, Manchester, 2000.
69. Karamuk E., Mayer J., Düring M., Wagner B., Bischoff B., Ferrario R., Billia M., Seidl R., Panizzon R., Wintermantel E., 'Embroidery technology for medical textiles' in *Medical Textiles*, editor Anand S., Woodhead Publishing Limited, Cambridge, 2001.
70. Goldsmith N., 'Solar tensile pavilion,' http://ndm.si.edu/EXHIBITIONS/sun/2/ obj_tent.mtm
71. Zabetakis A., Stamelaki A., Teloniati T., 'Solar textiles: Production and distribution of electricity coming from the solar radiation. Applications.' in *Proceedings of Conference on Fibrous Assemblies at the Design and Engineering Interface*, INTEDEC 2003, Edinburgh, 2003.
72. Muller, H.-F., 'Sailcloth arrangement for sails of water-going vessels,' US Patent Office, 6237521, May 2001.
73. Martin T., 'Sun shade canopy has solar energy cells within woven textile material that is wound onto a roller' (Warema Renkhoff GmbH), German Patent Office, DE10134314, January 2003.
74. Meoli, D., T. May-Plumlee, *Interactive Electronic Textile Development: A Review of Technologies*. Journal of Textile and Apparel, Technology and Management, 2002. 2(2).
75. Tibtech. *conductive yarns and fabrics for energy transfer and heating devices in SMART textiles and composites*. http://www.tibtech.com/metal_fiber_composition.php.
76. Skrifvars, M. and A. Soroudi, *Melt spinning of carbon nanotube modified polypropylene for electrically conducting nanocomposite fibers*. Solid State Phenomena, 2008. 151: p. 43–47.
77. Hegde, R.R., A. Dahiya, and M.G. Kamath. *Carbon Fibers*. 2004 http://web.utk.edu/~mse/Textiles/CARBON%20FIBERS.htm.

78. Dalmas, F., et al., *Carbon nanotube-filled polymer composites. Numerical simulation of electrical conductivity in three-dimensional entangled fibrous networks*. Acta Materialia, 2006. 54(11): p. 2923–2931.
79. Hunt, M.A., et al., *Patterned Functional Carbon Fibers from Polyethylene*. Advanced Materials, 2012. 24(18): p. 2386–2389.
80. Chung, D.D.L., ed. *Carbon Fiber Composites*. 1994, Elsevier. 227.
81. Pomfret, S.J., et al., *Electrical and mechanical properties of polyaniline fibers produced by a one-step wet spinning process*. Polymer, 2000. 41(6): p. 2265–2269.
82. Okuzaki, H., Y. Harashina, H. Yan, *Highly conductive PEDOT/PSS microfibers fabricated by wet-spinning and dip-treatment in ethylene glycol*. European Polymer Journal, 2009. 45(1): p. 256–261.
83. Woonphil, B., et al., *Synthesis of highly conductive poly (3,4-ethylenedioxythiophene) fiber by simple chemical polymerization*. Synthetic Metals, 2009. 159(13): p. 1244- 1246.
84. Saleem, A., L. Frormann, A. Iqbal, *Mechanical, Thermal and Electrical Resisitivity Properties of Thermoplastic Composites Filled with Carbon Fibers and Carbon Particles*. Journal of Polymer Research, 2007. 14(2): p. 121–127.
85. Bigg, D.M., *Mechanical and conductive properties of metal fiber-filled polymer composites*. Composites, 1979. 10(2): p. 95–100.
86. Show, Y. and H. Itabashi, *Electrically conductive material made from CNT and PTFE*. Diamond and Related Materials, 2008. 17(4–5): p. 602–605.
87. Shirakawa, H., et al., *Synthesis of electrically conducting organic polymers: Halogen derivatives of polyacetylene, (CH)*. Journal of the Chemical Society, Chemical Communications, 1977: p. 578–580.
88. Bigg, D.M., *Conductive polymeric compositions*. Polymer Engineering & Science, 1977. 17(12): p. 842–847.
89. Zabetakis, D., M. Dinderman, P. Schoen, *Metal-Coated Cellulose Fibers for Use in Composites Applicable to Microwave Technology*. Advanced Materials, 2005. 17(6): p. 734–738.
90. Xue, P. and X.M. Tao, *Morphological and Electromechanical Studies of Fibers Coated with Electrically Conductive Polymer*. Journal of Applied Polymer Science, 2005. 98: p. 1844–1854.
91. Negru, D., C.-T. Buda, and D. Avram, *Electrical Conductivity of Woven Fabrics Coated with Carbon Black Particles*. Fibers & Textiles in Eastern Europe, 2012. 20(1, 90)): p. 53–56.
92. Sen, A.k., *Coated Textiles Principles and Applications* J. Damewood, Editor 2001, Technomic Publishing Company, Inc.
93. Hohnholz, D., et al., *Uniform thin films of poly 3,4-ethylenedioxythiophene (PEDOT) prepared by in-situ deposition*. Chemical Communications, 2001(23): p. 2444–2445.
94. Rozek, Z., et al., *Potential applications of nanofiber textile covered by carbon coatings*. Journal of Achievements in Materials and Manufacturing Engineering, 2008. 27(1): p. 35–38.
95. Tao, X.M., *Smart Fibers, Fabrics and Clothing*. The Textile Institute, England, 2001.
96. Kaynak, A., et al., *Characterization of conductive polypyrrole coated wool yarns*. Fibers and Polymers, 2002. 3(1): p. 24–30.

97. Jönsson, S.K.M., et al., *The effects of solvents on the morphology and sheet resistance in poly(3,4-ethylenedioxythiophene)-polystyrene sulfonic acid (PEDOT-PSS) films.* Synthetic Metals, 2003. 139(1): p. 1–10.
98. Rajahn M., Rakhlin M., Schubert M B "Amorphous and heterogeneous silicone based films" MRS Proc. 664, 2001
99. Schubert M.B., Werner J. H., Mater. Today 9(42), 2006
100. Drew C., Wang X. Y., Senecal K., J. Macro. Mol. Sci Pure A 39, 2002, 1085
101. Baps B., Eder K. M., Konjuncu M., Key Eng. Mater. 206–213, 2002, 937
102. Gratzel M., Prog. Photovolt: Res. Appl. 8, 2000, 171
103. Verdenelli M., Parole S., Chassagneux F., Lettof J. M., Vincent H. Scharff J. P., J of Eur. Ceram. Soc. 23, 2003, 1207
104. Xie C., Tong W., Acta Mater, 53, 2005, 477
105. Muller D., Fromm E. Thin Mater. Solid Films 270, 1995, 411
106. Hu M. S. Evans A. G., Acta Mater, 37, 1998, 917
107. Yang Q. D., Thouless M. D., Ward S. M., J of Mech Phys Solids, 47, 1999, 1337
108. Agrawal D. C., Raj R., Acta Mater, 37, 1989, 1265
109. Rochal G., Leterrier Y., Fayet P., Manson J. Ae, Thin Solid Films 437, 2003, 204
110. Park S. H., Choi H. J., Lee S. B., Lee S. M. L., Cho S. E., Kim K. H., Kim Y. K., Kim M. R., Lee J. K. "Fabrications and photovoltaic properties of dye-sensitized solar cells with electrospun poly(vinyl alcohol) nanofibers containing Ag nanoparticles" Macromolecular Research 19(2), 2011, 142–146
111. Ramiera J., Plummera C.J.G., Leterriera Y., Mansona J.-A.E.,_, Eckertb B., and Gaudianab R. "Mechanical integrity of dye-sensitized photovoltaic fibers" Renewable Energy 33 (2008), 314–319
112. Hamedi M., Forchheimer R., Inganas O. Nat Mat. 6, 2007, 357
113. Bayindir M., Sorin F., Abouraddy A. F., Viens J., Hart S. D., Joannopoulus J. D., Fink Y. Nature 431, 2004, 826
114. Yadav A., Schtein M., Pipe K. P. J. of Power Sources 175, 2008, 909
115. Coner O., Pipe K. P. Shtein M. Fiber based organic PV devices Appl. Phys. Letters 92, 2008, 193306
116. Ghas A. P., Gerenser L. J., Jarman C. M., Pornailik J. E., Appl. Phys. Lett. 86, 2005, 223503
117. Wang P., Zakeeruddin S M, Gratzel M and Fluorine J Chem. 125, 2004, 1241
118. Yu J. W., Chin B. D., Kim J. K., Kang N. S. Patent IPC8 Class: AH01L3100FI, USPC Class: 136259, 2007
119. http://www.coatema.de/ger/downloads/veroeffentlichungen/news/0701_Textile%20 Month_GB.pdf
120. Krebs F. C., Fyenbo J., Jørgensen Mikkel "Product integration of compact roll-to-roll processed polymer solar cell modules: methods and manufacture using flexographic printing, slot-die coating and rotary screen printing" *J. Mater. Chem.*, 20, 2010, 8994–9001
121. Krebs F. C., Polymer solar cell modules prepared using roll-to-roll methods: Knife-overedge coating, slot-die coating and screen printing Solar Energy Materials and Solar Cells 93, (4), 2009, 465–475
122. Luo P., Zhu C., Jiang G. Preparation of CuInSe2 thin films by pulsed laser deposition the Cu–In alloy precursor and vacuum selenization, Solid State Communications, 146, (1–2), 2008, 57–60

123. Solar energy: the state of art Ed: Gordon J Pub: James and James Willium Road London, 2001
124. Loewenstein T., Hastall A., Mingebach M., Zimmermann Y., Neudeck A. and Schlettwein D., *Phys. Chem. Chem. Phys.*, 2008, 10, 1844.
125. Lincot D., Peulon S., *J. Electrochem. Soc.*, 1998, 145, 864.
126. Bedeloglua A., Demirb A., Bozkurta Y., Sariciftci N. S., Synthetic Metals 159 (2009) 2043–2048
127. Pettersson L.A.A., Ghosh S., Inganas O. Org Electron 2002; 3: 143.
128. Kim J.Y., Jung J.H., Lee D.E., Joo J. Synth Met 2002; 126: 311.
129. Kim W.H., Makinen A.J., Nikolov N., Shashidhar R., Kim H., Kafafi Z.H. Appl Phys Lett 80, (2002), 3844.
130. Jonsson S.K.M., Birgerson J., Crispin X., Greczynski G., Osikowicz W., van der Gon A.W.D., Salaneck W.R., Fahlman M. Synth. Met 1, 2003;139:
131. Ouyang J., Xu Q., Chu C. W., Yang Y., Lib G., Shinar Joseph S. On the mechanism of conductivity enhancement in poly(3,4-ethylenedioxythiophene): poly(styrene sulfonate) film through solvent treatment" Polymer 45, 2004, 8443–8450
132. Durisch W., Urban J., Smestad G. "Characterization of solar cells and modules under actual operating conditions" WERS 1996, 359–366
133. Wang W., Xia G., Zheng J., Feng L., Hao R. "Study of polycrystalline ZnTe(ZnTe: Cu) thin films for photovoltaic cells" Journal of Materials Science: Materials in Electronics 18(4), 2007, 427–431
134. Khanna, R.K. "Raman-spectroscopy of oligomeric SiO species isolated in solid methane."Journal of Chemical Physics 74 (4), (1981) 2108
135. Koch, C., Ito M., Schubert M., 'Low temperature deposition of amorphous silicon solar cells,' *Solar Energy Materials and Solar Cells*, 68, 2001, 227–236.
136. Aernouts, T. 19th European Photovoltaics Conference, June 7–11, Paris, France, 2004.
137. Lund P. D. Renewable energy 34, 2009, 53
138. Yaksel I. Renewable energy 4, 2008, 802
139. Gunes S., Beugebauer H., Saricftci N. S. Chem Rev. 107, 2007, 1324
140. Coakley KM,cGehee M D, Chem Mater. 16, 2004, 4533
141. Ji J.-Y., Shen T.-C., 'Low-temperature silicon epitaxy on hydrogen-terminated Si surfaces,' *Physical Review B* 70, 2004, 115–309.
142. Moncrieff R.W., *Man-made Fibres*, Newnes-Butterworths, 6th edn, London and Boston, 1975.
143. Hearle J.W.S., 'Introduction' in *High-performance Fibres*,' editor Hearle J.W.S., Woodhead Publishing Limited, Cambridge, 2001, pp. 1–22.
144. Jones F.R., 'Glass fiber' in *High-performance Fibres*, editor Hearle J.W.S., Woodhead Publishing Limited, Cambridge, 2001, pp. 191–238.
145. Abdelfattah M.S., 'Formation of textile structures for giant-area applications' in *Electronics on Unconventional Substrates – Electrotextiles and Giant-Area Flexible Circuits*, MRS Symposium proceedings Volume 736, Materials Processing Society, Warrendale, Pennsylvania, USA, 2003, pp. 25–36.
146. Bonderover E., Wagner S., Suo Z., 'Amorphous silicon thin transistors on kapton fibers,' in *Electronics on Unconventional Substrates – Electrotextiles and Giant-Area Flexible Circuits*, MRS Symposium Proceedings Volume 736, Materials Processing Society, Warrendale, Pennsylvania, USA, 2003, pp. 109–114.

CHAPTER 13

GEOMETRIC AND ELECTRONIC STRUCTURE OF THE MODELS OF DEKACENE AND EICOCENE WITHIN THE FRAMEWORK OF MOLECULAR GRAPHENE MODEL

V. A. BABKIN,[1] V. V. TRIFONOV,[1] V. YU. DMITRIEV,[1] D. S. ANDREEV,[1] A. V. IGNATOV,[1] E. S. TITOVA,[2] O. V. STOYANOV,[3] and G. E. ZAIKOV[4]

[1]*Volgograd State Architect-Build University Sebryakov Department, Russia,* [2]*Volgograd State Technical University, Russia,* [3]*Kazan State Technological University, Russia,* [4]*Institute of Biochemical Physics, Russian Academy of Sciences, Russian Federation*

CONTENTS

ABSTRACT

Quantum-chemical calculation of molecules dekacene, eicocene was done by method MNDO. Optimized by all parameters geometric and electronic structures of these compounds was received. Each of these molecular models has a universal factor of acidity equal to 33 (pKa=33). They all pertain to class of very weak H-acids (pKa>14).

13.1 AIMS AND BACKGROUNDS

The aim of this work is a study of electronic structure of molecules dekacene, eicocene and theoretical estimation its acid power by quantum-chemical method MNDO within the framework of molecular graphene model, which was discovered by Novoselov and Game in 2004 [1].

13.2 METHODICAL PART

The calculation was done with optimization of all parameters by standard gradient method built-in in PC GAMESS [2]. The calculation was executed in approach the insulated molecule in gas phase. Program MacMolPlt was used for visual presentation of the model of the molecule. [3].

13.3 THE RESULTS OF THE CALCULATION AND DISCUSSION

Geometric and electronic structures, general and electronic energies of molecules dekacene, eicocene was received by method MNDO and are shown on Figs. 13.1 and 13.2 and in Tables 13.1–13.3. The universal factor of acidity was calculated by formula: pKa = 49.4–134.61*$q_{max}H^+$ [4] (where, $q_{max}H^+$ – a maximum positive charge on atom of the hydrogen (by Milliken [1]) R=0.97, R– a coefficient of correlations, $q_{max}{}^{H-}$=+0.06). pKa=33. This formula was successfully used in the following articles: [5–7].

Quantum-chemical calculation of molecules dekacene, eicocene by method MNDO was executed for the first time. Optimized geometric and electronic structures of these compound was received. Acid power of molecules dekacene, eicocene was theoretically evaluated (pKa=33). These compounds pertain to class of very weak H-acids (pKa>14).

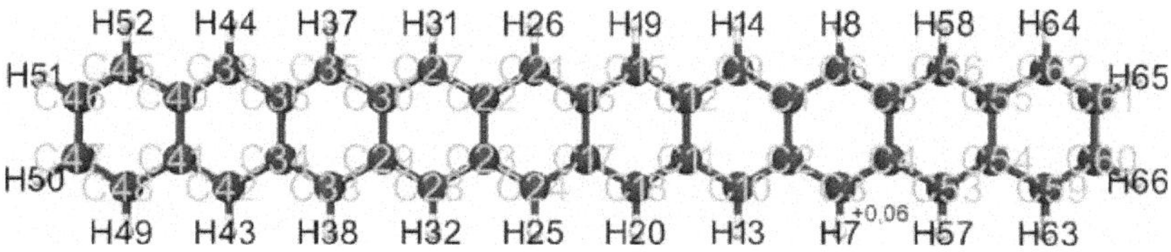

FIGURE 13.1 Geometric and electronic molecular structure of dekacene. (E_0= −550105 kDg/mol, E_{el}= −4850841 kDg/mol).

TABLE 13.1 Optimized Bond Lengths, Valent Corners and Atom Charges of Dekacene

Bond lengths	R,A	Valent corners	Grad	Atom	Charge (by Milliken)
C(1)-C(2)	1.46	C(3)-C(2)-C(1)	118	C(1)	–0.04
C(2)-C(3)	1.45	C(12)-C(9)-C(1)	122	C(2)	–0.04
C(3)-C(4)	1.38	C(4)-C(3)-C(2)	123	C(3)	–0.02
C(4)-C(5)	1.47	C(9)-C(1)-C(2)	119	C(4)	–0.06
C(5)-C(6)	1.38	C(5)-C(4)-C(3)	119	C(5)	–0.06
C(6)-C(1)	1.45	C(10)-C(2)-C(3)	123	C(6)	–0.02
H(7)-C(3)	1.09	C(6)-C(5)-C(4)	119	H(7)	0.06
H(8)-C(6)	1.09	C(54)-C(53)-C(4)	123	H(8)	0.06
C(9)-C(12)	1.44	C(1)-C(6)-C(5)	123	C(9)	–0.02
C(9)-C(1)	1.39	C(53)-C(4)-C(5)	118	C(10)	–0.02
C(10)-C(2)	1.39	C(2)-C(1)-C(6)	118	C(11)	–0.04
C(11)-C(10)	1.44	C(9)-C(1)-C(6)	123	C(12)	–0.04
C(12)-C(11)	1.46	C(4)-C(3)-H(7)	120	H(13)	0.06
H(13)-C(10)	1.38	C(1)-C(6)-H(8)	117	H(14)	0.06
H(14)-C(9)	1.45	C(11)-C(12)-C(9)	118	C(15)	–0.02
C(15)-C(16)	1.42	C(15)-C(12)-C(9)	123	C(16)	–0.04
C(15)-C(12)	1.41	C(1)-C(2)-C(10)	119	C(17)	–0.04
C(16)-C(17)	1.45	C(18)-C(11)-C(10)	123	C(18)	–0.02
C(17)-C(18)	1.42	C(2)-C(10)-C(11)	122	H(19)	0.06
C(18)-C(11)	1.41	C(15)-C(12)-C(11)	119	H(20)	0.06
H(19)-C(15)	1.09	C(10)-C(11)-C(12)	118	C(21)	–0.02

TABLE 13.1 (Continued)

Bond lengths	R,A	Valent corners	Grad	Atom	Charge (by Milliken)
H(20)-C(18)	1.09	C(16)-C(15)-C(12)	122	C(22)	−0.04
C(21)-C(22)	1.41	C(2)-C(10)-H(13)	120	C(23)	−0.04
C(21)-C(16)	1.42	C(12)-C(9)-H(14)	118	C(24)	−0.02
C(22)-C(23)	1.46	C(17)-C(16)-C(15)	119	H(25)	0.06
C(23)-C(24)	1.40	C(21)-C(16)-C(15)	123	H(26)	0.06
C(24)-C(17)	1.42	C(18)-C(17)-C(16)	119	C(27)	−0.02
H(25)-C(24)	1.09	C(22)-C(21)-C(16)	122	C(28)	−0.02
H(26)-C(21)	1.09	C(11)-C(18)-C(17)	122	C(29)	−0.04
C(27)-C(30)	1.39	C(21)-C(16)-C(17)	119	C(30)	−0.04
C(27)-C(22)	1.44	C(12)-C(11)-C(18)	119	H(31)	0.06
C(28)-C(23)	1.44	C(24)-C(17)-C(18)	123	H(32)	0.06
C(29)-C(28)	1.39	C(16)-C(15)-H(19)	118	C(33)	−0.02
C(30)-C(29)	1.46	C(11)-C(18)-H(20)	119	C(34)	−0.04
H(31)-C(27)	1.09	C(23)-C(22)-C(21)	119	C(35)	−0.02
H(32)-C(28)	1.09	C(27)-C(22)-C(21)	123	C(36)	−0.04
C(33)-C(29)	1.45	C(24)-C(23)-C(22)	119	H(37)	0.06
C(34)-C(33)	1.38	C(30)-C(27)-C(22)	122	H(38)	0.06
C(35)-C(36)	1.38	C(17)-C(24)-C(23)	122	C(39)	−0.02
C(35)-C(30)	1.45	C(27)-C(22)-C(23)	118	C(40)	−0.04
C(36)-C(34)	1.47	C(16)-C(17)-C(24)	119	C(41)	−0.04
H(37)-C(35)	1.09	C(28)-C(23)-C(24)	123	C(42)	−0.02
H(38)-C(33)	1.09	C(17)-C(24)-H(25)	119	H(43)	0.06
C(39)-C(40)	1.38	C(22)-C(21)-H(26)	119	H(44)	0.06
C(39)-C(36)	1.46	C(29)-C(30)-C(27)	119	C(45)	−0.04
C(40)-C(41)	1.48	C(35)-C(30)-C(27)	123	C(46)	−0.06
C(41)-C(42)	1.38	C(22)-C(23)-C(28)	118	C(47)	−0.06
C(42)-C(34)	1.46	C(33)-C(29)-C(28)	123	C(48)	−0.04
H(43)-C(42)	1.09	C(23)-C(28)-C(29)	122	H(49)	0.06
H(44)-C(39)	1.09	C(35)-C(30)-C(29)	118	H(50)	0.06
C(45)-C(46)	1.36	C(28)-C(29)-C(30)	119	H(51)	0.06
C(45)-C(40)	1.47	C(36)-C(35)-C(30)	124	H(52)	0.06
C(46)-C(47)	1.45	C(30)-C(27)-H(31)	120	C(53)	−0.02
C(47)-C(48)	1.36	C(23)-C(28)-H(32)	118	C(54)	−0.04

TABLE 13.1 (Continued)

Bond lengths	R,A	Valent corners	Grad	Atom	Charge (by Milliken)
C(48)-C(41)	1.47	C(30)-C(29)-C(33)	118	C(55)	−0.04
H(49)-C(48)	1.09	C(42)-C(34)-C(33)	123	C(56)	−0.02
H(50)-C(47)	1.09	C(29)-C(33)-C(34)	123	H(57)	0.06
H(51)-C(46)	1.09	C(39)-C(36)-C(34)	118	H(58)	0.06
H(52)-C(45)	1.09	C(34)-C(36)-C(35)	119	C(59)	−0.04
C(53)-C(54)	1.38	C(39)-C(36)-C(35)	123	C(60)	−0.06
C(53)-C(4)	1.46	C(33)-C(34)-C(36)	119	C(61)	−0.06
C(54)-C(55)	1.48	C(40)-C(39)-C(36)	123	C(62)	−0.04
C(55)-C(56)	1.38	C(36)-C(35)-H(37)	120	H(63)	0.06
C(56)-C(5)	1.46	C(29)-C(33)-H(38)	117	H(64)	0.06
H(57)-C(53)	1.09	C(41)-C(40)-C(39)	119	H(65)	0.06
H(58)-C(56)	1.09	C(45)-C(40)-C(39)	123	H(66)	0.06
C(59)-C(60)	1.36	C(42)-C(41)-C(40)	119		
C(59)-C(54)	1.47	C(46)-C(45)-C(40)	122		
C(60)-C(61)	1.45	C(34)-C(42)-C(41)	123		
C(61)-C(62)	1.36	C(45)-C(40)-C(41)	118		
C(62)-C(55)	1.47	C(36)-C(34)-C(42)	118		
H(63)-C(59)	1.09	C(48)-C(41)-C(42)	123		
H(64)-C(62)	1.09	C(34)-C(42)-H(43)	117		
H(65)-C(61)	1.09	C(40)-C(39)-H(44)	120		
H(66)-C(60)	1.09	C(47)-C(46)-C(45)	121		
		C(48)-C(47)-C(46)	121		
		C(41)-C(48)-C(47)	122		
		C(40)-C(41)-C(48)	118		
		C(41)-C(48)-H(49)	118		
		C(48)-C(47)-H(50)	121		
		C(47)-C(46)-H(51)	118		
		C(46)-C(45)-H(52)	120		
		C(55)-C(54)-C(53)	119		
		C(3)-C(4)-C(53)	123		
		C(56)-C(55)-C(54)	119		
		C(60)-C(59)-C(54)	122		
		C(5)-C(56)-C(55)	123		

TABLE 13.1 (Continued)

Bond lengths	R,A	Valent corners	Grad	Atom	Charge (by Milliken)
		C(59)-C(54)-C(55)	118		
		C(4)-C(5)-C(56)	118		
		C(6)-C(5)-C(56)	123		
		C(54)-C(53)-H(57)	120		
		C(5)-C(56)-H(58)	117		
		C(61)-C(60)-C(59)	121		
		C(53)-C(54)-C(59)	123		
		C(62)-C(61)-C(60)	121		
		C(55)-C(62)-C(61)	122		
		C(54)-C(55)-C(62)	118		
		C(56)-C(55)-C(62)	123		
		C(60)-C(59)-H(63)	120		
		C(55)-C(62)-H(64)	118		
		C(62)-C(61)-H(65)	121		
		C(61)-C(60)-H(66)	118		

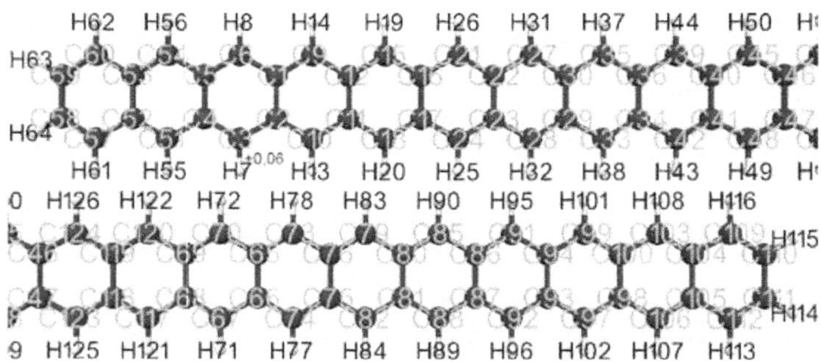

FIGURE 13.2 Geometric and electronic molecular structure of eicocene. (E_0= −1069853 kDg/mol, E_{el}= −11719827 kDg/mol).

TABLE 13.2 Optimized Bond Lengths, Valent Corners and Atom Charges of Eicocene

Bond lengths	R,A	Valence corners	Grad	Atom	Charge (by Milliken)
C(2)-C(1)	1.46	C(5)-C(6)-C(1)	122	C(1)	–0.03
C(3)-C(2)	1.44	C(11)-C(10)-C(2)	122	C(2)	–0.04

TABLE 13.2 (Continued)

Bond lengths	R,A	Valence corners	Grad	Atom	Charge (by Milliken)
C(4)-C(3)	1.39	C(1)-C(2)-C(3)	118	C(3)	–0.02
C(4)-C(5)	1.47	C(10)-C(2)-C(3)	123	C(4)	–0.04
C(5)-C(54)	1.46	C(5)-C(4)-C(3)	119	C(5)	–0.04
C(6)-C(5)	1.39	C(2)-C(3)-C(4)	122	C(6)	–0.02
C(6)-C(1)	1.44	C(54)-C(5)-C(4)	118	H(7)	+0.06
H(7)-C(3)	1.09	C(6)-C(5)-C(4)	119	H(8)	+0.06
H(8)-C(6)	1.09	C(53)-C(54)-C(5)	123	C(9)	–0.02
C(9)-C(1)	1.40	C(54)-C(5)-C(6)	123	C(10)	–0.02
C(10)-C(2)	1.40	C(2)-C(1)-C(6)	118	C(11)	–0.03
C(10)-C(11)	1.43	C(2)-C(3)-H(7)	118	C(12)	–0.03
C(11)-C(18)	1.42	C(5)-C(6)-H(8)	120	H(13)	+0.06
C(11)-C(12)	1.45	C(1)-C(6)-H(8)	118	H(14)	+0.06
C(12)-C(9)	1.43	C(2)-C(1)-C(9)	119	C(15)	–0.02
H(13)-C(10)	1.09	C(1)-C(2)-C(10)	119	C(16)	–0.04
H(14)-C(9)	1.09	C(18)-C(11)-C(10)	123	C(17)	–0.04
C(15)-C(12)	1.42	C(12)-C(11)-C(10)	119	C(18)	–0.02
C(16)-C(15)	1.41	C(17)-C(18)-C(11)	122	H(19)	+0.06
C(16)-C(17)	1.45	C(9)-C(12)-C(11)	119	H(20)	+0.06
C(17)-C(24)	1.44	C(15)-C(12)-C(11)	119	C(21)	–0.02
C(18)-C(17)	1.41	C(1)-C(9)-C(12)	122	C(22)	–0.04
H(19)-C(15)	1.09	C(18)-C(11)-C(12)	119	C(23)	–0.04
H(20)-C(18)	1.09	C(2)-C(10)-H(13)	119	C(24)	–0.02
C(21)-C(16)	1.44	C(11)-C(10)-H(13)	118	H(25)	+0.06
C(22)-C(21)	1.39	C(1)-C(9)-H(14)	119	H(26)	+0.06
C(23)-C(22)	1.46	C(9)-C(12)-C(15)	123	C(27)	–0.02
C(24)-C(23)	1.39	C(17)-C(16)-C(15)	119	C(28)	–0.02
H(25)-C(24)	1.09	C(12)-C(15)-C(16)	122	C(29)	–0.04
H(26)-C(21)	1.09	C(24)-C(17)-C(16)	118	C(30)	–0.04
C(27)-C(22)	1.45	C(18)-C(17)-C(16)	119	H(31)	+0.06
C(28)-C(23)	1.45	C(23)-C(24)-C(17)	122	H(32)	+0.06
C(29)-C(28)	1.38	C(24)-C(17)-C(18)	123	C(33)	–0.02
C(29)-C(30)	1.47	C(12)-C(15)-H(19)	119	C(34)	–0.04

TABLE 13.2 (Continued)

Bond lengths	R,A	Valence corners	Grad	Atom	Charge (by Milliken)
C(30)-C(27)	1.38	C(17)-C(18)-H(20)	119	C(35)	–0.02
H(31)-C(27)	1.09	C(15)-C(16)-C(21)	123	C(36)	–0.04
H(32)-C(28)	1.09	C(17)-C(16)-C(21)	118	H(37)	+0.06
C(33)-C(29)	1.46	C(16)-C(21)-C(22)	122	H(38)	+0.06
C(34)-C(33)	1.38	C(21)-C(22)-C(23)	119	C(39)	–0.02
C(34)-C(36)	1.48	C(27)-C(22)-C(23)	118	C(40)	–0.04
C(34)-C(42)	1.46	C(22)-C(23)-C(24)	119	C(41)	–0.04
C(35)-C(30)	1.46	C(28)-C(23)-C(24)	123	C(42)	–0.02
C(36)-C(35)	1.38	C(23)-C(24)-H(25)	120	H(43)	+0.06
H(37)-C(35)	1.09	C(16)-C(21)-H(26)	118	H(44)	+0.06
H(38)-C(33)	1.09	C(21)-C(22)-C(27)	123	C(45)	–0.02
C(39)-C(36)	1.46	C(22)-C(23)-C(28)	118	C(46)	–0.03
C(40)-C(39)	1.38	C(30)-C(29)-C(28)	119	C(47)	–0.03
C(40)-C(41)	1.48	C(23)-C(28)-C(29)	123	C(48)	–0.02
C(41)-C(48)	1.45	C(27)-C(30)-C(29)	119	H(49)	+0.05
C(42)-C(41)	1.38	C(35)-C(30)-C(29)	118	H(50)	+0.05
H(43)-C(42)	1.09	C(22)-C(27)-C(30)	123	C(51)	–0.02
H(44)-C(39)	1.09	C(22)-C(27)-H(31)	117	C(52)	–0.04
C(45)-C(40)	1.45	C(23)-C(28)-H(32)	117	C(53)	–0.04
C(46)-C(45)	1.39	C(28)-C(29)-C(33)	123	C(54)	–0.02
C(47)-C(46)	1.49	C(30)-C(29)-C(33)	118	H(55)	+0.06
C(48)-C(47)	1.39	C(36)-C(34)-C(33)	120	H(56)	+0.06
H(49)-C(48)	1.09	C(42)-C(34)-C(33)	123	C(57)	–0.04
H(50)-C(45)	1.09	C(29)-C(33)-C(34)	123	C(58)	–0.06
C(51)-C(4)	1.46	C(35)-C(36)-C(34)	120	C(59)	–0.06
C(52)-C(51)	1.38	C(41)-C(42)-C(34)	123	C(60)	–0.04
C(52)-C(53)	1.48	C(39)-C(36)-C(34)	118	H(61)	+0.06
C(53)-C(60)	1.47	C(27)-C(30)-C(35)	123	H(62)	+0.06
C(54)-C(53)	1.38	C(30)-C(35)-C(36)	123	H(63)	+0.06
H(55)-C(51)	1.09	C(42)-C(34)-C(36)	118	H(64)	+0.06
H(56)-C(54)	1.09	C(30)-C(35)-H(37)	117	C(65)	–0.04
C(57)-C(52)	1.47	C(29)-C(33)-H(38)	117	C(66)	–0.04

TABLE 13.2 (Continued)

Bond lengths	R,A	Valence corners	Grad	Atom	Charge (by Milliken)
C(58)-C(57)	1.36	C(35)-C(36)-C(39)	123	C(67)	–0.02
C(59)-C(58)	1.45	C(41)-C(40)-C(39)	119	C(68)	–0.04
C(60)-C(59)	1.36	C(36)-C(39)-C(40)	123	C(69)	–0.04
H(61)-C(57)	1.09	C(48)-C(41)-C(40)	118	C(70)	–0.02
H(62)-C(60)	1.09	C(42)-C(41)-C(40)	119	H(71)	+0.06
H(63)-C(59)	1.09	C(47)-C(48)-C(41)	124	H(72)	+0.06
H(64)-C(58)	1.09	C(48)-C(41)-C(42)	123	C(73)	–0.02
C(65)-C(70)	1.46	C(41)-C(42)-H(43)	120	C(74)	–0.02
C(66)-C(65)	1.47	C(36)-C(39)-H(44)	117	C(75)	–0.04
C(67)-C(66)	1.46	C(39)-C(40)-C(45)	123	C(76)	–0.04
C(68)-C(67)	1.37	C(41)-C(40)-C(45)	118	H(77)	+0.06
C(68)-C(69)	1.48	C(40)-C(45)-C(46)	124	H(78)	+0.06
C(69)-C(120)	1.46	C(45)-C(46)-C(47)	119	C(79)	–0.02
C(70)-C(69)	1.38	C(124)-C(46)-C(47)	118	C(80)	–0.04
H(71)-C(67)	1.09	C(46)-C(47)-C(48)	119	C(81)	–0.04
H(72)-C(70)	1.09	C(123)-C(47)-C(48)	123	C(82)	–0.02
C(73)-C(65)	1.38	C(47)-C(48)-H(49)	119	H(83)	+0.06
C(74)-C(66)	1.38	C(40)-C(45)-H(50)	117	H(84)	+0.06
C(75)-C(74)	1.46	C(3)-C(4)-C(51)	123	C(85)	–0.02
C(75)-C(76)	1.47	C(5)-C(4)-C(51)	118	C(86)	–0.03
C(76)-C(73)	1.46	C(53)-C(52)-C(51)	119	C(87)	–0.04
H(77)-C(74)	1.09	C(4)-C(51)-C(52)	123	C(88)	–0.02
H(78)-C(73)	1.09	C(60)-C(53)-C(52)	118	H(89)	+0.06
C(79)-C(76)	1.39	C(54)-C(53)-C(52)	119	H(90)	+0.06
C(80)-C(79)	1.45	C(59)-C(60)-C(53)	122	C(91)	–0.02
C(80)-C(81)	1.46	C(60)-C(53)-C(54)	123	C(92)	–0.02
C(81)-C(82)	1.45	C(4)-C(51)-H(55)	117	C(93)	–0.04
C(82)-C(75)	1.39	C(53)-C(54)-H(56)	120	C(94)	–0.04
H(83)-C(79)	1.09	C(51)-C(52)-C(57)	123	H(95)	+0.06
H(84)-C(82)	1.09	C(53)-C(52)-C(57)	118	H(96)	+0.06
C(85)-C(80)	1.40	C(52)-C(57)-C(58)	122	C(97)	–0.02
C(86)-C(85)	1.43	C(57)-C(58)-C(59)	121	C(98)	–0.04

TABLE 13.2 (Continued)

Bond lengths	R,A	Valence corners	Grad	Atom	Charge (by Milliken)
C(86)-C(87)	1.45	C(58)-C(59)-C(60)	121	C(99)	–0.02
C(87)-C(88)	1.43	C(52)-C(57)-H(61)	118	C(100)	–0.04
C(88)-C(81)	1.40	C(59)-C(60)-H(62)	120	H(101)	+0.06
H(89)-C(88)	1.09	C(58)-C(59)-H(63)	118	H(102)	+0.06
H(90)-C(85)	1.09	C(57)-C(58)-H(64)	121	C(103)	–0.02
C(91)-C(86)	1.42	C(69)-C(70)-C(65)	123	C(104)	–0.04
C(92)-C(87)	1.42	C(70)-C(65)-C(66)	118	C(105)	–0.04
C(93)-C(92)	1.41	C(73)-C(65)-C(66)	119	C(106)	–0.02
C(93)-C(94)	1.45	C(65)-C(66)-C(67)	118	H(107)	+0.06
C(94)-C(91)	1.41	C(74)-C(66)-C(67)	123	H(108)	+0.06
H(95)-C(91)	1.09	C(69)-C(68)-C(67)	120	C(109)	–0.04
H(96)-C(92)	1.09	C(66)-C(67)-C(68)	123	C(110)	–0.06
C(97)-C(93)	1.44	C(120)-C(69)-C(68)	118	C(111)	–0.06
C(98)-C(97)	1.39	C(70)-C(69)-C(68)	120	C(112)	–0.04
C(98)-C(100)	1.46	C(119)-C(120)-C(69)	123	H(113)	+0.06
C(98)-C(106)	1.45	C(120)-C(69)-C(70)	123	H(114)	+0.06
C(99)-C(94)	1.44	C(66)-C(67)-H(71)	117	H(115)	+0.06
C(100)-C(99)	1.39	C(69)-C(70)-H(72)	120	H(116)	+0.06
H(101)-C(99)	1.09	C(70)-C(65)-C(73)	123	C(117)	–0.02
H(102)-C(97)	1.09	C(65)-C(66)-C(74)	119	C(118)	–0.04
C(103)-C(100)	1.45	C(76)-C(75)-C(74)	118	C(119)	–0.04
C(104)-C(103)	1.38	C(66)-C(74)-C(75)	123	C(120)	–0.02
C(104)-C(105)	1.47	C(73)-C(76)-C(75)	118	H(121)	+0.05
C(105)-C(112)	1.47	C(79)-C(76)-C(75)	119	H(122)	+0.05
C(106)-C(105)	1.38	C(65)-C(73)-C(76)	123	C(123)	–0.02
H(107)-C(106)	1.09	C(66)-C(74)-H(77)	120	C(124)	–0.02
H(108)-C(103)	1.09	C(65)-C(73)-H(78)	120	H(125)	+0.05
C(109)-C(104)	1.47	C(73)-C(76)-C(79)	123	H(126)	+0.05
C(110)-C(109)	1.36	C(81)-C(80)-C(79)	118		
C(111)-C(110)	1.45	C(76)-C(79)-C(80)	122		
C(112)-C(111)	1.36	C(82)-C(81)-C(80)	118		
H(113)-C(112)	1.09	C(88)-C(81)-C(80)	119		
H(114)-C(111)	1.09	C(75)-C(82)-C(81)	122		

TABLE 13.2 (Continued)

Bond lengths	R,A	Valence corners	Grad	Atom	Charge (by Milliken)
H(115)-C(110)	1.09	C(74)-C(75)-C(82)	123		
H(116)-C(109)	1.09	C(76)-C(75)-C(82)	119		
C(117)-C(68)	1.46	C(76)-C(79)-H(83)	120		
C(118)-C(117)	1.38	C(75)-C(82)-H(84)	120		
C(118)-C(119)	1.49	C(79)-C(80)-C(85)	123		
C(118)-C(123)	1.44	C(81)-C(80)-C(85)	119		
C(119)-C(124)	1.44	C(87)-C(86)-C(85)	119		
C(120)-C(119)	1.38	C(80)-C(85)-C(86)	122		
H(121)-C(117)	1.09	C(88)-C(87)-C(86)	119		
H(122)-C(120)	1.09	C(92)-C(87)-C(86)	119		
C(123)-C(47)	1.42	C(81)-C(88)-C(87)	122		
C(124)-C(46)	1.42	C(82)-C(81)-C(88)	123		
H(125)-C(123)	1.09	C(81)-C(88)-H(89)	119		
H(126)-C(124)	1.09	C(80)-C(85)-H(90)	119		
		C(85)-C(86)-C(91)	123		
		C(87)-C(86)-C(91)	119		
		C(88)-C(87)-C(92)	123		
		C(94)-C(93)-C(92)	119		
		C(87)-C(92)-C(93)	122		
		C(91)-C(94)-C(93)	119		
		C(99)-C(94)-C(93)	118		
		C(86)-C(91)-C(94)	122		
		C(86)-C(91)-H(95)	119		
		C(87)-C(92)-H(96)	119		
		C(92)-C(93)-C(97)	123		
		C(94)-C(93)-C(97)	118		
		C(100)-C(98)-C(97)	119		
		C(106)-C(98)-C(97)	123		
		C(93)-C(97)-C(98)	122		
		C(99)-C(100)-C(98)	119		
		C(105)-C(106)-C(98)	123		
		C(103)-C(100)-C(98)	118		
		C(91)-C(94)-C(99)	123		

TABLE 13.2 (Continued)

Bond lengths	R,A	Valence corners	Grad	Atom	Charge (by Milliken)
		C(94)-C(99)-C(100)	122		
		C(106)-C(98)-C(100)	118		
		C(94)-C(99)-H(101)	118		
		C(93)-C(97)-H(102)	118		
		C(99)-C(100)-C(103)	123		
		C(105)-C(104)-C(103)	119		
		C(100)-C(103)-C(104)	123		
		C(112)-C(105)-C(104)	118		
		C(106)-C(105)-C(104)	119		
		C(111)-C(112)-C(105)	122		
		C(112)-C(105)-C(106)	123		
		C(105)-C(106)-H(107)	120		
		C(100)-C(103)-H(108)	117		
		C(103)-C(104)-C(109)	123		
		C(105)-C(104)-C(109)	118		
		C(104)-C(109)-C(110)	122		
		C(109)-C(110)-C(111)	121		
		C(110)-C(111)-C(112)	121		
		C(111)-C(112)-H(113)	120		
		C(110)-C(111)-H(114)	118		
		C(109)-C(110)-H(115)	121		
		C(104)-C(109)-H(116)	118		
		C(67)-C(68)-C(117)	123		
		C(69)-C(68)-C(117)	118		
		C(119)-C(118)-C(117)	119		
		C(123)-C(118)-C(117)	123		
		C(68)-C(117)-C(118)	123		
		C(124)-C(119)-C(118)	118		
		C(47)-C(123)-C(118)	124		
		C(120)-C(119)-C(118)	119		
		C(46)-C(124)-C(119)	124		
		C(123)-C(118)-C(119)	118		
		C(124)-C(119)-C(120)	123		

TABLE 13.2 (Continued)

Bond lengths	R,A	Valence corners	Grad	Atom	Charge (by Milliken)
		C(68)-C(117)-H(121)	117		
		C(119)-C(120)-H(122)	120		
		C(46)-C(47)-C(123)	118		
		C(45)-C(46)-C(124)	123		
		C(47)-C(123)-H(125)	118		
		C(46)-C(124)-H(126)	118		

TABLE 13.3 Total Energy (E_0), Maximal Charge on the Hydrogen Atom (q_{max}^{H+}) and Universal Factor of Acidity (pKa) of Molecules Dekacene and Eicocene

Molecules	E_0 (kDg/mol)	q_{max}^{H+}	pKa
Dekacene	–550,105	+0.06	33
Eicocene	–1,069,853	+0.06	33

KEYWORDS

- **acid strength**
- **dekacene**
- **eicocene**
- **method MNDO**
- **quantum chemical calculation**

REFERENCES

1. Novoselov, K. S. et al. Electric Field Effect in Atomically Thin Carbon Films, Science 306, 666 (2004); DOI:10.1126/science.1102896.
2. Shmidt, M. W., Baldrosge, K. K., Elbert, J. A., Gordon, M. S., Enseh, J. H., Koseki, S., Matsvnaga, N., Nguyen, K. A., Su, S. J. et al. J. Comput. Chem. 14, 1347–1363, (1993).

3. Bode, B. M. and Gordon, M. S. *J. Mol. Graphics Mod., 16*, 1998, 133–138
4. Babkin V.A., Fedunov R.G., Minsker K.S., et al.. Oxidation communication, 2002, №1, 25, 21–47.
5. Babkin, V. A., Ignatov, A. V., Ignatov, A. N., Gulyukin, M. N., Dmitriev, V. Yu., Stoyanov, O. V., Zaikov, G. E. Quantum-chemical Calculation of some molecules of triboratols. Kazan. "Herald" of Kazan State Technological University. 2013, Vol. 16, N2, 15–17
6. Babkin, V. A., Ignatov, A. V., Stoyanov, O. V., Zaikov, G. E. Quantum-chemical Calculation of some monomers of cationic polymerization with small cycles. Kazan. "Herald" of Kazan State Technological University. 2013, Vol. 16, N 4, 21–22
7. Babkin, V. A., Trifonov V. V., Lebedev, B. B. N. G., Dmitriev, V. Yu., Andreev, D. S., Stoyanov, O. V., Zaikov, G. E. Quantum-chemical Calculation of tetracene and pentacene by MNDO in approximation of the linear molecular model of graphene. Kazan. "Herald" of Kazan State Technological University. 2013, 16(7), 16–18.

CHAPTER 14

INTRODUCTION TO CERAMIZABLE POLYMER COMPOSITES*

R. ANYSZKA[1] and D. M. BIELIŃSKI[1,2]

[1]*Lodz University of Technology, Faculty of Chemistry, Institute of Polymer and Dye Technology, Stefanowskiego 12/16, 90-924 Lodz, Poland*

[2]*Institute for Engineering of Polymer Materials and Dyes, Division of Elastomers and Rubber Technology, Harcerska 30, 05-820 Piastow, Poland; Tel.: +4842 6313214; Fax: +4842 6362543; E-mail: dbielin@p.lodz.pl*

**The authors dedicate this work to Prof. Gennady Zaikov to commemorate his 80th anniversary.*

CONTENTS

ABSTRACT

In this chapter, ceramization (ceramification) of polymer composites is presented as a promising method for gaining flame retardancy of the materials. Because of its passive fire protecting character, ceramization effect can be applied in polymer composites, which are dedicated for work in public places like shopping centers, sport halls, galleries and museums, office buildings, theaters or cinemas and public means of transport. In case of fire, ceramizable polymer composites turn into barrier ceramic materials, ensuring integrity of objects like electrical cables, window frames, doors, ceilings, etc., exposed on flames and heat, preventing from collapsing of materials and making electricity working, enabling effective evacuation. Moreover, ceramization process decreases emission of toxic or harmful gaseous products of polymer matrix degradation as well as its smoke intensity. The paper describes mechanisms of ceramization for various polymer composites, especially focusing on silicone rubber-based ones, basic characteristics of the materials and ways of their testing.

14.1 INTRODUCTION

According to International Association of Fire and Rescue Service (CTIF) direct and indirect costs of fires can reach 1% of GDP in developed countries and even up to 2% of GDP in less-developed countries [1]. Important part of these costs is generated by fires of infrastructure, especially of buildings. Polymer materials and composites are nowadays commonly used as parts of wide range of civil engineering constructions, increasingly replacing conventional materials like concrete, wood and even metal, because of their good processability, low price, relatively high mechanical properties and corrosion resistance. However, polymer materials exhibit two major disadvantages, which are low fire resistance and low thermal stability. For example combustion heat of polyethylene or polypropylene is around 46–47 MJ/kg [2], in comparison to 32–37 MJ/kg reported for coal [3]. That's why their growing presence in building technology creates fire hazard. Moreover, lots of decorative parts and facilities are made of polymer materials, what significantly increases danger.

To eliminate this problem growing number of polymer flame retardants have been developed. But unfortunately, the most effective chemical compounds creates toxic, corrosive or harmful volatile products due to radical reactions present in gas or solid phase of fire space, which can cause even more harm for people health and infrastructure damage than flames or heat radiation [4]. Flame retardants based on physical mechanisms of combustibility lowering, like char formation, dissipation of heat from a fire zone or volatile fuel diluting are much safer, however unfortunately also less effective. Nevertheless, due to issue of toxicity, development in field of physical flame retardants has become one of the most important way to improve fire resistance of polymer composites due to their nontoxicity. According to Morgan and Gilman [5], application of ceramic amorphous oxides of low softening point temperature as a dispersed phase in polymer composites could strongly increase properties of char being formed under fire, being one of the most perspective way of developing new, flame retarded polymer materials. In fact, presence of this kind of glassy oxides in a composite results in possibility for creation of strong continuous ceramic phase on the surface of heat treated composite instead of rather weak and less continuous char. Thus, it begins development of new kind of passive flame retarded polymer materials – the so-called ceramizable (ceramifiable) composites.

14.2 PHENOMENON OF CERAMIZATION

Ceramization process bases on creation of mechanically strong, porous and insulating ceramic barrier layer on the surface of composite during fire or at high temperature. This is in fact an improved char forming mechanism, but mechanical and barrier properties of ceramic phase obtained are superior to char created after heat treatment. This ceramic, protective shield prevents from diffusion of volatile fuel, created during thermal decomposition of polymer matrix, to fire zone as well as diffusion of oxygen into condensed phase of the composite, where it accelerates degradation rate of polymer macromolecules. Moreover, stiff, continuous and thermally insulating layer decreases emission of harmful, corrosive and toxic product of firing and reduces production of smoke, what can significantly increase fire safety in public use buildings, especially from the point of view of possibility for evacuation of crowds of people from places endangered by fire.

In comparison to other flame retarded polymer materials, ceramizable composites exhibit higher thermal stability, nontoxicity, excellent barrier properties and ability to sustain their shape even after complete destruction of polymer matrix at elevated temperature.

14.3 ORGANIC POLYMER-BASED COMPOSITES

First information describing properties of ceramizable composites based on organic polymers was presented in Proceedings of "European Coatings Conference, 2006" [6]. The paper by Thomson et al. describes properties of composites based on poly(vinyl chloride) (PVC) and ethylene-propylene-diene rubber (EPDM). The authors have shown that their composites exhibit significantly lower heat release rate (HRR), smoke intensity and toxic carbon (II) oxide emission under cone calorimeter test. These studies have resulted in industrial application and now Australian company "Cerampolymerik" offers ceramizable PVC and EPDM composites for sale. However, this company seems to be the only manufacturer of organic polymer based ceramizable composites in the world.

In case of ceramizable composites based on organic polymers, dominating mechanism of ceramization is based on dispersed mineral phase playing the main role (Fig. 14.1). Because organic matrix degrades completely during heat treatment simulating fire conditions, it cannot participate in creation of ceramic phase. The most important component of dispersed phase is oxide amorphous frit called fluxing agent, which melts relatively quickly at elevated temperature and combine other, thermally stable mineral particles together, leading to formation of continuous, ceramic phase. This is physical type of ceramization mechanism, which theoretically can be applied for any kind of polymer.

Another mechanism, describing creation of continuous mineral phase on the surface of polymer composite during ceramization, is based on sintering process of refractory mineral filler particles due to condensation of hydroxyl groups present on their surface (Fig. 14.2). However, this process, originally proposed by Xiong et al. [7], plays minor role in forming brittle, porous ceramic phase of high endurance.

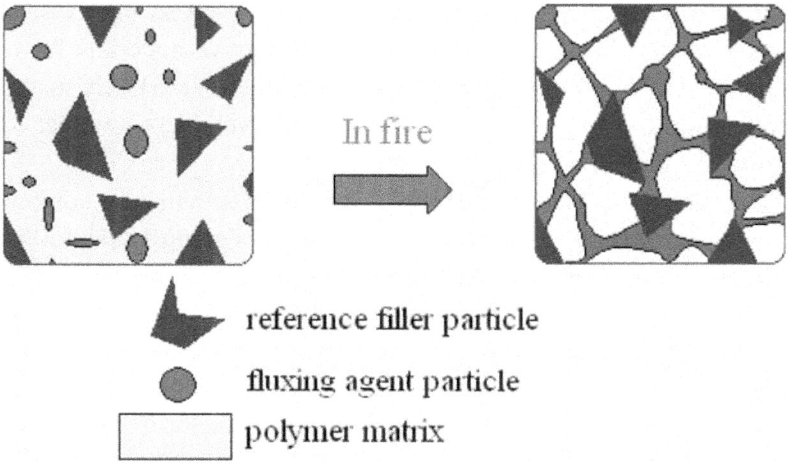

FIGURE 14.1 Scheme of physical mechanism of ceramization involving binding action of fluxing agent particles.

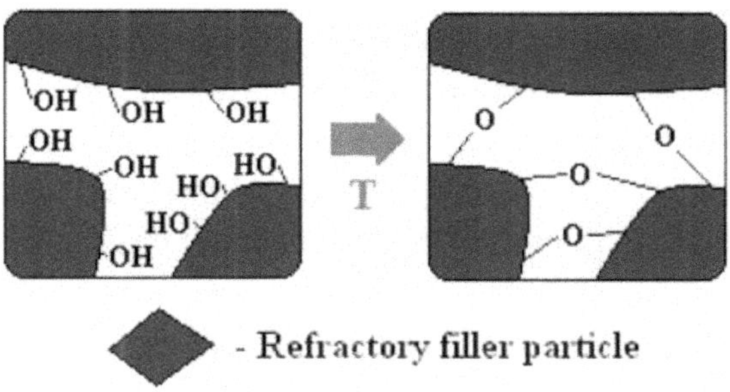

FIGURE 14.2 Mechanism of sintering mineral filler particles during heat treatment based on condensation of hydroxyl groups.

Ceramizable composites based on low density linear polyethylene (LLDPE) [8] and poly(vinyl acetate) (PVA) [9] have recently been described in scientific literature.

Polyethylene is one of the most representative polymer for a group of polyolefines. Currently, composites based on this material are increasingly used for cable covers production, replacing plasticized PVC, because of significantly lower toxicity level of their thermal degradation products. According to Wang et al. [8], ceramizable LLDPE composites exhibits good thermal stability and low flammability as well as high resistance against thermal shock, what make them perspective materials for cable insulations. Polyolefin-based ceramizable composites are likely to replace in cable industry commonly used more expensive silicone rubber-based composites which are much expensive.

According to Al-Hassany et al. [9], PVA based ceramizable composites can be used for sealing applications. The authors studied thermal stability and morphology of ceramizable composites after heat treatment. They have shown that composite containing kaolin particles as refractory filler exhibits better mechanical properties after ceramization than composite filled with talc particles. However, composite containing talc exhibit better thermal stability, probably due to high shape factor of talc plates. In both cases PVA based composites sustain their shape after heat treatment, which is one of the most important factors from the point of view of fire protection because ceramic phase being created, maintains barrier properties of seal preventing from oxygen diffusion to fire zone, what eventually results in extinction of fire.

According to Zhang et al. [10], polystyrene (PS) composites containing organically modified montmorillonite clays (OMMT) are able to create porous, ceramic residue, being composed of two different layered phases after heat treatment. This residue has thin "skin" layer containing a lot of small pores and thick "cellular" layer with relatively big pores (Fig. 14.3). We have obtained analogical structures after heat treatment of PVC ceramizable composites, prepared in our laboratory (Fig. 14.4). This result

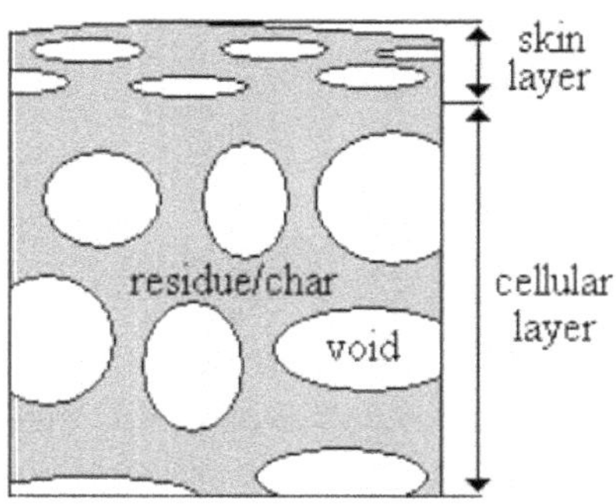

FIGURE 14.3 Scheme of porous structure of residue obtained after heat treatment of PS/OMMT composite.

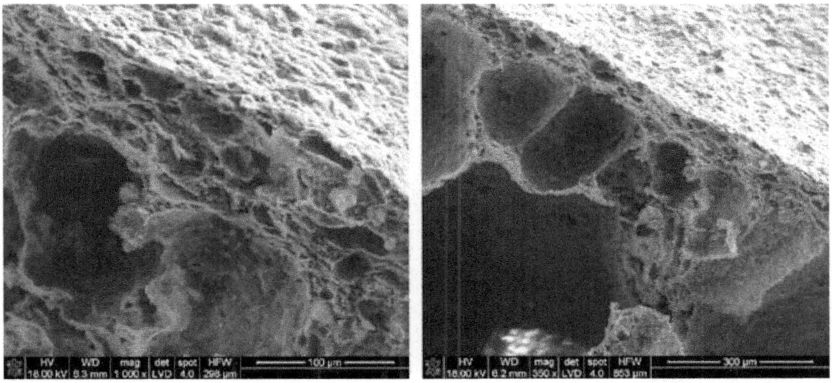

FIGURE 14.4 SEM pictures of two layered morphology of PVC ceramizable composite subjected to heat treatment.

proves on ceramization process being a developed version of char forming flame retardancy mechanism, acting similarly under fire conditions.

14.4 SILICONE RUBBER-BASED COMPOSITES

Despite a lot of research and development going on organic polymer-based composites, ceramizable composites based on silicone rubber still remains the most popular materials for cable industry. This is simply because of their ability to protect copper wire from heat melting and to sustain cable wires, maintaining integrity of electrical circuits even up to 120 min. in fire. During this time all important installation, like fire sprinklers, lamps, elevators, alarm systems etc. are able to work, increasing chance for evacuation people from dangerous zones. Polysiloxanes are the best kind of polymers which can be used as a continuous phase for ceramizable composites because they are able to create amorphous silica during thermal degradation under oxidizing atmosphere. Unfilled silicone degrades creating volatile cyclosiloxanes according to three mechanisms: unzipping reaction (Fig. 14.5), random scission reaction (Fig. 14.6) and externally catalyzed reaction (Fig. 14.7) [11]. After depolymerization, volatile cyclosiloxanes undergo burning producing silica, which can take part in creation of ceramic phase on the surface of composite.

Unzipping reactions take place when silicone macromolecules are terminated with a polar group like hydroxyl one (-OH). These kinds of

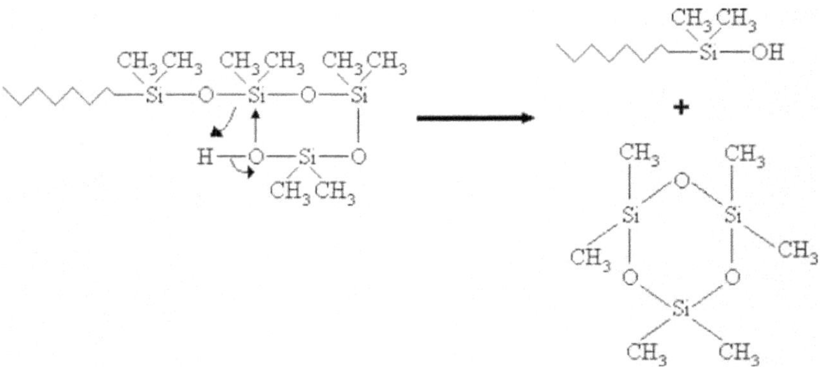

FIGURE 14.5 Scheme of unzipping reaction mechanism of hydroxyl-terminated PDMS macromolecule.

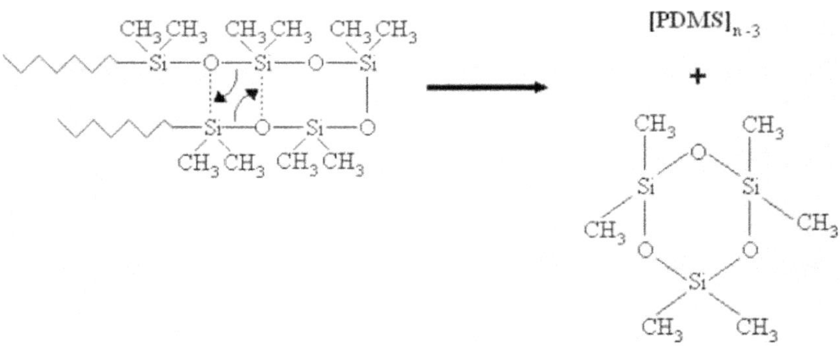

FIGURE 14.6 Scheme of random scission reaction mechanism of PDMS chain.

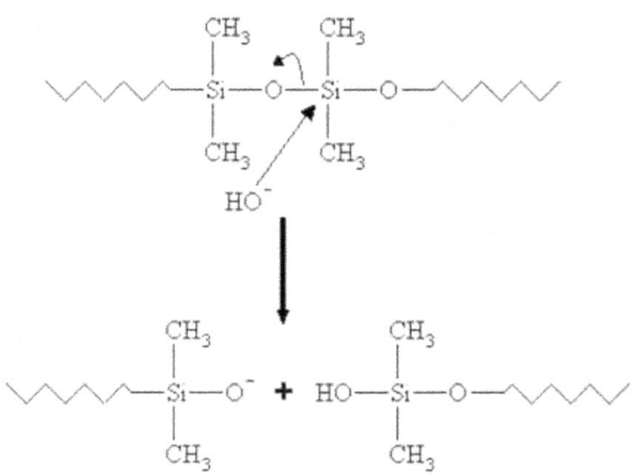

FIGURE 14.7 Scheme of externally catalyzed degradation of PDMS macromolecule.

reactive groups are present in room temperature vulcanizable silicones (RTV). They allow macromolecules to create cross-linked structure as a result of reaction with curing agents, for example, silanes.

Random scission reaction dominates when silicone macromolecules are terminated with nonpolar groups like methyl ($-CH_3$) or vinyl ($-CH=CH_2$) ones. These kinds of chains are characteristic for high temperature vulcanizable silicones (HTV), where cross-linking of polymer bases on radical reactions with peroxides or curing reactions involving platinum catalysts.

Impurities, which exhibit ion character can also take part in thermal degradation of silicones. For example hydroxyl anions present in moisture are able to catalyzed breaking of polysiloxane chain.

According to Camino and Lomakin [12, 13], thermal degradation of silicones occurs with competition between molecular mechanisms leading to creation of cyclic volatiles (described previously) and radical mechanism leading to cross-linking of macromolecules and creation of methane (Fig. 14.8).

The most important advantage of using silicone rubber as a continuous phase of ceramizable composites seems to be, that even without addition of fluxing agent, dispersed mineral particles are able to create ceramic continuous phase due to sintering via silica bridges created as a result of thermal degradation of silicone macromolecules bonded to their surfaces (Fig. 14.9) [7, 14–17].

Based on SEM analysis of ceramizable silicone composites after heat treatment, Hanu et al. [17], proposed scheme describing process of silica bridges creation on the surface of muscovite mica filler, leading to obtain continuous ceramic phase (Fig. 14.10). They have concluded

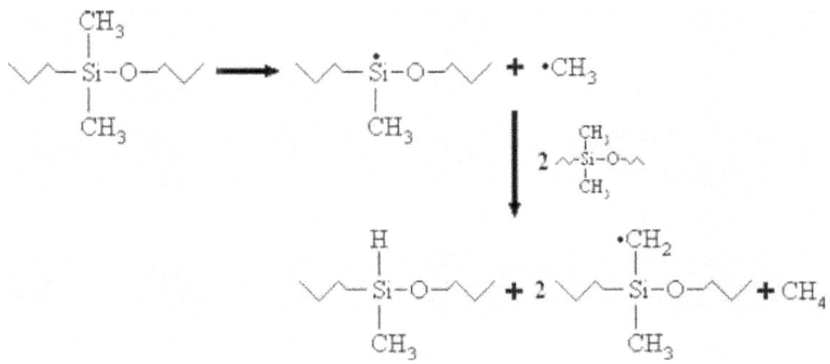

FIGURE 14.8 Radical mechanism of thermal degradation of silicones leading to obtain cross-linked structure and methane.

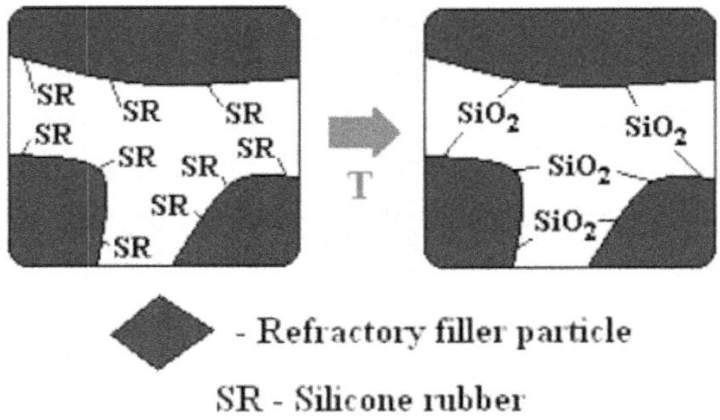

FIGURE 14.9 Mechanism of sintering mineral filler particles during heat treatment, based on creation of silica bridges, as proposed by Xiong et al. [7].

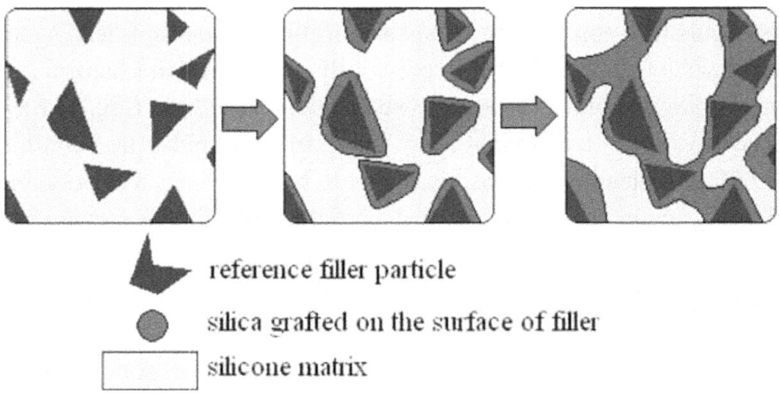

FIGURE 14.10 Scheme of ceramic phase creation via silica bridges.

from morphology of the system, that character of filler surface and size of its particles are crucial factors from the point of view of ceramic shield properties.

Nevertheless, incorporation of fluxing agent into silicone rubber-based ceramizable composites significantly improves mechanical and barrier properties of ceramic phase obtain after their heat treatment [18–25]. Moreover, chemical reactions taking place between mineral particles of dispersed phase or products of their thermal oxidation, may result in creation of even more durable ceramic phase (Fig. 14.11). For example, calcium oxide (II) created after decarboxylation of calcium carbonate filler can react with silica obtain after thermal degradation of silicone rubber, what leads to creation of wollastonite or larnite particles.

$$T > 600\,^{\circ}C$$

$$CaO + SiO_2 \rightarrow CaSiO_3 \quad \text{Wollastonite}$$

$$2CaO + SiO_2 \rightarrow Ca_2SiO_4 \quad \text{Larnite}$$

FIGURE 14.11 Chemical reactions between mineral compounds leading to creation of more complex mineral phases.

14.4.1 PREPARATION AND PROCESSING

Simple way of preparation of ceramizable polymer composites is their huge advantage. Elastomer-based composites can be prepared using classic internal or external mixers used in rubber technology. For preparation of thermoplastic composites the best way is to apply twin-screw extruder, which also allows for final forming of product shape. Major problem faced during preparation of ceramizable composites is associated with high amount of refractory mineral fillers, what strongly increases viscosity of compositions, adversely affecting their processing and forming.

Due to high amount of different mineral fillers, rheological properties of composites could be very different, even if the same polymer play role of matrix (Fig. 14.12). In our previous study [26] we have compared rheological properties of various commercially available silicone rubber-based composites.

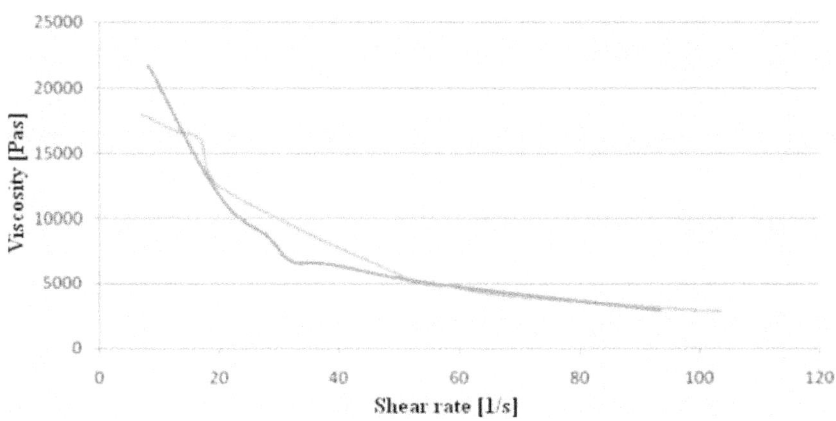

FIGURE 14.12 Viscosity in function of shear rate of silicone rubber-based ceramizable composites originated from various producers.

Curing of ceramizable composite mixes can be also quite challenging. Strong acidic or alkaline character of fillers can negatively influence efficiency of cross-linking, what could lead even to prevent creation of chemical bonds between macromolecules of elastomer matrix, what especially concerns peroxide cross-linking reactions. However, our recent work have shown, that high addition of alkaline mineral fillers, like MgO, CaO, or organofilized montmorillonite, can solve the curing problem of silicone composites [27].

14.4.2 PROPERTIES AND TESTING

High amount of mineral fillers strongly influences mechanical properties of ceramizable polymer composites. Their viscosity becomes significantly higher and their stiffness increases, what altogether may lead to lower impact resistance of the materials. Of course, it is possible to test ceramizable composites as any other kind of polymer materials, determining their rheological, mechanical and tribological properties, etc., but the most important types of studies are related to morphology of ceramic phase being created, its thermal properties and fire resistance of composites heat insulation and thermal stability, respectively).

For studying thermal properties of composites thermogravimetry, combined with differential scanning calorimetry (TG-DSC) can be applied. Another valuable source of information is cone calorimetry analysis, which can provide data on heat release rate (HRR) in function of temperature or time, smoke intensity or concentration of toxic products of polymer matrix degradation. In our previous work [28], we performed cone calorimetry studies for some commercially available silicone rubber-based composites They showed superior HRR characteristic of ceramizable composites over conventional silicone rubber.

For information on flame retardancy oxygen index (OI) or limited oxygen index (LOI) is commonly determined. No less popular is vertical burning test described by UL-94 standard.

For ceramizable electrical wire covers a number of standards have been exclusively prepared, for example EN 50264-1 and EN 50382-1 – defining requirements for cables destined for railway applications or EN 60332–1-2 describing method of fire testing for cables (Fig. 14.13).

FIGURE 14.13 Pictures representing cable test under fire conditions, according to EN 60332-1-2 standard.

In our previous study, we have described that Fourier transform infrared spectroscopy (FT-IR) of smoke and volatiles produced during ceremization process of composites can be also concern as very accurate tool for analyzing thermal properties of ceramizable composites [29].

A new challenge is to prepare suitable methodology for testing properties of ceramic phase created after composite heat/flame treatment. For qualitative analysis of its morphology scanning electron microscopy equipped with energy-dispersive X-ray spectroscope (SEM-EDS) seems to be sufficient. More difficult are analysis of thermal or electrical conductivity or even mechanical properties of created ceramic phase being created, because of sample deformation accompanying ceramization, making practically impossible to maintaining required specimen shape. However, indirect information of this kind of properties can be evaluated from porosimetry analysis or simple mechanical determinations like three point bending test of ceramic phase resulted from ceramization.

Porosimetric characteristics of various commercially available ceramizable silicone rubber-based composites [28] point on possibility to meet engineering requirements for ceramized phase by controlling morphology of materials, being related to composition, processing and anticipated thermal conditions during fire (temperature and heating rate).

14.5 FUTURE DIRECTIONS

Future studies will probably lead to obtain new kinds of glassy amorphous frits, acting as fluxing agents, which could provide more and more effective ceramization process of composites. Nowadays, this special kind of fillers exhibit insufficient low value of softening point temperature and additionally size of their particles still remains far too big. Moreover, very often they exhibit quite high hygroscopicity, which disturbs processing of composites. First studies describing comilling of fluxing agent with refractory mineral fillers before compounding process are promising [30]. This is one of potential ways to decrease average size of primary particles of fluxing agent as well as to obtain its better dispersion and distribution in polymer matrix.

Another direction for development of ceramizable polymer composites will surely originate from application of various organic polymers as a composite matrix. Wide range of polymers or copolymers can be used due to their different fields of application. Already only civil engineering presents an enormous potential market. Even commonly used so far polymers, like, for example, PVC, which generate toxic or harmful products under their thermal degradation or characterizing themselves by high smoke intensity can be considered as matrices for ceramizable composites.

Another very important market seems to be cable industry, requiring permanent progress in technology and properties of silicone rubber-based composites destined for wire covers. Our recent studies have shown, that rather low mechanical properties of produced so far silicone ceremizable composites can be increased by incorporation of mineral or polyamide fibers [31].

ACKNOWLEDGEMENTS

This work was supported by the EU Integrity Fund, project POIG 01.03.01-00-067/08-00.

KEYWORDS

- **ceramization (ceramification)**
- **flame retardancy**
- **poly(vinyl chloride)**
- **polymer composites**

REFERENCES

1. Brushlinsky N.N., Hall J.R., Sokolov S.V., Wagner P.: CTIF – World Fire Statistics. Report No. 11 (2006).
2. Vasile C.: Handbook of polyolefins. New York USA: Marcell Dekker Inc. (2000).
3. Ulanovskii M.L.: Heat of Combustion of Coal: Basic Principles and New Calculation Methods. Coke and Chemistry, 53, 322–329 (2010).
4. Shaw S.D., Blum A., Weber R., Kannan K., Rich D., Lucas D., Koshland C.P., Dobraca D., Hanson S., Birnbaum L.S.: Halogenated flame retardants: do the fire safety benefits justify the risks? Reviews on Environmental Health, 25, 261–305 (2010).
5. Morgan A.B., Gilman J.W.: An overview of flame retardancy of polymeric materials: application, technology and future directions. Fire and Materials, 37, 259–279 (2013).
6. Thomson K.W., Rodrigo P.D.D., Preston C.M., Griffin G.J.: Ceramifying Polymers for Advanced Fire Protection Coatings. Berlin Germany: Proceedings of European Coatings Conference, 99–110 (2006).
7. Xiong Y., Shen Q., Chen F., Luo G., Yu K., Zhang L.: High strength retention and dimensional stability of silicone/alumina composite panel under fire. Fire and Materials, 36, 254–263 (2012).
8. Wang T., Shao H., Zhang Q.: Ceramifying fire-resistant polyethylene composites. Advanced Composites Letters, 19, 175–179 (2010).
9. Al-Hassany Z., Genovese A., Shanks R. A.: Fire-retardant and fire-barrier poly(vinyl acetate) composites for sealant application. eXPRESS Polymer Letters, 4, 79–93 (2010).
10. Zhang J., Bai M., Wang Y., Xiao F.: Featured structures of fire residue of high-impact polystyrene/organically modified montmorillonite nanocomposites during burning. Fire and Materials, 36, 661–670 (2012).
11. Dvornic P.R. In: Jones RG, Ando W, Chojnowski J, editors: Thermal stability of polysiloxanes: silicone–containing polymers. Dordrecht, the Netherlands: Kluwer Academic Publisher, 185–212 (2000).

12. Camino G., Lomakin S., Lazzari M.: Polydimethylsiloxane thermal degradation. Part 1. Kinetic aspects. Polymer, 42, 2395–2402 (2001).
13. Camino G., Lomakin S., Lageard M.: Thermal polydimethylsiloxane degradation. Part 2. The degradation mechanisms. Polymer, 43, 2011–2015 (2002).
14. Mansouri J., Burford R.P., Cheng Y.B., Hanu L.: Formation of strong ceramified ash from silicone-based composites. Journal of Materials Science, 40, 5741–5749 (2005).
15. Mansouri J., Burford R.P., Cheng Y.B.: Pyrolysis behavior of silicone-based ceramifying composites. Materials Science and Engineering A, 425, 7–14 (2006).
16. Hamdani S., Longuet C., Perrin D., Lopez-Cuesta J-M., Ganachaud F.: Flame retardancy of silicone-based materials. Polymer Degradation and Stability, 94, 465–495 (2009).
17. Hanu L.G., Simon G.P., Cheng Y.B.: Preferential orientation of muscovite in ceramifiable silicone composites. Materials Science and Engineering A, 398, 180–187 (2005).
18. Marosi G., Márton A., Anna P., Bertalan G., Marosfői B., Szép A.: Ceramic precursor in flame retardant systems. Polymer Degradation and Stability, 77, 259–265 (2002).
19. Mansouri J., Wood C.A., Roberts K., Cheng Y.B., Burford R.P.: Investigation of the ceramifying process of modified silicone-silicate compositions. Journal of Materials Science, 42, 6046–6055 (2007).
20. Hanu L.G., Simon G.P., Mansouri J., Burford R.P., Cheng Y.B.: Development of polymer-ceramic composites for improved fire resistance. Journal of Materials Processing Technology, 153–154, 401–407 (2004).
21. Bieliński D.M., Anyszka R., Pędzich Z., Dul J.: Ceramizable silicone rubber-based composites. International Journal of Advanced Materials Manufacturing and Characterization, 1, 17–22 (2012).
22. Hanu L.G., Simon G.P., Cheng Y.B.: Thermal stability and flammability of silicone polymer composites. Polymer Degradation and Stability, 91, 1373–1379 (2006).
23. Pędzich Z., Bukanska A., Bieliński D.M., Anyszka R., Dul J., Parys G.: Microstructure evolution of silicone rubber-based composites during ceramization at different conditions. International Journal of Advanced Materials Manufacturing and Characterization, 1, 29–35 (2012).
24. Pędzich Z., Bieliński D.M.: Microstructure of silicone composites after ceramization. Composites, 10, 249–254 (2010).
25. Dul J., Parys G., Pędzich Z., Bieliński D.M., Anyszka R.: Mechanical properties of silicone-based composites destined for wire covers. International Journal of Advanced Materials Manufacturing and Characterization, 1, 23–28 (2012).
26. Bieliński D.M., Anyszka R., Masłowski M., Pingot T., Pędzich Z.: Rheology, extrudability and mechanical properties of ceramizable silicone composites. Composites Theory and Practice, 11, 252 – 257 (2011).
27. Anyszka R., Bieliński D.M., Kowalczyk M.: Influence of dispersed phase selection on ceramizable silicone composites cross-linking. Elastomers, 17, 16–20 (2013).
28. Composites Theory and Practice, 11, 252 – 257 (2011).

29. Anyszka R., Bieliński D.M., Jędrzejczyk M.: Thermal behavior of silicone rubber-based ceramizable composites characterized by Fourier transform infrared (FT-IR) spectroscopy and microcalorimetry. Applied Spectroscopy, 67, 1437–1440 (2013).
30. Bieliński D.M., Anyszka R. In: Jurkowski B. Rydarowski H. editors: Silicone ceramizable composites protecting against fire. Radom Poland. WNITE-PIB. 389–412 (2012).
31. Pędzich Z., Anyszka R., Bieliński D.M., Ziąbka M., Lach R., Zarzecka-Napierała M.: Silicon-basing ceramizable composites containing long fibers. Journal of Materials Science and Chemical Engineering, 1, 43–48 (2013).

CHAPTER 15

THE RHEOLOGICAL BEHAVIOR OF LIQUID TWO-PHASE GELATIN-LOCUST BEAN GUM SYSTEMS

YURIJ A. ANTONOV[1] and M. P. GONÇALVES[2]

[1]*N.M. Emanuel Institute of Biochemical Physics, Russian Academy of Sciences, Kosigin Str. 4. 119334 Moscow, Russia*

[2]*CEQUP/Departamento de Engenharia Química, Faculdade de Engenharia, Universidade do Porto, Rua dos Bragas, 4099 Porto Codex, Portugal*

CONTENTS

ABSTRACT

In this chapter, we have attempted to establish relationships between the phase viscosity ratios of liquid gelatin (2%)–LBG (0.8%) systems and their rheological properties. To do that, LBG samples with different degrees of thermodegradation were taken.

15.1 INTRODUCTION

The behavior of aqueous biopolymer systems in a flow is of interest for both the academic and the applied science and also in many industrial applications. The possibility of formation of a fiber-like structure in concentrated biopolymer emulsions under shear and their consequent gelation was shown many years ago [1, 2] for different water-protein-polysaccharide mixtures at definite conditions. However, until the present time, it is not quite clear why similar liquid biopolymer's mixtures, at the same shear rate, can form or not form anisotropic structures, and why their rheological properties are so complicated [3]. To answer to these questions, the establishment of the relationship between local change of selected parameters of the systems and their macrostructure and rheological properties are essential.

The rheological behavior of liquid two phase gelatin-locust bean gum (LBG) systems, comprising liquid LBG enriched continuous phase and deformable in a flow gelatin enriched dispersed particles seems to be determined, at the same phase composition, by phase viscosity ratio (μ). In the μ range from 0.03 to 0.21, viscosity dropped to values noticeably lower (13–40 times) than those of the corresponding LBG solution. Downfall in the viscosity of the mixtures was not observed at μ=0.5–0.6, corresponding that to the maximum energy scatter inside the droplets, in agreement with Mason's conception of droplets' deformation and disruption of liquid Newtonian emulsions.

15.2 METHODS

Gelatin, type B, from bovine skin, ($[\eta]_{298}$ = 0.633 $dL{\cdot}g^{-1}$ in 2 M KSCN, M_η = 243,000 Da) was purchased from Sigma Chemical CO. (Lot 63H0312), and LBG ($[\eta]_{298}$ = 10.35 $dL{\cdot}g^{-1}$ in water, $M_\eta = 1.51\times10^6$ Da), was kindly obtained from Systems Bio-Industries, France, and additionally purified as described before [4].

Thermally degraded LBG samples denoted as A ($[\eta]_{298}$ = 9.36 $dL{\cdot}g^{-1}$, $M_v = 1.36\times10^6$ Da); B ($[\eta]_{298}$ = 8.87 $dL{\cdot}g^{-1}$, $M_v = 1.29\times10^6$ Da); C ($[\eta]_{298}$ = 7.11 $dL{\cdot}g^{-1}$, $M_v = 1.03\times10^6$ Da) and D ($[\eta]_{298}$ = 4.69 $dL{\cdot}g^{-1}$, $M_v = 0.673\times10^6$) were obtained from purified LBG according to Chikamai et al.[5].

Morphology of two-phase gelatin-LBG systems was estimated using the optical microscope AXIOSKOP (Zeiss, Germany).

Rheological measurements of aqueous LBG and LBG-gelatin mixed solutions in the two-phase region were performed at 313±0.1 K with a controlled stress rheometer, Carri-Med (CSL 50) fitted with parallel plate geometry (gap 1 mm, 40 mm Ø).

15.3 RESULTS

In Fig. 15.1(a,b), the flow behavior of gelatin (2.0%)–LBG (0.8%) mixtures with different μ, containing the initial and the thermally degraded LBG samples, is compared to that of solutions (0.8%) of the same LBG

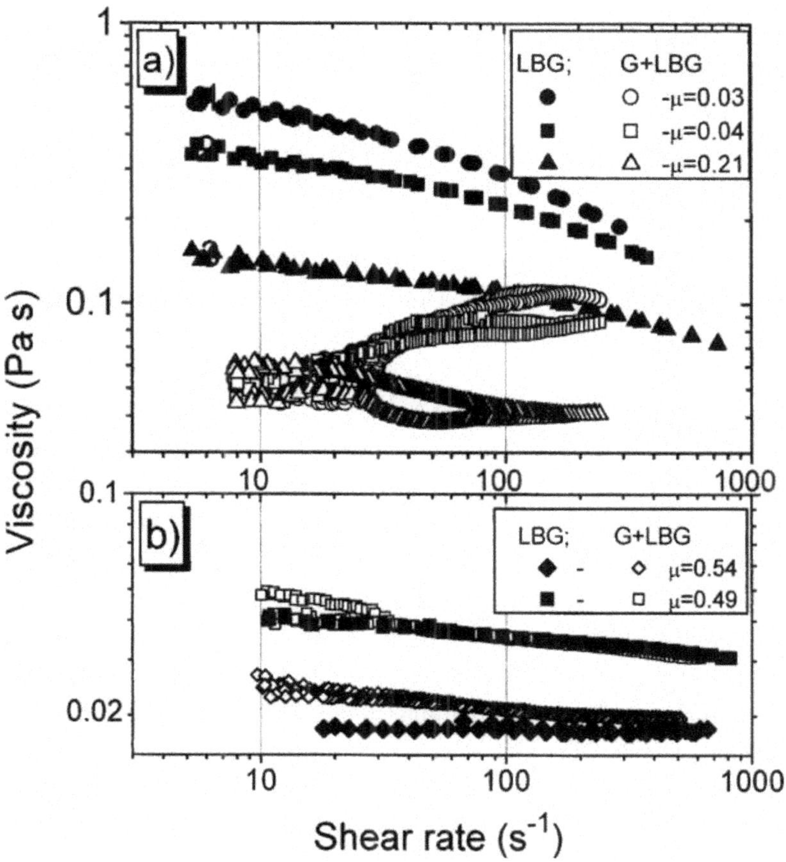

FIGURE 15.1 Flow curves of solutions (0.8%) of the thermally degradaded LBG samples and of their two-phase systems with gelatin (2.0%) with different phase viscosity ratios (m). m was taken at shear rate of 100 s^{-1}. T=313 K.

samples. This composition was selected on the basis of knowledge of phase diagrams [4], first, because it corresponds to a point far from the binodal (in these conditions, all the mixtures containing degraded LBG samples were two-phase ones) and second, the phase volume ratio for all these systems was constant and equal to ~0.64 (that gives the possibility of excluding the effect of the phase volume ratio on the rheology of these mixtures). Also, phase composition was practically the same for all the systems studied; only a small difference was observed for the system where LBG with $M_v = 0.673 \times 10^6$ Da was used. As expected, solutions of degraded LBG samples exhibit rheological properties typical of polymer solutions. The flow behavior of LBG solutions remains absolutely insensitive to shear history. The only difference observed between undegraded an degraded LBG samples was a considerable decrease in the viscosity and in the dynamic moduli (data are not presented) of the degraded LBG. For a 0.8% solution, the viscosity at low shear rate of the most degraded LBG sample (hydrolysis' time = 96 h) was 35 times less than that of the initial LBG sample.

The flow behavior of the mixed systems studied depended qualitatively on the phase viscosity ratio. This ratio was determined from the flow curves of the equilibrium phases obtained by centrifugation of the mixtures at 380,000 g for 1 h, at 313 K. The μ values reported here were obtained at three different shear rate ($10\ s^{-1}$, $20\ s^{-1}$ and $100\ s^{-1}$). The two-phase systems with $0.03 < \mu < 0.21$, containing LBG with a relatively low degree of degradation (3–7 h), exhibited a rheological behavior similar to that of the mixture with the undegraded LBG sample: viscosity much lower than that of an LBG solution with the same concentration (Fig. 15.1a). The flow curves obtained were non-quantifiable after multiple repetition of the shear test. On the contrary, for systems with $\mu = 0.49$ and $\mu = 0.54$ (containing high degraded LBG samples), the flow curves were quite similar to those of LBG solutions of the same concentration (Fig. 15.1b). On the other hand, it was not observed a noticeable decrease of the viscoelasticity of the system, as it was observed for the casein-guar gum system in the two-phase region, when the guar gum enriched phase was the continuous one [3].

Viscosity of all the mixtures with degraded LBG was less than the viscosity of the continuous LBG enriched phase and higher than the viscosity of gelatin enriched dispersed phase. This is illustrated in Fig. 15.2 for mixtures with $\mu = 0.49$ and $\mu = 0.54$, containing LBG samples with two degrees of thermal degradation (20 h and 96 h). The viscosity of these

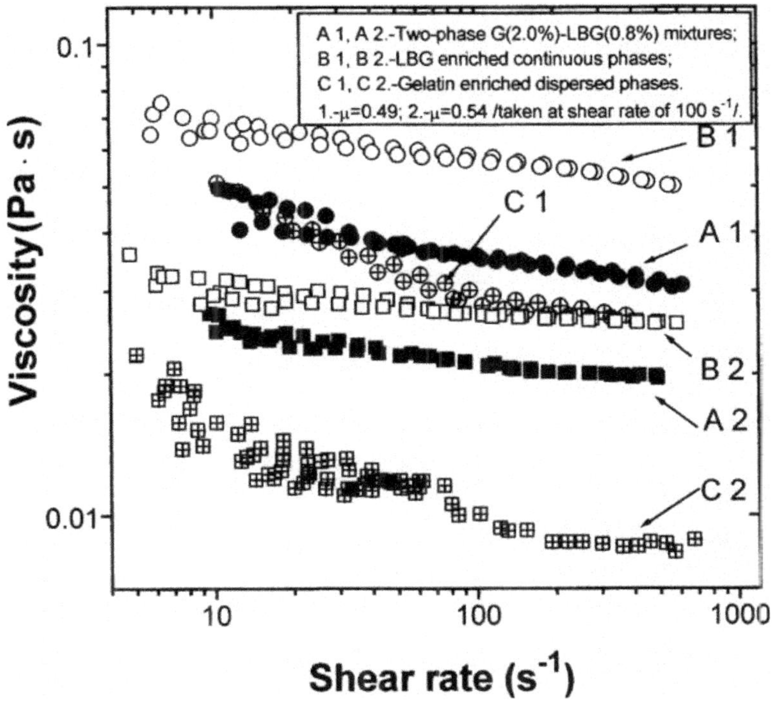

FIGURE 15.2 Flow curves of selected two-phase gelatin (2.0%)–LBG (0.8%)systems, containing thermally degraded LBG samples (curves A1 and A2) and their coexisting gelatin enriched dispersed (curves C1 and C2) and LBG enriched continuous (curves B1 and B2) media. T = 313 K.

mixtures as well as the viscosity of LBG enriched continuous phases practically don't depend on shear rate.

15.4 DISCUSSION

Decrease in viscosity of the ternary water-gelatin-LBG systems with $0.03<\mu<0.21$ and their shear sensitivity are a clear indication of liquid-liquid phase separation and of the droplets' elongation in a flow [3, 6]. Aqueous two-phase protein-polysaccharide systems have a very low surface tension ($\leq 1.0 \times 10^{-5}$ N/m) [7], and they should be very prone to these phenomena. Similar but more significant effects, that is, viscosity of the mixture lower than the sum of the viscosities of the individual components, were described recently for the water-casein-guar gum mixture [3].

Therefore we can regard the ratio of the mixture's viscosity to the viscosity of the LBG solution as a characteristic of droplets' elongation and attempt to analyze these systems by comparison with the main conclusions of the general theory [7] which describes the behavior of two unmixed Newtonian liquids. Theory of droplet's deformation for two unmixed liquids when one or both liquids are non-Newtonian is absent. Therefore, the behavior of such systems can be estimated by comparison with the Newtonian one [9]. It has been established [9, 10] that the main features of the behavior of such systems are very similar to those obtained for Newtonian systems. Quantitative analogy is absent.

Goldsmith and Mason [8] described four types of a droplet's disruption:

First type, at $\mu<0.2$. When a critical shear rate value is attained, the droplet takes a sigmoidal form and is partially disrupted, forming some small droplets.

Second type, at $0.2 \leq \mu < 0.8$. The central part of the droplet under shear suddenly thins, resulting in its disruption and formation of two relatively large droplets and three small droplets. The easiest droplet's disruption takes place when $0.3 < \mu < 0.6$.

Third type, at $0.8 \leq \mu < 2.0$. The droplet keeps a large elongation, and is disrupted instantaneously at a critical elongation, forming a lot of small droplets.

Fourth type, at $\mu > 2.0$. The droplet is not disrupted even at a shear rate of 40 s^{-1}, orienting itself along the flow.

Measurements on a single droplet deformation and burst by Grace [11] also revealed that the critical capillary number at breakup reaches its minimum value for viscosity ratios between 0.1 and 1. Breakup processes are very effective in this range. At higher viscosity ratios, fracture of the droplets under shear flow becomes more difficult, eventually even impossible when μ is larger than 4.

The dependence of η_{LBG+G}/η_{LBG} from the phase viscosity ratio, determined at shear rate values of 10 s^{-1}, 20 s^{-1} and 100 s^{-1} (in conditions of constant pH, ionic strength, and phase volume ratios), are shown in Fig. 15.3. All the data obtained can be described by a single curve. The maximum η_{LBG+G}/η_{LBG} value is equal to 1 at μ ~0.54–0.71, and the minimum η_{LBG+G}/η_{LBG} values (~0.1) are located at $\mu \cong 0.012$. Since fall in viscosity of ternary two-phase solvent-polymer 1-polymer 2 systems,

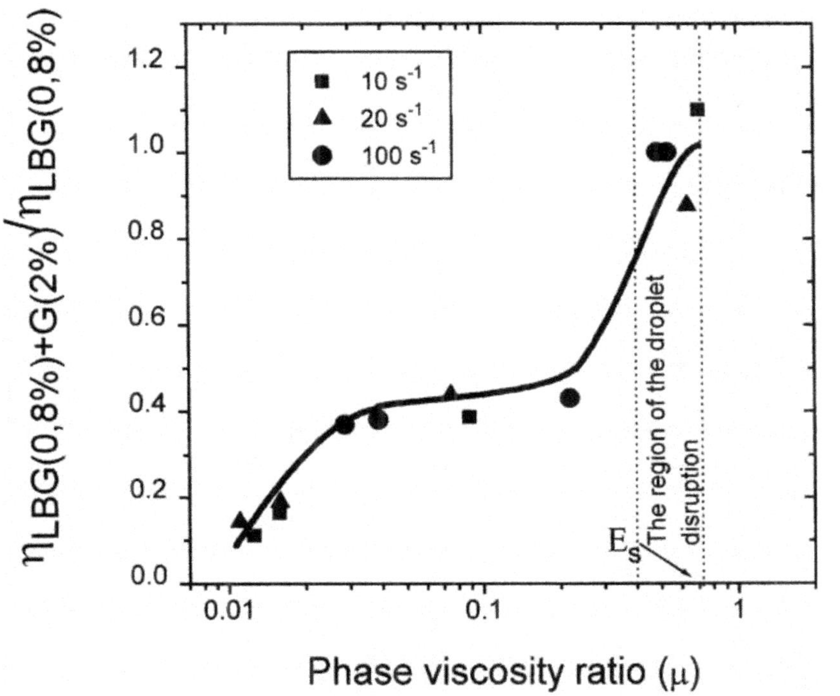

FIGURE 15.3 The dependence of η_{LBG} + G/η_{LBG} of the gelatin (2.0%)–LBG (0.8%) system from the phase viscosity ratio. m was taken at three different shear rates (■ 10 s^{-1}; ▲ 20 s^{-1}; ▲100 s^{-1}).

normally, is connected with formation of anisotropic structures in a flow [7], it is possible to suppose, for 0.012<μ<0.21 the presence of easily deformable in a flow liquid dispersed particles, and for μ ~0.54–0.7, a full disruption, of the droplets. In fact, we observed these phenomena by the microscope after shearing the mixtures between glass plates. The easy disruption of the droplets at μ=0.49–0.7 is in agreement with the main results of Goldsmith and Mason described above, and is in agreement with theory [12], because maximum energy scatter inside the droplet (E_s), that is, the maximum of droplets' disruption takes place at μ=0.632.

The fact that the viscosity of the two-phase mixtures with μ< 0.3 after the change in the sign of the shear rate gradient is normally less (at the same shear rate value) than the viscosity before change, can be probably due to the sharp increase in the total surface volume of the droplets after their disruption at high shear rate, resulting in considerable loss of the system's thermal energy (change of the system's enthalpy).

REFERENCES

1. Tolstoguzov, V.B. Mzel'sky, A.I., Gulov, V.Ya. (1974) *Colloid and Polymer Sci.*, 252, 124.
2. Antonov, Yu.A., Grinberg, V.Ya., Zhuravskaya, N.A., Tolstoguzov, V.B. (1980), *J. Text. Studies.*, 11, 199.
3. Lefebvre, J., Doublier, J.-L., Y.A. Antonov, Yu.A. (1996) *Composite Mechanics and Design,* 2, 121.
4. Alves, M., Antonov, Yu.A., Gonçalves, M.P. (1999), *Food Hydrocoll,* 13, 77.
5. Chikamai, B.N., Banks, W.B., Anderson, D.M.W. and Weiping, W. (1996), *Food Hydrocoll.*, 10, 309.
6. Paul, D.R., Newman, S. (Eds.) "Polymer Blends," Academic Press, New York, 1978.
7. Albertsson, P.-A., "Partition of Cell Particles and Macromolecules," 3rd ed, Almgvist And Wiksel, Stockholm, 1986.
8. Goldsmith, H.L., Mason, S.G. in Eirich, F.R., editor, "Rheology-theory and applications," Vol. 4., Academic Press, New York, 1967, Chapter 2.
9. Karnis, A., Mason, S.G. (1967), *J. Coll. Interface Sci.*, 24,164.
10. Gauthier, F., Goldsmith, H.L., Mason, S.G. (1971), *Trans. Soc. Rheol.*,15, 297.
11. Grace, H.P. (1982), *Chem. Eng.Commun.*,14, 225.
12. Chaffey, C. E., Mason, S.G. (1966) *J. Coll. Interface Sci.*, 21, 254.

CHAPTER 16

THE KINETIC DPPH-METHOD OF ANTIRADICAL ACTIVITY ANALYSIS OF THE MATERIALS OF PLANT ORIGIN

V. A. VOLKOV and V. M. MISIN

N.M. Emanuel Institute of Biochemical Physics, 4, Kosygin street, Moscow, 119334, Russian Federation; E-mail: vl.volkov@mail.ru

CONTENTS

ABSTRACT

In this study, a kinetic method of analysis of compounds from edible and medicinal plant extracts activity against stable radical 2,2-diphenyl-1-picrylhydrazyl (DPPH) is developed. The initial rate of DPPH's decay in standard conditions is suggested and theoretically explained as a kinetic parameter to compare the extract antiradical activity. A 10–150-fold decrease of the DPPH's reaction rate with plant extract antioxidants is achieved by the addition of acids into the reaction medium. Such results were explained by changes of input of different mechanisms into the whole process of scavenge of DPPH radical. A decrease of the reaction rate for the optimum of added acid's concentration with the acid strength increasing is also observed. The influence of the acids extracted from plant material on the results is excluded by this method because of the stronger acid addition. It is found that the linear interval of the dependence of DPPH's conversion degree after the first 30 min from the start of the reaction vs. the initial antioxidant concentration lies from 0 to 60%.

16.1 INTRODUCTION

Direct methods for evaluation of the antioxidant activity (AOA) of individual chemical compounds and compositions of complex composition are based on studying the effect of antioxidants on the kinetics of model of oxidation of hydrocarbons, fatty acids, or biological materials [1]. In practice, however, very often used indirect methods, which look at the parameters that correlate with the antioxidant activity of antiradical antioxidants. Among them is a method based on the interaction of AO with a stable chromogen-radical 2,2-diphenyl-1-pikrilgidrazilom (DPPH).

A number of studies [2–5] showed a correlation between the results obtained by this method, and the results obtained by direct methods. Its advantages include high reproducibility, ease of operations performed, the general availability of the necessary equipment, high sensitivity, high selectivity for antiradical AO. In the visible region of the spectrum of DPPH in organic solvents has a maximum absorption at a wavelength of 517 nm (Fig. 16.1), which vanishes in the interaction of the radical with substances – donors of hydrogen atoms or free radicals of different structure [6–8].

Until now DPPH radical used almost exclusively for the quantitative analysis of antioxidants in natural objects. Attempts to create methods of kinetic analysis of antiradical activity of antioxidants taken previously by some authors, cannot be called successful. In these studies, the activity of antioxidants analyzed at significant conversion depths of substances,

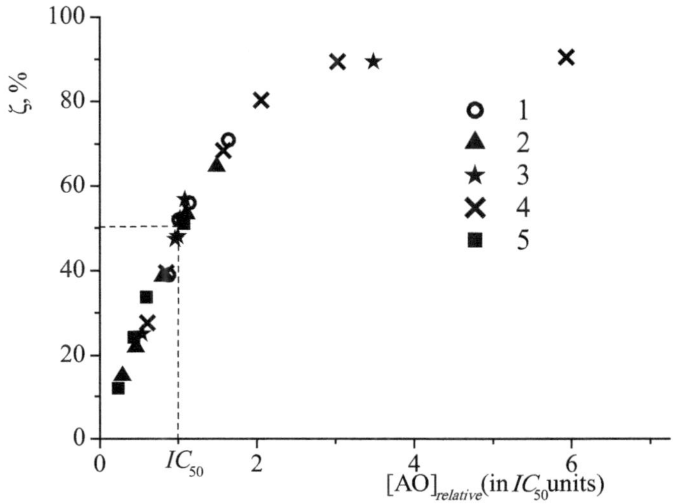

FIGURE 16.1 Dependence of the DPPH conversion degree for the first 30 min of reaction (ζ) on the relative AO concentration in the extracts of mugwort (1), red clover (2), tansy (3), St. John's wort (4) and yarrow (5).

that is, when the most active components has already been consumed. No detailed studies were also carried out according to the dependence of kinetic characteristics of the interaction of DPPH with extractive substances from plants on the reaction medium.

According to the published data [9–11], the transfer of a hydrogen atom from a phenolic AO on DPPH radical can occur for at least two independent competing mechanisms. Radical mechanism of HAT (hydrogen atom transfer), based on the direct abstraction of a hydrogen atom from a molecule of AO, proceeds with the highest rate in nonpolar solvents:

$$\text{ArOH} + \text{DPPH}^{\bullet} \rightarrow \text{ArO}^{\bullet} + \text{DPPH-H} \quad (1)$$

Question of a second (ionic) mechanism remains controversial. Several researchers [10] believe that it is based on electron transfer molecule of ionized phenolic AO molecule of DPPH. This mechanism is given the name SPLET (sequential proton loss – electron transfer), prevails in solvents having high proton affinity:

$$\text{ArOH} \underset{+H^+}{\overset{-H^+}{\rightleftarrows}} \text{ArO}^- + \text{DPPH}^{\bullet} \longrightarrow \text{ArO}^{\bullet} + \text{DPPH}^- \xrightarrow{+H^+} \text{DPPH-H} \quad (2)$$

Other authors [11] give the name of the second mechanism ET – PT (electron transfer – proton transfer) and believe that it proceeds as follows:

$$ArOH + DPPH \rightleftarrows ArOH^{+\bullet} + DPPH^{-} \longrightarrow ArO^{\bullet} + DPPH\text{-}H \quad (3)$$

In almost all organic solvents the reaction proceeds at a high rate that complicates kinetic studies, forcing many authors confined to finding the amount of stock in the plant extract. Furthermore, the most commonly used organic solvents (methanol, ethanol), has the fastest reaction mechanism SPLET, while in the lipid phase in the process of lipid peroxidation inhibition Antioxidants react with radicals by the mechanism HAT. This casts doubt on the adequacy of any comparison of antioxidant activity of extracts from plants with their antiradical activity against DPPH [10].

When determining the kinetic parameters of the samples of plant origin is always necessary to know the concentration of AO in these samples. In the literature there are no data about the range in which dependence of the conversion of DPPH on the concentration of AO entered into the system is linear.

In this chapter, the authors set the task of finding solutions to these problems.

16.2 MATERIALS AND METHODS

Aerial parts of medicinal plants were collected during the flowering period in the Tver region and dried according to [12]. Extraction ball milled material was made by 95% ethanol for 30 min with continuous shaking at 120 min^{-1}. Sample and solvent ratio 1:120 (w/v). Apples, onions and garlic bulbs premilled knife. Juice was mechanically squeezed from citrus pulp, after which AO from the remainder where further extracted.

Kinetic curves of DPPH radical interaction with plant extracts were recorded on a spectrophotometer "Specord M40" equipped with computer software Soft Spectra 5.0, thermostat and thermostatically controlled cuvette holder. The reaction is carried out in quartz cuvettes with a layer thickness of 10 mm of the sample at 293 ± 1 K by pouring 2.4 mL of 8.7×10^{-5} M DPPH solution in ethanol to 0.8 mL of the extract, the concentration of which is equal to IC_{50}. IC_{50} – the concentration of the extract, at which the degree of conversion of DPPH $\zeta = 50\%$ at a given initial concentration of the radical $[DPPH]_0$, fixed volume ratio of the mixed solutions and the exposure time

[13]. Processing curves performed on a PC using the software Microcal Origin 7.0 and SYSTAT TableCurve 2Dv 5.1.

16.3 RESULTS AND DISCUSSION

16.3.1 STUDY OF THE DEPENDENCE OF *Z* FROM $[AO]_0$

Experimental points obtained from experiments for determining the ζ of series of consecutive dilutions of five different extracts of medicinal plants are placed in a coordinate plane in Fig. 16.1.

The graph shows that all the points fit well on a single curve, the range of linear dependence of ζ from $[AO]_0$ is 0 – 60%. This allows a linear interpolation and extrapolation of data in the specified range of ζ for establishing extract concentration corresponding to IC_{50}.

16.3.2 KINETIC STUDIES

When using DPPH method we propose to use the initial reaction rate under certain fixed values of the initial concentration of DPPH and the amount of antioxidants as a kinetic parameter to compare APA extracts of various plants.

In accordance with the laws of formal kinetics

$$w_0 = (k_{eff})_0[DPPH]_0^m[AO]_0^n, \tag{4}$$

where w_0 – the initial reaction rate, $(k_{eff})_0$ – effective (apparent) reaction rate constant at the initial time, m and n – effective (apparent) partial orders of reaction on DPPH and AO respectively. Thus, w_0 is linearly related to $(k_{eff})_0$.

By its physical meaning $(k_{eff})_0$ is the sum of the rate constants for the interaction of individual antioxidants with DPPH in the first stage, multiplied by their mole fraction in the amount of AO of the extract:

$$(k_{eff})_0 = k_1\frac{(C_1)_0}{[AO]_0} + k_2\frac{(C_2)_0}{[AO]_0} + \cdots + k_n\frac{(C_n)_0}{[AO]_0}, \tag{5}$$

where $(C_1)_0$, $(C_2)_0$, …, $(C_n)_0$; k_1, k_2, …, k_n – the initial concentrations of individual AO and the corresponding rate constant for the reaction with

DPPH. k_{eff} decreases monotonically during the reaction, since changing the composition of AO to decrease the proportion of the most active compounds, which are rapidly consumed at the beginning of the process.

The value of w_0, as $(k_{eff})_0$, is mainly determined by the values of the rate constants and the mole fraction of the most active components in the amount of AO, which contribute most to the antioxidant properties of the extract.

The fact that the calculation of w_0 allows to "see" the most active AO, gives the benefits of this value in comparison with the half-time of DPPH ($t_{1/2}$) proposed as the kinetic parameter in [14]. Because of the change in the AO during the reaction half-time is inversely proportional not to $(k_{eff})_0$, but a certain average value $k_{eff} < (k_{eff})_0$ for a period of mixing the reactants to ($t_{1/2}$). This value k_{eff} corresponds to the time when the most active antioxidants has been consumed.

Using software Microcal Origin 7.0 and (independently) SYSTAT TableCurve 2D v5.1 an equation describing the interaction of all the kinetic curves of DPPH with AO plant extracts with high precision ($r^2 > 99{,}9\%$) is derived:

$$D_{DPPH} = D_{\infty} + A_1 \exp(-t/a_1) + A_2 \exp(-t/a_2), \quad (6)$$

where D_{DPPH} – absorbance of the solution at л = 517 nm, t – time from the start of the reaction, D_{∞}, A_1, A_2, a_1, a_2 – parameters chosen by the program. In the program menu Origin this equation is given the name "Exponential Decay 2"; lettering parameters we changed to reflect the physical sense.

Thus, knowing the time from mixing the reactants to registration of the kinetic curve by the device and initial optical density of DPPH, the curve could be approximated to the initial moment and the initial velocity can be calculated:

$$w_0 = (A_1/a_1 + A_2/a_2) \cdot e_{DPPH}^{-1}, \quad (7)$$

wherein e_{DPPH} – the molar extinction coefficient of DPPH in ethanol, equal to 1.15±0.02)×10^4 $mol^{-1} \cdot cm^{-1}$[15][1].

[1]This value is determined for $\lambda = 516$ nm, however, due to the large width of the peak extinction of DPPH at $\lambda = 517$ nm, according to our data, different from that at 516 nm of less than 0.1%, much less than the error in determining eDPPH.

Applying Eq. (6) allows to smooth noise in the kinetic curves and thus reduce the error in their differentiation, as well as to approximate curves at time t = 0 if not applicable stopped-flow equipment.

As mentioned above, plant antioxidants react with of DPPH in ethanol at a high rate (Fig. 16.4, curves 1–3), which hampers the correct definition of w_0 without the use of stopped-flow equipment. W_0 obtained underestimated of the true value, especially in the case of the most active extracts, as the record of the kinetic curve begins at an essential degree of transformation of the initial components.

Introduction of the acetic acid in the reaction mixture causes a sharp drop in the reaction rate (Fig. 16.2, curves 1'–3'). Changes in the UV spectrum of the mixture are shown on Fig. 16.3. For all the studied extracts minimum reaction rate was observed when the acetic acid concentration of about 50 mM (Fig. 16.4). In this case the initial reaction rate was 10–40 times lower compared with the experiments without the addition of acid. With increasing power of injected acids (monochloroacetic, trichloroacetic acid, hydrochloric acid) a minimum rate of reaction shifted towards

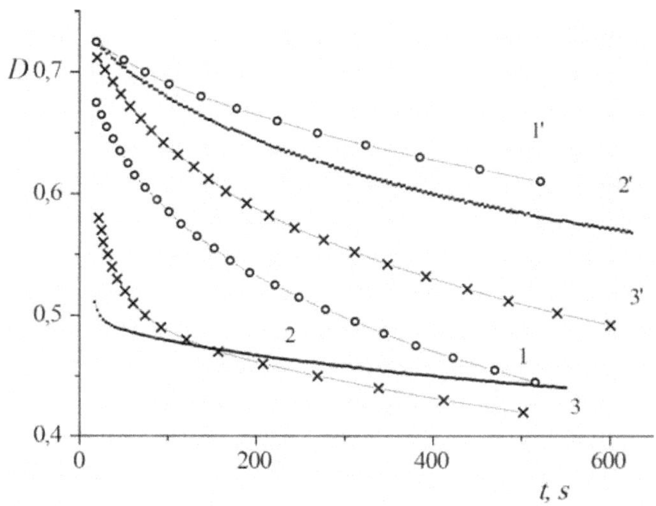

FIGURE 16.2 Kinetic curves of the DPPH absorbance decrease at 517 nm during it's interaction with AO of the *Hypericum perforatum* (1), tansy (2) and yarrow (3) extracts. 1', 2', 3'–the same experiments in the presence of 50 mM acetic acid. $[DPPH]_0 = 6.5 \times 10^{-5}$ M, $[AO]_0 = IC_{50}$.

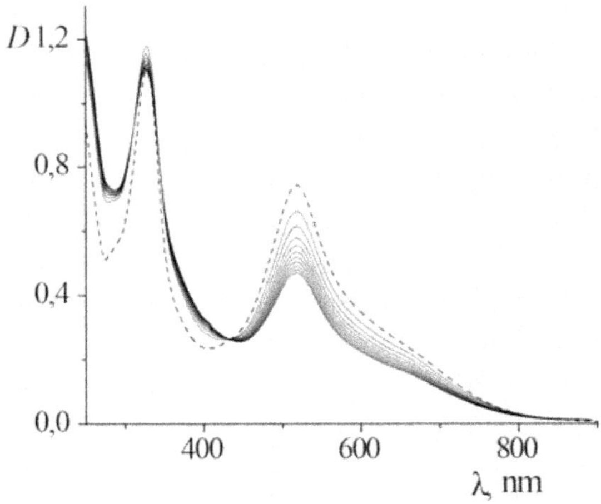

FIGURE 16.3 UV spectrum of the DPPH radical in ethanol (dotted line) and the dynamics it's change in presence of 10 mM of acetic acid after addition of the *Hypericum perforatum* extract (solid line; conducted through 3, 6, 9, 12, 15, 18, 21, 24, 27, and 30 min after the start of the reaction). DPPH taken in excess with respect to the antiradical components of the extract. Along with the gradual decrease in the intensity of the absorption band of the radical at 517 nm an abrupt increase in optical density in the range of 350–420 nm and 315 nm is observed when adding the extract in the reaction medium. The latter is caused by the presence of phenolic compounds in the extract, which absorb in those wavelength ranges.

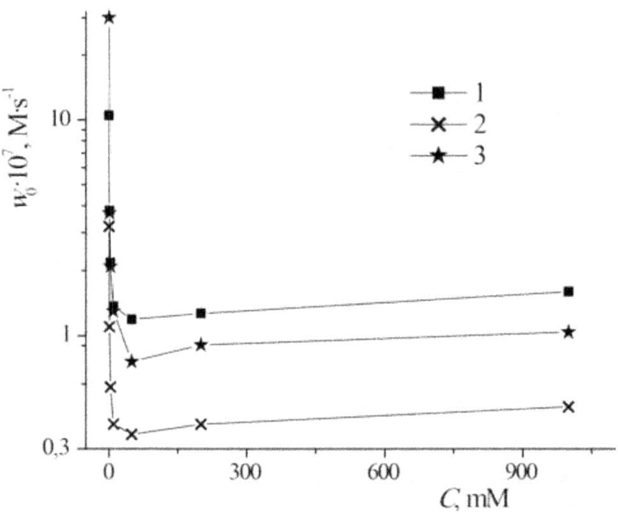

FIGURE 16.4 Changes in initial DPPH reaction rate with antioxidants of yarrow (1), St. John's wort (2), tansy (3) extracts in ethanol in the presence of acetic acid at concentrations of 0, 1, 3, 10, 50, 200, 1000 mM (semilogarithmic scale).

smaller values acid concentration (3.3 mM for monochloroacetic, 0.1 mM for trichloroacetic and hydrochloric acid, Fig. 16.5).

The rate of reaction at acid concentrations corresponding to the minimal rates also decreased with increasing acid strength. In experiments with tansy extract w_0 decreased 1.4 fold in the presence of monochloroacetic, 2.1 times – trichloroacetic acid and 3.4 times – hydrochloric acid compared to the rate in the presence of acetic acid. Figure 16.7 shows the kinetic curves of absorbance drop of DPPH by the interaction with tansy extract antioxidants without the addition of acids (1) and at acid concentration in the reaction medium corresponds to the minimum speed of the reaction (2–5).

In ethanol, acidified with 0.1 mM HCl, slowest process flow is observed.

However, AO of grapefruit, oranges, garlic (Fig. 16.6–16.9), as well as other members of the citrus and onion families (lemon, onion) interact with a minimum rate with DPPH at the concentration of hydrogen chloride in ethanol for about 3–10 mM.

In the reaction of DPPH with orange fruit AO 0.1 mM solution of HCl in ethanol inhibits the mechanism SPLET so weak, that by means of used equipment the initial portion of the kinetic curve could not be written (Fig. 16.7, curve 1).

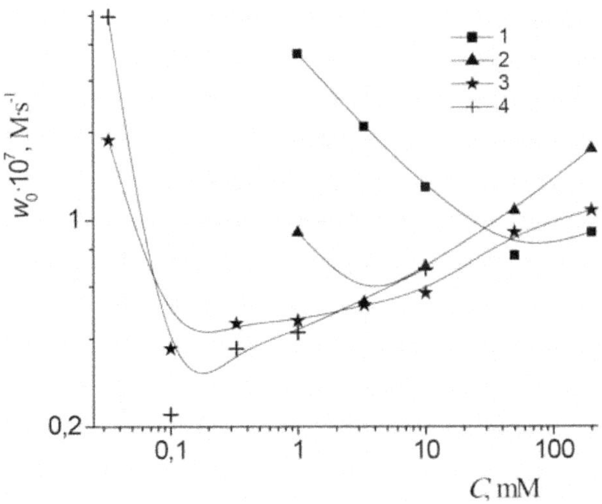

FIGURE 16.5 Changes in initial rate of the DPPH reaction with AO of tansy extract (on a logarithmic scale) by the introduction of various concentrations of acids into the reaction system: acetic (1) monochloroacetic (2), trichloroacetic (3), hydrochloric (4).

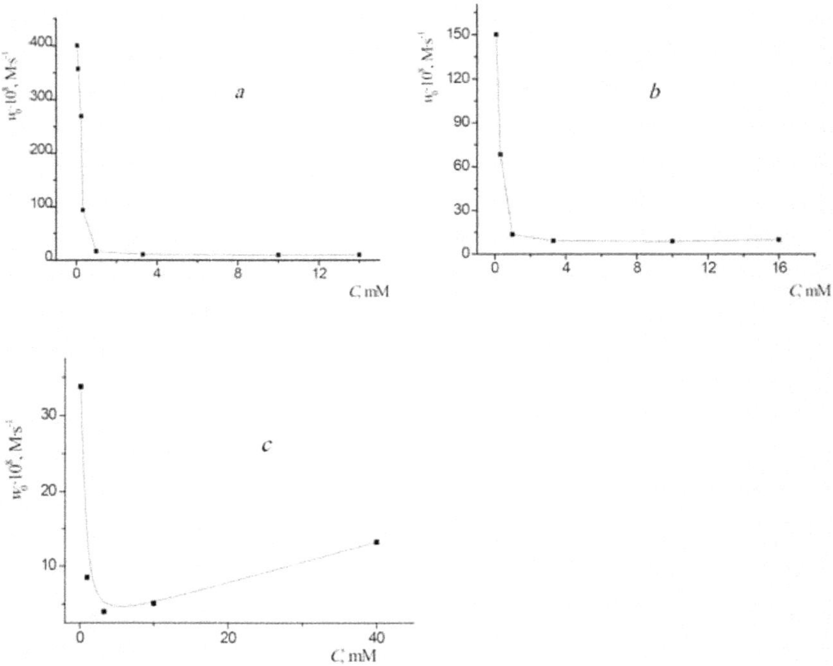

FIGURE 16.6 The dependence of initial reaction rate of DPPH radical w_0 with antioxidants from extract of orange fruit (s), grapefruit (b) and garlic bulbs (c) on the concentration of hydrochloric acid C in the reaction medium.

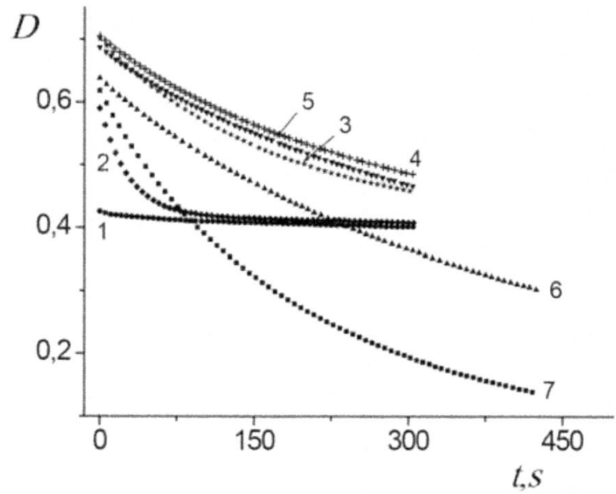

FIGURE 16.7 Decrease in absorbance of DPPH solution D at its interaction with the orange fruit antioxidants in the presence of various concentrations of hydrochloric acid in ethanolic medium: 0.1 мM (1); 0.33 мM (2); 1 мM (3); 3.3 мM (4); 10 мM (5); 30 мM (6); 100 мM (7).

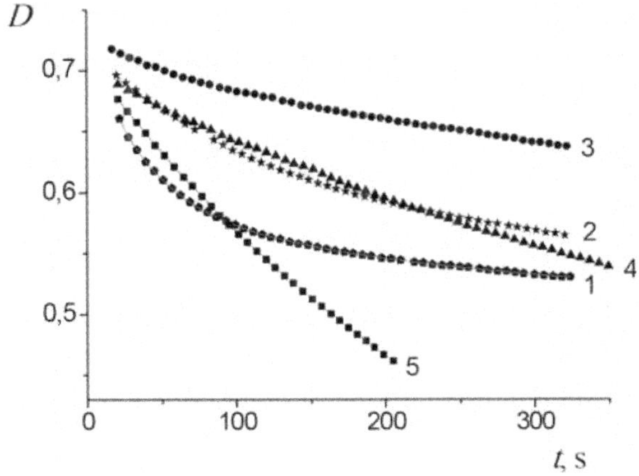

FIGURE 16.8 Decrease in absorbance of DPPH solution D at its interaction with the garlic bulb antioxidants in the presence of various concentrations of hydrochloric acid in ethanolic medium: 0.1 мM (1); 1 мM (2); 3.3 мM (3); 10 мM (4); 40 мM (5).

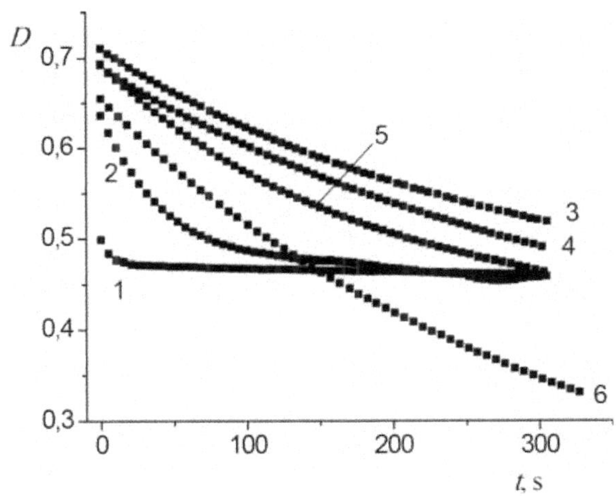

FIGURE 16.9 Decrease in absorbance of DPPH solution D at its interaction with the grapefruit fruit antioxidants in the presence of various concentrations of hydrochloric acid in ethanolic medium: 0.1 мM (1); 0.33 мM (2); 1 мM (3); 3.3 мM (4); 10 мM (5); 40 мM (6).

As can be seen from the figures, despite the differences in the position of the minimum rate of the reaction remains unchanged general character of changes of its velocity with increasing acid concentration. Due to the presence of the lone electron pairs on the oxygen atom of ethanol molecule phenolic AO can dissociate, giving proton to solvent molecule and turning into phenoxide anions. As the result, in this reaction medium high rate of the process by the mechanism SPLET occurs:

$$ArOH \underset{+H^+}{\overset{-H^+}{\rightleftarrows}} ArO^- \xrightarrow{+DPPH^\bullet} ArO^\bullet + DPPH^- \xrightarrow{+H^+} ArO^\bullet + DPPH\text{-}H$$

Hydrogen chloride dramatically slows the reaction by the mechanism SPLET, as the increase in the concentration of protonated solvent molecules (*solvH*$^+$) due to the introduction of a strong acid leads to a drastic shift of the equilibrium towards the undissociated molecules of a phenolic AO:

$$ArOH + solv \rightleftarrows ArO^- + solvH^+$$

Increasing the speed of the reaction at a further increase of the concentration of HCl indicates that process proceeds for at least one another mechanism sensitive to acid concentration in parallel with the mechanism SPLET. DPPH consumption rate by the mechanism HAT (purely radical, non-ions and charge transfer) may decrease, but not increase with increasing concentration of HCl, accompanied by increase of the permittivity of the medium. On the other hand, the latter may assist in the separation of ionic pairs when the reaction by the mechanism ET – PT [11], which is accompanied by increase of DPPH consumption rate:

$$ArOH + DPPH^\bullet \rightleftarrows ArOH^{+\bullet} + DPPH^- \longrightarrow ArO^\bullet + DPPH\text{-}H$$

Thus, the dependence of the reaction rate of the DPPH radical with extractives of plants is the sum of two functions – increasing and decreasing, respectively, reflecting the dependence of the rates of SPLET and ET – PT on the acid concentration. At the minimum of the total reaction rate decrease of process SPLET rate per unit change in acid concentration is equal to increase of the rate of the process ET – PT. As can be seen from Figs. 16.5 and 16.7–16.9, the minimum rate reaction for various multicomponent systems of plant origin is in HCl concentration range of 0.1–10 mM. Thus, in order to ensure the maximum of the SPLET process rate reduction in all cases, it is necessary to carry out the reaction at a concentration of HCl in the reaction system about ≈ 3–10 mM.

Increasing of this concentration does not seem appropriate, since this leads to a decrease of the HAT radical mechanism relative contribution to the total reaction rate. This mechanism of interaction of AO with radicals prevails if the process flows in hydrophobic environment. It is based on the homolytic cleavage of OH bond, so that the reaction rate by this mechanism characterizes the strength of this bond. Besides this, at HCl concentrations of 30 and 100 mM, rapid DPPH degradation accompanied with bleaching of solution is observed (curves 6 and 7 in Fig. 16.7). This way D quickly go to optical densities significantly lower than the baseline of the curve 1 when it reaches a plateau, corresponding to 0.1 mM HCl. The same behavior is demonstrated by curve 5 in Fig. 16.8 and the curve 6 in Fig. 16.9 describing the decrease in absorbance of DPPH solution containing 40 mM HCl during the interaction with AO of garlic and grapefruit. Individual selection of concentration HCl, corresponding to the minimum reaction rate for each object also seems unreasonable. Rate of the process ET – PT depends on the concentration of hydrogen chloride. Changing of this concentration makes the results not comparable. In addition, the determination of the position of this minimum for each object – a very laborious and lengthy process.

An example of using the method for determining the activity of antioxidants of some food and medicinal plants based on the analysis of the kinetics of their interaction with DPPH radical is shown in Table 16.1.

TABLE 16.1 Antiradical Activity of Antioxidants in Alcoholic Extracts of Some Food and Medicinal Plants ($[HCl]$ = 3.3 мM)

No	Source of AO	$w_0 \times 10^8$, M·s^{-1}	k_{eff}, M^{-1}·s^{-1}
1	Lemon (fruit pulp)	13	260
2	Orange (fruit pulp)	11	220
3	Grapefruit (fruit pulp)	9.1	190
4	Onion (*Allium cepa L.*)	10	210
5	Garlic (bulb)	4.0	81
6	Tansy (leaves)	5.0	100
7	Tea "Hibiscus"	24	500
8	Apple (sort of "Golden")	11.5	230
9	Red grapes (fruit)	7.5	150

It follows from the table that the highest antiradical activity have tea AO "Hibiscus," citrus fruits, apples, and onion bulbs of *Allium cepa L.*

Antiradical activity of many plant extracts (Table 16.2) can also be analyzed at 0.1 mM HCl concentration in the reaction medium. However, AO, which minimum of the reaction rate is observed near this value of hydrogen chloride concentration in this case reacts with DPPH at a rate of 1.75–2.5 times less than in the presence of 3.3 mM HCl. As a result, the relative contribution of radical and ionic mechanisms is quite different. So, such data are not comparable with the results of kinetic analysis of antiradical activity of AO which minimum rate of interaction with of DPPH lies in the concentration of hydrogen chloride in excess of 0.1 mM.

According to our data, in the above mentioned acid concentrations have no significant effect on either the extinction coefficient of DPPH in ethanol, or the stability of the radical components and antiradical extracts.

Relative error of w_0 determination does not exceed 15%, which is approximately equal to the error of calculation of the kinetic parameters for the reactions of DPPH with individual substances, where is possible to calculate the rate constant using kinetic equations [15].

Since, in accordance with the theory of Ingold et al. [10, 16], reducing the reaction rate when acid is introduced into system occurs due to the suppression of SPLET mechanism, it can be assumed that the data obtained in this way should be better correlated with experiments based on free

TABLE 16.2 Antiradical Activity of Antioxidants in Alcoholic Extracts of Some Food and Medicinal Plants ([HCl] = 0.1 mM)

No	Source of AO	$w_0 \times 10^8$, $M \cdot s^{-1}$	k_{eff}, $M^{-1} \cdot s^{-1}$
1	Peppermint (*Mentha piperita L.*)	2.0	41
2	St. John's wort (*Hypericum perforatum L.*)	2.1	43
3	*Melissa officinalis L.*	2.4	49
4	Yarrow (*Achillea millefolium L.*)	2.3	48
5	Chamomile (*Matricaria chamomilla L.*)	2.6	53
6	Tansy (*Tanacetum vulgare L.*)	2.0	41
7	Tea "Hibiscus"	12	250
8	Apple (sort of "Golden")	5.0	100
9	Red grapes (*Vitis vinifera L.*)	4.3	87

radical chain oxidation, which requires further experimental verification. Besides, alcoholic extracts from plants always contain organic acids having different power in different concentrations that can significantly affect the results obtained without the addition of external acid. By suppressing mechanism SPLET this effect is excluded.

16.4 CONCLUSIONS

1. Kinetic parameters based on measurement of the initial reaction rate, the most adequately characterize the antiradical activity of antioxidants in the objects of vegetable origin.
2. In quantitative analysis of antioxidants by the DPPH-method it is necessary to consider that the decrease in absorbance dependence of the solution of the radical on the concentration of AO linear in the range of DPPH conversion degree from 0 to 60%.
3. Carrying out the reaction low molecular weight AO of plant origin with a stable radical DPPH in the environment of 3.3–10 mM solution of HCl in ethanol allows to reduce its rate by about two orders of magnitude by suppressing SPLET mechanism. This enables observation of the process, starting from small conversion degrees, high precision of calculation of the kinetic parameters and elimination of the influence of organic acids contained in the extracts of plants on kinetic parameters. In addition, the dramatic increase in the contribution of radical mechanism HAT to the observed rate of the process closer processes in a system based on the use of DPPH to systems based on the radical chain oxidation of hydrocarbons.
4. The similarity of kinetic parameters change patterns of the reaction of DPPH with plants extractive substances and pure phenolic AO when hydrochloric acid introduced into the reaction medium confirms the decisive role of phenols in of the total antiradical activity of alcoholic extracts of plants.
5. The experiments confirms that the ionic mechanism of DPPH interaction with phenolic AO in ethanolic medium can be carried out in parallel by the scheme SPLET (electron transfer from anion phenolic AO) and by the scheme ET – PT (electron transfer from un-ionized phenolic AO).

KEYWORDS

- **2,2-diphenyl-1-picrylhydrazyl**
- **antioxidants**
- **antiradical activity**
- **DPPH**
- **plant extracts**

REFERENCES

1. V. A. Roginsky, E. A. Lissi. Review of methods to determine chain-breaking antioxidant activity in food. Food Chem., 92, 235–254 (2005).
2. B. A. Silva, F. Ferreres, J. O. Malva, and A. C. P. Dias. Phytochemical and antioxidant characterization of *Hypericum perforatum* alcoholic extracts. Food Chem., 90 (1–2), 157–167 (2005).
3. C. Imark, M. Kneubbhl, and S. Bodmer. Occurrence and activity of natural antioxidants in herbal spirits. Innovative Food Science & Emerging Technologies, 1(4), 239–243 (2000).
4. W. Bondet, W. Brand-Williams, and C. Berset. Kinetics and Mechanisms of Antioxidant Activity using the DPPH$^{\bullet}$ Free Radical Method. Lebensm.-Wiss u.-Technol., 30, 609–615 (1997).
5. K. Schwarz, G. Bertelsen, L. R. Nissen et al. Investigation of plant extracts for the protection of processed foods against lipid oxidation. Comparison of antioxidant assays based on radical scavenging, lipid oxidation and analysis of the principal antioxidant compounds. Eur. Food Res. Technol., 212, 319–328 (2001).
6. A. L. Buchachenko, A. M. Wasserman. Stable Radicals. Electronic structure, reactivity and application. Khimiya, Moscow, 1973, 408 P. [in Russian].
7. E. G. Rozantsev. Free iminoxyl radicals. Khimiya, Moscow, 1970, 216 P. [in Russian].
8. N. Nishimura and T. Moriya. Reaction between 2,2-Diphenyl-1-picrilhydrazyl and Phenols. Substituent and Solvent Effects. Bull. Chem. Soc. Jap., 50(8), 1969–1974 (1977).
9. M. Musialik, G. Litwinienko. Scavenging of dpph Radicals by Vitamin E is Accelerated by Its Partial Ionization: the Role of Sequential Proton Loss Electron Transfer. Organic Letters, 7 (22), 4951–4954 (2005)
10. G. Litwinienko and K. U. Ingold. Abnormal Solvent Effects on Hydrogen Atom Abstractions. 1. The Reactions of Phenols with 2,2-Diphenyl-1-picrylhydrazyl (dpph) in Alcohols. J. Org. Chem., 68, 3433–3438 (2003).
11. M. Leopoldini, T. Marino, N. Russo, and M. Toscano. Antioxidant Properties of Phenolic Compounds: H-Atom versus Electron Transfer Mechanism. J. Phys. Chem. A., 108, 4916–4922 (2004).
12. USSR State Pharmacopoeia, 11th ed., Issue 2, Meditsina, Moscow (1990).

13. O. I. Aruoma. Neuroprotection by Bioactive Components in Medicinal and Food Plant Extracts. Mutation Research/Fundamental and Molecular Mechanisms of Mutagenesis, 523–524, 9–20 (2003).
14. T. A. Filippenko, N. I. Belaya, F. N. Nikolaevsky. Phenolic Compounds of Plant Extracts and Their Activity in the Reaction with Diphenylpicrylhydrazyl. Khim.-pharm. zhurn., 38 (8), 34–36 (2004).
15. M. C. Foti, C. Daquino and C. Geraci. Electron-Transfer Reaction of Cinnamic Acids and Their Methyl Esters with the DPPH˙ Radical in Alcoholic Solutions. J. Org. Chem., 69(7), 2309–2314 (2004).
16. G. Litwinienko and K. U. Ingold. Effects on Hydrogen Atom Abstraction. 2. Resolution of the Curcumin Antioxidant Controversy. The Role of Abnormal Solvent Sequential Proton Loss Electron Transfer. J. Org. Chem., 69, 5888–5896 (2004).

CHAPTER 17

UPDATE ON MODERN FIBERS, FABRICS AND CLOTHING

A. K. HAGHI

University of Guilan, Rasht, Iran

CONTENTS

17.1 INTRODUCTION

The fast development and changes in life style has attracted peoples towards a more comfort and luxurious life. People are moving towards small, safer, cheaper and fast working products which not only reduces the work load but also help them to carry out their works at a much greater pace with minimum efforts. There have been development of gadgets that are much smaller in size like microchip, nano capsules, carbon tubes, memory cards, pen drives etc. which reduces the problems of transport, and storage and are also much faster and reliable by which we can carry out more of our work in less time. In the formation and development of such products nanotechnology plays a very important and vital role [1, 2].

The term nanotechnology (sometimes shortened to "nanotech") comes from nanometer – a unit of measure of 1 billionth of a meter of length. The concept of Nanotechnology was given by Nobel Laureate Physicist Richard Feynman, in 1959. Nanotechnology is defined as the understanding, manipulation, and control of matter at the length scale on nanometer, such that the physical, chemical, and biological properties of materials (individual atoms, molecules and bulk matter) can be engineered, synthesized or altered to develop the next generations of improved materials, devices, structures, and systems [3].

Generally, nanotechnology deals with structures that are sized between 1 to 100 nm in at least one dimension and involves developing materials or devices possessing dimension within size. Nanotechnology creates structure that have excellent properties by controlling atoms and molecules, functional materials, devices and systems on the nanometer scale by involving precise placement of individual atoms [4]. Although nanotechnology is a relatively recent development in scientific research, the development of its central concepts happened over a longer period of time. The emergence of nanotechnology in the 1980s was caused by the convergence of experimental advances. The early 2000s also saw the beginnings of commercial applications of nanotechnology, although these were limited to bulk applications of nanomaterials, such as the silver nano platform for using silver nanoparticles as an antibacterial agent, nanoparticles-based transparent sunscreens, and carbon nanotubes for stain-resistant textiles [5, 6].

Throughout history, the textiles sectors have been used worldwide in a wide range of consumer applications. Natural fibers, such as cotton, silk, and wool, along with synthetic fibers, such as polyester and nylon, continue to be the most widely used fibers for apparel manufacturing. Synthetic fibers are mostly suitable for domestic and industrial applications, such as carpets, tents, tires, ropes, belts, cleaning cloths, and medical products. Natural and synthetic fibers generally have different characteristics, which make them ideally suitable mainly for apparel. Depending on the end-use application, some of those characteristics may be good, while the others may not be as good to contribute to the desired performance of the end product [7, 8].

As stated previously, nanotechnology brings the possibility of combining the merits of natural and synthetic fibers, such that advanced fabrics that complement the desirable attributes of each constituent fiber can be produced. In the last decade, the advent of nanotechnology has spurred significant developments and innovations in this field of textile technology. By using nanotechnology, there have been developments of several fabric treatments to achieve certain enhanced fabric attributes, such as superior durability, softness, tear strength, abrasion resistance, durable-press and wrinkle-resistance. The (nano)-treated core component of a core-wrap bicomponent fabric provides high strength, permanent antistatic behavior, and durability, while the traditionally treated wrap component of the fabric provides desirable softness, comfort, and esthetic characteristics [9, 10].

17.2 THE TEXTILE INDUSTRY BACKGROUND

The origin of the textile industry is lost in the past. Fine cotton fabrics have been found in India, dating from some 6–7000 years ago, and fine and delicate linen fabrics have been found from two to three thousand years ago, at the height of the Egyptian civilizations. More recent archaeological excavations, among some of Europe's oldest Stone Age sites, have found imprints of textile structures, dating back some 25,000 years, but in the humid conditions obtaining in these more northerly areas, all traces of the actual textiles have long disappeared, unlike those from the dry areas of India and Egypt [11].

Until more recent times, the spinning of the yarns and the weaving of the fabrics were generally undertaken by small groups of people, working together – often as a family group. However, during the Roman occupation of England, the Romans established a factory' at Winchester, for the production, on a larger scale, of warm Woolen blankets, to help reduce the impact of the British weather on the soldiers from southern Europe [12].

In the family context, it generally fell to the female side to undertake the spinning, while the weaving was the domain of the men. Spinning was originally done using the distaff to hold the unspun fibers, which were then teased out using the fingers and twisted into the final yarn on the spindle. In the 1530s, in Brunswick, a 'spinning wheel' was invented, with the wheel driven by a foot pedal, giving better control and uniformity to the yarns produced. Often, great skill was developed, as shown by the records of a woman in Norwich, who spun one pound of combed wool into a single yarn measuring 168,000 yards, and from the same weight of cotton, spun a yarn of 203,000 yards. In today's measures this is equivalent to a cotton count of 240, or approximately 25 decitex. Cotton count is the number of hanks of 840 yards (768 meters) giving a total weight of 1 lb (453.6 g). A Tex is a measurement of the linear density of a yarn or cord, being the weight in grams of a 1,000 m length; a decitex is the weight in grams of a 10,000 m length [13, 14].

By the eighteenth century, small cooperatives were being formed for the production of textiles, but it was really only with the mechanization of spinning and weaving during the Industrial Revolution, that mass production started.

Up to this time, both spinning and weaving were essentially hand operations. Handlooms were operated by one person, passing the weft (the transverse threads) by hand, and performing all the other stages of weaving manually. In 1733, John Kay invented the 'flying shuttle,' which enabled a much faster method for inserting the weft into the fabric at the loom and greatly increased the productivity of the weavers [15].

Until the advent of the flying shuttle, the limiting factor in the production chain for fabrics was the output of the individual weaver, but this now changed and with the more rapid use of the yarns, their production became the limiting factor in the total process. In 1764, this was partly resolved by the invention, by James Hargreaves, of the 'Spinning Jenny,' which was developed further by Sir Richard Arkwright, with his water spinning frame, in 1769, and then in 1779, by Samuel Crompton, with his 'spinning mule' [16].

Alongside these developments in spinning, similar changes were taking place in the weaving field, with the invention of the power loom by Edmund Cartwright, in 1785.

With this increase in mechanization of the whole industry, it was logical to bring the production together, rather than keeping it widely spread throughout the homes of the producers. Accordingly, factories were established [17]. The first of such was in Doncaster in 1787, with many power looms powered by one large steam engine. Unfortunately, this was not a financial success, and the mill only operated for about 3 or 4 years [18].

Meanwhile, other mills were being established, in Glasgow, Dumbarton and Manchester. A large mill was erected at Knott Mill, Manchester, although this burnt down after only about 18 months. The first really successful mill was opened in Glasgow in 1801 [19].

However, this industrialization was not to everyone's liking; many individuals were losing their livelihoods to the mass production starting to come from the increasing number of mills. This led to a backlash from the general public, resulting in the Luddite Riots in 1811–1812, when bands of masked people under the leadership of 'King Ludd' attacked the new factories, smashing all the machinery therein. It was only after very harsh suppression, resulting in the hanging or deportation of convicted Luddites in 1813, that this destruction was virtually stopped. However, there were still some outbreaks of similar actions in 1816, during the depression following the end of the British war with France, and this intermittent action only finally stopped when general prosperity increased again in the 1820s [20].

Following this, the textile industry expanded considerably, particularly in the areas where the raw materials were readily available. For example, the Woolen mills in East Anglia, where there was good grazing for the sheep, and in West Yorkshire and Eastern Lancashire, where either coal was available for powering the new steam engines, or where fast flowing streams existed to provide the energy source for water-powered mills. The main Woolen textile production developed in Yorkshire, as it was easier and cheaper to transport the raw wool there, than to carry the large quantities of coal required to power the mills to the wool growing areas [21]. In Lancashire, with the ports of Liverpool and Manchester close by for the importation of cotton from America, the cotton industry grew and flourished. However, in the 1860s, due to the American Civil War, the supply of cotton from America dried up and caused great hardship among the cotton towns of south and east Lancashire [22].

On account of this, and with the great strides being made in chemistry, research was begun to find ways of making artificial yarns and fibers. The first successful artificial yarn was the Chardonnet 'artificial silk,' a cellulosic fiber regenerated from spun nitrocellulose. Further developments lead to the cuprammonium process and then to the viscose process for the production of another cellulosic, rayon [23]. This latter viscose was fully commercialized by Courtaulds in 1904, although it was not widely used in rubber reinforcement until the 1920s, with the development of the balloon type [24].

Research continued into fiber-forming polymers, but the next new fully synthetic yarn was not discovered until the 1930s, when Wallace Hume Carothers, working for DuPont, discovered and developed nylon. This was first commercialized in 1938 and was widely developed during the 1940s to become one of the major yarn types used. Continuing research led to the discovery of polyester in 1941, and over the ensuing decades, polyolefin fibers (although because of their low melting/softening temperatures, these are not used as reinforcing fibers in rubbers) and aramids [25].

As the chemical industry greatly increased the types of yarns available for textile applications, so the machinery used in the industry was being developed. Whereas the basic principles of spinning and weaving have not significantly changed over the millennia, the speed and efficiency of the equipment used for this has been vastly been improved. In weaving, the major changes have been related to the method of weft insertion; the conventional shuttle has been replaced by rapiers, air and water jets, giving far higher speeds of weft insertion [26].

Other methods of fabric formation have similarly been developed, such as the high speed knitting machines and methods for producing fabric webs [27].

The use of nanotechnology in the textile industry made it multifunctional which can produce fibers with variable functions and applications such as UV protection, antiodor, antimicrobial etc. In many cases smaller amounts of the additive are required, for the saving on resources. The success of Nanotechnology and its potential applications in textiles lies in various fields where new methods are combined with multifunctional textile systems, durable etc. without affecting the inherent properties of the textiles including softness, flexibility, washability, etc. Keeping the above factors in consideration, the present review highlighted the use of nanotechnology in textile industry and textile engineering, the types and methods of preparation of different nano composites used in textiles [28, 29].

17.3 TEXTILE TYPES

Textiles can be made from many materials. They are classified on the basis of their component fibers into, animal (wool, silk), plant (cotton, flax, jute), mineral (asbestos, glass fiber), and synthetic (nylon, polyester, acrylic). They are also classified as to their structure or weave, according to the manner in which warp and weft cross each other in the loom. Textiles are made in various strengths and degrees of durability, from the finest gossamer to the sturdiest canvas. The relative thickness of fibers in cloth is measured in deniers. Microfiber refers to fibers made of strands thinner than one denier [30, 31].

17.3.1 ANIMAL TEXTILES

The main animal fiber used for textiles is wool. Animal textiles are commonly made from hair or fur of animals. Silk is another animal fiber produces one of the most luxurious fabrics. Sheep supply most of the wool, but members of the camel family and some goats also furnish wool. Wool, commonly used for warm clothing, refers to the hair of the domestic goat or sheep and it is coated with oil known as lanolin, which is water-proof and dirt-proof making a comfortable fabric for dresses, suits, and sweaters. The term woolen refers to raw wool, while *worsted* refers to the yarn spun from raw wool. Cashmere, the hair of the Indian Cashmere goat, and mohair, the hair of the North African Angora goat, are types of wool known for their softness. Other animal textiles made from hair or fur is alpaca wool, vicuña wool, llama wool, and camel hair. They are generally used in the production of coats, jackets, ponchos, blankets, and other warm coverings. Angora refers to the long, thick, soft hair of the Angora rabbit [32, 33].

Silk is an animal textile made from the fibers of the cocoon of the Chinese silkworm. It is spun into a smooth, shiny fabric prized for its sleek texture. Silk comes from cocoons spun by silkworms. Workers unwind the cocoons to obtain long, natural filaments. Fabrics made from silk fibers have great luster and softness and can be dyed brilliant colors. Silk is especially popular for saree, scarfs and neckties [34].

17.3.2 PLANT TEXTILES

Plants provide more textile fibers than do animals or minerals. Cotton fibers produce soft, absorbent fabrics that are widely used for clothing, sheets, and towels. Fibers of the flax plant are made into linen. The strength and beauty of linen have made it a popular fabric for fine tablecloths, napkins, and handkerchiefs [35].

Grass, rush, hemp, and sisal are all used in making rope. In the first two cases, the entire plant is used for this purpose, while in the latter two; only fibers from the plant are used. Coir (coconut fiber) is used in making twine, floor mats, door mats, brushes, mattresses, floor tiles, and saking. Straw and bamboo are both used to make hats. Straw, a dried form of grass, is also used for stuffing, as is kapok. Fibers from pulpwood trees, cotton, rice, hemp, and nettle are used in making paper. Cotton, flax, jute, and modal are all used in clothing. Piña (pineapple fiber) and ramie are also fibers used in clothing, generally with a blend of other fabrics such as cotton. Seaweed is sometimes used in the production of textiles. A water soluble fiber known as *alginate* is produced and used as a holding fiber. When the cloth is finished, the alginate is dissolved, leaving an open area [36, 37].

17.3.3 MINERAL TEXTILES

Asbestos and basalt fiber are used for vinyl tiles, sheeting, and adhesives, "transite" panels and siding, acoustical ceilings, stage curtains, and fire blankets. Glass fiber is used in the production of spacesuits, ironing board and mattress covers, ropes and cables, reinforcement fiber for motorized vehicles, insect netting, flame-retardant and protective fabric, soundproof, fireproof, and insulating fibers. Metal fiber, metal foil, and metal wire have a variety of uses, including the production of "cloth-of-gold" and jewelry [38, 39].

17.3.4 SYNTHETIC TEXTILES

Most manufactured fibers are made from wood pulp, cotton linters, or petrochemicals. Petrochemicals are chemicals made from crude oil and natural gas. The chief fibers manufactured from petrochemicals include nylon, polyester, acrylic, and olefin. Nylon has exceptional strength, wears well, and is easy to launder. It is popular for hosiery and other clothing and for carpeting and upholstery. Such products as conveyor belts and fire hoses are also made of nylon. All synthetic textiles are used primarily in the production of clothing [37, 40].

- Polyester fiber is used in all types of clothing, either alone or blended with fibers such as cotton.
- Acrylic is a fiber used to imitate wools, including cashmere, and is often used in place of them.
- Nylon is a fiber used to imitate silk and is tight-fitting; it is widely used in the production of pantyhose.
- Lycra, spandex, and tactel are fibers that stretch easily and are also tight-fitting. They are used to make active wear, bras, and swimsuits.
- Olefin (Polypropylene or Herculon) fiber is a thermal fiber used in active wear, linings, and warm clothing.
- Lurex is a metallic fiber used in clothing embellishment.
- Ingeo is a fiber blended with other fibers such as cotton and used in clothing. It is prized for its ability to wick away perspiration.

17.4 TEXTILE PRODUCTION

17.4.1 PRODUCTION METHODS

Most textiles are produced by twisting fibers into yarns and then knitting or weaving the yarns into a fabric. This method of making cloth has been used for thousands of years. But throughout most of that time, workers did the twisting, knitting, or weaving largely by hand. With today's modern machinery, textile mills can manufacture as much fabric in a few seconds as it once took workers weeks to produce by hand [41]. The production of textiles are done by different methods and some of the common production methods are listed as follows [42]: (i) *Weaving* (by machine as well as by hand); (ii) *Knitting;* (iii) *Crochet;* (iv) *Felt* (fibers are matted together to produce a cloth); (v) *Braiding;* (vi) *Knotting*. *Weaving* is a textile production method that involves interlacing a set of vertical threads (called the warp) with a set of horizontal threads (called the weft). This is done on a machine known as a loom, of which there are a number of types. Some weaving is still done by hand, but a mechanized process is used most often. Tapestry, sometimes classed as embroidery, is a modified form of plain cloth weaving [43].

Knitting and *crocheting* involve interlacing loops of yarn, which are formed either on a knitting needle or crochet hook, together in a line. The two processes differ in that the knitting needle has several active loops at one time waiting to interlock with another loop, while crocheting never has more than one active loop on the needle [44].

Other specially prepared fabrics not woven are *felt* and *bark* (or tapa) cloth, which are beaten or matted together, and a few in which a single thread is looped or plaited, as in crochet and netting work and various laces. *Braiding* or *plaiting* involves twisting threads together into cloth. Knotting involves tying threads together and is used in making macramé [45].

Most textiles are now produced in factories, with highly specialized power looms, but many of the finest velvets, brocades, and table linens are still made by hand. Lace is made by interlocking threads together independently, using a backing and any of the methods described above, to create a fine fabric with open holes in the work. Lace can be made by hand or machine. The weaving of carpet and rugs is a special branch of the textile industry.

Carpets, rugs, velvet, velour, and velveteen are made by interlacing a secondary yarn through woven cloth, creating a tufted layer known as a nap or pile [46].

17.4.2 PRODUCTION OF COTTON CLOTHES

In the 1700s, English textile manufacturers developed machines that made it possible to spin thread and weave cloth into large quantities. Today, the

United States, Russia, China and India are major producers of cotton. When cotton arrives at a textile mill, several blenders feed cotton into *cleaning machines*, which mix the cotton, break it into smaller pieces and remove trash. The cotton is sucked through a pipe into *picking machines*. Beaters in these machines strike the cotton repeatedly to knock out dirt and separate lumps of cotton into smaller pieces. Cotton then goes to the *carding machine*, where the fibers are separated. Trash and short fibers are removed. Some cotton goes through a *comber* that removes more short fibers and makes a stronger, more lustrous yarn. This is followed by spinning processes which do three jobs: *draft* the cotton, or reduce it to smaller structures, straighten and parallel the fibers and lastly, put twist into the yarn. The yarns are then made into cloth by weaving, knitting or other processes [47, 48].

Some of the properties of cotton are discussed as follows: (i) soft and comfortable, (ii) wrinkles easily, (iii) absorbs perspiration quickly, (iv) good color retention and good to print on, and (v) strong and durable [49].

17.4.3 PRODUCTION OF WOOL

The processing of wool involves four major steps. First comes shearing, followed by sorting and grading, making yarn and lastly, making fabric. This is followed by grading and sorting, where workers remove any stained, damaged or inferior wool from each fleece and sort the rest of the wool according to the quality of the fibers. Wool fibers are judged not only on the basis of their strength but also by their *fineness* (diameter), length, *crimp* (waviness) and color [50].

The wool is then scoured with detergents to remove the *yolk* and such impurities as sand and dust. After the wool dries, it is *carded*. The carding process involves passing the wool through rollers that have thin wire teeth. The teeth untangle the fibers and arrange them into a flat sheet called a *web*. The web is then formed into narrow ropes known as *silvers*. After carding, the processes used in making yarn vary slightly, depending on the length of the fibers. Carding length fibers are used to make woolen yarn. Combing length fibers and French combing length fibers are made into *worsted yarn* [51].

Woolen yarn, which feels soft, has a fuzzy surface and it is heavier than worsted. While worsted wool is lighter and highly twisted, it is also smoother, and is not as bulky, thus making it easier to carry or transport about. Making worsted wool requires a greater number of processes, during which the fibers are arranged parallel to each other. The smoother the hard-surface worsted yarns, the smoother the wool it produces, meaning, less fuzziness. Fine worsted wool can be used in the making of athletics attire, because it is not as hot as polyester, and the weave of the fabric allows wool to absorb perspiration, allowing the body to "breathe." Wool manufacturers knit or weave yarn into a variety of fabrics. Wool may also be dyed at various stages of the manufacturing process and undergo finishing processes to give them the desired look and feel [52, 53].

The finishing of fabrics made of woolen yarn begins with *fulling*. This process involves wetting the fabric thoroughly with water and then passing it through the rollers. Fulling makes the fibers interlock and mat together. It shrinks the material and gives it additional strength and thickness. Worsteds go through a process called *crabbing* in which the fabric passes through boiling water and then cold water. This procedure strengthens the fabric.

The exclusive features of cotton fabric are (i) hard wearing and absorbs moisture, (ii) does not burn over a flame but smolders instead, (iii) lightweight and versatile, (iv) does not wrinkle easily, and (v) resistant to dirt and wear and tear [54].

17.4.4 PRODUCTION OF SILK

Silkworms are cultivated and fed with mulberry leaves. Some of these eggs are hatched by artificial means such as an incubator, and in the olden times, the people carried it close to their bodies so that it would remain warm. Silkworms that feed on smaller, domestic tree leaves produce the finer silk, while the coarser silk is produced by silkworms that have fed on oak leaves. From the time they hatch to the time they start to spin cocoons, they are very carefully tended to. Noise is believed to affect the process, thus the cultivators try not to startle the silkworms. Their cocoons are spun from the tops of loose straw. It will be completed in two to three days' time. The cultivators then gather the cocoons and the chrysales are killed by heating and drying the cocoons. In the olden days, they were packed with leaves and salt in a jar, and then buried in the ground, or else other insects might bite holes in it. Modern machines and modern methods can be used to produce silk but the old-fashioned hand-reels and looms can also produce equally beautiful silk [55, 56].

The properties of silk includes: (i) versatile and very comfortable, (ii) absorbs moisture, (iii) cool to wear in the summer yet warm to wear in winter, (v) easily dyed, (v) strongest natural fiber and is lustrous, and (vi) poor resistance to sunlight exposure [57].

17.4.5 PRODUCTION OF NYLON MATERIALS

Nylon is made by forcing molten nylon through very small holes in a device called 'spinneret.' The streams of nylon harden into filament once they come in contact with air. They are then wound onto bobbins. These fibers are drawn (stretched) after they cool. Drawing involves unwinding the yarn or filaments and then winding it around another spool. Drawing makes the molecules in each filament fall into parallel lines. This gives the nylon fiber strength and elasticity. After the whole drawing process, the yarn may be twisted a few turns per yard or meters as it is wound onto spools. Further treatment to it can give it a different texture or bulk [58].

Properties of the nylon: (i) it is strong and elastic, (ii) it is easy to launder, (iii) it dries quickly, (iv) it retains its shape, and (v) it is resilient and responsive to heat setting [59].

17.4.6 PRODUCTION OF POLYESTER

Polyesters are made from chemical substances found mainly in petroleum. Polyesters are manufactured in three basic forms – *fibers*, *films* and *plastics*. Polyester fibers are used to make fabrics. Poly ethylene terephthalate or simply PET is the most common polyester used for fiber purposes. This is the polymer used for making soft drink bottles. Recycling of PET by remelting and extruding it as fiber may saves much raw materials as well as energy. PET is made by ethylene glycol with either terephthalic acid or its methyl ester in the presence of an antimony catalyst. In order to achieve high molecular weights needed to form useful fibers, the reaction has to be carried out at high temperature and in a vacuum [60].

17.5 BASIC NANOTECHNOLOGY

Two main approaches are used in nanotechnology that is, the bottom up approach and the top-down approach. In case of the "bottom-up" approach, the different type of materials and the instruments are made up from different types of molecular components which combine themselves by chemical ways basing on the mechanism of molecular recognition. In case of the "top-down" approaches, various nano-objects are made from various types of components without atomic-level control. Materials reduced to the nanoscale can show different properties compared to what they exhibit on a macro scale, enabling unique applications. The basic premise is that properties can dramatically change when a substance's size is reduced to the nanometer range. For instance, ceramics which are normally brittle can be deformable when their size is reduced, opaque substances become transparent (copper); stable materials turn combustible (aluminum); insoluble materials become soluble (gold) [61, 62].

Nanoparticles can be prepared from a variety of materials such as proteins, polysaccharides and synthetic polymers. The selection of materials in mainly depended on factors like size of the nanoparticles required, inherent properties such as aqueous, solubility and stability, surface characteristics i.e. charge and permeability, degree of biodegradability, biocompatibility and toxicity, release of the desired product, antigenicity of the final product etc. Polymeric nanoparticles have been prepared most frequently be three methods: (1) Dispersion of the performed polymers; (2) Polymerization of the monomers; and (3) Ionic gelation of the hydrophobic or hydrophilic polymers. However, techniques like supercritical fluid technology and particle repulsion in non wetting templates (PRINT) have been also used in modern days [63–65].

17.6 NANOTECHNOLOGY IN TEXTILE INDUSTRY AND TEXTILE ENGINEERING

Of the many applications of nanotechnology, textile industry has been currently added as one of the most benefited sector. Application of nanotechnology in textile industry has tremendously increased the durability of fabrics, increase its comfortness, hygienic properties and have also reduces its production cost. Nanotechnology also offers many advantages as compare to the conventional process in term of economy, energy saving, ecofriendliness, control release of substances, packaging, separating and storing materials on a microscopic scale for later use and release under control condition [66]. The unique and new properties of nanotechnology have attracted scientists and researchers to the textile industry and hence the use of nanotechnology in the textile industry has increased rapidly. This may be due to the reason that textile technology is one of the best areas for development of nanotechnology. The textile fabrics provide best suitable substrates where a large surface area is present for a given weight or a given volume of fabric. The synergy between nanotechnology and textile industry uses this property of large interfacial area and a drastic change in energetic is experienced by various macromolecules or super molecules in the vicinity of a fiber when changing from wet state to a dry state [67].

The application of nanoparticles to textile materials have been the objective of several studies aimed at producing finished fabrics with different functional performances. Nanoparticles can provide high durability for treated fabrics as they posse large surface area and high surface energy that ensure better affinity for fabrics and led to an increase in durability of the desired textile function. The particle size also plays a primary role in determining their adhesion to the fibers. It is reasonable to except that the largest particle agglomerates will be easily removed from the fiber surface, while the smallest particle will penetrate deeper and adhere strongly into the fabric matrix [68, 69]. Thus, decreasing the size of particles to nanoscale dimensional, fundamentally changes the properties of the material and indeed the entire substance.

A whole variety of novel nanotech textiles are already on the market at this moment. Areas where nanotech enhanced textiles are already seeing some applications include sporting industry, skincare, space technology and clothing as well as materials technology for better protection in extreme environments. The use of nanotechnology allows textiles to become multifunctional and produce fabrics with special functions, including antibacterial, UV-protection, easy-clean, water- and stain repellent and antiodor [68].

17.7 TECHNOLOGICAL USES

Textile materials are materials for the daily use. Besides, textiles are also play a vital role in fashion shows. Technically, they are applied in variety of our life savers including safety belt, and the airbags in the cars, bullet-proof vests protect against weapons, used as Implant material in medical applications. Recently, PU foams that can be combined with TPU films are used as excellent material for functional medical wound dressing which are not only help wounds to heal but also allow the wound to breathe by permeate the water-vapor [70, 71].

17.7.1 ADVANCEMENTS IN TEXTILE PRODUCTION

Technological advances during the past decade have opened many new doors for the Textile and Apparel industries, especially in the area of rapid prototyping and related activities. During the past decade, the textile and apparel complex has been scrambling to adjust to a rapidly changing business environment. Textiles and yarns have been around for thousands of years but in the last 50 years, progress in the technology has been most remarkable. The application of textiles and yarns have move beyond clothing and fabrics and they are increasingly used in high value-added applications such as composites, filtration media, gas separation, sensors and biomedical engineering. With the emergence of nanotechnology, the users of textiles and yarns are switching their attention to the production of nanometer diameter fibers [72, 73].

17.7.2 BODY SCANNING

The development of three dimensional body scanning technologies may have significant potential for use in the apparel industry. First, this technology has the potential of obtaining an unlimited number of linear and nonlinear measurements of human bodies (in addition to other objects) in a matter of seconds. Because an image of the body is captured during the scanning process, the location and description of the measurements can be altered as needed in mere seconds, as well. Second, the measurements obtained using this technology has the potential of being more precise and reproducible than measurements obtained through the physical measurement process. Third, with the availability of an infinite number of linear and nonlinear measurements the possibility exists for garments to be created to mold to the 3 dimensional shapes of unique human bodies. Finally, the scanning technology allows measurements to be obtained in a digital format that could integrate automatically into apparel CAD systems without the human intervention that takes additional time and can introduce error [74, 75].

17.7.3 APPAREL CAD

Adoption of CAD/CAM technology over the past few decades has increased the speed and accuracy of developing new products, reducing the manpower required to complete the development process. Unfortunately, this technology has also encouraged manufacturers to simplify the design of garments, allowing a more efficient use of materials and making mass production much easier. These systems initially only made an effort to adapt traditional manual methods instead of encouraging innovation in design or fit adaptations. Current developments in the area of information technology help build on the traditional CAD/CAM functions and offer a new way of looking at and using the systems for design and product development [76, 77].

17.7.4 NANOFIBER FABRICATION

As with all new technologies, polymeric nanofibers have brought with it a new beginning to the understanding of polymeric fibers. One apparent advantage of nanofibers is the huge increase in the surface area to volume ratio. Given the huge potential of nanofibers, the key is to use a technique that is able to easily fabricate nanofibers out of most, if not all the different type of polymers. A number of processing techniques such as drawing [78], template synthesis [79], phase separation [80], self-assembly [81], electrospinning [82], etc. have been used to prepare polymer nanofibers in recent years. The drawing is a process similar to dry spinning in fiber industry, which can make one-by-one very long single nanofibers. The template synthesis, as the name suggests, uses a nanoporous membrane as a template to make nanofibers of solid (a fibril) or hollow (a tubule) shape. The phase separation consists of dissolution, gelation, and extraction using a different solvent, freezing, and drying resulting in nanoscale porous foam. The process takes relatively long period of time to transfer the solid polymer into the nano-porous foam. The self-assembly is a process in which individual, preexisting components organize themselves into desired patterns and functions.

17.7.5 ELECTROSPINNING

Electrospinning is a process where continuous fibers with diameters in the submicron range are produced through the action of an applied electric field imposed on a polymer solution. Textiles made from these fibers have high surface area and small pore size, making them ideal materials for use in protective clothing. There are currently several applications of electrospinning that are being investigated, including: The fabrication of transparent composites reinforced with nanofibers. The effect of processing conditions on the morphology of polymers has been investigated. The scaling up of current production techniques is one of the main areas of research in which we have ongoing interest is the Adhesives, Permselective membranes, Anti-fouling coatings, Active protective barriers against chemical and biological threats are few examples of nanofibers produced through electrospinning [83].

17.7.5.1 Using Electrospun Nanofibers: Background and Terminology

The three inherent properties of nanofibrous materials that make them very attractive for numerous applications are their high specific surface area (surface area/unit mass), high aspect ratio (length/diameter) and their biomimicking potential. These properties lead to the potential application of electrospun fibers in such diverse fields as high-performance filters, absorbent textiles, fiber-reinforced composites, biomedical textiles for wound dressings, tissue scaffolding and drug-release materials, nano- and microelectronic devices, electromagnetic shielding, photovoltaic devices and high-performance electrodes, as well as a range of nanofiber-based sensors [84, 85].

In many of these applications the alignment, or controlled orientation, of the electrospun fibers is of great importance and large-scale commercialization of products will become viable only when sufficient control over fiber orientation can be obtained at high production rates. In the past few years research groups around the world have been focusing their attention on obtaining electrospun fibers in the form of yarns of continuous single nanofibers or uniaxial fiber bundles. Succeeding in this will allow the processing of nanofibers by traditional textile processing methods such as weaving, knitting and embroidery. This, in turn, not only will allow the significant commercialization of several of the applications cited above, but will also open the door to many other exciting new applications [86, 87].

Incorporating nanofibers into traditional textiles creates several opportunities. In the first instance, the replacement of only a small percentage of the fibers or yarns in a traditional textile fabric with yarns of similar diameter, but now made up of several thousands of nanofibers, can significantly increase the toughness and specific surface area of the fabric without increasing its overall mass [88]. Alternatively, the complete fabric can even be made from nanofiber yarns. This has important implications in protective clothing applications, where lightweight, breathable fabrics with protection against extreme temperatures, ballistics, and chemical or biological agents are often required. On an esthetic level, nanofiber textiles also exhibit extremely soft handling characteristics and have been proposed for use in the production of artificial leather and artificial cashmere [85]. In biomedical applications the similarity between certain electrospun polymeric nanofibers and the naturally occurring nanofibrous structures of connective tissues such as collagen and elastin gives rise to the opportunity of creating artificial biomimicking wound dressings and tissue engineering scaffolds. Several studies on nonwoven nanofiber webs of biocompatible polymers have already shown potential in this area. Simple three-dimensional constructs for vascular prostheses have also been manufactured by electrospinning onto preformed templates. Although these initial studies show that enhanced cell adhesion, cell proliferation and scaffold vascularization can be obtained on porous, nonwoven nanofiber webs, the simplicity of the constructs and the fragile nature of nonwoven webs still limit their applicability to small areas [89, 90].

Creating complex three-dimensional scaffold structures with fibers aligned in a controlled fashion along the directions of the forces that are usually present in dynamic tissue environments, as for instance in muscles and tendons, will lead to significant improvements in the performance of tissue engineering scaffolds. With continuous nanofiber yarns it will become possible to create such aligned fiber structures on a large scale, simply by weaving. In addition, the age-old techniques of knitting and embroidery can then be applied to create very complicated, three-dimensional scaffolds, with precisely controlled porosity, and yarns placed exactly along the lines of dynamic force [91].

Several other fields will also benefit from the availability of continuous yarns from electrospun fibers. Owing to the high fiber-aspect ratios and increased fiber–matrix adhesion caused by the high specific surface areas, aligned nanofiber yarns can lead to stronger and tougher, lightweight, fiber reinforced composite materials. The incorporation of nanofiber-based sensors into textiles can lead to new opportunities in the fields of smart and electronic textiles. Aligned nanofiber yarns of piezo-electric polymers and other microactuator materials may lead to better performance in advanced robotics applications [92].

Since the revival of electrospinning in the early 1990s, several research groups have worked on controlling the orientation of electrospun fibers. Those who have worked in the field of electrospinning over the past decade have come from various disciplinary backgrounds, including physics, chemistry and polymer science, chemical and mechanical engineering, and also from the traditional textiles field. The result of this has been that literature on the topic of electrospinning, and especially yarns from electrospun fibers, is plagued with terminology from different disciplines, which often leads to misunderstanding and even self-contradictory statements. So, for instance, in paper on the electrospinning of individual fibers of a novel polymer some authors might use the term yarn when they are actually referring to an individual fiber, or authors might refer to spinning a filament when they are actually spinning a yarn [93].

- **Fiber** – a single piece of a solid material, which is flexible and fine, and has a high aspect ratio (length/diameter ratio).

- **Filament** – a single fiber of indefinite length.
- **Tow** – an untwisted assembly of a large number of filaments; tows are cut up to produce staple fibers.
- **Sliver** – an assembly of fibers in continuous form without twist. The assembly of staple fibers, after carding but before twisting, is also known as a sliver.
- **Yarn** – a generic term for a continuous strand of textile fibers or filaments in a form suitable for knitting, weaving or otherwise intertwining to form a fabric.
- **Staple fiber** – short-length fibers, as distinct from continuous filaments, which are twisted together (spun) to form a coherent yarn. Most natural fibers are staple fibers, the main exception being silk which is a filament yarn. Most artificial staple fibers are produced in this form by slicing up a tow of continuous filaments.
- **Staple fiber yarn** – a yarn consisting of twisted together (spun) staple fibers.
- **Filament yarn** – a yarn normally consisting of a bundle of continuous filaments. The term also includes monofilaments.
- **Core-spun yarn** – a yarn consisting of an inner core yarn surrounded by staple fibers. A core-spun yarn combines the strength and/or elongation of the core thread and the characteristics of the staple fibers that form the surface.
- **Denier** – a measure of linear density: the weight in grams of 9000 meters of yarn.
- **Tex** – another measure of linear density: the weight in grams of 1000 meters of yarn.

17.7.5.2 Controlling Fiber Orientation

As stated in the previous section, achieving control over the orientation of electrospun fibers is an important step towards many of their potential applications. However, if one considers the fact that fiber formation occurs at very high rates (several hundreds of meters of fiber per second) and that the fiber formation process coincides with a very complicated three-dimensional whipping of the polymer jet (caused by electrostatic bending instability), it becomes clear that controlling the orientation of fibers formed by electrospinning is no simple task [94].

Various mechanical and electrostatic approaches have been taken in efforts to control fiber alignment. The two most successful methods are the following:

- Spinning onto a rapidly rotating surface – Several research groups have been routinely using this technique to obtain reasonably aligned fibers. The rapid rotation of a drum or disk and coinciding high linear velocity of the collector surface allows fast take-up of the electrospun fibers as they are formed. The 'point-to-plane' configuration of the electric field does, however, lead to fiber orientations that deviate from the preferred orientation. A special instance of the rotating drum set-up involves spinning onto a rapidly rotating sharp-edged wheel, which uses an additional electrostatic effect, since the sharp edge of the wheel creates a stronger converging electrostatic field, or a 'point-to-point' configuration, which has a focusing effect on the collected fibers (Fig. 17.1). This in turn leads to better alignment of the fibers [95, 96].
- The gap alignment effect – uniaxially aligned arrays of electrospun fibers can be obtained through the gap alignment effect, which occurs when charged electrospun fibers are deposited onto a collector that consists of two electrically conductive substrates, separated by an insulating gap. This electrostatic effect (Fig. 17.2) has been observed by various groups. Recently this was investigated in more detail. Briefly, the lowest energy configuration for an array of highly charged fibers between two conductive substrates, separated by an insulating gap, is obtained when fibers align parallel to each other [97].

17.7.5.3 Producing noncontinuous or short yarns

Both spinning onto a rapidly rotating collector and the gap alignment effect have been used to obtain short yarns for experimental purposes [98]:

Rotating collector method

Fibers can be spun onto a rapidly rotating disk, where the shearing force of the rotating disk led to aligned fibrous assemblies with good orientation.

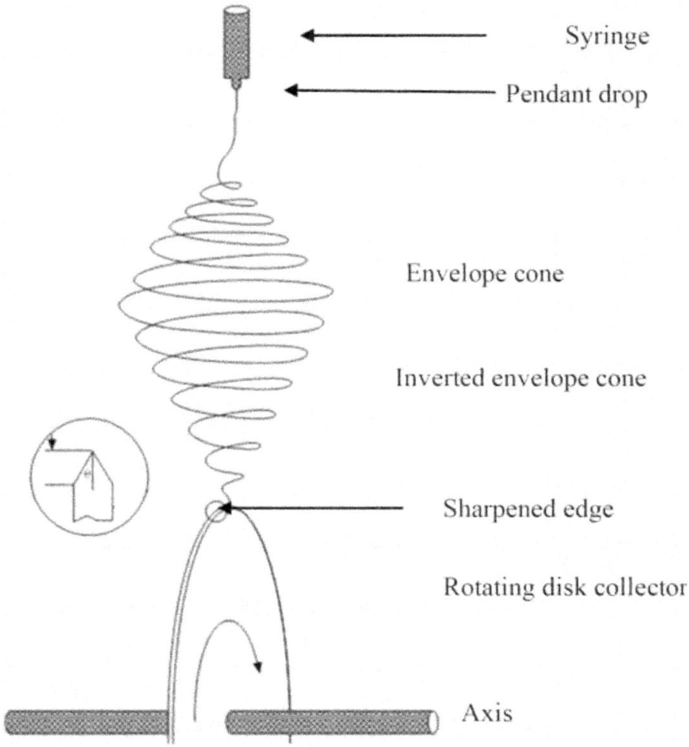

FIGURE 17.1 Converging electrostatic field on sharp-edged wheel electrode.

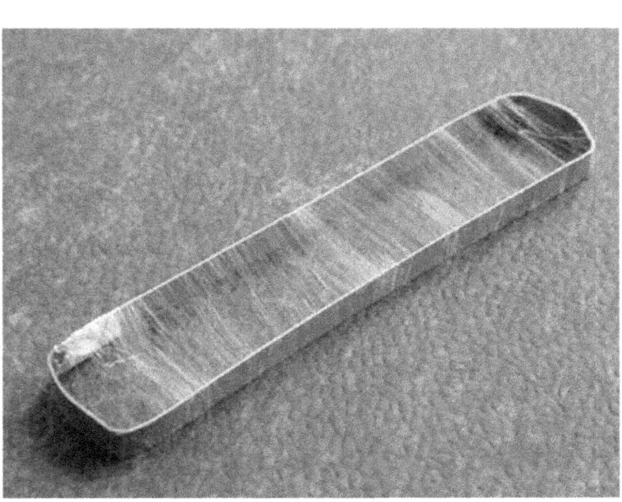

FIGURE 17.2 Aligned fibers obtained through the gap alignment effect.

These oriented fibers could then be collected and manually twisted into a yarn [99]. Twisted yarns obtained from tows spun on a rotating disk collector. The stress–strain behavior of the yarns can be examined and the modulus, ultimate strength and elongation at the ultimate strength can be measured as a function of twist angle.

Gap alignment method

Short yarns can be made by quickly passing a wooden frame through the electrospinning jet several times (for up to an hour), in a process also known as 'combing,' resulting in a tow of reasonably aligned fibers, which were then 'gently twisted' to form a yarn [100] (Figs. 17.3 and 17.4).

17.7.5.4 Producing Continuous Yarns

A common misconception in recent electrospinning literature is that the first literature on electrospinning dates back to 1934 when Formhals patented a method for manufacturing yarns from electrospun fibers. Some of the first publications on electrospinning date back as far as 1902, when first Cooley and then Morton patented processes for dispersing fluids. In both patents, the authors describe processes for producing very fine artificial fibers by delivering a solution of a fiber-forming material, such as

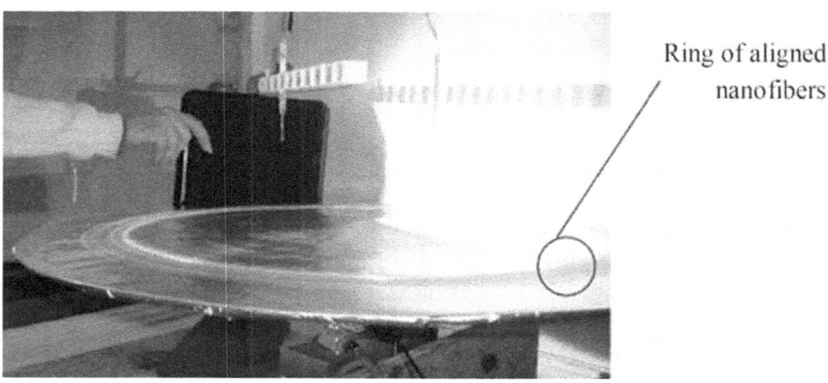

FIGURE 17.3 Aligned fiber tows on rotating disk collector.

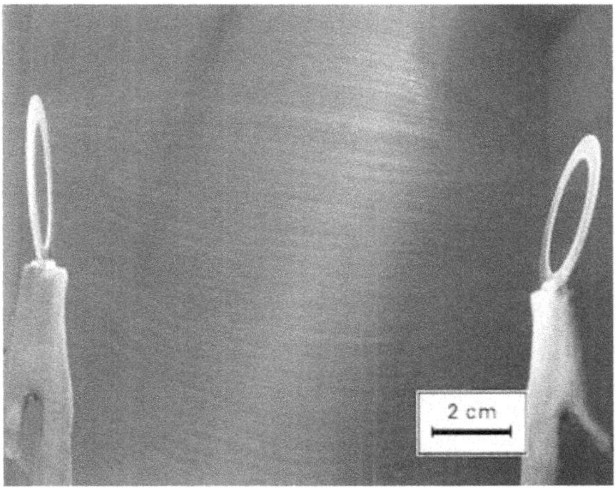

FIGURE 17.4 Fibers aligned between two parallel ring collectors.

pyroxylin, a nitrated form of cellulose, dissolved in alcohol or ether, into a strong electric field [101, 102].

The reason for this oversight is unclear, but it could possibly be blamed on differences in terminology since the term 'electrospinning' has only become popular with the revival of the process in the mid-1990s [100]. Another more puzzling oversight, which has recently led several authors to bemoan the lack of processes for making continuous yarns from electrospun fibers, is that Formhals actually registered a series of seven patents over a period of ten years between 1934 and 1944 and that all these patents describe processes and/or improvements to processes for the manufacture of continuous yarns from electrospun fibers. Since the youngest of these patents is more than 60 years old, one could speculate that these processes did not really work, which would explain the absence of commercially available electrospun fiber yarns. An alternative explanation could be that, since Formhals lived in Mainz, Germany, and since the last patent application was filed in 1940, the disruption of World War II and the ensuing years simply led to the processes being forgotten. Closer inspection of the patents, aided by more recent knowledge of the electrospinning process, also leads us to believe that at least some of the described processes are viable and that they deserve further consideration. The patents of Formhals show a gradual evolution of his yarn production process over time and in many instances he applied the same fundamental practical aspects of electrospinning that have reoccurred in the recent literature. These included obtaining aligned fibers by spinning onto conductive strips or rods that were separated in space from each other by an insulating material (gap alignment effect), increasing production rates by using multiple spinnerets, regulating the electric field between the source and the collector by adding additional electrostatic elements, using corona discharge to discharge the electrospun fibers, and posttreating the electrospun fibers by submerging them in a liquid bath [103].

Rotating dual-collector yarn

Formhals's original patent relates to the manufacture of slivers of cellulose acetate fibers by electrospinning from a cogwheel source onto various collector set-ups. In these collector set-ups fibers are first spun onto a rotating collector and then removed in a continuous fashion, onto a second take-up roller. The first of these collector set-ups consisted of a solid conductive wheel or ring with string attached to the edge. In this set-up the wheel was rotated for a short period while fibers were spun onto its edge. The process was then stopped, the string was loosened and then drawn over rollers and/or through twist-imparting rings to a second take-up roller, and the spinning process restarted. The newly spun sliver or semitwisted yarn was then drawn off continuously onto the second take-up roller. Another collector set-up consisted of a looped metal belt with fixtures to push or blow the fiber sliver off the belt before the fiber sliver was collected on a second take-up roller. The concept of using multiple spinnerets for increasing production rates was also introduced in this patent.

Formhals later identified several problems related to his first design and hence in subsequent patents he made various additions and/or alterations to the original design, which were intended to eliminate these problems. These problems and their solutions included the following [92, 104]:

- Problem: Fibers flying to-and-fro between the source and the collector

  Solution: In his second patent, Formhals claimed that one cause of the fibers flying to-and-fro between the jet and collector was that the collector was at too high a voltage of opposite polarity and that resulting corona discharge from the collector reversed the charge of the fibers while they were passing between the jet and the collector. This in turn caused them to change direction and fly back to the source. He proposed to eliminate this problem by adding a voltage regulator on the collector-side of the circuit in order to down-regulate the voltage of the collector.

- Problem: Fibers not drying sufficiently between source and collector.

  Solution: Formhals later designed various additions to his spinning system for regulating the shape and intensity of the electrical field in the vicinity of the spinning source.30 This was done in order to direct the formed fibers along a longer, predetermined and constant path towards the collecting electrode and was achieved by placing, in close proximity to the fiber stream, conductive strips, wires, plates and screens, which were connected to the same potential as the fiber source. These additions allowed a more thorough drying of fibers before they were deposited on the collector.

An additional problem, which is not specifically discussed in Formhals's patents, but which can be foreseen when examining his first collector design, is that fiber alignment would be less than ideal when spinning onto a solid wheel or belt. It appears, however, that Formhals did encounter this problem and that he overcame it by using the gap alignment effect in the design of subsequent collectors. The design consisted of a picket-fence-like belt, with individual, pointed electrodes, separated from each other by an air gap.

Multi-collector yarn

In this patented process from the Korea Research Institute of Chemical Technology, continuous slivers or twisted yarns of different polymers, but especially of polyamide–polyimide copolymers, are claimed to be obtained by electrospinning first onto one stationary or rotating plate or conductive mesh collector, where the charges on the fibers are neutralized, and then continuously collecting the fibers from the first onto a second rotating collector.

A diagram depicting the process is given in Fig. 17.5. The underlying principle of this process closely resembles the rotating dual-collector yarn process patented by Formhals in 1934 [105].

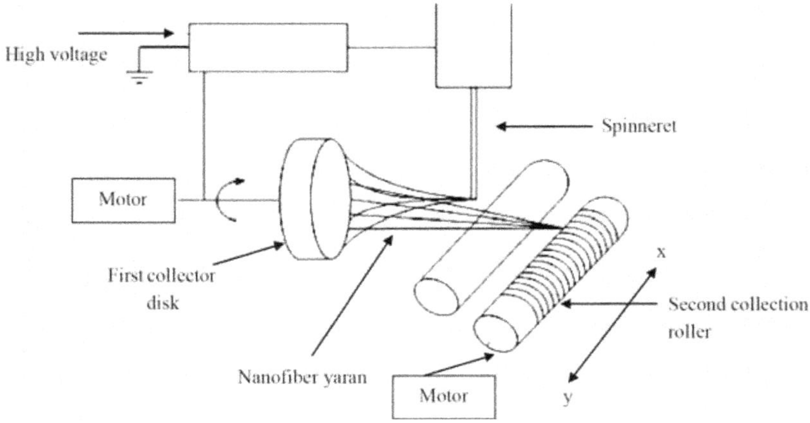

FIGURE 17.5 Multi-collector yarn process diagram.

Core-spun yarn

In his 1940 patent, Formhals described a method for making composite yarns by electrospinning onto existing cotton, wool or other preformed yarn. It was also proposed that a sliver of fibers, such as wool, could be coated with the electrospun fibers before twisting the product into an intimately blended yarn [103].

Staple fiber yarn

Formhals developed a method for controlling the length of the electrospun fibers with the main objective being to manufacture fibers with a controlled and comparatively short length. This goal was achieved by modulating the electric field using a spark gap. In this modulation, it was preferable to periodically switch the field strength to at least 35% and preferably 20% of its original voltage in order to interrupt the electrospinning process for a short period and thereby create a sliver of short fibers, which could then be spun into a staple fiber yarn [103].

Continuous filament yarn

Instead of spinning directly onto the counter electrode, Formhals altered the process so the polarity of the charge on the fibers was changed before reaching the counter electrode. This was achieved by using high voltage of opposite polarity on a sharp-edged or thin wire counter electrode. The high voltage led to corona discharge, which initially reduced and eventually inverted the charge on the fiber while it was traveling from the source to the collector [106].

This caused the fibers to turn away from the collector electrode and they could then be intercepted at a point below the counter electrode and rolled up as a continuous filament yarn on a take-up roller. In a final improvement on the system, the entire spinning apparatus was encased in a box with earthed conductive siding. This avoided build-up of charges in the panels, which could lead to disturbance of the electric field inside the spinning chamber and disruption of the spinning process. In addition, variable voltage power on the source and collector electrodes allowed the tuning and moving of the position of the neutral zone in which the yarn formation process takes place, which in turn allowed better control over the continuity of the spinning process [107, 108].

Self-assembled yarn

The self-assembled yarn process was developed by Ko et al. at Drexel University. When a solution of pure polymer, or a polymer-containing polymer blend, was electrospun onto a solid conductive collector under appropriate conditions, the fibers did not deposit on the collector in the form of a flat nonwoven web as is usually observed. Instead, initial fibers deposited on a relatively small area of the collector and then subsequent fibers started accumulating on top of them and then on top of each other, forming a self-assembled yarn structure that rapidly grew upwards from the collector towards the spinneret. The formation of a self-assembled yarn is illustrated in Fig. 17.6. The self-assembling yarn, suspended in the space between the Spinneret and the collector, continued to grow in this fashion until it reached a critical point somewhere in the vicinity of the spinneret. At this critical point, a branched tree-like fiber structure formed and newly formed fibers deposited on the branches of the tree. The yarn

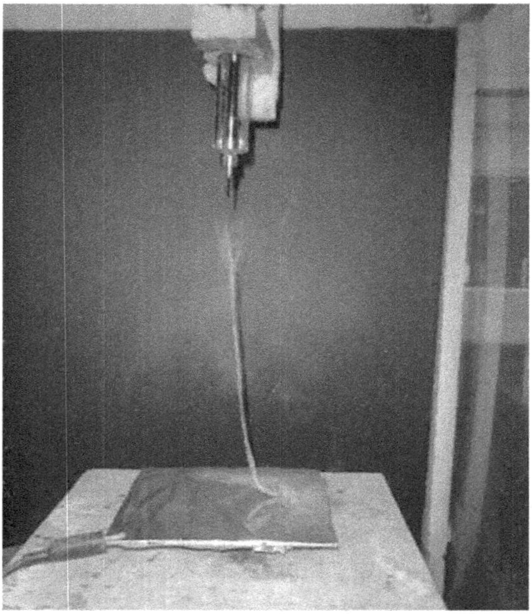

FIGURE 17.6 Self-assembled yarn formation.

could then be collected by slowly taking up the fibers collected on one of the tree branch structures, or by slowly moving the target electrode away from the spinneret. Post-processing of the yarn, including twisting, could be done in a second step [109].

It was proposed that the charge on the electrospun fibers, which is induced through the high voltage in the spinneret, is dissipated through the evaporation of the solvent during the electrospinning process, so that the fibers are essentially neutral when they reach the collector electrode. This could explain why the initial fibers deposit on such a small area on the collector. If the fibers on the collector are charged, they repel incoming fibers leading to an expanding random web. Neutral fibers would not have the same repelling effect on incoming fibers and so the fibers would collect on a smaller area. Neutral fibers, deposited on top of each other, and therefore closer to the spinneret than the target electrode surface, also form an attractive target for incoming fibers. This would explain why subsequent fibers selectively deposit on the tip of the self-assembling yarn [85].

Conical collector yarn

A method for the production of hollow fibers by the electrospinning process was reported by Kim et al. The conventional electrospinning device was modified to include a conical collector and an air-suction orifice to generate hollow and void-containing, uniaxially aligned electrospun fibers. Use of the conical collector allowed for the collection of aligned yarns with diameters of approximately 157 nm [110].

Spin-bath collector yarn

In this recently published method, developed by our group at Stellenbosch University, continuous uniaxial fiber bundle yarns are obtained by electrospinning onto the surface of a liquid reservoir counter electrode. The web of electrospun fibers, which forms on the surface of the spin-bath, is drawn at low linear velocity (ca. 0.05 m/s) over the liquid surface and onto a take-up roller. A diagrammatic representation of the electrospinning set-up is given in Fig. 17.7. All the yarns obtained using this method exhibit very high degrees of fiber alignment and bent fiber loops are observed in all the yarns [47].

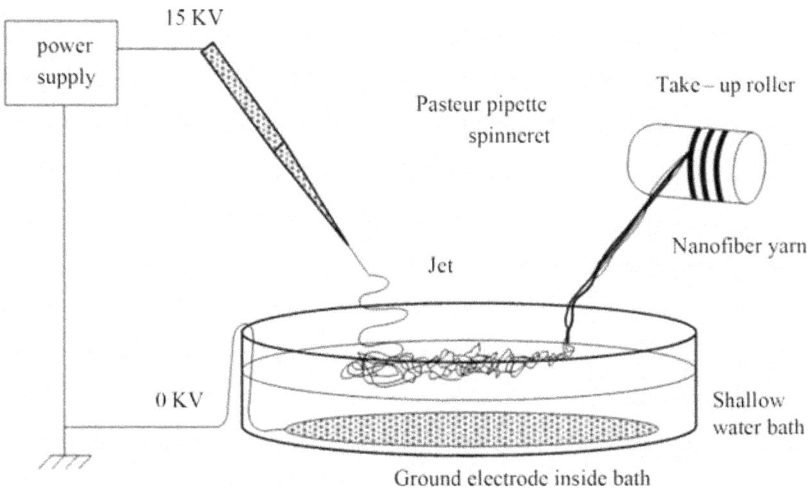

FIGURE 17.7 Yarn-spinning set-up with grounded spin-bath collector electrode.

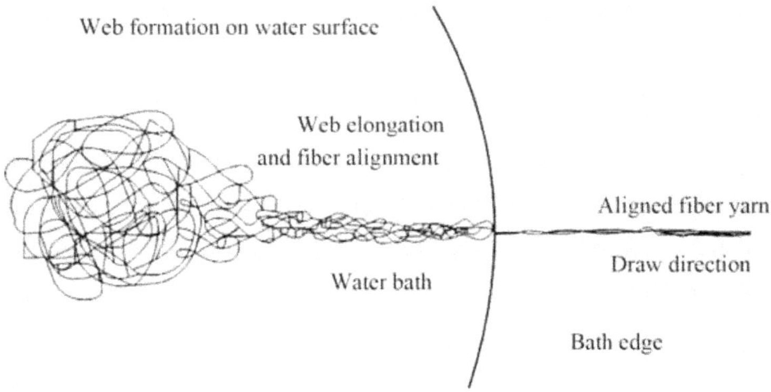

FIGURE 17.8 Top view of the yarn formation process.

The process of yarn formation is illustrated in Fig. 17.8. It can be described in three phases. In the first phase, a flat web of randomly looped fibers forms on the surface of the liquid. In the second phase, when the fibers are drawn over or through the liquid, the web is elongated and alignment of the fibers takes place in the drawing direction. The third phase consists of drawing the web off the liquid and into air. The surface tension of the remaining liquid on the web pulls the fibers together into a three-dimensional, round yarn structure.

The average yarn obtained in a single-spinneret electrospinning set-up contains approximately 3720 fibers per cross-section and approximately 180 m of yarn can be spun per hour. The yarns obtained are very fine, with calculated linear densities in the order of 10.1 denier. Although higher linear densities can be obtained by reducing the yarn take-up rate, this is accompanied by a decrease in fiber alignment within the yarn. Currently investigations are focused on various options to overcome these challenges by, for instance, combining aligned yarns from multiple spinnerets into single yarns.

Twisted nonwoven web yarn

This method, patented by Raisio Chemicals Korea Inc. involves electrospinning nanofibers through multiple nozzles to obtain a nonwoven nanofiber web, either directly in a ribbon form or in a larger form, which is

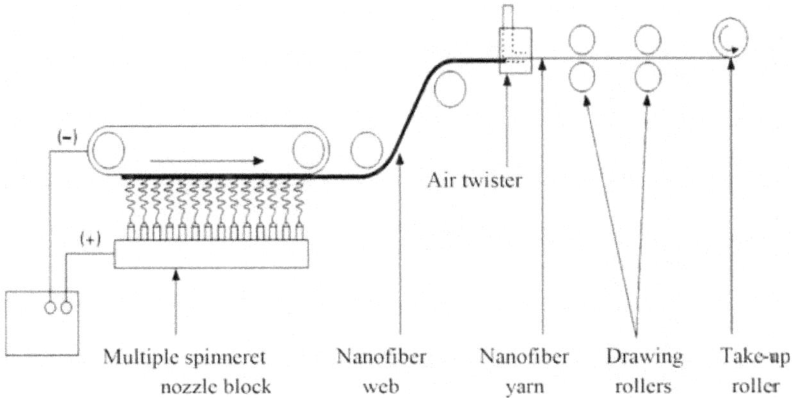

FIGURE 17.9 Spinning process used to prepare a twisted nonwoven web yarn.

then cut into ribbons, and subsequently passing the nanofiber web ribbons through an air twister to obtain a twisted nanofiber yarn. A diagram depicting the process is given in Fig. 17.9 [92].

Grooved belt collector yarn

In a recent patent by Kim and Park, a ribbon-shaped nanofiber web is prepared by electrospinning onto a collector consisting of an endless belt-type nonconductive plate with grooves formed at regular intervals along a lengthwise direction and a conductive plate inserted into the grooves of the nonconductive plate. The nanofiber webs are electrospun onto the conductive plates in the grooves and later separated from the collector, focused, drawn and wound into a yarn [92].

Vortex bath collector yarn

In this patent pending process developed at the National University of Singapore, a basin with a hole at the bottom is used to allow water to flow out in such a manner that a vortex is created on the water surface.

Electrospinning is carried out over the top of the basin so that electrospun fibers are continuously deposited on the surface of the water. Owing to the presence of the vortex, the deposited fibers are drawn into a bundle

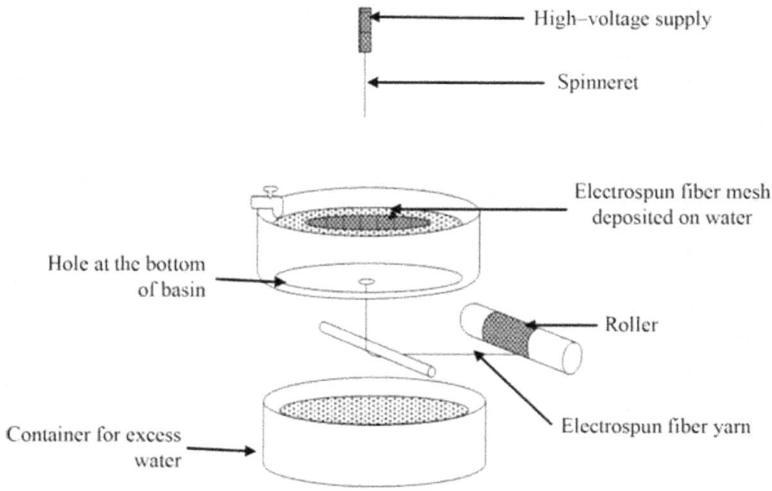

FIGURE 17.10 Vortex bath collector yarn process.

as they flow through the water vortex. Generally, a higher feed rate or multiple spinnerets are required to deliver sufficient fibers on the surface of the water so that the resultant fiber yarn has sufficient strength to withstand the drawing and winding process. Figure 17.10 shows the set-up used for the yarn drawing process.

Yarn drawing speeds as high as 80 m/min have been achieved and yarns made of poly(vinylidene fluoride) (PVDF) and polycaprolactone have been fabricated using this process [103].

Gap-separated rotating rod yarn

This method developed by Doiphode and Reneker at the University of Akron uses the gap alignment effect in a very similar way to the work published by Dalton et al. discussed before. The process is described with reference to Fig. 17.11. Fibers are electrospun between a 2 mm metal rod on the right and a hollow 25 mm metal rod with a hollow hemisphere attached to its end on the left. Both the geometries are grounded and placed at a distance of a few centimeters. Fibers are collected across the

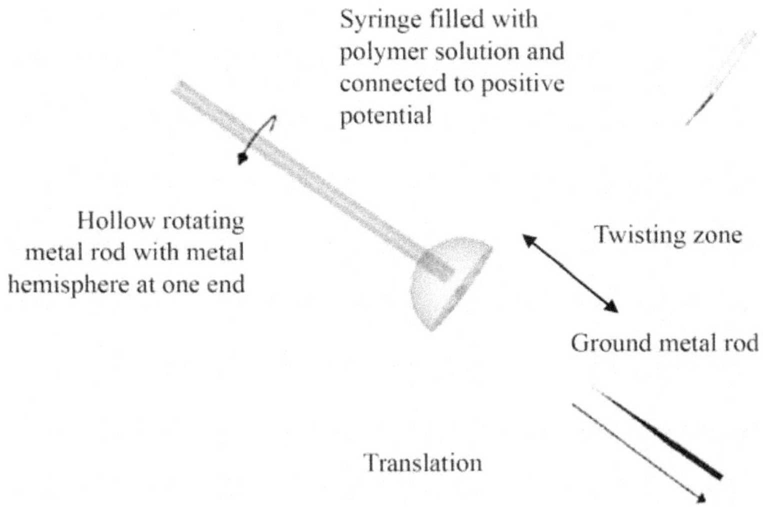

FIGURE 17.11 Schematic diagram for gap-separated rotating rod yarn set-up.

gap between these two collector surfaces and are given a twist by rotating the hemispherical collector. Yarn collected in this manner on the tip of the metal rod can be translated away from the rotating collector, thereby drawing the yarn and producing yarn continuously. Yarns with lengths up to 30 cm were produced by this method and the creators of the process believe that optimizing the winding mechanism can lead to production of continuous yarns [111].

Conjugate electrospinning yarn

Methods for making continuous nanofiber yarns based on the principle of conjugate electrospinning were recently published and patented by Xinsong Li et al. at South-east University in Nanjing as well as Luming Li and co-workers [29] at Tsinghua University in Beijing. In conjugate electrospinning, two spinnerets or two groups of spinnerets are placed in an opposing configuration and connected to high voltage of positive and negative polarity respectively. The process is presented diagrammatically in Fig. 17.12. Oppositely charged fiber jets are ejected from the spinnerets

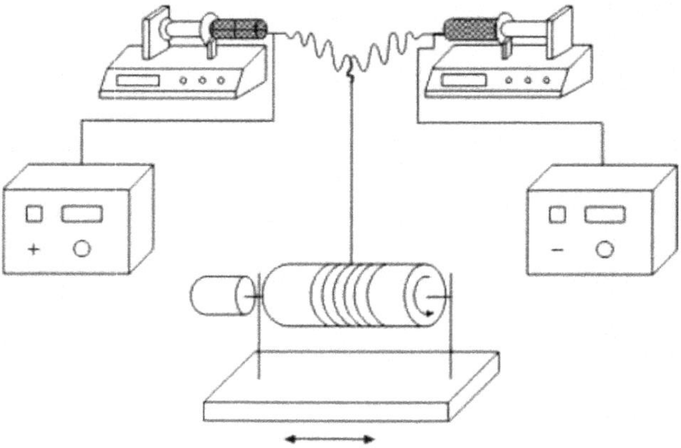

FIGURE 17.12 Conjugate electrospinning set-up.

and Coulombic attraction leads the oppositely charged fibers to collide with each other. The collision of the fibers leads to rapid neutralization of the charges on the fibers and rapid decrease in their flying speeds. In the processes described by both groups, the neutralized fibers are then collected onto take-up rollers to form yarns. Each continuous yarn contains a large quantity of nanofibers, which are well aligned along the longitudinal axis of the yarn. Conjugate electrospinning works for a variety of polymers, composites and ceramics.

17.8 NANOPARTICLES IN TEXTILES FINISHING

Fabric treated with nanoparticles of TiO_2 and MgO replaces fabrics with active carbon, previously used as chemical and biological protective materials. The photocatalytic activity of TiO_2 and MgO nanoparticles can break harmful and toxic chemicals and biological agents. These nanoparticles can be pre-engineered to adhere to textile substrates via spray coating or electrostatic methods. Textiles with nanoparticles finishing are used to convert fabrics into sensor-based materials which has numerous applications. If nanocrystalline epiezoceramic particles are incorporated into fabrics, the finished fabric can convert exerted mechanical forces into electrical signals enabling the monitoring of bodily functions such as heart rhythm and pulse if they are worn next to skin [112–114].

17.9 FABRIC FINISHING BY USING NANOTECHNOLOGY

Finishing of textile fabrics is made of natural and synthetic fibers to achieve desirable surface texture, color and other special esthetic and functional properties, has been a primary focus in textile manufacturing industries. In the last decade, the advent of nanotechnology has spurred significant developments and innovations in this field of textile technology [115]. Fabric finishing has taken new routes and demonstrated a great potential for significant improvements by applications of nanotechnology. The developments in the areas of surface engineering and fabric finishing have been highlighted in several researches. There are many ways in which the surface properties of a fabric can be manipulated and enhanced, by implementing appropriate surface finishing, coating, and/or altering techniques, using nanotechnology. A few representative applications of fabric finishing using nanotechnology are schematically displayed in the Fig. 17.13.

Nanotechnology provides plenty of efficient tools and techniques to produce desirable fabric attributes, mainly by engineering modifications of the fabric surface. For example, the prevention of fluid wetting towards

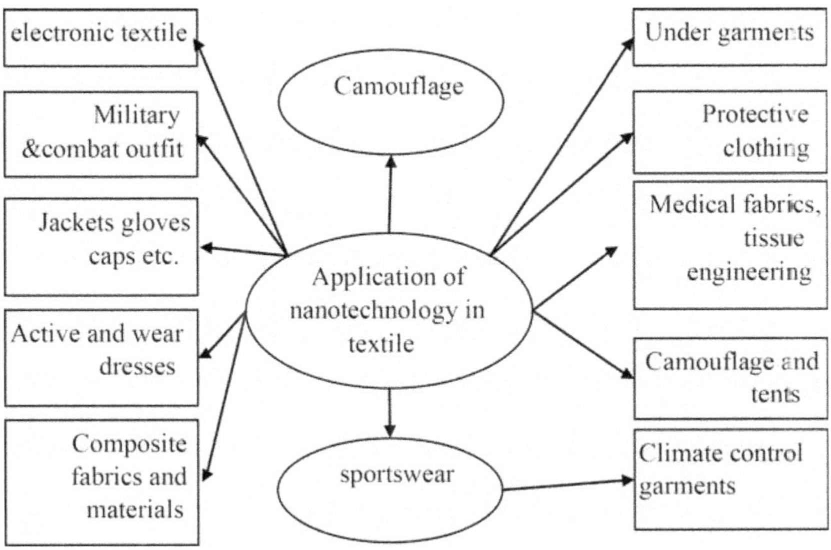

FIGURE 17.13 Some representative applications of nanotechnology in textiles.

the development of water or stain-resistant fabrics has always been of great concerning textile manufacturing.

The basic principles and theoretical background of "fluid-fabric" surface interaction are well described in recent manuscript. It has been demonstrated that by altering the micro and nano-scale surface features on a fabric surface, a more robust control of wetting behavior can be attained. The alteration in the fabric's surface properties enables to exhibit the "Lotus-Effect," which demonstrates the natural hydrophobic behavior of a leaf surface. This sort of surface engineering, which is capable of replicating hydrophobic behavior, can be used in developing special chemical finishes for producing water and/or stain- resistant fabrics [116, 117]. In recent years, several attempts have been made by researchers and industries to use similar concepts of surface-engineered modifications through nanotechnology to develop high performance textile and smart textile. The concept of surface engineering and nano-textile develops hydrophobic fabric surfaces that are capable of repelling liquids and resisting stains, while complementing the other desirable fabric attributes, such as breathability, softness, and comfort [118].

17.10 TEXTILE MODIFICATION

Considering special advantages and high potentialities of the application of nanostructured materials in textile industry, especially for producing high performance textiles, here we reviewed the application of nano-structured materials for antibacterial modification of textile and polymeric materials. The modification of textile fibers is carried out by commonly used chemical or electro-chemical application methods. Many of the classical textile finishing techniques (e.g., hydrophobization, easy-care finishing) that are already used since decades are among these methods. Modification of textiles via producing polymeric nano-composites and also surface modification of textiles with metallic and inorganic nanostructured materials are developed due to their unique properties. Considering the fact that fiber and film processing are the most difficult procedures of molding polymeric materials, bulk modification of continuous multifilament yarns is an extremely sensitive process. However, achieving optimum process conditions will present an economical technique [119, 120].

Different methods have been used for surface modifications of textiles by using poly carboxylic acids as spacers for attaching TiO_2 nanoparticles to the fabrics [121] and argon plasma grafting nano-particles on wool surface [122]. Plasma pretreatment has been used for the generation of active groups on the surface to be combined with TiO_2 nanoparticles [123]. The radical groups on the surface have also been generated using irradiation of the textile surfaces with UV light to bond the nano-particles [124]. Deposition of nanoparticles from their metallic salt solution on the surface pretreated with RF-plasma and vacuum-UV [125].

Nanotechnology holds great potential in the textile and clothing industry offering enhanced performance of textile manufacturing machines and processes so as to overcome the limitations of conventional methods. Nanofibers have good properties such as high surface area, a small fiber diameter, good filtration properties and high permeability. Nanofibers can be obtained via electro-spinning application or bicomponent extrusion (islands in the sea technique) [126]. One of the interesting areas for the application of nanotechnology in the textile industry are coating and finishing processes of textiles which is done by the techniques like sol-gel [127] and plasma [128]. Nanotech enhanced textiles include sporting industry, skincare, space technology and clothing as well as material technology exhibiting better healthcare systems, protective clothing and integrated electronics. By using nanotechnology, textiles with self-cleaning surfaces have attracted much attention which is created by the lotus effect. In brief, nanoscaled structures similar to those of a lotus leaf create a surface that causes water and oil to be repelled, forming droplets, which will simply roll of the surface, taking any with them [62].

With the advent of nanoscience and technology, a new area has developed in the area of textile finishing called "Nanofinishing." Growing awareness of health and hygiene has increased the demand for bioactive or antimicrobial and UV-protecting textiles. Coating the active surfaces cause UV blocking, antimicrobial, flame retardant, water repellant and self-cleaning properties. The UV-blocking property of a fabric is enhanced when a dye, pigment, delustrant, or ultraviolet absorber finish is present that absorbs ultraviolet radiation and blocks its transmission through a fabric to the skin. Metal oxides like ZnO as UV-blocker are more stable when compared to organic UV-blocking agents. For antibacterial finishing, ZnO nanoparticles scores over nano-silver in cost-effectiveness, whiteness, and UV-blocking property [129, 130].

17.11 TYPES OF NANOMATERIALS

17.11.1 NANOCOMPOSITE FIBERS

A composite is a material that combines one or more separate components. Composites are designed to exhibit the best properties of each component. A large variety of systems combining one, two and three dimensional materials with amorphous materials mixed at the nanometer scale [131]. Nanostructure composite fibers are intensively used in automotive, aerospace and military applications. Nanocomposite fibers are produced by dispersing nanosize fillers into a fiber matrix. Due to their large surface area and high aspect ratio, nanofillers interact with polymer chain movement and thus reduce the chain mobility of the system. Being evenly distributed in polymer matrices, nanoparticles can carry load and increase the toughness and abrasion resistance. Most of the nanocomposite fibers use fillers such as nanosilicates, metal oxide nanoparticles, GNF as well as single-wall and multiwall CNT [132, 133]. Some novel CNT reinforced polymer composite materials have been developed, which can be used for developing multifunctional textiles having superior strength, toughness, lightweight, and high electrical conductivity [72].

17.11.2 CARBON NANOFIBERS AND CARBON NANOPARTICLES

Carbon nanofibers and carbon black nanoparticles are among the most commonly used nanosize filling materials. Nanofibers can be defined as fibers with a diameter of less than 1 mm or 1000 nm and are characterized as having a high surface area to volume ratio and a small pore size in fabric form [66]. Carbon nanofibers can effectively increase the tensile strength of composite fibers due to its high aspect ratio, while carbon black nanoparticles can improve their abrasion resistance and toughness. Several fiber-forming polymers used as matrices have been investigated including polyester, nylon and polyethylene with the weight of the filler from 5 to 20% [134].

There are numerous applications in which nanofibers could be suited. The high surface area to volume ratio and small pore size allows viruses and spore-forming bacterium such as anthrax to be trapped. Filtration devices and wound dressings are just some of the applications in which nanofibers could be used. Researchers are investigating textile materials made from nanofibers which can act as a filter for pathogens (bacteria, viruses), toxic gasses, or poisonous or harmful substances in the air. Medical staff, fire fighters, the emergency services or military personnel could all benefit from protective garments made from nanofibers materials [135].

17.11.3 CLAY NANOPARTICLES

Clay nanoparticles are resistant to heat, chemicals and electricity, and have the ability to block UV light. Incorporating clay nanoparticles into a textile can result in a fabric with improved tensile strength, tensile modulus, flexural strength and flexural modulus. Nanocomposite fibers which use clay nanoparticles can be engineered to be flame, UV light resistant and anti-corrosive. Although there have been a number of flame retardant finishes available since the 1970s, the emission of toxic gasses when set ablaze make them somewhat hazardous. Clay nanoparticles have been incorporated into nylon to impart flame retardant characteristics to the textile without the emission of toxic gas. The addition of clay nanoparticles has made polypropylene dyeable. Metal oxide nanoparticles of TiO_2, Al_2O_3, ZnO and MgO exhibit photocatalytic ability, electrical conductivity, UV absorption and photo-oxidizing capacity against chemical and biological species. The main research efforts involving the use of nanoparticles of metal oxides have been focused on antimicrobial, self-decontaminating and UV blocking applications for both military protection gears and civilian health products [66, 72]. Nylon fibers filled with ZnO nanoparticles can provide UV shielding function and reduce static electricity on nylon fibers. A composite fiber with nanoparticle of TiO_2 or MgO can provide self-sterilizing function [134].

17.11.4 CARBON NANOTUBES

Carbon Nanotube is a tubular form of carbon with diameter as small as nanometer (nm). A carbon nanotube is configurationally equivalent to a two dimensional graphene sheet rolled into a tube. They can be metallic or semiconducting, depending on chirality. CNT are one of the most promising materials due to their high strength and high electrical conductivity. CNT consists of tiny shell(s) of graphite rolled up into a cylinder(s) [69, 136]. CNT exhibit 100 times the tensile strength of steel at one-sixth weight, thermal conductivity better than all but the purest diamond, and an electrical conductivity similar to copper, but with the ability to carry much higher currents. The potential applications of CNTs include conductive and high-strength composite fibers, energy storage and energy conversion devices, sensors, and field emission displays. Possible applications include screen displays, sensors, aircraft structures, explosion-proof blankets and electromagnetic shielding. The composite fibers have potential applications in safety harnesses, explosion-proof blankets, and electromagnetic shielding applications. Continuing research activities on CNT fibers involve study of different fiber polymer matrices such as polymethylmethacrylate (PMMA) and polyacrylonitrile (PNA) as well as CNT dispersion and orientation in polymers [137].

17.11.5 NANOCELLULAR FOAM STRUCTURE

Polymeric materials with nanosize porosity exhibit lightweight, good thermal insulation, as well as high cracking resistance at high temperature without sacrifices in mechanical strength. By choosing the pretreatment condition to the fiber, the transverse mechanical properties of the composite can be also enhanced through the molecular diffusion across the interface between the fiber and the matrix. The nanocomposites clearly surpass the mechanical properties of most comparable cellulosic materials, their greatest advantage being the fact that they are fully bio-based and biodegradable, but also of relatively high strength. A potential application of cellular structure is to encapsulate functional compounds such as pesticides and drugs inside of the nanosize cells. One of the approaches to fabricate nanocellular fibers is to make use of a thermodynamic instability during supercritical carbon dioxide extrusion and reduce the size of the cellular fibers that can be used as high-performance composite fibers as well as for sporting and aerospace materials [72, 136].

17.12 PROPERTIES OF NANO-TEXTILE FIBERS

17.12.1 WATER REPELLENCE

The water-repellent property of fabric created by nano-whiskers, which are hydrocarbons and 1/1000 of the size of a typical cotton fiber, when added to the fabric create a peach fuzz effect without lowering the strength of cotton. The spaces between the whiskers on the fabric are smaller than the typical drop of water, but still larger than water molecules; water thus, remains on the top of the whiskers and above the surface of the fabric. However, liquid can still pass through the fabric, if pressure is applied to it [117, 118].

Nanosphere impregnation involving a three-dimensional surface structure with gel forming additives which repel water and prevent dirt particles from attaching themselves are also used. Once water droplets fall onto them, water droplets bead up and, if the surface slopes slightly, will roll off. As a result, the surfaces stay dry even during a heavy shower. Furthermore, the droplets pick up small particles of dirt as they roll, and so the leaves of the lotus plant keep clean even during light rain. By altering the micro and nano-scale surface features on a fabric surface, a more robust control of wetting behavior can be attained. It has been demonstrated that by combining the nanoparticles of hydroxyapatite, TiO_2, ZnO and Fe_7O_3 with other organic and inorganic substances, the audio frequency plasma of fluorocarbon chemical was applied to deposit a nanoparticulate hydrophobic film onto a cotton fabric surface to improve its water repellent property. This sort of surface engineering, which is capable of replicating hydrophobic behavior, can be used in developing special chemical finishes for producing water-and/or stain- resistant fabrics while complementing the other desirable fabric attributes, such as breathability, softness and comfort. The surfaces of the textile fabrics can be appreciably modified to achieve considerably greater abrasion resistance, UV resistance, electromagnetic and infrared protection properties [138, 139].

17.12.2 UV PROTECTION

Inorganic UV blockers are more preferable to organic UV blockers as they are nontoxic and chemically stable under exposure to both high temperatures and UV [140, 141]. Inorganic UV blockers are usually certain semiconductor oxides such as TiO_2, ZnO, SiO_2 and Al_2O_3. Among these semiconductor oxides, TiO_2 and ZnO are commonly used. It was determined that nano-sized titanium dioxide and zinc oxide are more efficient at absorbing and scattering UV radiation than the conventional size, and are thus better to provide protection against UV rays. This is due to the fact that nano-particles have a larger surface area per unit mass and volume than the conventional materials, leading to the increase of the effectiveness of blocking UV radiation [68, 140]. Various researchers have worked on the application of UV blocking treatment to fabric using nanotechnology. UV blocking treatment for cotton fabrics are developed using the sol-gel method. A thin layer of titanium dioxide is formed on the surface of the treated cotton fabric which provides excellent UV protection; the effect can be maintained after 50 home launderings [141]. Apart from titanium dioxide, zinc oxide nano rods of 10 to 50 nm in length are also applied to cotton fabric to provide UV protection. According to the studies on the UV blocking effect, the fabric treated with zinc oxide nanorods were found to have demonstrated an excellent UV protective factor (UPF) rating. This effect can be further enhanced by using a different procedure for the application of nanoparticles on the fabric surface. When the process of padding is used for applying the nanoparticles on to the fabric, the nanoparticles get applied not only on the surface alone but also penetrates into the interstices of the yarns and the fabric, that is, some portion of the nanoparticles get penetrate into the fabric structure. Such Nanoparticles which do not stay on the surface may not be very effective in shielding the UV rays. It is worthwhile that only the right (face) side of the fabric gets exposed to the rays and therefore, this surface alone needs to be covered with the nanoparticles for better UV protection. Spraying (using compressed air and spray gun) the fabric surface with the nanoparticles can be an alternate method of applying the nanoparticles [118].

17.12.3 ANTIMICROBIAL

Although many antimicrobial agents are already in used for textile, the major classes of antimicrobial for textile include organo-silicones, organo-metallics, phenols and quaternary ammonium salts. The bi- phenolic compounds exhibit a broad spectrum of antimicrobial activity. For imparting antibacterial properties, nano-sized silver, titanium dioxide, zinc oxide, triclosan and chitosan are used [118]. Nano-silver particles have an extremely large relative surface area, thus increasing their contact with bacteria or fungi and vastly improving their bactericidal and fungicidal effectiveness. Nano-silver is very reactive with protein and shows antimicrobial properties at concentrations as low as 0.0003 to 0.0005%. When contacting bacteria and fungi, it will adversely affect cellular metabolism and inhibits cell growth. It also suppresses respiration, the basal metabolism of the electron transfer system, and the transport of the substrate into the microbial cell membrane. Furthermore, it inhibits the multiplication and growth of those bacteria and fungi which cause infection, odor, itchiness and sores [140]. Some synthetic antimicrobial nano particles which are used in textiles are as follows. Triclosan, a chlorinated bi- phenol, is a synthetic, nonionic and broad spectrum antimicrobial agent possessing mostly antibacterial alone with some antifungal and antiviral properties. Chitosan, a natural biopolymer, is effectively used as antibacterial, antifungal, antiviral, nonallergic and biocompatible. ZnO nanoparticles have been widely used for their antibacterial and UV-blocking properties.

17.12.4 ANTISTATIC

An antistatic agent is a compound used for treatment of materials or their surfaces in order to reduce or eliminate buildup of static electricity generally caused by the triboelectric effect. The molecules of an antistatic agent often have both hydrophilic and hydrophobic areas, similar to those of a surfactant; the hydrophobic side interacts with the surface of the material, while the hydrophilic side interacts with the air moisture and binds the water molecules [133]. As synthetic fibers provide poor antistatic properties, research work concerning the improvement of the antistatic properties of textiles by using nanotechnology has been at large. It was determined that nano-sized particles like titanium dioxide, zinc oxide whiskers, nano-antimony-doped tin oxide and silanenanosol could impart antistatic properties to synthetic fibers. Such material helps to effectively dissipate the static charge which is accumulated on the fabric [118]. On the other hand, silanenanosol improves antistatic properties, as the silica gel particles on fiber absorb water and moisture in the air by amino and hydroxyl groups and bound water. Electrically conductive nano-particles are durably anchored in the fibrils of the membrane of Teflon, creating an electrically conductive network that prevents the formation of isolated chargeable areas and voltage peaks commonly found in conventional antistatic materials. This method can overcome the limitation of conventional methods, which is that the antistatic agent is easily washed off after a few laundry cycles [62].

17.12.5 WRINKLE RESISTANCE

To impart wrinkle resistance to fabric, resin is commonly used in conventional methods. However, there are limitations to applying resin, including a decrease in the tensile strength of fiber, abrasion resistance, water absorbency and dye-ability, as well as breathability. To overcome the limitations of using resin, some researchers employed nano-titanium dioxide and nano-silica to improve the wrinkle resistance of cotton and silk respectively [118]. Nano-titanium dioxide employed with carboxylic acid as a catalyst under UV irradiation to catalyzes the cross-linking reaction between the cellulose molecule and the acid. On the other hand, nano-silica when applied with maleic anhydride as a catalyst could successfully improve the wrinkle resistance of silk [116]. More over the wrinkle recovery of the fabrics can also be improved to a great extent by imparting techniques like padding and exhaustion beside the use of nanomaterials to the fabrics. Studies also have suggest that treatment of fabrics with microwaves are more wrinkle resistant as comparable to oven curing, because it generates higher frequency and volumetric heating which minimizes the damage from over drying.

17.13 SMART TEXTILES

All applications textile materials have a vast number of clear advantages:

- They are omnipresent, everybody is familiar with them
- They are easy to use and to maintain
- Clothes have a large contact with the body
- They make us look nice
- They are extremely versatile in terms of raw materials used, arrangement of the fibers, finishing treatments, shaping etc.

They can be made to fit Typical applications where textile structures are to be preferred are:

- Long term or permanent contact without skin irritation,
- Home applications,
- Applications for children: in a discrete and careless way,
- Applications for the elderly: discretion, comfort and esthetics are important.

The multifunctional textiles such as fashion and environmental protection, ballistic and chemical protection, flame protection are all passive systems. The smart textiles are a new generation of fibers, yarns, fabrics and garments that are able to sense stimuli and changes in their environments, such as mechanical, thermal, chemical, electrical, magnetic and optical changes, and then respond to these changes in predetermined ways. They are multifunctional textile systems that can be classified into three categories of passive smart textiles, active smart textiles and very smart textiles [142].

The functionalities of smart textiles can be classified in five groups: sensoring, data processing, actuation, communication, energy.

At this moment, most of the progress has been achieved in the area of sensoring. Many type of parameters can be measured:

- Temperature

- Biopotentials: cardiogram, myographs, encephalographs
- Acoustic: heart, lungs, digestion, joints
- Ultrasound: blood flow
- Biological, chemical
- Motion: respiration, motion
- Pressure: blood
- Radiation: IR, spectroscopy
- Odor, sweat
- Mechanical skin parameters
- Electric (skin) parameters

Some of these parameters are well known, like cardiogram and temperature. Nevertheless, permanent monitoring also opens up new perspectives for these traditional parameters too. Indeed today evaluation is usually based on standards for global population groups. Permanent monitoring supported by self learning devices will allow the set up of personal profiles for each individual, so that conditions deviating from normal can be traced the soonest possible. Also diagnosis can be a lot more accurate.

Apart from the actual measuring devices data processing is a key feature in this respect. These types of data are new. They are numerous with multiple complex interrelationships and time dependent. New self learning techniques will be required.

Actuation is another aspect. Identification of problems only makes sense when followed by an adequate reaction. This reaction can consist of reporting or calling for help, but also drug supply and physical treatment. A huge challenge in this respect is the development of high performance muscle like materials.

Smart textiles is a new aspect in textile that is a multidiscipline field of research in many sciences and technologies such as textile, physics, chemistry, medicine, electronics, polymers, biotechnology, telecommunications, information technology, microelectronics, wearable computers, nanotechnology and microelectromechanical machines. Shape memory materials, conductive materials, phase change materials, chromic materials, photonic fibers, mechanical responsive materials, intelligent coating/membranes, micro and nanomaterials and piezoelectric materials are applied in smart textiles [143].

The objective of smart textile is to absorb a series of active components essentially without changing its characteristics of flexibility and comfort. In order to make a smart textile, firstly, conventional components such as sensors, devices and wires are being reshaped in order to fit in the textile, ultimately the research activities trend to manufacture active elements made of fibers, yarns and fabrics structures. Smart textiles are ideal vehicle for carrying active elements that permanently monitor our body and the environment, providing adequate reaction should something happen [144].

The smart textiles have some of the capabilities such as biological and chemical sensing and responding, power and data transmission from wearable computers and polymeric batteries, transmitting and receiving RF signals and automatic voice warning systems as to 'dangers ahead' that may be appropriate in military applications. Other than military applications of smart textiles, mountain climbers, sportsmen, businessmen, healthcare and medical personnel, police, and firemen will be benefitted from the smart textiles technologies.

A smart textile can be active in many other fields. Smart textiles as a carrier of sensor systems can measure heart rate, temperature, respiration, gesture and many other body parameters that can provide useful information on the health status of a person. The smart textiles can support the rehabilitation process and react adequately on hazardous conditions that may have been detected. The reaction can consist of warning, prevention or active protection. After an event has happened, the smart textile is able to analyze the situation and to provide first aid [145].

Wearable electronics and photonics, adaptive and responsive structures, biomimetics, bioprocessing, tissue engineering and chemical/drug releasing are some of the research areas in integrated processes and products of smart textiles. There are some areas that the research activities have reached the industrial application. Optical fibers, shape memory polymers, conductive polymers, textile fabrics and composites integrated with optical fiber sensors have been used to monitor the health of major bridges and buildings. The first generation of wearable motherboards has been developed, which has sensors integrated inside garments and is capable of detecting injury and health information of the wearer and transmitting such information remotely to a hospital. Shape memory polymers have been applied to textiles in fiber, film and foam forms, resulting in a range of high performance fabrics and garments, especially sea-going garments. Fiber sensors, which are capable of measuring temperature, strain/stress, gas, biological species and smell, are typical smart fibers that can be directly applied to textiles. Conductive polymer-based actuators have achieved very high levels of energy density. Clothing with its own senses and brain, like shoes and snow coats which are integrated with Global positioning system and mobile phone technology, can tell the position of the wearer and give him/her directions. Biological tissues and organs, like ears and noses, can be grown from textile scaffolds made from biodegradable fibers [146, 147].

17.13.1 PHASE CHANGE MATERIALS

Phase change materials are thermal storage materials that are used to regulate temperature fluctuations. The thermal energy transfer occurs when a material changes from a solid to a liquid or from a liquid to a solid. This is called a change in state, or phase.

Incorporating microcapsules of PCM into textile structures improves the thermal performance of the textiles. Phase change materials store energy when they change from solid to liquid and dissipate it when they change back from liquid to solid. It would be most ideal, if the excess heat a person produces could be stored intermediately somewhere in the clothing system and then, according to the requirement, activated again when it starts to get chilly.

TABLE 17.1 Phase Change Materials

Phase change material	Melting temperature °C	Crystallization temperature °C	Heat storage capacity in J/g
Eicosane	36.1	30.6	246
Nonadecane	32.1	26.4	222
Octadecane	28.2	25.4	244
Heptadecane	22.5	21.5	213
Hexadecane	18.5	16.2	237

The most widespread PCMs in textiles are paraffin-waxes with various phase change temperatures (melting and crystallization) depending on their carbon numbers. The characteristics of some of these PCMs are summarized in Table 17.1. These phase change materials are enclosed in microcapsules, which are 1–30 μm in diameter. Hydrated inorganic salts have also been used in clothes for cooling applications. PCM elements containing Glauber's salt (sodium sulfate) have been packed in the pockets of cooling vests.

PCM can be applied to fibers in a wet-spinning process, incorporated into foam or embedded into a binder and applied to fabric topically, or contained in a cell structure made of a textile reinforced synthetic material [148, 149].

In manufacturing the fiber, the selected PCM microcapsules are added to the liquid polymer or polymer solution, and the fiber is then expanded according to the conventional methods such as dry or wet spinning of polymer solutions and extrusion of polymer melts.

Fabrics can be formed from the fibers containing PCM by conventional weaving, knitting or nonwoven methods, and these fabrics can be applied to numerous clothing applications.

In this method, the PCMs are permanently locked within the fibers, the fiber is processed with no need for variations in yarn spinning, fabric knitting or dyeing and properties of fabrics (drape, softness, tenacity, etc.) are not altered in comparison with fabrics made from conventional fibers. The microcapsules incorporated into the fibers in this method have an upper loading limit of 5–10% because the physical properties of the fibers begin to suffer above that limit, and the finest fiber is available. Due to the small content of microcapsules within the fibers, their thermal capacity is rather modest, about 8–12 J/g.

Usually PCM microcapsules are coated on the textile surface. Microcapsules are embedded in a coating compound such as acrylic, polyurethane and rubber latex, and applied to a fabric or foam. In lamination of foam containing PCMs onto a fabric, the selected PCMs microcapsules can be mixed into a polyurethane foam matrix, from which moisture is removed, and then the foam is laminated on a fabric. Typical concentrations of PCMs range from 20% to 60% by weight. Microcapsules should be added to the liquid polymer or elastomer prior to hardening. After foaming (fabricated from polyurethane) microcapsules will be embedded within the base material matrix. The application of the foam pad is particularly recommended because a greater amount of microcapsules can be introduced into the smart textile. In spite of this, different PCMs can be used, giving a broader range of regulation temperatures. Additionally, microcapsules may be anisotropically distributed in the layer of foam. The foam pad with PCMs may be used as a lining in a variety of clothing such as gloves, shoes, hats and outerwear. Before incorporation into clothing or footwear the foam pad is usually attached to the fabric, knitted or woven, by any conventional means such as glue, fusion or lamination [150, 151].

The PCM microcapsules are also applied to a fibrous substrate using a binder (e.g., acrylic resin). All common coating processes such as knife over roll, knife over air, screen-printing, gravure printing, dip coating may be adapted to apply the PCM microcapsules dispersed throughout a polymer binder to fabric. The conventional pad–mangle systems are also suitable for applying PCM microcapsules to fabrics. The formulation containing PCMs can be applied to the fabric by the direct nozzle spray technique.

There are many thermal benefits of treating textile structures with PCM microcapsules such as cooling, insulation and thermo regulating effect. Without phase change materials the thermal insulation capacity of clothing depends on the thickness and the density of the fabric (passive insulation). The application of PCM to a garment provides an active thermal insulation effect acting in addition to the passive thermal insulation effect of the garment system. The active thermal insulation of the PCM controls the heat flux through the garment layers and adjusts the heat flux to the thermal circumstances. The active thermal insulation effect of the PCM results in a substantial improvement of the garment's thermo-physiological wearing comfort [142]. Intensity and duration of the PCM's active thermal insulation effect depend mainly on the heat-storage capacity of the PCM microcapsules and their applied quantity. In order to ensure a suitable and durable effect of the PCM, it is necessary to apply proper PCM in sufficient quantity into the appropriate fibrous substrates of proper design.

The PCM quantity applied to the active wear garment should be matched with the level of activity and the duration of the garment use. Furthermore, the garment construction needs to be designed in a way which assists the desired thermo-regulating effect. Thinner textiles with higher densities readily support the cooling process. In contrast, the use of thicker and less dense textile structures leads to a delayed and therefore more efficient heat release of the PCM. Further requirements on the textile substrate in a garment application include sufficient breathability, high flexibility, and mechanical stability [152].

In order to determine a sufficient PCM quantity, the heat generated by the human body has to be taken into account carrying out strenuous activities under which the active wear garments are worn. The heat generated by the body needs to be entirely released through the garment layers into the environment. The necessary PCM quantity is determined according to the amount of heat which should be absorbed by the PCM to keep the heat balance equalized. It is mostly not necessary to put PCM in all parts of the garment. Applying PCM microcapsules to the areas that provide problems from a thermal standpoint and thermoregulating the heat flux through these areas is often enough. It is also advisable to use different PCM microcapsules in different quantities in distinct garment locations [151].

17.13.1.1 Applications of Textiles Containing PCMS

Fabrics containing PCMs have been used in a variety of applications including apparel, home textiles and technical textiles (Table 17.2).

Phase change materials are used both in winter and summer clothing. PCM is used not only in high-quality outerwear and footwear, but also in the underwear, socks, gloves, helmets and bedding of world-wide

TABLE 17.2 Application of PMs in Textiles

Casual clothing	Underwear, Jackets, sport garments
Professional clothing	Fire fighters protective clothing, Bullet proof fabrics, Space suits, Sailor suits
Medical uses	Surgical gauze, Bandage, Nappies, Bed lining, Gloves, Gowns, Caps, Blankets
Shoe linings	Ski boots, Golf shoes
Building materials	IInd proof. In concrete
Life style apparel	Elegant fleece vests
Other uses	Automotive interiors, Battery warmers

brand leaders. Seat covers in cars and chairs in offices can consist of phase change materials.

Currently, phase change materials are being used in a variety of outdoor apparel items such as smart jackets, vests, men's and women's hats and rainwear, outdoor active-wear jackets and jacket lining, golf shoes, trekking shoes, ski and snowboard gloves, boots, earmuffs and protective garments. In protective garments, the absorption of body heat surplus, insulation effect caused by heat emission of the PCM into the fibrous structure and thermo-regulating effect, which maintains the microclimate temperature nearly constant are the specified functions of PCM contained smart textile [153].

The addition of PCMs to fabric-backed foam significantly increases the weight, thickness, stiffness, flammability, insulation value, and evaporative resistance value of the material. It is more effective to have one layer of PCM on the outside of a tight-fitting, two layer ensemble than to have it as the inside layer. This may be because the PCMs closest to the body did not change phase [154].

PCM protective garments should improve the comfort of workers as they go through these environmental step changes (e.g., warm to cold to warm, etc.). For these applications, the PCM transition temperature should be set so that the PCMs are in the liquid phase when worn in the warm environment and in the solid phase in the cold environment. The effect of phase change materials in clothing on the physiological and subjective thermal responses of people would probably be maximized if the wearer was repeatedly going through temperature transients (i.e., going back and forth between a warm and cold environment) or intermittently touching hot or cold objects with PCM gloves [154].

One example of practical application of PCM smart textile is cooling vest. This is a comfort garment developed to prevent elevated body temperatures in people who work in hot environments or use extreme physical exertion. The cooling effect is obtained from the vest's 21 PCM elements containing Glauber's salt which start absorbing heat at a particular temperature (28°C). Heat absorption from the body or from an external source continues until the elements have melted. After use the cooling vest has to be charged at room temperature (24°C) or lower.

When all the PCMs are solidified the cooling vest is ready for further use [155].

A new generation of military fabrics feature PCMs which are able to absorb, store and release excess body heat when the body needs it resulting in less sweating and freezing, while the microclimate of the skin is influenced in a positive way and efficiency and performance are enhanced [156].

In the medical textiles field, a blanket with PCM can be useful for gently and controllably reheating hypothermia patients. Also, using PCMs in bed covers regulates the micro climate of the patient [157].

In domestic textiles, blinds and curtains with PCMs can be used for reduction of the heat flux through windows. In the summer months large amounts of heat penetrate the buildings through windows during the day. At night in the winter months the windows are the main source of thermal loss. Results of the test carried out on curtains containing PCM have indicated a 30% reduction of the heat flux in comparison to curtains without PCM [158].

17.13.2 SHAPE MEMORY MATERIALS

Shape memory materials are able to 'remember' a shape, and return to it when stimulated, for example, with temperature, magnetic field, electric field, pH-value and UV light. An example of natural shape memory textile material is cotton, which expands when exposed to humidity and shrinks back when dried. Such behavior has not been used for esthetic effects because the changes, though physical, are in general not noticeable to the naked eye. The most common types of such SMMs materials are shape memory alloys and polymers, but ceramics and gels have also been developed. When sensing this material specific stimulus, SMMs can exhibit dramatic deformations in a stress free recovery. On the other hand, if the SMM is prevented from recovering this initial strain, a recovery stress (tensile stress) is induced, and the SMM actuator can perform work. This situation where SMM deforms under load is called restrained recovery [159].

Because of the wide variety of different activation stimuli and the ability to exhibit actuation or some other predetermined response, SMMs can be used to control or tune many technical parameters in smart material systems in response to environmental changes –such as shape, position, strain, stiffness, natural frequency, damping, friction and water vapor penetration. Both the fundamental theories and engineering aspects of SMMs have been investigated extensively and a rather wide variety of different SMMs are presently commercial materials. Commercialized shape memory products have been based mainly on metallic SMAs, either taking advantage of the shape change due the shape memory effect or the superelasticity of the material, the two main phenomena of SMAs. SMPs and shape memory gels are developed at a quick rate, and within the last few years also some products based on magnetic shape memory alloys have been commercialized. SMC materials, which can be activated not only by temperature but also by elastic energy, electric or magnetic field, are mainly at the research stage [160].

17.13.2.1 Applications of Textiles Containing SMMs

There are many potential applications of shape memory polymers in industrial components like automotive parts, building and construction products, intelligent packing, implantable medical devices, sensors and actuators, etc. SMPs are used in toys, handgrips of spoons, toothbrushes, razors and kitchen knives, also as an automatic choking device in small-size engines. One of the most well-known examples of SMP is a clothing application, a membrane called Diaplex. The membrane is based on polyurethane based shape memory polymers developed by Mitsubishi Heavy Industries [161].

Polyurethane is an example of shape memory polymers which is based on the formation of a physical cross-linked network as a result of entanglements of the high molecular weight linear chains, and on the transition from the glassy state to the rubber-elastic state. Shape memory polyurethane is a class of polyurethane that is different from conventional polyurethane in that these have a segmented structure and a wide range of Tg. The long polymer chains entangle each other and a three-dimensional

TABLE 17.3 Some of the Shape Memory Polymers for Textiles Applications

Polymer	Physical interactions	Form
	Original shape	**Transient shape**
Polynorbomene	Chain entanglement	Glassy state
Polyurethane	Microcrystal	Glassy state
Polyethylene/nylon-6 graft copolymer	Crosslinking	Microcrystal
Styrene-1,4-butadiiene block copolymer	Microcrystal/Glassy state of polystyrene	Microcrystal of poly (1,4-butadiene)
Ethylene oxide-ethylene terphethalate block copolymer	Microcrystal of PET	Microcrystal of PEO
Polymethylene-1, 3-cyclopentane) polyethylene block copolymer	Microcrystal of PE	Microcrystal/Glassy state of pmcp

network is formed. The polymer network keeps the original shape even above Tg in the absence of stress. Under stress, the shape is deformed and the deformed shape is fixed when cooled below Tg. Above the glass transition temperature polymers show rubber-like behavior [162].

The material softens abruptly above the glass transition temperature Tg. If the chains are stretched quickly in this state and the material is rapidly cooled down again below the glass transition temperature the polynorbornene chains can neither slip over each other rapidly enough nor become disentangled. It is possible to freeze the induced elastic stress within the material by rapid cooling. The shape can be changed at will. In the glassy state the strain is frozen and the deformed shape is fixed. The decrease in the mobility of polymer chains in the glassy state maintains the transient shape in polynorbornene. The recovery of the material's original shape can be observed by heating again to a temperature above Tg. This occurs because of the thermally induced shape-memory effect. The disadvantage of this polymer is the difficulty of processing because of its high molecular weight [162, 163]. Some of the shape memory polymers are suitable for textiles applications are shown in Table 17.3.

Shape memory polymers can be laminated, coated, foamed, and even straight converted to fibers. There are many possible end uses of these smart textiles. The smart fiber made from the shape memory polymer can be applied as stents, and screws for holding bones together.

Shape memory polymer coated or laminated materials can improve the thermophysiological comfort of surgical protective garments, bedding and incontinence products because of their temperature adaptive moisture management features.

Films of shape memory polymer can be incorporated in multilayer garments, such as those that are often used in the protective clothing or leisurewear industry. The shape memory polymer reverts within wide range temperatures. This offers great promise for making clothing with adaptable features. Using a composite film of shape memory polymer as an interliner in multilayer garments, outdoor clothing could have adaptable thermal insulation and be used as protective clothing. A shape memory polymer membrane and insulation materials keep the wearer warm. Molecular pores open and close in response to air or water temperature to increase or minimize heat loss. Apparel could be made with shape memory fiber. Forming the shape at a high temperature provides creases and pleats in such apparel as slacks and skirts. Other applications include fishing yarn, shirt neck bands, cap edges, casual clothing and sportswear. Also, using a composite film of shape memory polymers as an interlining provides apparel systems with variable tog values to protect against a variety of weather conditions [163].

17.13.3 CHROMIC MATERIALS

Chromic materials are the general term referring to materials, which their color changes by the external stimulus. Due to color changing properties, chromic materials are also called chameleon materials. This color changing phenomenon is caused by the external stimulus and chromic materials can be classified depending on the external stimulus of induction.

Photochromic, thermochromic, electrochromic, piezochromic, solvatechromic and carsolchromic are chromic materials that change their color by the external stimulus of heat, electricity, pressure, liquid and an electron beam, respectively. Photochromic materials are suitable for sun lens applications. Most photochromic materials are based on organic materials or silver particles. Thermochromic materials change color reversibly with changes in temperature. The liquid crystal type and the molecular rearrangement type are thermochromic systems in textiles. The thermochromic materials can be made as semiconductor compounds, from liquid crystals or metal compounds. The change in color occurs at a predetermined temperature, which can be varied. Electrochromic materials are capable of changing their optical properties (transmittance and/or reflectance) under applied electric potentials. The variation of the optical properties is caused by insertion/extraction of cations in the electrochromic film. Piezochromism is the phenomenon where crystals undergo a major change of color due to mechanical grinding. The induced color reverts to the original color when the fractured crystals are kept in the dark or dissolved in an organic solvent [164].

Solvatechromism is the phenomenon, where color changes when it makes contact with a liquid, for example, water. Materials that respond to water by changing color are also called hydrochromic and this kind of textile material can be used, for example, for swimsuits.

17.13.3.1 Applications of Textiles Containing Chromic Materials

The majority of applications for chromic materials in the textile sector today are in the fashion and design area, in leisure and sports garments. In workwear and the furnishing sector a variety of studies and investigations are in the process by industrial companies, universities and research centers. Chromic materials are one of the challenging material groups when thinking about future textiles. Color changing textiles are interesting, not only in fashion, where color changing phenomena will exploit for fun all the rainbow colors, but also in useful and significant applications in soldier and weapons camouflage, workwear and in technical and medical textiles. The combination of SMM and thermochromic coating is an interesting area which produces shape and color changes of the textile material at the same time [165].

17.13.4 DESIGNING THE SMART TEXTILE SYSTEMS

Comfort is very important in textiles because stresses lead to increased fatigue. The potential of smart textiles is to measure a number of body parameters such as skin temperature, humidity and conductivity and show the level of comfort through the textile sensors. To keep the comfort of textiles, adequate actuators are needed that can heat, cool, insulate, ventilate and regulate moisture. The use of the smart system should not require any additional effort. The weight of a smart textile system should not reduce operation time of the rescue worker.

Other key issues for the design of a smart textile system are [166]:

- Working conditions – relevant parameters: only relevant information should be provided in order to avoid additional workload; this includes indication of danger and need for help.
- Effective alarm generation: the rescue worker or a responsible person should be informed adequately on what needs to be done.
- System maintenance: it must be possible to treat the suit using usual maintenance procedures.
- Cost must be justified
- Robustness
- Energy constraints: energy requirements must be optimized
- Long range transmission: transmission range must be adjusted to the situation of use.

Fighting a fire in a building is different from fighting one in an open field.

A wearable smart textile system basically comprises following components [36]:

- Sensors to detect body or environmental parameters;
- A data processing unit to collect and process the obtained data;
- An actuator that can give a signal to the wearer;
- An energy supply that enables working of the entire system;
- Interconnections that connect the different components;
- A communication device that establishes a wireless communication link with a nearby base station.

The main layers concerned with smart clothing are the skin layer and two clothing layers. Physically the closest clothing layer for a human user is an underwear layer, which transports perspiration away from the skin area. The function of this layer is to keep the interface between a user and the clothes comfortable and thus improve the overall wearing comfort. The second closest layer is an intermediate clothing layer, which consists of the clothes that are between the underclothes and outdoor clothing. The main purpose of this layer is considered to be an insulation layer for warming up the body. The outermost layer is an outerwear layer, which protects a human against hard weather conditions.

The skin layer is located in close proximity to the skin. In this layer we place components that need direct contact with skin or need to be very close to the skin. Therefore, the layer consists of different user interface devices and physiological measurement sensors. The number of the additional components in underwear is limited owing to the light structure of the clothing.

An inner clothing layer contains intermediate clothing equipped with electronic devices that do not need direct contact with skin and, on the other hand, do not need to be close to the surrounding environment. These components may also be larger in size and heavier in weight compared to components associated with underclothes. It is often beneficial to fasten components to the inner clothing layer, as they can be easily hidden. Surrounding clothes also protect electronic modules against cold, dirt and hard knocks.

Generally, the majority of electronic components can be placed on the inner clothing layer. These components include various sensors, a central processing unit and communication equipment. Analogous to ordinary clothing, additional heating to warming up a person in cold weather conditions is also associated with this layer. Thus, the inner layer is the most suitable for batteries and power regulating equipment, which are also sources of heat.

The outer clothing layer contains sensors needed for environment measurements, positioning equipment that may need information from the surrounding environment and numerous other accessories. The physical surroundings of smart clothing components measure the environment and the virtual environment accessed by communication technologies. Soldier and weapons camouflage is possible by using chromic materials in outer layer of smart textiles [10].

17.13.5 DATA TRANSFER REQUIREMENTS IN SMART TEXTILES

The data transfer requirements can be divided into internal and external. The internal transfer services are divided into local health and security related measurements. Many of the services require or result in external communications between the smart clothing and its environment. Wired data transfer is in many cases a practical and straightforward solution. Thin wires routed through fabric are an inexpensive and high capacity medium for information and power transfer. The embedded wires inside clothing do not affect its appearance. However, wires form inflexible parts of clothing and the detaching and reconnecting of wires decrease user comfort and the usability of clothes. The cold winter environment especially stiffens the plastic shielding of wires. In hard usage and in cold weather conditions, cracking of wires also becomes a problem [167]. The connections between the electrical components placed on different pieces of clothing are another challenge when using wires. During dressing and undressing, connectors should be attached or detached, decreasing the usability of clothing. Connectors should be easily fastened, resulting in the need for new connector technologies.

A potential alternative to plastic shielded wires is to replace them with electrically conductive fibers. Conductive yarns twisted from fibers form a soft cable that naturally integrates in the clothing's structure keeping the system as clothing-like as possible. Fiber yarns provide durable, flexible and washable solutions. Also lightweight optical fibers are used in wearable applications, but their function has been closer to a sensor than a communication medium [168, 169]. The problem of conductive fibers is due to the reliable connections of them. Ordinary wires can be soldered directly to printed circuit boards, but the structure of the fiber yarn is more sensitive to breakage near the solder connections.

Protection materials that prevent the movement of the fiber yarn at the interface of the hard solder and the soft yarn must be used. Optical fibers are commonly used for health monitoring applications and also for lighting purposes.

Low-power wireless connections provide increased flexibility and also enable external data transfer within the personal space. Different existing and emerging WLAN and WPAN types of technologies are general purpose solutions for the external communications, providing both high speed transfer and low costs. For wider area communications and full mobility, cellular data networks are currently the only practical possibility [170].

17.13.6 OPTICAL FIBERS IN SMART TEXTILES

Optical fibers are currently being used in textile structures for several different applications. Optic sensors are attracting considerable interest for a number of sensing applications [171]. There is great interest in the multiplexed sensing of smart structures and materials, particularly for the real-time evaluation of physical measurements (e.g., temperature, strain) at critical monitoring points. One of the applications of the optical fibers in textile structures is to create flexible textile-based displays based on fabrics made of optical fibers and classic yarns [172]. The screen matrix is created during weaving, using the texture of the fabric. Integrated into the system is a small electronics interface that controls the LEDs that light groups of fibers. Each group provides light to one given area of the matrix. Specific control of the LEDs then enables various patterns to be displayed in a static or dynamic manner. This flexible textile-based displays are very thin size and ultra-light weight. This leads one to believe that such a device could quickly enable innovative solutions for numerous applications. Bending in optical fibers is a major concern since this causes signal attenuation at bending points. Integrating optical fibers into a woven perform requires bending because of the crimping that occurs as a result of weave interlacing. However, standard plastic optical fiber materials like poly methyl methacrylate, polycarbonate and polystyrene are rather stiff compared to standard textile fibers and therefore their integration into textiles usually leads to stiffen of the woven fabric and the textile touch is getting lost [173]. Alternative fibers with appropriate flexibility and transparency are not commercially available yet.

17.13.7 CONDUCTIVE MATERIALS IN SMART TEXTILES

Several conductive materials are in use in smart textiles. Conductive textiles include electrically conductive fibers fabrics and articles made from them. Flexible electrically conducting and semiconducting materials, such as conductive polymers, conductive fibers, threads, yarns, coatings and ink are playing an important role in realizing lightweight, wireless and wearable interactive electronic textiles. Generally, conductive fibers can be divided into two categories such as naturally conductive fibers and treated conductive fibers.

Naturally, conductive fibers can be produced purely from inherently conductive materials, such as metals, metal alloys, carbon sources, and conjugated polymers [174].

However, it is important to point out that nanofibers produced from polymers do not in general show quantum effects usually associated with nanotechnology, with spatially confined matter. The reason simply is first of all that the nature of the electronic states of organic materials including polymer materials that control optical and electronic properties does not resemble the one known for semiconductors or conductors.

Electronic states that are not localized but rather extend throughout the bulk material are characteristics of such nonorganic materials, with the consequence that modifications first of all of the absolute size and secondly of the geometry of a body made from them have strong effects on properties particularly as the sizes approach the few tens to a few nanometers scale. Organic materials, on the other hand, display predominantly localized states for electronic excitations, electronic transport with the states being defined by molecular groups such as chromophore groups or complete molecules. The consequence is that the electronic states are not affected as the dimensions of the element such as a fiber element are reduced down into the nanometer scale [175].

Furthermore, both amorphous polymers and partially crystalline polymers have structures anyway even in the bulk, in macroscopic bodies that are restricted to the nanometer scale. The short – range order of amorphous polymers, as represented by the pair correlation function, does not extend beyond about 2 nm and the thickness of crystalline lamellae is typically in the range of a few nanometers or a few tens of nanometer, respectively. So, the general conclusion is that the reduction of the diameter of fibers made from polymers or organic materials for that matter will affect neither optical nor electronic properties to a significant degree or the intrinsic structure. This will, of course, be different if we are concerned with fibers composed of metals, metal oxides, semiconductors as accessible via the precursor route. In such cases, one is well within the range of quantum effects and the properties of such fibers have to be discussed along the lines spelt out in textbooks on quantum effects [176].

Now, staying with the subject of polymer fibers the discussion given above should certainly not lead to the conclusion that nanofibers and nonwovens composed of nanofibers do not display unique properties and functions of interest both in the areas of basic science but also technical applications, just the contrary.

Taking the reduction of the fiber diameter into the nm scale as a first example it is readily obvious that the specific surface area increases dramatically as the fiber diameter approaches this range. In fact, it increases with the inverse of the fiber radius.

It is obvious that the specific surface increases from about 0.1 m^2/g for fibers with a thickness of about 50 micrometers (diameter of a human hair) to about 300 m^2/g for fibers with a thickness of 10 nm.

Secondly, the strength of fibers also scales inversely with the fiber diameter, thus increasing also very strongly with decreasing fiber diameter, following the Griffith criteria. The reason is that the strength tends to be controlled by surface flaws, the probability of which will decrease along a unit length of the fiber as the surface area decreases. So, a decrease of the fiber diameter from about 50 micrometer to about 10 nm is expected to increase the strength from about 300 N/mm 2 by a factor of about 1000 and more. Thirdly, the pores of nonwoven membranes composed of nanofibers reach the nm range as the fibers get smaller and smaller. A reduction of the fiber diameter from say about 50 micrometer, for which pores with diameter of about 500 micrometer are expected to about 100 nm will cause the pore diameter to decrease to about 1 micrometer. In a similar way, a further reduction to fiber diameters to about 10 nm will cause the average pores to display pore diameters around 100 nm. This will certainly show up in the selectivity of the filters with respect to solid particles, aerosols, etc. to be discussed further below [175].

The reduction in pore diameter is, furthermore, connected with strong modifications of the dynamics of gases and fluids within the nonwoven, located within or flowing through the nonwoven. Thermal insulations, for instance, in nonwovens containing a gas is controlled for larger pores – larger than the average free path length of the molecules – by just this free path length. However, as the pores get smaller the collision of the molecules with the pore walls – fiber surface for fiber -based nonwovens – takes over the control of thermal insulation with an increase of the thermal insulation that can amount to several orders of magnitude. This aspect will also be discussed below [177].

Finally, the flow of gases or fluids around a fiber changes very strongly in nature as the fiber diameter goes down to the nanometer range with a transition of the flow regime from the conventional one to the so – called Knudsen regime, to be discussed below in more detail.

All these effects are classical ones yet have major impact on nanofiber properties and applications. These examples show that nanofibers and nonwovens composed of them display unique properties of functions already based on classical phenomena. This suggests that such fibers/nonwovens can be used with great benefits in various types of applications. The spectrum of applications that can be envisioned for electrospun nanofibers is extremely broad due to their unique intrinsic structure, surface properties and functions [175].

Highly conductive flexible textiles can be prepared by weaving thin wires of various metals such as brass and aluminum. These textiles have been developed for higher degrees of conductivity.

Metal conductive fibers are very thin filaments with diameters ranging from 1 to 80 μm produced from conductive metals such as ferrous alloys, nickel, stainless steel, titanium, aluminum and copper. Since they are different from polymeric fibers, they may be hard to process and have problems of long-term stability. These highly conductive fibers are expensive, brittle, heavier and lower processability than most textile fibers.

Treated conductive fibers can be produced by the combination of two or more materials, such as nonconductive and conductive materials. These conductive textiles can be produced in various ways, such as by impregnating textile substrates with conductive carbon or metal powders, patterned printing, and so forth. Conducting polymers, such as polyacetylene, polypyrrole, polythiophene and polyaniline offer an interesting alternative. Among them, polypyrrole has been widely investigated owing to its easy preparation, good electrical conductivity, good environmental stability in ambient conditions and because it poses few toxicological problems. PPy is formed by the oxidation of pyrrole or substituted pyrrole monomers. Electrical conductivity in PPy involves the movement of positively charged carriers or electrons along polymer chains and the hopping of these carriers between chains. The conductivity of PPy can reach the range 102 S cm-1, which is next only to PA and PAn. With inherently versatile molecular structures, PPys are capable of undergoing many interactions [178, 179].

The conductive fibers obtained through special treatments such as mixing, blending, or coating are also known as CPCs, can have a combination of the electrical and mechanical properties of the treated materials. Fibers containing metal, metal oxides and metal salts are a proper alternative for metal fibers. Polymer fibers may be coated with a conductive layer such as polypyrrole, copper or gold. The conductivity will be maintained as long as the layer is intact and adhering to the fiber. Chemical plating and dispersing metallic particles at a high concentration in a resin are two general methods of coating fibers with conductive metals [180].

The brittleness of PPy has limited the practical applications of it. The processability and mechanical properties of PPy can be improved by incorporating some polymers into PPy [181]. However, the incorporation of a sufficient amount of filler generally causes a significant deterioration in the mechanical properties of the conducting polymer, in order to exceed the percolation threshold of conductivity [182]. Another route to overcoming this deficiency is by coating the conducting polymer on flexible textile substrates to obtain a smooth and uniform electrically conductive coating that is relatively stable and can be easily handled [183].

Thus, PPy-based composites may overcome the deficiency in the mechanical properties of PPy, without adversely affecting the excellent physical properties of the substrate material, such as its mechanical strength and flexibility. The resulting products combine the usefulness of a textile substrate with electrical properties that are similar to metals or semiconductors.

Due to electron-transport characteristics of Conjugated polymers or ICPs, they are regarded as semi conductors or even sometimes conductors. Due to their high conductivity, lower weight, and environmental stability, they have a very important place in the field of smart and interactive textiles [184].

The conductivity of materials is often affected by several parameters which may be exploitable mechanisms for use as a sensor. Extension, heating, wetting and absorption of chemical compounds in general may increase or decrease conductivity. Swelling or shrinkage of composite fibers of carbon nanotubes alters the distance between the nanoparticles in the fibers, causing the conductivity to change.

Fibers containing conductive carbon are produced with several methods such as loading the whole fibers with a high concentration of carbon, incorporating the carbon into the core of a sheath–core bicomponent fiber, incorporating the carbon into one component of a side–side or modified side–side bicomponent fiber, suffusing the carbon into the surface of a fiber.

Nanoparticles such as carbon nanotubes can be added to the matrix for achieving conductivity. Semi-conducting metal oxides are often nearly colorless, so their use as conducting elements in fibers has been considered likely to lead to fewer problems with visibility than the use of conducting carbon. The oxide particles can be embedded in surfaces, or incorporated into sheath–core fibers, or react chemically with the material on the surface layer of fibers.

Conductive fibers can also be produced by coating fibers with metal salts such as copper sulfide and copper iodide. Metallic coatings produce highly conductive fibers; however adhesion and corrosion resistance can present problems. It is also possible to coat and impregnate conventional fibers with conductive polymers, or to produce fibers from conductive polymers alone or in blends with other polymers.

Conductive fibers/yarns can be produced in filament or staple lengths and can be spun with traditional nonconductive fibers to create yarns that possess varying wearable electronics and photonics degrees of conductivity. Also, conductive yarns can be created by wrapping a non-conductive yarn with metallic copper, silver or gold foil and be used to produce electrically conductive textiles.

Conductive threads can be sewn to develop smart electronic textiles. Through processes such as electrodeless plating, evaporative deposition, sputtering, coating with a conductive polymer, filling or loading fibers and carbonizing, a conductive coating can be applied to the surface of fibers, yarns or fabrics. Electrodeless plating produces a uniform conductive coating, but is expensive. Evaporative deposition can produce a wide range of thicknesses of coating for varying levels of conductivity. Sputtering can achieve a uniform coating with good adhesion. Textiles coated with a conductive polymer, such as, polyaniline and polypyrrole, are more conductive than metal and have good adhesion, but are difficult to process using conventional methods.

Adding metals to traditional printing inks creates conductive inks that can be printed onto various substrates to create electrically active patterns. The printed circuits on flexible textiles result in improvements in durability, reliability and circuit speeds and in a reduction in the size of the circuits. The printed conductive textiles exhibit good electrical properties after printing and abrading. The inks withstand bending without losing conductivity. However, after 20 washing cycles, the conductivity decreases considerably. Therefore, in order to improve washability, a protective polyurethane layer is put on top of the printed samples, which resulted in the good conductivity of the fabrics, even after washing. Currently, digital printing technologies promote the application of conductive inks on textiles [185].

17.13.7.1 Applications of Conductive Smart Textile

Electrically conductive textiles make it possible to produce interactive electronic textiles. They can be used for communication, entertainment, health care, safety, homeland security, computation, thermal purposes, protective clothing, wearable electronics and fashion. The application of conductive smart textile in combination with electronic advices is very widespread. In location and positioning, they can be used for child monitoring, geriatric monitoring, integrated GPS monitoring, livestock monitoring, asset tracking, etc.

In infotainment, they can be used for integrated compact disc players, MP3 players, cell phones and pagers, electronic game panels, digital cameras, and video devices, etc.

In health and biophysical monitoring, they can be used for cardiovascular monitoring, monitoring the vital signs of infants, monitoring clinical trials, health and fitness, home healthcare, hospitals, medical centers, assisted-living units, etc.

They can be used for soldiers and personal support of them in the battlefield, space programs, protective textiles and public safety (fire-fighting, law enforcement), automotive, exposure-indicating textiles, etc.

They can be also used to show the environmental response such as color change, density change, heating change, etc. Fashion, gaming, residential interior design, commercial interior design and retail sites are other application of conductive smart textiles.

17.13.8 SMART FABRICS FOR HEALTH CARE

The continuous monitoring of vital signs of some patients and elderly people is an emerging concept of health care to provide assistance to patients as soon as possible either online or offline. A wearable smart textile can provide continuous remote monitoring of the health status of the patient. Wearable sensing systems will allow the user to perform everyday activities without discomfort. The simultaneous recording of vital signs would allow parameter extrapolation and intersignal elaboration, contributing to the generation of alert messages and synoptic patient tables. In spite of this, a smart fabric is capable of recording body kinematic maps with no discomfort for several fields of application such as rehabilitation and sports [186].

17.13.9 ELECTRONIC SMART TEXTILES

The components of an electronic smart textile that provide several functions are sensors unit, network unit, processing unit, actuator unit and power unit. On the smart textile, several of these functions are combined to form services. Providing information, communication or assistance are possible services. Because mobility is now a fundamental aspect of many services and devices, these smart textiles can be used for health applications such as monitoring of vital signs of high-risk patients and elderly people, therapy and rehabilitee, knowledge applications such as instructions and navigation and entertainment applications such as audio and video devices. For communication between the different components of smart textile applications, both wired and wireless technologies are applicable. An applied solution for data transferring is often a compromise based on application requirements, operational environment, available and known technologies, and costs [146].

17.13.10 ELECTRICAL CIRCUITS IN SMART TEXTILE STRUCTURES

In order to form flexible circuit boards, printing of circuit patterns is carried out on polymeric substrates such as films. Fabric based circuits potentially offer additional benefits of higher flexibility in bending and shear, higher tear resistance, as well as better fatigue resistance in case of repeated deformation. Different processes that have been described in literature for the fabrication of fabric based circuits include embroidery of conductive threads on fabric substrates, weaving and knitting of conductive threads along with nonconductive threads, printing or deposition and chemical patterning of conductive elements on textile substrates.

The insulating fabric could be woven, nonwoven, or knitted [187]. The conductive threads can be embroidered in any shape on the insulating fabric irrespective of the constituent yarn path in a fabric. One of the primary disadvantages of embroidery as a means of circuit formation is that it does not allow formation of multilayered circuits involving conductive threads traversing through different layers as is possible in the case of woven circuits. Conductive threads can be either woven or knitted into a fabric structure along with nonconductive threads to form an electrical circuit. One of the limitations of using weaving for making electrical circuits is that the conductive threads have to be placed at predetermined locations in the warp direction while forming the warp beam or from a creel during set up of the machine. Different kinds of conductive threads can be supplied in the weft or filling direction and inserted using the weft selectors provided on a weaving machine. Some modifications to the yarn supply system of the machine may be needed in order to process the conductive threads that are more rigid [188].

In most conventional weft knitting machines, like a flatbed machine, the conductive threads can be knitted in the fabric only in one direction, i.e., the course (or cross) direction. In order to keep the conductive element in a knit structure straight, one can insert a conductive thread in the course direction such that the conductive thread is embedded into the fabric between two courses formed from nonconductive threads.

Processes that have been employed to form a patterned conductive path on fabric surfaces include deposition of polymeric or nonpolymeric conducting materials and subsequent etching, reducing, or physical removal of the conductive materials from certain regions. Thus, the conductive material that is not removed forms a patterned electrical circuit or a region of higher conductivity. The biggest problem associated with patterning of circuits from thin conductive films (polymeric or metallic) deposited on fabric substrates is that use of an etching agent for forming a circuit pattern leads to nonuniform etching, as some of the etching liquid is absorbed by the threads of the underlying substrate fabric [189]. Another problem with deposition of conductive films on fabric substrates is that bending the fabric may lead to discontinuities in conductivity at certain points.

There are different device attachment methods like raised wire connectors, solders, snap connectors, and ribbon cable connectors in electronic smart textiles. Soldering produces reliable electrical connections to conductive threads of an electronic textile fabric but has the disadvantage of not being compatible with several conductive threads or materials like stainless steel. Moreover, soldering of electronic devices to threads that are insulated is a more complex process involving an initial step of removal of insulation from the conductive threads in the regions where the device attachments are desired and insulation of the soldered region after completion of the soldering process. The main advantage of employing snap connectors is the ease of attachment or removal of electronic devices from these connectors, whereas the main disadvantages are the large size of the device and the weak physical connection formed between the snap connectors and the devices. Ribbon cable connectors employ insulation displacement in order to form an interconnection with insulated conductor elements integrated into the textiles. A v-shaped contact cuts through the insulation to form a connection to the conductor. Firstly, the ribbon cable connector is attached to the conductive threads in an e-textile fabric and subsequent electronic devices and printed circuit boards are attached to the ribbon cable connector. One of the advantages of employing ribbon cable connectors for device attachment is the ease of attachment and removal of the electronic devices to form the electronic textiles [190, 191].

17.13.11 NEW TEXTILE MATERIALS FOR ENVIRONMENTAL PROTECTION

From prehistoric times till now, air pollution from hazardous chemical and biological particles is an essential threat to humans' health. Together with the development of civilization and escalation of the conflicts between nations, the risk of loss of health and even life due to polluted air increases considerably. Therefore the continuous development of the new materials used for protection of human respiratory tracts against hazardous particles is observed. The fibrous materials play a special role in this subject. Davies in his work 'Air Filtration' has presented an interesting review of the earliest literature considering problems connected with filtering polluted air [192].

For centuries, miners have used special clothes to protect nose and mouth against dust. Bernardino amazzini, who lived on the turn of the seventeenth century, in his work 'De morbis artificum' indicated the need for protection of the respiratory tracts against dusts of workers laboring in various professions listed by him. Brise Fradin developed in 1814 the first device, which provided durable protection of the respiratory tracts. It was composed of a container filled with cotton fibers which was connected by a duct with the user's mouth. The first filtration respiratory mask was designed at the beginning of the nineteenth century with the aim of protecting the users against diseases transmitted by the breathing system. In these times, firemen began to use masks specially designed for them. The first construction of such a 'mask' was primitive: a leather helmet was connected with a hose which supplied air from the ground level.

The construction was based on the observation that during fire, fewer amounts of toxic substances were at the ground level than at the level of the fireman's mouth. In addition, a layer of fibers protected the lower air inlet. John Tyndall, in 1868 designed a mask which consisted of some layers of differentiated structure. A clay layer separated the first two layers of dry cotton fibers. Between the two next cotton fiber layers can be inserted charcoal, and the last two cotton fiber layers were separated by a layer of wool fibers saturated with glycerine. The history of the development of filtration materials over the nineteenth century has been described in a work elaborated by Feldhaus [193].

The twentieth century left a lasting impression of the First World War, during which toxic gases were used for the first time. This was the reason that after 1914, the further history of the development of filtration materials was connected with absorbers of toxic substances manufactured with the use of charcoal and fibrous materials. The next discovery, which changed the approach to the designing of filtration materials, was done in 1930. Hansen, in his filter applied a mixture of fibers and resin as filtration materials. This caused an electrostatic field being created inside the material. The action of electrostatic forces

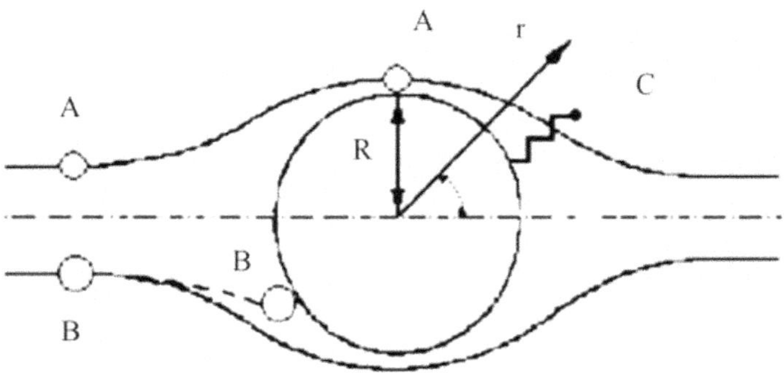

FIGURE 17.14 Particle capture mechanism: A – particle captured by interception; B – particle captured by inertial impaction; C – particle captured by diffusive deposition.

on dust particles significantly increases the filtration efficiency of the materials manufactured.

The brief historical sketch presented above indicates that textile fibers were one of the material components, which protect the respiratory tracts, and have been applied from the dawn of history. From the beginning they had been used intuitively, without understanding the mechanism of filtration. The first attempts of scientific description of the filtration mechanism were presented by Albrecht [194], Kaufman [195], Langmuir [196], and recently by Brown [6] who characterized the four basic physical phenomena of mechanical deposition in the following way:

- direct interception occurs when a particle follows a streamline and is captured as a result of coming into contact with the fiber;
- inertial impaction is realized when the deposition is effected by the deviation of a particle from the streamline caused by its own inertia; in diffusive deposition the combined action of airflow and Brownian motion brings a particle into contact with the fiber; gravitational settling resulting from gravitation forces.

Illustration of the above mechanisms of filtration is presented in Fig. 17.14.

The analysis of equations determining the filter efficiency governed by the mechanisms specified above indicates that the most important parameters deciding about filtering efficiency are the thickness of filters, the diameter of fibers and the porosity of the filter. Identification of these phenomena was the basis for development of new technologies for filtering materials composed from ultra thin fibers. These technologies mainly are based on manufacturing the nonwovens directly from the dissolved or melted polymers using melt-blown technique, flash spinning and electrospinning.

Additionally theoretical consideration also indicates that the activity of fibers on particles significantly increases if an electrostatic field is formed inside the nonwoven. This is the reason that nonwovens are additional modified. Three following groups of fibrous electrostatic materials used can be distinguished, based upon their ability to generate an electrostatic charge:

- materials in which the charge is generated by corona discharge after fiber or web formation,
- materials in which the charge is generated by induction during spinning in an electrostatic field, and

- materials in which the charge is induced as the result of the tribo-electric effect.

KEYWORDS

- **applications**
- **electrospun nanofibers**
- **modern fabrics**
- **modern fibers**
- **nanotechnology**
- **textile industry**
- **update**

REFERENCES

1. Selin, C., *Expectations and the Emergence of Nanotechnology.* Science, Technology & Human Values, 2007, 32(2), 196–220.
2. Mansoori, G.A., *Principles of Nanotechnology: Molecular-Based Study of Condensed Matter in Small Systems.* 2005: World Scientific.
3. Peterson, C.L., *Nanotechnology: From Feynman to the Grand Challenge of Molecular Manufacturing.* Technology and Society Magazine, IEEE, 2004, 23(4), 9–15.
4. Gleiter, H., *Nanostructured Materials: Basic Concepts and Microstructure.* Acta Materialia, 2000, 48(1), 1–29.
5. Peterson, C.L., *Nanotechnology-Evolution of the Concept.* Journal of the British Interplanetary Society, 1992, 45, 395–400.
6. Yadugiri, V.T. and R. Malhotra, *'Plenty of Room'-Fifty Years after the Feynman Lecture.* Current Science, 2010, 99(7), 900–907.
7. Wilson, K., *History of Textiles.* 1979: Westview Press.
8. Harris, J., *Textiles, 5,000 Years: An International History and Illustrated Survey.* 1995, New York City, United States: Harry N. Abrams, Inc.
9. Sawhney, A.P.S., et al., *Modern Applications of Nanotechnology in Textiles.* Textile Research Journal, 2008, 78(8), 731–739.
10. Tao, X., *Smart Fibers, Fabrics and Clothing.* Vol. 20. 2001: Woodhead publishing.
11. Tyrer, R.B., *The Demographic and Economic History of the Audiencia of Quito: Indian Population and the Textile Industry, 1600–1800,* 1976: University of California, Berkeley.

12. Chapman, S.D. and S. Chassagne, *European Textile Printers in the Eighteenth Century: A Study of Peel and Oberkampf.* 1981: Heinemann Educational Books, Pasold Fund.
13. Jeremy, D.J., *Transatlantic Industrial Revolution: The Diffusion of Textile Technologies between Britain and America, 1790–1830 s.* 1981: MIT Press.
14. Hekman, J.S., *The Product Cycle and New England Textiles.* The Quarterly Journal of Economics, 1980, 94(4), 697–717.
15. Mantoux, P., *The Industrial Revolution in the Eighteenth Century: An Outline of the Beginnings of the Modern Factory System in England.* 2013: Routledge.
16. French, G.J., *Life and Times of Samuel Crompton of Hall-in-the-Wood: Inventor of the Spinning Machine Called the Mule.* 1862: Charles Simms and Co.
17. O'brien, P., *The Micro Foundations of Macro Invention: The Case of the Reverend Edmund Cartwright.* Textile History, 1997, 28(2), 201–233.
18. Aspin, C., *The Cotton Industry.* 1981: Osprey Publishing.
19. Smout, T.C., *The Development and Enterprise of Glasgow 1556-1707,* Scottish Journal of Political Economy, 1959, 6(3), 194–212.
20. Sale, K., *Rebels against the Future: The Luddites and their War on the Industrial Revolution: Lessons for the Computer Age.* 1995: Basic Books.
21. Aspin, C., *The Woollen Industry.* 1982: Osprey Publishing.
22. Beckert, S., *Emancipation and Empire: Reconstructing the Worldwide Web of Cotton Production in the Age of the American Civil War.* The American Historical Review, 2004, 109(5), 1405–1438.
23. Heberlein, G., *Georges Heberlein,* 1935, Google Patents.
24. Townswnd, B.A., *Full Circle in Cellulose.* 1993: New York: Elsevier Applied Science.
25. Hicks, E.M., et al., *The Production of Synthetic-Polymer Fibers.* Textile Progress, 1971, 3(1), 1–108.
26. Cook, J.G., *Handbook of Textile Fibers: Man-Made Fibers.* Vol. 2. 1984: Elsevier.
27. Conrad, D.J., et al., *Ink-Printed, Low Basis Weight Nonwoven Fibrous Webs and Method,* 1995, Google Patents.
28. Sawhney, A.P.S., et al., *Modern Applications of Nanotechnology in Textiles.* Textile Research Journal, 2008, 78(8), 731–739.
29. Brown, P. and K. Stevens, *Nanofibers and Nanotechnology in Textiles.* 2007: Elsevier.
30. Valko, E.I., *Textile Material,* 1966, Google Patents.
31. Higgins, L. and M.E. Anand, *Textiles Materials and Products for Activewear and Sportswear.* Technical Textile Market, 2003, 52, 9–40.
32. Hunter, L., *Mohair, Cashmere and other Animal Hair Fibers,* in *Handbook of Natural Fibers,* R.M. Kozłowski, Editor. 2012, Woodhead Publishing. p. 196–290
33. Kuffner, H. and C. Popescu, *Wool Fibers,* in *Handbook of Natural Fibers,* R.M. Kozłowski, Editor. 2012, Woodhead Publishing. p. 171–195.
34. Roff, W.J. and J.R. Scott, *Silk and Wool,* in *Fibers, Films, Plastics and Rubbers,* W.J. Roff and J.R. Scott, Editors. 1971, Butterworth-Heinemann. p. 188–196.
35. Gordon, S., *Identifying Plant Fibers in Textiles: The Case of Cotton,* in *Identification of Textile Fibers,* M.M. Houck, Editor. 2009, Woodhead Publishing. p. 239–258.
36. Shahid ul, I., M. Shahid, and F. Mohammad, *Perspectives for Natural Product Based Agents Derived from Industrial Plants in Textile Applications – A Review.* Journal of Cleaner Production, 2013, 57, 2–18.

37. Ansell, M.P. and L.Y. Mwaikambo, *The Structure of Cotton and other Plant Fibers*, in *Handbook of Textile Fiber Structure*, S.J. Eichhorn, et al., Editors. 2009, Woodhead Publishing. p. 62–94.
38. Brown, R.C., et al., *Pathogenetic Mechanisms of Asbestos and other Mineral Fibers.* Molecular Aspects of Medicine, 1990, 11(5), 325–349.
39. Porter, R.M., *Glass Fiber Compositions*, 1991, Google Patents.
40. Milašius, R. and V. Jonaitienė, *Synthetic Fibers for Interior Textiles*, in *Interior Textiles*, T. Rowe, Editor. 2009, Woodhead Publishing. p. 39–46.
41. Rugeley, E.W., T.A. Feild Jr, and J.L. Petrokubi, *Synthetic Textile Articles*, 1947, US Patent
42. Woolman, M.S. and E.B. McGowan, *Textiles: A Handbook for the Student and the Consumer*. 1915: Macmillan.
43. Todd, M.P., *Hand-loom Weaving: A Manual for School and Home*. 1902: Rand, McNally.
44. Horne, P. and S. Bowden, *Knitting and Crochet.* Design, 1974, 75(3), 38–38.
45. Burt, E.C., *Bark-Cloth in East Africa.* Textile History, 1995, 26(1), 75–88.
46. Halls, Z., *Machine-Made Lace 1780–1820 (Pillow and Bobbin to Bobbin and Carriage).* Costume, 1970, 4(Supplement-1), 46–50.
47. Smith, C.W. and J.T. Cothren, *Cotton: Origin, History, Technology, and Production.* Vol. 4. 1999: John Wiley & Sons.
48. Kriger, C.E., *Guinea Cloth: Production and Consumption of Cotton Textiles in West Africa before and during the Atlantic Slave Trade.* The Spinning World. A Global History of Cotton Textiles, 2009, 1200–1850.
49. Betrabet, S.M., K.P.R. Pillay, and R.L.N. Iyengar, *Structural Properties of Cotton Fibers: Part II: Birefringence and Structural Reversals in Relation to Mechanical Properties.* Textile Research Journal, 1963, 33(9), 720–727.
50. Alston, J.M. and J.D. Mullen, *Economic Effects of Research into Traded Goods: The Case of Australian Wool.* Journal of Agricultural Economics, 1992, 43(2), 268–278.
51. Company, I.T. and I.C. Schools, *Wool, Wool Scouring, Wool Drying, Burr Picking, Carbonizing, Wool Mixing, Wool Oiling: Woolen Carding, Woolen Spinning, Woolen and Worsted Warp Preparation.* Vol. 79. 1905: International Textbook Company.
52. Parkes, C., *The Knitter's Book of Wool: The Ultimate Guide to Understanding, Using, and Loving this Most Fabulous Fiber*. 2011: Random House LLC.
53. D. Arden, L.D. and A. Pfister, *Woolen and Worsted Yarn for an Elizabethan Knitted Suite.* 2003.
54. Knapton, J.J.F., et al., *The Dimensional Properties of Knitted Wool Fabrics Part I: The Plain-Knitted Structure.* Textile Research Journal, 1968, 38(10), 999–1012.
55. Oikonomides, N., *Silk Trade and Production in Byzantium from the Sixth to the Ninth Century: The Seals of Kommerkiarioi.* Dumbarton Oaks Papers, 1986, 40, 33–53.
56. Lock, R.L., *Process for Making Silk Fibroin Fibers*, 1993, Google Patents.
57. Kaplan, D.L. *Silk Polymers*. in *Workshop on Silks: Biology, Structure, Properties, Genetics (1993: Charlottesville, Va.)*. 1994, American Chemical Society.
58. Bolton, E.K., *Chemical Industry Medal. Development of Nylon.* Industrial & Engineering Chemistry, 1942, 34(1), 53–58.

59. Li, L., et al., *Formation and Properties of Nylon-6 and Nylon-6/Montmorillonite Composite Nanofibers.* Polymer, 2006, 47(17), 6208–6217.
60. Harazoe, H., M. Matsuno, and S. Noda, *Method of Manufacturing Polyesters*, 1996, Google Patents.
61. Mijatovic, D., J.C.T. Eijkel, and A. V. D. Berg, *Technologies for Nanofluidic Systems: Top-Down vs. Bottom-Up—A Review.* Lab on a Chip, 2005, 5(5), 492–500.
62. Patra, J.K. and S. Gouda, *Application of Nanotechnology in Textile Engineering: An Overview.* Journal of Engineering and Technology Research, 2013, 5(5), 104–111.
63. Mohanraj, V.J. and Y. Chen, *Nanoparticles-A Review.* Tropical Journal of Pharmaceutical Research, 2007, 5(1), 561–573.
64. Mehnert, W. and K. Mäder, *Solid Lipid Nanoparticles: Production, Characterization and Applications.* Advanced Drug Delivery Reviews, 2001, 47(2), 165–196.
65. Roney, C., et al., *Targeted Nanoparticles for Drug Delivery through the Blood–Brain Barrier for Alzheimer's Disease.* Journal of Controlled Release, 2005, 108(2), 193–214.
66. Joshi, M. and A. Bhattacharyya, *Nanotechnology–A New Route to High-Performance Functional Textiles.* Textile Progress, 2011, 43(3), 155–233.
67. Kaounides, L., H. Yu, and T. Harper, *Nanotechnology Innovation and Applications in Textiles Industry: Current Markets and Future Growth Trends.* Materials Science and Technology, 2007, 22(4), 209–237.
68. Kathiervelu, S.S., *Applications of Nanotechnology in Fiber Finishing.* Synthetic Fibers, 2003, 32(4), 20–22.
69. Wang, C.C. and C.C. Chen, *Physical Properties of the Crosslinked Cellulose Catalyzed with Nanotitanium Dioxide under UV Irradiation and Electronic Field.* Applied Catalysis A: General, 2005, 293, 171–179.
70. Price, D., et al., *Burning Behavior of Foam/Cotton Fabric Combinations in the Cone Calorimeter.* Polymer Degradation and Stability, 2002, 77(2), 213–220.
71. Lamba, N.M.K., K.A. Woodhouse, and S.L. Cooper, *Polyurethanes in Biomedical Applications*. 1997: CRC press.
72. Qian, L. and J.P. Hinestroza, *Application of Nanotechnology for High Performance Textiles.* Journal of Textile and Apparel, Technology and Management, 2004, 4(1), 1–7.
73. Singh, V.K., et al. *Applications and Future of Nanotechnology in Textiles*. in *National Cotton Council Beltwide Cotton Conference*. 2006.
74. Gibson, P., H. S. Gibson, and C. Pentheny, *Electrospinning Technology: Direct Application of Tailorable Ultrathin Membranes.* Journal of Industrial Textiles, 1998, 28(1), 63–72.
75. Snyder, R.G., *The Bionic Tailor: TC2's 3-D Body Scanner*, 2001, IEEE-INST Electrical Electronics Engineers INC 345 E 47TH ST NY 10017–2394 USA.
76. Okabe, H., et al. *Three Dimensional Apparel CAD System*. in *ACM SIGGRAPH Computer Graphics*. 1992, ACM.
77. Yan, H. and S.S. Fiorito, *Communication: CAD/CAM Adoption in US Textile and Apparel Industries.* International Journal of Clothing Science and Technology, 2002, 14(2), 132–140.
78. Xing, X., Y. Wang, and B. Li, *Nanofibers Drawing and Nanodevices Assembly in Poly (Trimethylene Terephthalate).* Optics Express, 2008, 16(14), 10815–10822.

79. Ikegame, M., K. Tajima, and T. Aida, *Template Synthesis of Polypyrrole Nanofibers Insulated within One-Dimensional Silicate Channels: Hexagonal versus Lamellar for Recombination of Polarons into Bipolarons.* Angewandte Chemie International Edition, 2003, 42(19), 2154–2157.
80. He, L., et al., *Fabrication and Characterization of Poly (L-Lactic Acid) 3D Nanofibrous Scaffolds with Controlled Architecture by Liquid–Liquid Phase Separation from a Ternary Polymer–Solvent System.* Polymer, 2009, 50(16), 4128–4138.
81. Hartgerink, J.D., E. Beniash, and S.I. Stupp, *Self-Assembly and Mineralization of Peptide-Amphiphile Nanofibers.* Science, 2001, 294(5547), 1684–1688.
82. Li, D. and Y. Xia, *Fabrication of Titania Nanofibers by Electrospinning.* Nano Letters, 2003, 3(4), 555–560.
83. Ramakrishna, S., et al., *Electrospun Nanofibers: Solving Global Issues.* Materials Today, 2006, 9(3), 40–50.
84. Reneker, D.H. and I. Chun, *Nanometer Diameter Fibers of Polymer, Produced by Electrospinning.* Nanotechnology, 1996, 7(3), 216–233.
85. Frenot, A. and I.S. Chronakis, *Polymer Nanofibers Assembled by Electrospinning.* Current Opinion in Colloid & Interface Science, 2003, 8(1), 64–75.
86. Dersch, R., et al., *Electrospun Nanofibers: Internal Structure and Intrinsic Orientation.* Journal of Polymer Science Part A: Polymer Chemistry, 2003, 41(4), 545–553.
87. Yang, F., et al., *Electrospinning of Nano/Micro Scale Poly (L-Lactic Acid) Aligned Fibers and their Potential in Neural Tissue Engineering.* Biomaterials, 2005, 26(15), 2603–2610.
88. Moroni, L., et al., *Fiber Diameter and Texture of Electrospun PEOT/PBT Scaffolds Influence Human Mesenchymal Stem Cell Proliferation and Morphology, and the Release of Incorporated Compounds.* Biomaterials, 2006, 27(28), 4911–4922.
89. Jayakumar, R., et al., *Novel Chitin and Chitosan Nanofibers in Biomedical Applications.* Biotechnology Advances, 2010, 28(1), 142–150.
90. Zhang, Y., et al., *Recent Development of Polymer Nanofibers for Biomedical and Biotechnological Applications.* Journal of Materials Science: Materials in Medicine, 2005, 16(10), 933–946.
91. Nazarov, R., H.-J. Jin, and D.L. Kaplan, *Porous 3-D scaffolds from regenerated silk fibroin.* Biomacromolecules, 2004, 5(3), 718–726.
92. Smit, E.A., *Studies towards High-throughput Production of Nanofiber Yarns*, 2008, Stellenbosch University: Stellenbosch.
93. Subbiah, T., et al., *Electrospinning of Nanofibers.* Journal of Applied Polymer Science, 2005, 96(2), 557–569.
94. Murugan, R. and S. Ramakrishna, *Design Strategies of Tissue Engineering Scaffolds with Controlled Fiber Orientation.* Tissue Engineering, 2007, 13(8), 1845–1866.
95. Badrossamay, M.R., et al., *Nanofiber Assembly by Rotary Jet-Spinning.* Nano letters, 2010, 10(6), 2257–2261.
96. Srary, J., *Method of and Apparatus for Ringless Spinning of Fibers*, 1970, Google Patents.
97. Katta, P., et al., *Continuous Electrospinning of Aligned Polymer Nanofibers onto a Wire Drum Collector.* Nano Letters, 2004, 4(11), 2215–2218.
98. M. Creight, D.J., et al., *Short Staple Yarn Manufacturing*. Vol. 700. 1997, NC, USA: Carolina Academic Press Durham,.

99. Pan, H., et al., *Continuous Aligned Polymer Fibers Produced by a Modified Electrospinning Method.* Polymer, 2006, 47(14), 4901–4904.
100. Dzenis, Y.A., *Spinning Continuous Fibers for Nanotechnology.* Science 2004, 304 p. 1917–1919.
101. Jacobs, V., R.D. Anandjiwala, and M. Maaza, *The Influence of Electrospinning Parameters on the Structural Morphology and Diameter of Electrospun Nanofibers.* Journal of Applied Polymer Science, 2010, 115(5), 3130–3136.
102. Sarkar, K., et al., *Electrospinning to Forcespinning.* Materials Today, 2010, 13(11), 12–14.
103. Smit, A.E., U. Buttner, and R.D. Sanderson, *Continuous Yarns from Electrospun Nanofibers*. Nanofibers and Nanotechnology in Textiles. 2007, 45.
104. Hong, Y.T., et al., *Filament Bundle Type Nano Fiber and Manufacturing Method Thereof*, 2010, Google Patents.
105. Zhou, F.L. and R.H. Gong, *Manufacturing Technologies of Polymeric Nanofibers and Nanofiber Yarns.* Polymer International, 2008, 57(6), 837–845.
106. Teo, W.E. and S. Ramakrishna, *A Review on Electrospinning Design and Nanofiber Assemblies.* Nanotechnology, 2006, 17(14), p. R89-R106.
107. Kleinmeyer, J., J. Deitzel, and J. Hirvonen, *Electro Spinning of Submicron Diameter Polymer Filaments*, 2003, Google Patents.
108. Childs, H.R., *Process of Electrostatic Spinning*, 1944, US Patent
109. Ko, F.K., *Nanofiber Technology: Bridging the Gap between Nano and Macro World*, in *Nanoengineered Nanofibrous Materials*. 2004, p. 1–18.
110. Murray, M.P., *Cone Collecting Techniques for Whitebark Pine.* Western Journal of Applied Forestry, 2007, 22(3), 153–155.
111. Hermes, J., *Apparatus for Winding Yarn*, 1986, Google Patents.
112. Yadav, A., et al., *Functional Finishing in Cotton Fabrics Using Zinc Oxide Nanoparticles.* Bulletin of Materials Science, 2006, 29(6), 641–645.
113. Perelshtein, I., et al., *A One-Step Process for the Antimicrobial Finishing of Textiles with Crystalline TiO_2 Nanoparticles.* Chemistry-A European Journal, 2012, 18(15), 4575–4582.
114. Jiu, J.T., et al., *The Preparation of MgO Nanoparticles Protected by Polymer.* Chinese Journal of Inorganic Chemistry, 2001, 17(3), 361–365.
115. Soane, D.S., et al., *Nanoparticle-Based Permanent Treatments for Textiles*, 2003, Google Patents.
116. Song, X.Q., et al., *The Effect of Nano-Particle Concentration and Heating Time in the Anti-Crinkle Treatment of Silk.* Journal of Jilin Institute of Technology, 2001, 22, 24–27.
117. Russell, E., *Nanotechnologies and the Shrinking World of Textiles.* Textile Horizons, 2002, 9(10), 7–9.
118. Wong, Y.W.H., et al., *Selected Applications of Nanotechnology in Textiles.* AUTEX Research Journal, 2006, 6(1), 1–8.
119. Dastjerdi, R. and M. Montazer, *A Review on the Application of Inorganic Nano-Structured Materials in the Modification of Textiles: Focus on Anti-Microbial Properties.* Colloids and Surfaces B: Biointerfaces, 2010, 79(1), 5–18.

120. Dastjerdi, R., M. Montazer, and S. Shahsavan, *A New Method to Stabilize Nanoparticles on Textile Surfaces.* Colloids and Surfaces A: Physicochemical and Engineering Aspects, 2009, 345(1–3), 202–210.
121. Nazari, A., M. Montazer, and M.B. Moghadam, *Introducing Covalent and Ionic Cross-linking into Cotton through Polycarboxylic Acids and Nano TiO2.* Journal of The Textile Institute, 2012, 103(9), 985–996.
122. Gorjanc, M., et al., *The Influence of Water Vapor Plasma Treatment on Specific Properties of Bleached and Mercerized Cotton Fabric.* Textile Research Journal, 2010, 80(6), 557–567.
123. Mihailović, D., et al., *Functionalization of Cotton Fabrics with Corona/Air RF Plasma and Colloidal TiO_2 Nanoparticles.* Cellulose, 2011, 18(3), 811–825.
124. Karimi, L., et al., *Effect of Nano TiO_2 on Self-cleaning Property of Cross-linking Cotton Fabric with Succinic Acid Under UV Irradiation.* Photochemistry and photobiology, 2010, 86(5), 1030–1037.
125. Yuranova, T., et al., *Antibacterial Textiles Prepared by RF-Plasma and Vacuum-UV Mediated Deposition of Silver.* Journal of Photochemistry and Photobiology A: Chemistry, 2003, 161(1), 27–34.
126. M. Quaid, M. and P. Beesley, *Extreme Textiles: Designing for High Performance.* 2005: Princeton Architectural Press.
127. Mahltig, B., H. Haufe, and H. Böttcher, *Functionalization of Textiles by Inorganic Sol–Gel Coatings.* Journal of Materials Chemistry, 2005, 15(41), 4385–4398.
128. Kang, J.Y. and M. Sarmadi, *Textile Plasma Treatment Review–Natural Polymer-Based Textiles.* American Association of Textile Chemists and Colorists Review, 2004, 4(10), 28–32.
129. Rahman, M., et al., *Tool-Based Nanofinishing and Micromachining.* Journal of Materials Processing Technology, 2007, 185(1), 2–16.
130. Wang, R.H., J.H. Xin, and X.M. Tao, *UV-Blocking Property of Dumbbell-Shaped ZnO Crystallites on Cotton Fabrics.* Inorganic Chemistry, 2005, 44(11), 3926–3930.
131. Lee, H.J., S.Y. Yeo, and S.H. Jeong, *Antibacterial Effect of Nanosized Silver Colloidal Solution on Textile Fabrics.* Journal of Materials Science, 2003, 38(10), 2199–2204.
132. Sennett, M., et al., *Dispersion and Alignment of Carbon Nanotubes in Polycarbonate.* Applied Physics A, 2003, 76(1), 111–113.
133. Weiguo, D., *Research on Properties of Nano Polypropylene/TiO_2 Composite Fiber.* Journal of Textile Research, 2002, 23(1), 22–23.
134. Meier, U., *Carbon Fiber-Reinforced Polymers: Modern Materials in Bridge Engineering.* Structural Engineering International, 1992, 2(1), 7–12.
135. Huang, Z.M., et al., *A Review on Polymer Nanofibers by Electrospinning and their Applications in Nanocomposites.* Composites Science and Technology, 2003, 63(15), 2223–2253.
136. Daoud, W.A. and J.H. Xin, *Low Temperature Sol-Gel Processed Photocatalytic Titania Coating.* Journal of Sol-Gel Science and Technology, 2004, 29(1), 25–29.
137. Hartley, S.M., et al. *The next Generation of Chemical and Biological Protective Materials Utilizing Reactive Nanoparticles.* in *24th Army Science Conference.* 2004, Orlando, Florida, USA.

138. Wang, R.H., et al., *ZnO Nanorods Grown on Cotton Fabrics at Low Temperature.* Chemical Physics Letters, 2004, 398(1), 250–255.
139. Zhang, J., et al., *Hydrophobic Cotton Fabric Coated by a Thin Nanoparticulate Plasma Film.* Journal of Applied Polymer Science, 2003, 88(6), 1473–1481.
140. Yang, H., S. Zhu, and N. Pan, *Studying the Mechanisms of Titanium Dioxide as Ultraviolet-Blocking Additive for Films and Fabrics by an Improved Scheme.* Journal of Applied Polymer Science, 2004, 92(5), 3201–3210.
141. El-Molla, M.M., et al., *Nanotechnology to Improve Coloration and Antimicrobial Properties of Silk Fabrics.* Indian Journal of Fiber and Textile Research, 2011, 36(3), 266–271.
142. Mondal, S., *Phase Change Materials for Smart Textiles–An Overview.* Applied Thermal Engineering, 2008, 28(11), 1536–1550.
143. Hu, J., *Advances in Shape Memory Polymers.* 2013: Elsevier.
144. Spencer, B.F., M.E. R. Sandoval, and N. Kurata, *Smart Sensing Technology: Opportunities and Challenges.* Structural Control and Health Monitoring, 2004, 11(4), 349–368.
145. Diamond, D., et al., *Wireless Sensor Networks and Chemo-/Biosensing.* Chemical Reviews, 2008, 108(2), 652–679.
146. Tao, X.M., *Wearable Electronics and Photonics.* 2005: Elsevier.
147. Koncar, V., *Optical Fiber Fabric Displays.* Optics and Photonics News, 2005, 16(4), 40–44.
148. Sharma, A., et al., *Review on Thermal Energy Storage with Phase Change Materials and Applications.* Renewable and Sustainable Energy Reviews, 2009, 13(2), 318–345.
149. Zalba, B., et al., *Review on Thermal Energy Storage with Phase Change: Materials, Heat Transfer Analysis and Applications.* Applied Thermal Engineering, 2003, 23(3), 251–283.
150. Shin, Y., D.I. Yoo, and K. Son, *Development of Thermoregulating Textile Materials with Microencapsulated Phase Change Materials (PCM). IV. Performance Properties and Hand of Fabrics Treated with PCM Microcapsules.* Journal of Applied Polymer Science, 2005, 97(3), 910–915.
151. B. García, L., et al., *Phase Change Materials (PCM) Microcapsules with Different Shell Compositions: Preparation, Characterization and Thermal Stability.* Solar Energy Materials and Solar Cells, 2010, 94(7), 1235–1240.
152. Tyagi, V.V., et al., *Development of Phase Change Materials Based Microencapsulated Technology for Buildings: A Review.* Renewable and Sustainable Energy Reviews, 2011, 15(2), 1373–1391.
153. Nelson, G., *Application of Microencapsulation in Textiles.* International Journal of Pharmaceutics, 2002, 242(1), 55–62.
154. Shim, H., E.A. M. Cullough, and B.W. Jones, *Using Phase Change Materials in Clothing.* Textile Research Journal, 2001, 71(6), 495–502.
155. Gao, C., K. Kuklane, and I. Holmer, *Cooling Vests with Phase Change Material Packs: The Effects of Temperature Gradient, Mass and Covering Area.* Ergonomics, 2010, 53(5), 716–723.
156. Tang, S.L.P. and G.K. Stylios, *An Overview of Smart Technologies for Clothing Design and Engineering.* International Journal of Clothing Science and Technology, 2006, 18(2), 108–128.

157. Buckley, T.M., *Flexible Composite Material with Phase Change Thermal Storage*, 1999, Google Patents.
158. Soares, N., et al., *Review of Passive PCM Latent Heat Thermal Energy Storage Systems towards Buildings' Energy Efficiency.* Energy and Buildings, 2013, 59, 82–103.
159. Park, J.Y., et al., *Growth kinetics of nanograins in SnO_2 fibers and size dependent sensing properties.* Sensors and Actuators B: Chemical, 2011, 152(2), 254–260.
160. Cho, C.G., *Shape Memory Material*, in *Smart Clothing.* 2010, p. 189–221.
161. Wang, M., X. Luo, and D. Ma, *Dynamic Mechanical Behavior in the Ethylene Terephthalate-Ethylene Oxide Copolymer with Long Soft Segment as a Shape Memory Material.* European Polymer Journal, 1998, 34(1), 1–5.
162. Mother, P.T., H.G. Jeon, and T.S. Haddad, *Strain Recovery in POSS Hybrid Thermoplastics.* Polymer Preprints (USA), 2000, 41(1), 528–529.
163. Otsuka, K. and C.M. Wayman, *Shape Memory Materials.* 1999: Cambridge University Press.
164. B. Laurent, H. and H. Dürr, *Organic Photochromism (IUPAC Technical Report).* Pure and Applied Chemistry, 2001, 73(4), 639–665.
165. Mattila, H.R., *Intelligent Textiles and Clothing.* Vol. 3. 2006: CRC press England.
166. Kiekens, P. and S. Jayaraman, *Intelligent Textiles and Clothing for Ballistic and NBC Protection.* 2011: Springer.
167. Rantanen, J., et al., *Smart Clothing Prototype for the Arctic Environment.* Personal and Ubiquitous Computing, 2002, 6(1), 3–16.
168. Lind, E.J., et al. *A Sensate Liner for Personnel Monitoring Applications.* in *First International Symposium on Wearable Computers.* 1997, IEEE.
169. Lee, K. and D.S. Kwon. *Wearable Master Device Using Optical Fiber Curvature Sensors for the Disabled.* in *IEEE International Conference on Robotics and Automation.* 2001, IEEE.
170. Thomas, H.L., *Multi-Structure Ballistic Material*, 1998, Google Patents.
171. Rao, Y.J., *Fiber Bragg Grating Sensors: Principles and Applications*, in *Optical Fiber Sensor Technology.* 1998, Springer. p. 355–379.
172. Deflin, E. and V. Koncar. *For Communicating clothing: The Flexible Display of Glass fiber Fabrics is Reality.* in *2nd International Avantex Symposium.* 2002.
173. Rothmaier, M., M. Luong, and F. Clemens, *Textile Pressure Sensor Made of Flexible Plastic Optical Fibers.* Sensors, 2008, 8(7), 4318–4329.
174. Marchini, F., *Advanced Applications of Metallized Fibers for Electrostatic Discharge and Radiation Shielding.* Journal of Industrial Textiles, 1991, 20(3), 153–166.
175. Wendorff, J.H., S. Agarwal, and A. Greiner, *Electrospinning: Materials, Processing, and Applications.* 2012: John Wiley & Sons.
176. Brotin, T., et al., *[n]-Polyenovanillins (n= 1–6) as New Push-Pull Polyenes for Nonlinear Optics: Synthesis, Structural Studies, and Experimental and Theoretical Investigation of Their Spectroscopic Properties, Electronic Structures, and Quadratic Hyperpolarizabilities.* Chemistry of materials, 1996, 8(4), 890–906.
177. C. Jr, P.H., *Nonwoven Thermal Insulating Stretch Fabric and Method for Producing Same*, 1985, Google Patents.

178. Omastová, M., et al., *Synthesis, Electrical Properties and Stability of Polypyrrole-Containing Conducting Polymer Composites.* Polymer International, 1997, 43(2), 109–116.
179. Thieblemont, J.C., et al., *Thermal Analysis of Polypyrrole Oxidation in Air.* Polymer, 1995, 36(8), 1605–1610.
180. Bashir, T., *Conjugated Polymer-based Conductive Fibers for Smart Textile Applications*. 2013: Chalmers University of Technology.
181. Ruckenstein, E. and J.H. Chen, *Polypyrrole Conductive Composites Prepared by Coprecipitation.* Polymer, 1991, 32(7), 1230–1235.
182. Chen, Y., et al., *Morphological and Mechanical Behavior of an in Situ Polymerized Polypyrrole/Nylon 66 Composite Film.* Polymer Communications, 1991, 32(6), 189–192.
183. Gregory, R.V., W.C. Kimbrell, and H.H. Kuhn, *Electrically Conductive Non-Metallic Textile Coatings.* Journal of Industrial Textiles, 1991, 20(3), 167–175.
184. Batchelder, D.N., *Color and Chromism of Conjugated Polymers.* Contemporary Physics, 1988, 29(1), 3–31.
185. Kazani, I., et al., *Electrical Conductive Textiles Obtained by Screen Printing.* Fibers & Textiles in Eastern Europe, 2012, 20(1), 57–63.
186. Pacelli, M., et al., *Sensing Threads and Fabrics for Monitoring Body Kinematic and Vital Signs*, in *Fibers and Textiles for the Future*2001, p. 55–63.
187. Post, E.R., et al., *Electrically Active Textiles and Articles Made Therefrom*, 2001, Google Patents.
188. Jachimowicz, K.E. and M.S. Lebby, *Textile Fabric with Integrated Electrically Conductive Fibers and Clothing Fabricated Thereof*, 1999, Google Patents.
189. Marculescu, D., et al., *Electronic Textiles: A Platform for Pervasive Computing.* Proceedings of the IEEE, 2003, 91(12), 1995–2018.
190. Child, A.D. and A.R. D. Angelis, *Patterned Conductive Textiles*, 1999, Google Patents.
191. K. Jr, W.C. and H.H. Kuhn, *Electrically Conductive Textile Materials and Method for Making Same*, 1990, Google Patents.
192. Davies, C.N., *Air Filtration.* 1973, New York: Academic Press.
193. Kruiēska, I., E. Klata, and M. Chrzanowski, *New Textile Materials for Environmental Protection*, in *Intelligent Textiles for Personal Protection and Safety*. 2006, p. 41–53.
194. Albrecht, F., *Theoretische Untersuchungen* über *die Ablagerung von Staub aus strömender Luft und ihre Anwendung auf die Theorie der Staubfilter.* Physikalische Zeitschrift, 1931, 23, 48–56.
195. Walkenhorst, W., *Physikalische Eigenschaften von Stäuben sowie Grundlagen der Staubmessung und Staubbekämpfung*, in *Pneumokoniosen*. 1976, Springer. p. 11–70.
196. Langmuir, I., *Report on Smokes and Filters*, in *Section I*1942: US Office of Scientific Research and Development.
197. Wente, V.A., *Superfine Thermoplastic Fibers.* Industrial & Engineering Chemistry, 1956, 48(8), 1342–1346.
198. Buntin, R.R. and D.T. Lohkamp, *Melt Blowing-One-Step WEB Process for New Nonwoven Products.* Tappi, 1973, 56(4), 74–77.
199. Angadjivand, S., R. Kinderman, and T. Wu, *High Efficiency Synthetic Filter Medium*, 2000, Google Patents.

200. Zeleny, J., *The electrical discharge from Liquid Points, and a Hydrostatic Method of measuring the electric intensity at their surface.* Physical Review, 1914, 3(2), 69–91.
201. Formhals, A., *Process and Apparatus Fob Pbepabing*, 1934, Google Patents.
202. Taylor, G., *Disintegration of Water Drops in an Electric Field.* Proceedings of the Royal Society of London. Series A. Mathematical and Physical Sciences, 1964, 280(1382), 383–397.
203. Gibson, P., H. S. Gibson, and D. Rivin, *Transport Properties of Porous Membranes Based on Electrospun Nanofibers.* Colloids and Surfaces A: Physicochemical and Engineering Aspects, 2001, 187, 469–481.
204. Patarin, J., B. Lebeau, and R. Zana, *Recent Advances in the Formation Mechanisms of Organized Mesoporous Materials.* Current Opinion in Colloid & Interface Science, 2002, 7(1), 107–115.
205. Weghmann, A. *Production of Electrostatic Spun Synthetic Microfiber Nonwovens and Applications in Filtration.* in *Proceedings of the 3rd World Filtration Congress, Filtration Society*. 1982, London.
206. Yarin, A.L. and E. Zussman, *Upward Needleless Electrospinning of Multiple Nanofibers.* Polymer, 2004, 45(9), 2977–2980.
207. Majeed, S., et al., *Multi-Walled Carbon Nanotubes (MWCNTs) Mixed Polyacrylonitrile (PAN) Ultrafiltration Membranes.* Journal of Membrane Science, 2012, 403, 101–109.
208. Macedonio, F. and E. Drioli, *Pressure-Driven Membrane Operations and Membrane Distillation Technology Integration for Water Purification.* Desalination, 2008, 223(1), 396–409.
209. Merdaw, A.A., A.O. Sharif, and G.A.W. Derwish, *Mass Transfer in Pressure-Driven Membrane Separation Processes, Part II.* Chemical Engineering Journal, 2011, 168(1), 229–240.
210. Van Der Bruggen, B., et al., *A Review of Pressure-Driven Membrane Processes in Wastewater Treatment and Drinking Water Production.* Environmental Progress, 2003, 22(1), 46–56.
211. Cui, Z.F. and H.S. Muralidhara, *Membrane Technology: A Practical Guide to Membrane Technology and Applications in Food and Bioprocessing.* 2010: Elsevier. 288.
212. Shirazi, S., C.J. Lin, and D. Chen, *Inorganic Fouling of Pressure-Driven Membrane Processes — A Critical Review.* Desalination, 2010, 250(1), 236–248.
213. Pendergast, M.M. and E.M.V. Hoek, *A Review of Water Treatment Membrane Nanotechnologies.* Energy & Environmental Science, 2011, 4(6), 1946–1971.
214. Hilal, N., et al., *A comprehensive review of nanofiltration membranes: Treatment, pretreatment, modeling, and atomic force microscopy.* Desalination, 2004, 170(3), 281–308.
215. Srivastava, A., S. Srivastava, and K. Kalaga, *Carbon Nanotube Membrane Filters*, in *Springer Handbook of Nanomaterials*. 2013, Springer. p. 1099–1116.
216. Colombo, L. and A.L. Fasolino, *Computer-Based Modeling of Novel Carbon Systems and Their Properties: Beyond Nanotubes*. Vol. 3. 2010: Springer. 258.
217. Polarz, S. and B. Smarsly, *Nanoporous Materials.* Journal of Nanoscience and Nanotechnology, 2002, 2(6), 581–612.
218. Gray-Weale, A.A., et al., *Transition-state theory model for the diffusion coefficients of small penetrants in glassy polymers.* Macromolecules, 1997, 30(23), 7296–7306.

219. Rigby, D. and R. Roe, *Molecular Dynamics Simulation of Polymer Liquid and Glass. I. Glass Transition.* The Journal of chemical physics, 1987, 87, 7285.
220. Freeman, B.D., Y.P. Yampolskii, and I. Pinnau, *Materials Science of Membranes for Gas and Vapor Separation.* 2006: Wiley. com. 466.
221. Hofmann, D., et al., *Molecular Modeling Investigation of Free Volume Distributions in Stiff Chain Polymers with Conventional and Ultrahigh Free Volume: Comparison Between Molecular Modeling and Positron Lifetime Studies.* Macromolecules, 2003, 36(22), 8528–8538.
222. Greenfield, M.L. and D.N. Theodorou, *Geometric Analysis of Diffusion Pathways in Glassy and Melt Atactic Polypropylene.* Macromolecules, 1993, 26(20), 5461–5472.
223. Baker, R.W., *Membrane Technology and Applications.* 2012: John Wiley & Sons. 592
224. Strathmann, H., L. Giorno, and E. Drioli, *Introduction to Membrane Science and Technology.* 2011: Wiley-VCH Verlag & Company. 544.
225. Chen, J.P., et al., *Membrane Separation: Basics and Applications*, in *Membrane and Desalination Technologies*, L.K. Wang, et al., Editors. 2008, Humana Press. p. 271–332.
226. Mortazavi, S., *Application of Membrane Separation Technology to Mitigation of Mine Effluent and Acidic Drainage.* 2008: Natural Resources Canada. 194.
227. Porter, M.C., *Handbook of Industrial Membrane Technology.* 1990: Noyes Publications. 604.
228. Naylor, T.V., *Polymer Membranes: Materials, Structures and Separation Performance.* 1996: Rapra Technology Limited. 136.
229. Freeman, B.D., *Introduction to Membrane Science and Technology. By Heinrich Strathmann.* Angewandte Chemie International Edition, 2012, 51(38), 9485–9485.
230. Kim, I., H. Yoon, and K.M. Lee, *Formation of Integrally Skinned Asymmetric Polyetherimide Nanofiltration Membranes by Phase Inversion Process.* Journal of applied polymer science, 2002, 84(6), 1300–1307.
231. Khulbe, K.C., C.Y. Feng, and T. Matsuura, *Synthetic Polymeric Membranes: Characterization by Atomic Force Microscopy.* 2007: Springer. 198.
232. Loeb, L.B., *The Kinetic Theory of Gases.* 2004: Courier Dover Publications. 678.
233. Koros, W.J. and G.K. Fleming, *Membrane-Based Gas Separation.* Journal of Membrane Science, 1993, 83(1), 1–80.
234. Perry, J.D., K. Nagai, and W.J. Koros, *Polymer membranes for hydrogen separations.* MRS bulletin, 2006, 31(10), 745–749.
235. Hiemenz, P.C. and R. Rajagopalan, *Principles of Colloid and Surface Chemistry, revised and expanded.* Vol. 14. 1997: CRC Press.
236. McDowell-Boyer, L.M., J.R. Hunt, and N. Sitar, *Particle transport through porous media.* Water Resources Research, 1986, 22(13), 1901–1921.
237. Auset, M. and A.A. Keller, *Pore-scale processes that control dispersion of colloids in saturated porous media.* Water Resources Research, 2004, 40(3).
238. Bhave, R.R., *Inorganic membranes synthesis, characteristics, and applications.* Vol. 312. 1991: Springer.
239. Lin, V.S.-Y., et al., *A porous silicon-based optical interferometric biosensor.* Science, 1997, 278(5339), 840–843.

240. Hedrick, J., et al. *Templating nanoporosity in organosilicates using well-defined branched macromolecules.* in *MATERIALS RESEARCH SOCIETY SYMPOSIUM PROCEEDINGS.* 1998, Cambridge University Press.
241. Hubbell, J.A. and R. Langer, *Tissue engineering* Chem. Eng. News 1995, 13, 42–45.
242. Schaefer, D.W., *Engineered porous materials* MRS Bulletin 1994, 19, 14–17.
243. Hentze, H.P. and M. Antonietti, *Porous Polymers and Resins.* Handbook of Porous Solids, 1964–2013.
244. Endo, A., et al., *Synthesis of ordered microporous silica by the solvent evaporation method.* Journal of materials science, 2004, 39(3), 1117–1119.
245. Sing, K., et al., *Physical and biophysical chemistry division commission on colloid and surface chemistry including catalysis.* Pure and Applied Chemistry, 1985, 57(4), 603–619.
246. Kresge, C., et al., *Ordered mesoporous molecular sieves synthesized by a liquid-crystal template mechanism.* nature, 1992, 359(6397), 710–712.
247. Yang, P., et al., *Generalized syntheses of large-pore mesoporous metal oxides with semicrystalline frameworks.* nature, 1998, 396(6707), 152–155.
248. Jiao, F., K.M. Shaju, and P.G. Bruce, *Synthesis of Nanowire and Mesoporous Low-Temperature $LiCoO_2$ by a Post-Templating Reaction.* Angewandte Chemie International Edition, 2005, 44(40), 6550–6553.
249. Ryoo, R., et al., *Ordered mesoporous carbons.* Advanced Materials, 2001, 13(9), 677–681.

INDEX

D

Q

R

S

For Product Safety Concerns and Information please contact our EU
representative GPSR@taylorandfrancis.com
Taylor & Francis Verlag GmbH, Kaufingerstraße 24, 80331 München, Germany

www.ingramcontent.com/pod-product-compliance
Ingram Content Group UK Ltd.
Pitfield, Milton Keynes, MK11 3LW, UK
UKHW022208190926
475702UK00003B/259
* 9 7 8 1 7 7 1 8 8 0 8 5 5 *